Principles of
Linear Algebra
With *Maple*™

Principles of Linear Algebra With *Maple*™

Kenneth Shiskowski
Eastern Michigan University

Karl Frinkle
Southeastern Oklahoma State University

A JOHN WILEY & SONS, INC., PUBLICATION

Published by John Wiley & Sons, Inc., Hoboken, New Jersey.

Published simultaneously in Canada.

For general information on our other products and services or for technical support, please contact our Customer Care Department within the United States at (800) 762-2974, outside the United States at (317) 572-3993 or fax (317) 572-4002.

Wiley also publishes its books in a variety of electronic formats. Some content that appears in print may not be available in electronic formats. For more information about Wiley products, visit our web site at www.wiley.com.

Library of Congress Cataloging-in-Publication Data:

Shiskowski, Kenneth, 1954–
 Linear algebra using Maple / Kenneth Shiskowski, Karl Frinkle.
  p. cm. — (Pure and applied mathematics)
 On t.p. the registered trademark symbol "TM" is superscript following "Maple" in the title.
 Includes index.
 ISBN 978-0-470-63759-3 (cloth)
 1. Algebras, Linear—Data processing. 2. Maple (Computer file) I. Frinkle, Karl, 1977– II. Title.
 QA185.D37S45 2010
 512'.50285--dc22   2010013923

Printed in the United States of America.

10 9 8 7 6 5 4 3 2 1

Contents

Preface

This book is an attempt to cross the gap between beginning linear algebra and the computational linear algebra one encounters more frequently in applied settings. The underlying theory behind many topics in the field of linear algebra is relatively simple to grasp, however, to actually apply this knowledge to nontrivial problems becomes computationally intensive. To do these computations by hand would be tedious at best, and many times simply unrealistic. Furthermore, attempting to solve such problems by the old pencil and paper method does not give the average reader any extra insight into the problem. *Maple*^{TM} allows the reader to overcome these obstacles, giving them the power to perform complex computations that would take hours by hand, and can help to visualize many of the geometric interpretations of linear algebra topics in two and three dimensions in a very intuitive fashion. We hope that this book will challenge the reader to become proficient in both theoretical and computational aspects of linear algebra.

Overview of the Text

Chapter 1 of this book is a brief introduction to *Maple* and will help the reader become more comfortable with the program. This chapter focuses on the commands and packages most commonly used when studying linear algebra and its applications. *Maple* commands will always be preceded by an arrow: >, while the output will be displayed in a centered fashion below the *Maple* command. The reader can enter these commands and obtain the same results, assuming that they have entered the commands correctly. Note also that all of the images in this book were produced with *Maple*. The overall intent of this book is to use *Maple* to enhance the concepts of linear algebra, and therefore *Maple* will be integrated into this book in a very casual manner. Where one normally explains how to perform some operation by hand in a standard text, many of the times, we simply use *Maple* commands to perform the same task. Thus, the reader should attempt to become as familiar with the *Maple* syntax as quickly as possible.

At the end of each section, you will find two types of problems: *Home-*

work Problems and *Maple Problems*. The former consists of strictly pen and pencil computation problems, inquiries into theory, and questions about concepts discussed in the section. The idea behind these problems is to ensure that the reader has an understanding of the concepts introduced and can put them to use in problems that can be worked out by hand. For example, *Maple* can multiply matrices together much faster than any person can and without any algebraic mistakes, so why should the reader ever perform these tasks by hand? The answer is simple: In order to fully grasp the mechanics of matrix multiplication, simple problems must be worked out by hand. This manual labor, although usually deemed tedious, is an important tool in learning reinforcement. The *Maple Problems* portion of the homework typically involve problems that would take too long, or would be too computationally complex, to solve by hand. There are many problems in the *Maple Problems* portion that simply ask you to verify your answers to questions from the preceding *Homework Problems*, implying that you can think of *Maple* as a "solutions manual" for a large percentage of this text. You will also notice that several sections are missing the *Homework Problems* section. These sections correspond to special topics that are discussed because they can only be explored in detail with *Maple*.

Website and Supplemental Material

We suggest students and instructors alike visit the book's companion website, which can be found at either of the following addresses:

http://carmine.se.edu/kfrinkle/PrinciplesOfLinearAlgebraWithMaple
http://people.emich.edu/kshiskows/PrinciplesOfLinearAlgebraWithMaple

At the above location, you can download *Maple* worksheets, corresponding to each section's *Maple* commands, along with many other resources. These files can be used with the book so that the reader does not have to retype all of the *Maple* code in order to do problems or practice the material. We highly suggest that all readers unfamiliar with *Maple* (and even those who are) read over the relevant sections of the "Introduction to *Maple*" worksheet before they get too far into the book in order to understand the book's *Maple* code better. Specifically, we suggest looking at plotting/graphing material, differences between sets, lists and strings, and expressions versus functions and how *Maple* uses each. *Homework Problems* solutions and *Maple Problems* worksheets are also available for download.

We should also mention the *Maple* Adoption Program allows an instructor to register a course with *Maple*, so that students can get the student version of *Maple* at a discount. For more information, visit the Maplesoft™ website here:

http://www.maplesoft.com/academic/adoption/

Suggested Course Outlines

It would be nice if we could always cover all of the topics that we wanted to in a given course. This rarely happens, but there are obviously core topics that should be covered. Furthermore, some of the advanced topics require knowledge beyond what students in a basic linear algebra course may have. The appropriate prerequisites for this course would be trigonometry, a precalculus course in algebra, and trig. Also, a computer programming course would be helpful because we are using *Maple*. A year-long course in calculus would also be beneficial in regards to several topics. Here is a list of sections that require advanced knowledge:

- Section 7.2 - Differentiation
- Section 7.3 - Multivariable calculus
- Section 10.1 - Green's theorem
- Section 10.3 - Divergence theorem and double integrals
- Section 11.3 - Gradients and Lagrange multipliers
- Section 12.4 - Linear differential equations

We suggest that as much of the book be covered as possible, but here is the minimum suggested course outline:

Chapter 1	Sections $1-2$	1 lecture
Chapter 2	Sections $1-3$	4 lectures
Chapter 3	Sections $1-2$	3 lectures
Chapter 5	Sections $1-6$	8 lectures
Chapter 6	Sections $1-4$	5 lectures
Chapter 8	Sections $1-5$	7 lectures
Chapter 9	Sections $1-4$	5 lectures
Chapter 11	Sections $1-3$	3 lectures
Chapter 12	Sections $1-3,5$	3 lectures
	Total	39 lectures

Upon inspection of the above outline, you will notice that Chapters 4, 7, and 10 have been completely omitted. Chapter 4 has interesting applications of matrix multiplication to geometry, business, finance, and curve fitting, and we highly suggest covering Sections 4.1 and 4.4. Curve fitting is covered in Section 4.2, but is covered more in-depth in Chapter 11, where pseudoinverses and the method of least-squared deviation are introduced. Chapter 7 contains applications of the information learned about vectors in Chapter 6. If you wish to cover any of the topics in Chapter 10, we highly suggest that you cover Section 7.1. Chapter 10 is a fun chapter on linear maps and how they affect geometric objects. Affine maps are included in this chapter and should be given serious consideration as a topic to cover.

Final Remarks

We hope that both students and instructors will find this book to be a unique read. Our goal was to tell a story, rather than follow the standard textbook formula of: definition, theorem, example, and then repeat for 500 pages. We also hope that you really enjoy using *Maple*, both to explore the geometric and computational aspects of linear algebra, and to verify your pencil and paper work. We very much would like to hear your comments. Some of the questions we would always like answered, both from the student and the instructor, follow:

1. Were there topics that were difficult to grasp from the explanation and examples given? If so, what would you suggest we add/change to help make comprehension easier?

2. How did you enjoy the mixture of homework and *Maple* problems? Did you gain anything from verifying your answers to the homework problems with *Maple*?

3. What were some of your favorite/least favorite sections, and why?

4. Do you feel there were important topics, integral to a first semester course in linear algebra, that were missing from this text?

5. Did embedding *Maple* commands and output within the actual explanation of topics help to illustrate the topics?

6. Overall, what worked the best for you in this text, and what really did not work?

It would be wonderful if this text, in its first edition, was free of errors: both grammatical and mathematical. However, no matter how many times we read and proofread this text, it is a certainty that something will be missed. We hope you contact us with any and all mistakes that you have found, along with comments and suggestions that you may have.

Acknowledgments

First, we would like to express our thanks to Jacqueline Palmieri, Christine Punzo, Stephen Quigley, and Susanne Steitz-Filler of John Wiley & Sons, Inc. for making the entire process, from the original proposal, to project approval, to final submission, incredibly smooth. The four of you were supportive, encouraging, very enthusiastic, and quick to respond to any questions that arose over the course of this project. We appreciate this very much. We would also like to thank the following individuals who were involved in the original peer review process:

Derek Martinez, Central New Mexico University
Dror Varolin, Stony Brook University
Gian Mario Besana, DePaul University

Chris Moretti, Southeastern Oklahoma State University
Andrew Ross, Eastern Michigan University

Special thanks go to Bobbi Page, who took the time to read large portions of early drafts of this book, pored over the copious copyedits, and made many invaluable suggestions. The two successive spring semester Linear Algebra students at Southeastern Oklahoma State University deserve a warm round of applause for being guinea pigs and error hunters. Thanks also goes out to the countless students from the many courses Dr. Shiskowski has taught at Eastern Michigan University, with the help of *Maple*, specifically the Linear Algebra courses. We would also like to thank Mark Bickham, whose idea for a title to this book finally made both authors happy. Thanks again to everyone who was involved in this project, at any point, at any time. If we forgot to add your name this time around, perhaps you will make it into the second edition.

Kenneth Shiskowski   Karl Frinkle
Eastern Michigan University   Southeastern Oklahoma State University
kenneth.shiskowski@emich.edu   kfrinkle@se.edu

With all textbooks, one should attempt to be consistent with notation, not only within the text, but within the field of mathematics upon which it is based. For the most part, we have done this.

Table of Symbols and Notation

$\mathbf{B}, \mathbf{K_1}, \mathbf{Q}$	Bold capital letters designate sets of objects, usually vectors, or a field
$\mathbb{R}, \mathbb{C}$	Real and complex numbers, respectively
$\mathbb{R}^n, \mathbb{C}^n$	n-tuples of real and complex numbers, respectively
$\mathbb{R}^{m \times n}, \mathbb{C}^{m \times n}$	$m \times n$ matrices with real and complex entries
$\mathbb{S}, \mathbb{T}, \mathbb{R}$	Math script capital letters denote vector spaces
$\dim(\mathbb{S})$	Dimension of a vector space $\mathbb{S}$
$u,\ x\ e_k$	Lower-case letters are designated as scalars
$\vec{u}, \vec{v}, \vec{e_k}$	Lower-case letters with arrows over them are vectors, or column matrices
$\langle 1, 2, -1 \rangle, \langle x_1, x_2 \rangle$	Vectors expressed in component form
$\vec{u} \cdot \vec{y}$	Dot product of two vectors
$\vec{u} \times \vec{y}$	Cross product of two vectors
$\text{proj}_{\vec{v}}(\vec{w})$	Projection of $\vec{w}$ onto $\vec{v}$
$\text{comp}_{\vec{v}}(\vec{w})$	Component of $\vec{w}$ onto $\vec{v}$
A, C, X	Single capital letters represent matrices
AB, AX	Matrix multiplication has no symbol, two matrices in sequence implies multiplication
A^T, A^{-1}	Transpose, inverse of a matrix
$(A\|B)$	Augmented matrix, with A on the left, B on the right
$A_{2,3}, B_{j,k}$	Entries of a matrix are indexed by row,column
$\det(A), \text{adj}(A)$	Determinant, adjoint of a matrix
$p(A)$	Pseudoinverse of a matrix
$T : \mathbb{R} \to \mathbb{S}$	Linear map from vector space $\mathbb{R}$ to vector space $\mathbb{S}$
$\text{Ker}(T), \text{Im}(T)$	The kernel, image of a linear map T
$\int_a^b f(u)\ du$	Integral of $f(u)$ with respect to u on $[a, b]$
$\displaystyle\prod_{j=1}^{n} x_j$	The product $x_1 x_2 \cdots x_n$
$\displaystyle\sum_{j=1}^{n} x_j$	The sum $x_1 + x_2 + \cdots + x_n$
$\nabla D(\vec{a})$	Gradient of vector valued function D
$\dfrac{dF}{dx}, \dfrac{\partial F}{\partial x}$	Derivative, partial derivative, of F with respect to x

Chapter 1

An Introduction To Maple

This chapter presents a very brief introduction to the *Maple* computer algebra system that will be used throughout this book for the production of all of our computations and graphics. *Maple* is an amazingly powerful piece of software that is specifically designed to do all things mathematical, this also makes it extremely useful to the general users of mathematics in all the sciences. By using *Maple*, we can increase our mathematical productivity at least 10-fold since *Maple* is much faster and more accurate than we mere humans will probably ever be. You will have to be patient in the beginning when you are first learning *Maple*, since like all software (*Excel*, *Word*, etc.), it will take some time to get used to the syntax and idiosyncracies of *Maple*.

As soon as you have read this short chapter, you should immediately download the *Maple* file (worksheet) called "Introduction to *Maple*" from this books companion website, whose address can be found in the preface. Please download this *Maple* file to your home computer so that you can play with it in order to master some of the basics of *Maple* programming. Of course, you will also need to have a copy of *Maple* on your computer in order for you to actually use the worksheet. If you are using this text for a course in linear algebra, then hopefully your instructor has registered the course with the *Maple* adoption program so that you can get a discounted copy of the latest student version of *Maple* from the *Maple* website at www.maplesoft.com. The "Introduction to *Maple*" worksheet gives a much more thorough overview of *Maple* and its inner mysteries than could be given in this chapter, but this file also goes into some areas of mathematics other than those directly relevant to linear algebra, as such, you can ignore the sections on statistics, financial mathematics, and calculus, unless you have some interest and prior knowledge in these areas.

One last word about *Maple* that is also generically true about all software, the Help leaves much to be desired, although *Maple*'s Help has gotten significantly better with each new version, which is at *Maple* 14 with the writing of this book. By this, we do not mean that *Maple*'s Help is useless, merely that it

1

does not often provide the needed examples of how to use *Maple*'s commands in more than the most basic of ways. Hopefully, our "Introduction to *Maple*" file will provide the needed assistance that *Maple*'s Help can not give you in using this text. You should also feel free to check out the Internet for other sources of information on how to use *Maple* beyond this introductory file. And a good *Maple*ing to you all!

If you are using an older version of *Maple* (e.g., *Maple* 9.5), then we suggest using the classic worksheet version. However, if you have a newer version of *Maple* (e.g., *Maple* 14), once the program has been opened, go to *File* → *New* → *Worksheet Mode*. The other type of document style under the *New* setting is *Document Mode*, and will still perform all the same commands, but it is geared more toward presentation than utility. The main reason that we focus on the worksheet versus the document mode is that we are attempting to make as much of the *Maple* code found in this book as backward compatible as possible. There have been many new and wonderful enhancements and additions to *Maple* since both of the authors started using it (long, long ago), so feel free to explore!

1.1 The Commands

Before we start, be sure you have opened up a new worksheet document. Once this is done, you will notice a bunch of commands on the left, as well as various options above the worksheet window in the toolbar. Feel free to explore these, paying particular attention to the expressions icon on the left. It has several shortcuts for popular commands you may want to use on a regular basis that allow you to simply fill in variables, limits, and other entries into standard mathematical operators. The newer versions of *Maple*, including *Maple* 13, have two different input types on the command line. For the purpose of this book, we will assume that you are using the "Text" version of the command line. The reason is that if you select the "Math" option (which is the standard) your input will be displayed in nice formula fashion, which looks good, but for the purposes of entering commands can be potentially burdensome. In "Text" mode, you can type commands character for character as they appear in this book.

No matter which version of *Maple* you use, one of the most important commands to remember is the *restart* command, which clears the values of all the currently used variables. We highly suggest that the *restart* command be placed at the very beginning of each problem you do, especially if there is more than one problem per worksheet file. The next most important thing to learn is how to load packages. Not every command is immediately accessible upon startup. This feature reduces the total amount of memory and CPU cycles that the program uses. Most of the basic commands that a user would encounter

do not require a package to be loaded. However, there are many linear algebra concepts (e.g., computing the determinant of a matrix), which require the
linear algebra package *linalg* to be loaded. To load the linear algebra package,
the following command is used.

> with(linalg):

If you wish to know what commands are loaded with the *linalg* package,
simply use a semicolon instead. Ending a command with a colon suppresses
the output, while ending a command with a semicolon displays the output. In
Maple 14, you do not even need to end the command with a semicolon; simply
press "Enter" after typing in your command.

> with(linalg);

[*BlockDiagonal, GramSchmidt, JordanBlock, LUdecomp, QRdecomp, Wronskian, addcol, addrow, adj, adjoint, angle, augment, backsub, band, basis, be-
zout, blockmatrix, charmat, charpoly, cholesky, col, coldim, colspace, colspan,
companion, concat, cond, copyinto, crossprod, curl, definite, delcols, delrows,
det, diag, diverge, dotprod, eigenvals, eigenvalues, eigenvectors, eigenvects, en-
termatrix, equal, exponential, extend, ffgausselim, fibonacci, forwardsub, frobenius, gausselim, gaussjord, geneqns, genmatrix, grad, hadamard, hermite, hes-
sian, hilbert, htranspose, ihermite, indexfunc, innerprod, intbasis, inverse,
ismith, issimilar, iszero, jacobian, jordan, kernel, laplacian, leastsqrs, linsolve,
matadd, matrix, minor, minpoly, mulcol, mulrow, multiply, norm, normalize,
nullspace, orthog, permanent, pivot, potential, randmatrix, randvector, rank,
ratform, row, rowdim, rowspace, rowspan, rref, scalarmul, singularvals, smith,
stackmatrix, submatrix, subvector, sumbasis, swapcol, swaprow, sylvester,
toeplitz, trace, transpose, vandermonde, vecpotent, vectdim, vector, wronskian*]

As you can see from the above output, the list of commands loaded with
the *linalg* package is quite extensive and we will not cover all of these spe-
cialized commands in this text. Throughout the text, you will notice that a
similar package, *LinearAlgebra*, is also used. This one is used most often when
dealing with vectors and linear algebra plotting constructs. *Maple* considers
the *LinearAlgebra* package to have superseded the earlier *linalg* package, although *linalg* is still available for use, and we primarily use the *linalg* package
throughout this text. Also, the datatype of a matrix is different between the
two packages, and so not always interchangeable. The reader should perform
the command: with(LinearAlgebra); in order to see the difference in command
names for essentially the same operations as well as the new commands not
found in the *linalg* package. Next, we will move on to some simple computational commands. As a first example, we can add two numbers together.

> 5+3;

$$8$$

If we want to multiply the result of the previous example by 12, *Maple* has a shortcut that comes in handy if the previous result is quite complicated. The percent sign, %, is used to represent the previous computed output. It is very important to remember that this command refers to the previously computed output, not necessarily the output from the line previous, as *Maple* will allow you to execute command lines in any order desired.

> %*12;

$$96$$

The program also contains all of the trigonometric functions, as well as many other special functions. You can also enter several commands on the same line when separated by semicolons, but pay special attention to the syntax required in the newer version of *Maple*. You will need to separate the commands by semicolons in any version, otherwise *Maple* will read it as multiplication in the newer versions and/or an error in older versions.

> sin(Pi/4); ln(3); exp(2); 2/3;

$$\frac{1}{2}\sqrt{2}$$

$$\ln(3)$$

$$e^2$$

$$\frac{2}{3}$$

The standard operation symbols for addition, subtraction, multiplication, and division are given by +, -, *, and /, respectively. *Maple* does exact computations unless otherwise told to do approximations involving integers and fractions. You can also place a decimal point after an integer to make it approximate instead of exact. If we wish to express the expressions above in floating point form, we can use the *evalf* command.

> evalf(sin(Pi/4)); evalf(ln(3)); evalf(exp(2)); 2./3;

$$0.7071067810$$

$$1.098612289$$

$$7.389056099$$

$$0.6666666667$$

Notice that the exponential function is defined by *exp*. To compute e^2, we used the command *exp(2)*, which brings up the concept of variables and functions and how they translate to *Maple*. To store a value in a variable, *Maple* needs to know the difference between an equality and an assignment and thus requires that you use an ':=' to actually store a value to a variable.

> f:= 12+sqrt(3)*sin(Pi/4);

$$f := 12 + \frac{1}{2}\sqrt{3}\sqrt{2}$$

We have also introduced the *sqrt* command as well. From algebra, we can also represent the square root as a fractional exponent.

> 12+(3^(1/2))*sin(Pi/4);

$$12 + \frac{1}{2}\sqrt{3}\sqrt{2}$$

Now one can also define an expression involving variables:

> f:= 12+sqrt(3+x);

$$f := 12 + \sqrt{3+x}$$

If you wish to evaluate the expression at $x = 3$, in essence looking for $f(3)$, you may be tempted to simply type in the command $f(3)$. If we execute this command, notice what happens:

> f(3);

$$12 + \sqrt{3 + x(3)}$$

To actually evaluate this expression at $x = 3$, we need to use the *subs* command.

> subs({x=3},f);

$$12 + \sqrt{6}$$

To make things easier and to avoid using the *subs* command for such a simple concept, we now show how to define functions.

> g:= x -> 12+sqrt(4+x);

$$g := x \rightarrow 12 + \sqrt{4+x}$$

> g(4);

$$12 + 2\sqrt{2}$$

The use of the arrow, ->, creates a variable that *Maple* now recognizes as a function, with the letter x as the place holder for the independent variable. We can simply replace the variable x with another expression. For instance, if we want to know what $g(3 - t)$ is, we use the following command:

> g(3-t);

$$12 + \sqrt{7 - t}$$

The *unapply* command can be used to achieve the same results as the arrow -> described previous.

> f2:= unapply(12+sqrt(3+x), x);

$$f2 := x \rightarrow 12 + \sqrt{3 + x}$$

> f2(3);

$$12 + \sqrt{6}$$

> f2(3.);

$$14.44948974$$

If f is an expression, then instead of using the *subs* command to evaluate f, we can also use *eval*, which seems to work better than *subs* in some cases, especially when the rule is quite complicated.

> eval(f2(x), x = 3.);

$$14.44948974$$

> eval(f2(x), x = 3);

$$12 + \sqrt{6}$$

The example below better illustrates the difference between *subs* and *eval* for evaluating expressions.

> g2:= x^2 + sin(x) - exp(x);

$$g2 := x^2 + \sin(x) - e^x$$

> eval(g2, x = 3.);

$$-10.94441691$$

> subs(x = 3., g2);

$$9. + \sin(3.) - e^{3.}$$

Plotting a function is quite simple, as can be seen in Figure 1.1. For example, we can plot f using the following *plot* command:

> plot(f,x=-5..5);

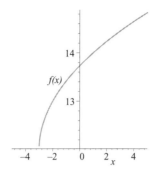

Figure 1.1: The graph of f.

To plot g, as seen in Figure 1.2, we need to modify the syntax slightly. Notice that the independent variable is explicitly written. We could also use a completely different variable (e.g., t), and let t range from -5 to 5 instead. The only difference in the resulting graph would be that the horizontal axis would be labeled t instead of x.

> plot(g(x),x=-5..5);

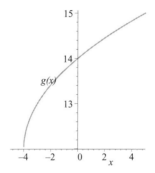

Figure 1.2: The graph of g.

Now, we can now plot the expression f and the function $g(x)$ together in the same *plot* command, as seen in Figure 1.3. The "thickness" option has also been used so that we can give thickness 1 to f and thickness 2 to $g(x)$. Notice that $[f, g(x)]$ is the list of the expression f and function $g(x)$, which orders them with f first and $g(x)$ second. This also allows us to give the two thicknesses as the list $[1, 2]$, which assigns the thicknesses to f and $g(x)$ in the same order. If the thickness 2 was to be the same for both, then only one thickness option would be needed given as "thickness = 2", and we could then give $\{f, g(x)\}$ instead of $[f, g(x)]$. $\{f, g(x)\}$ is the set containing f and $g(x)$ that does not order them in any way.

> plot([f,g(x)], x = -5..5, thickness = [1,2]);

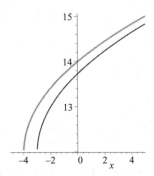

Figure 1.3: The graphs of f and g together.

We can also have *Maple* plot functions of two variables in the *xyz*-coordinate system. For example, consider the expression $h = e^{-(x^2+y^2)}$ (Figure 1.4). Plotting a function of two variables is similar to plotting a function of one variable, except the *plot3d* command is used instead of *plot*. We must also be sure to define the domain in terms of the two independent variables.

> h:= exp(-(x^2+y^2));

$$h := e^{-x^2-y^2}$$

> plot3d(h, x=-3..3, y=-3..3);

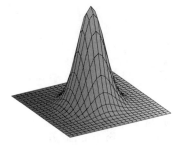

Figure 1.4: Plot of the surface defined by h.

To learn more about the *plot* and *plot3d* commands, on a new command line, enter either *?plot* or *?plot3d*. To learn more about the possible options, try *?plot, options* and *?plot3d, options*. Throughout this book, you will find many examples of plot commands and the various options that can be used in conjunction with it. *Maple* also has a *plots* package that you should load into *Maple* like the *linalg* package before doing any sophisticated plotting since it contains many useful commands. If you forget something about a *Maple* command, another way to access the Help file for this command is to place your cursor in the middle of the command name, and then go up to Help. You should then see Help on "command name", which you can select to quickly get to this command's Help file.

Now, we will move on to one of the more important commands that one can use in *Maple*. The *solve* command can be used to solve very simple equations (e.g., $x^2 - 2x + 1 = x$) and is implemented as follows:

> solve(x^2-2*x+1=x);

$$\frac{3}{2} + \frac{\sqrt{5}}{2}, \frac{3}{2} - \frac{\sqrt{5}}{2}$$

Note that both solutions were returned. Since there was only one unknown variable in the expression, it is understood what variable needs to be isolated. The *solve* command can also be used to solve more complex equations and also systems of equations, which will be of great importance in our exploration of linear algebra. Consider the following system of equations (notice that this system is nonlinear):

$$4xy - 3x + 6y = 0$$
$$4x - 3y + 23xy = 0 \tag{1.1}$$

We can solve this system using the *solve* command similar to before, but now we must specify which variables we are solving for.

> ANS:= solve({4*x*y-3*x+6*y=0, 4*x-3*y+23*x*y=0}, {x,y});

$$ANS := \{x = 0, y = 0\}, \left\{x = -\frac{3}{17}, y = -\frac{1}{10}\right\}$$

Notice that *Maple* gave two solutions to the system and the solutions were stored in the variable *ANS*. To access the second solution, we use the following command:

> ANS[2];

$$\left\{x = -\frac{3}{17}, y = -\frac{1}{10}\right\}$$

To access the first coordinate in the second solution, we have to perform the following command:

> ANS[2][1];

$$x = -\frac{3}{17}$$

However, this does not give us that actual value, it gives us the equality $x = -\frac{3}{17}$. To get only the value of $-\frac{3}{17}$, we use the *rhs* command, which tells *Maple* to grab the (r)ight (h)and (s)ide of the equation. The *lhs* command will grab only the left-hand side of an equation in a similar manner.

> rhs(ANS[2][1]);

$$-\frac{3}{17}$$

Now we well plot the two implicitly defined functions given in equation (1.1) in a square interval around the two solution points that *Maple* found, as depicted in Figure 1.5. To do this, we introduce the *implicitplot* command. Notice that the solutions to the nonlinear system (1.1) are the intersections of the two curves. The *plots* package must also be loaded to use the *display* command.

> with(plots):
> f1:= 4*x*y-3*x+6*y=0; f2:=4*x-3*y+23*x*y=0;

$$f1 := 4xy - 3x + 6y = 0$$
$$f2 := 4x - 3y + 23xy = 0$$

> plotf1:= implicitplot(f1, x=-2..2, y=-2..2, color=BLUE, thickness=2):
> plotf2:= implicitplot(f2, x=-2..2, y=-2..2, color=RED, thickness=2, numpoints = 5000):
> display({plotf1,plotf2}, view=[-1..1,-1..1]);

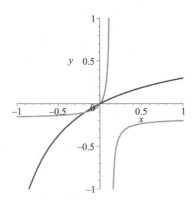

Figure 1.5: Intersection of the two implicitly defined curves from equation (1.1).

1.2   Programming

Maple would be nothing more than a glorified graphing calculator if it did not have a built-in programming structure. We already have been introduced to the concept of variables, which is one aspect of programming. Now, we introduce some commands to manipulate variables. The first command we will consider is the *for* loop command. We will do a simple example first.

```
> for k from 1 to 5 do
    k;
  od;
```

$$1$$
$$2$$
$$3$$
$$4$$
$$5$$

If you are having problems entering the second line without executing the first, it can all be scrunched onto one line, however, pressing "Shift" and "Enter" at the same time will take you to the next line without executing a single command. Inserting a colon at the end of the *od*, instead of a semicolon, will suppress output. Also notice that a *for* loop must end with the *od* command.

```
> for k from 1 to 5 do
    k;
  od:
```

However, the colon cannot suppress all output if it is overridden by the *print* command, as can be seen in the following loop:

```
> for k from 1 to 5 do
    print(k);
  od:
```

$$1$$
$$2$$
$$3$$
$$4$$
$$5$$

Like most programming languages, there are many counting options in the *for* loop command. The general syntax for the *for* loop is as follows:

```
> for variable from start by step to stop do
    place commands here
  od;
```

Note that to end a *for* loop you can also use the *end do* command. So the syntax could also look like the following:

```
> for variable from start by step to stop do
     place commands here
  end do;
```

We now do another simple example, counting backward:

```
> for k from 5 by -1 to 1 do
     print(k);
  od:
```

$$5$$
$$4$$
$$3$$
$$2$$
$$1$$

Or we can count by 2's.

```
> for k from 1 by 2 to 5 do
     print(k);
  od:
```

$$1$$
$$3$$
$$5$$

After the *for* loop, the next statement to learn is the *if* statement. The *if* statement is a conditional statement that verifies whether an expression is true or false. If true, the statement is then executed. There are many variations of the *if* statement, with extra options (e.g., *else* and *else if*). Next, we execute a simple *if* statement inside a *for* loop. The *if* statement will check to see if the loop variable j satisfies the inequality $j \leq 5$. If it does, then the *print* command will be executed, otherwise the *if* statement will not be executed on the current loop of the *for* statement.

```
> for j from 1 to 20 do
     if j <= 5 then
        print(j);
  fi:
```

od:

$$1$$
$$2$$
$$3$$
$$4$$
$$5$$

We do another example with an *elif* and *else* added to the *if* statement. The *elif* command is short for *else if*, which is the standard way to add an extra condition to the *if* command, while the *else* command is used only if none of the conditions in the *else* and *else if* structures are satisfied. We are also going to terminate the *if* statement with the *end if* command, instead of using the shorthand notation of *fi*.

```
> for j from 1 to 10 do
    if j<= 5 then
        print(j);
    elif j >=9 then
        print(j);
    else
        print('no');
    end if:
end do:
```

$$1$$
$$2$$
$$3$$
$$4$$
$$5$$
$$no$$
$$no$$
$$no$$
$$9$$
$$10$$

And remember, before you go on to the next chapter, make your way through the "Introduction to *Maple*" worksheet we told you about at the beginning of this chapter.

Chapter 2

Linear Systems of Equations and Matrices

2.1 Linear Systems of Equations

The basic idea behind a general linear system of equations is that of a system of two xy-plane line equations $ax + by = c$ and $dx + ey = f$, for real constants a through f. Such a system of two line equations in the xy-plane has a *simultaneous solution*, the intersection set of the two lines. In general, we consider a simultaneous solution to be the collection of points that simultaneously satisfies all equations in a given system of equations. In the two-line equation system example we have only three possibilities. If the two lines are not parallel, then their intersection set is a single point P, hence the simultaneous solution will simply be $\{P\}$. If the two lines are parallel and distinct, then their intersection set is empty and there is no simultaneous solution. Last, if the two lines are the same, the simultaneous solution is simply the entire line, which consists of an infinite number of points. We now give an example of each of these three situations. As you will see, the solution points of each of these three systems are where each pair of lines intersect. In these systems, in order to solve by hand for the solution points, we must try to solve for one of the variables x or y by itself without the other variable. In essence, we are assuming each pair of lines intersect at a single point unless we discover otherwise.

Example 2.1.1. Our first example is to solve the following system:

$$5x + 2y = 9$$
$$8x - 3y = -4 \tag{2.1}$$

In order to solve for one of the variables x or y by itself without the other variable, we can multiply the first equation by 3 and multiply the second equation by 2, which makes the coefficients of y negatives of each other. Now, we can add the two equations together eliminating the variable y, and so solving for x. Similarly, we can solve for y alone by multiplying the first equation by 8 and multiplying the second equation by 5 and then subtracting the second equation from the first:

$$
\begin{array}{cc}
3(5x + 2y \; = \; 9) & 8(5x + 2y \; = \; 9) \\
+2(8x - 3y \; = -4) & -5(8x - 3y \; = -4) \\
\hline
31x + 0y = 19 & 0x + 31y = 92 \\
x = \frac{19}{31} & y = \frac{92}{31}
\end{array}
$$

These two equations tell us that the system of two lines intersects at the point $\left(\frac{19}{31}, \frac{92}{31}\right)$. So the solution to this system is a single point. Now, we want the *solve* command of *Maple* to give us the same result.

> solve({5*x+2*y=9, 8*x-3*y=-4}, {x,y});

$$
\left\{ x = \frac{19}{31}, y = \frac{92}{31} \right\}
$$

Last, for this example we plot these two lines and see they intersect at this point, as depicted in Figure 2.1. The command *implicitplot* in the *plots* package will plot equations or systems of equations in two variables.

> with(plots):
> implicitplot([5*x+2*y=9, 8*x-3*y=-4], x=-5..5, y=0..5, thickness = 2, color = [blue, red]);

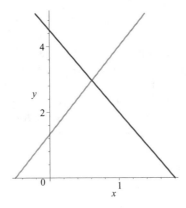

Figure 2.1: The intersection of system (2.1) is the single point $\left(\frac{19}{31}, \frac{92}{31}\right)$.

Example 2.1.2. Our second example is to solve the following system:

$$5x + 2y = 9$$
$$10x + 4y = -7 \tag{2.2}$$

Notice that the first equation for system (2.2) is the same as that of system (2.1). Applying the same ideas of Example 2.1.1, we multiply the first equation by 2 and then subtract the equations:

$$2(5x + 2y = 9)$$
$$-(10x + 4y = -7)$$
$$\overline{0x + 0y = 11}$$
$$0 = 11$$

The equation $0 = 11$ is impossible and so our system of two lines has no solution. This should not be surprising since these two lines are not identical, but are parallel since they have the same slope $m = -\frac{5}{2}$. The *solve* command gives no result indicating that the system has no solution or intersection point. This situation is illustrated in Figure 2.2.

> solve({5*x+2*y=9, 10*x+4*y=-7}, {x,y});
> implicitplot([5*x+2*y=9, 10*x+4*y=-7], x=-5..5, y=-5..5, thickness = 2, color=[blue, red]);

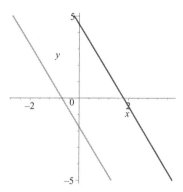

Figure 2.2: The parallel lines of system (2.2) have no solution set.

So we have done an example of two lines with one intersection point, two lines with no intersection point, and all that is left is to do an example that results in an infinite number of solutions.

Example 2.1.3. Now, we consider the following system of equations:

$$5x + 2y = 9$$
$$-5x - 2y = -9$$

(2.3)

Upon inspecting system (2.3), it should be clear that the line equations are really the same since the second is the negative of the first. Thus, there is truly only one equation in this system, the first equation. So the solution points of this system are all of the points that lie on this first equation's line. Notice that if we add the equations of this system we get the equation $0 = 0$, that tells us that one of our equations is not needed.

> solve({5*x+2*y=9, -5*x-2*y=-9}, {x,y});

$$\left\{ x = x, y = -\frac{5}{2}x + \frac{9}{2} \right\}$$

The *solve* command has told us that the solutions of this system of two lines are all of the points $\left(x, -\frac{5}{2}x + \frac{9}{2}\right)$, as x varies over all real numbers. In this solution x is the variable that determines the solutions as x changes its value. This is an infinite number of solution points, one for each value of x.

The situation for a system of two lines in the xy-plane turns out to be typical for general linear systems, where simultaneous solutions contain no points, a single point, or an infinite number of points. Next, we introduce some definitions that will help us to understand the more general situation.

Definition 2.1.1. A *scalar* is defined to be a real or complex constant.

In linear algebra, a scalar is essentially a synonym for a constant, and this terminology originally was used in physics to distinguish between constants versus vectors. For simplicity, and so that we can plot our results, we shall generally assume that a scalar is a real constant. However, sometimes we shall take our scalars to be complex, since real-life problems usually need only real scalars, but not always, as we will see in Chapter 12 with eigenvalues and eigenvectors. Recall that the complex numbers $\mathbb{C}$ are an extension of the real numbers $\mathbb{R}$ where

$$\mathbb{C} = \left\{ a + bi \mid a, b \in \mathbb{R},\ i^2 = -1 \right\}$$

The complex numbers $\mathbb{C}$ were created in order to solve algebra problems (e.g., solving polynomial equations and factoring polynomials where their coefficients are all real).

For clarity, we shall often say real scalar although you should not be surprised if we use or say complex scalar. The basic facts of how linear systems of equations work do not depend on whether our scalars are real or complex, as we shall see throughout the rest of this book.

Definition 2.1.2. A *linear equation* in the variables $x_1, x_2, \ldots, x_n$ is of the form

$$a_1 x_1 + a_2 x_2 + \cdots + a_n x_n = b$$

for scalars $a_1, a_2, \ldots, a_n$ and b. If the scalars in a linear equation are all real numbers, then the values of its variables are also all real, but if at least one scalar is a complex number, then the values of the linear equation's variables are also complex.

These definitions help to generalize the equation of a line in the xy-plane with variables x and y. If a linear equation has only three variables, then they are usually denoted by x, y, and z with the linear equation written as $ax + by + cz = d$ for real constants a through d. The plot of such a three variable linear equation is a plane in space with coordinates x, y, and z. Now, let us have *Maple* plot the three variable linear equation $5x - 2y + 7z = 15$ in order to see that this is true.

```
> f:= 5*x-2*y+7*z=15:
> g:= solve(f,z);
```

$$g := -\frac{5}{7}x + \frac{2}{7}y + \frac{15}{7}$$

```
> plot3d(g, x=-7..7, y=-7..7, axes = boxed);
```

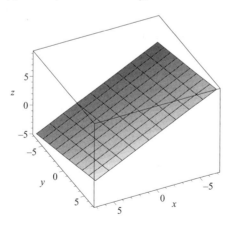

Figure 2.3: The plane defined by g.

To plot the plane, as seen in Figure 2.3, notice that we first solved for the variable z as a function of x and y using the *solve* command, and then plotted the resulting expression g. Now let us discuss why the plot of the linear equation $ax + by = c$ is a line (one-dimensional) while the plot of the linear equation $ax + by + cz = d$ is a plane (two-dimensional). The reason lies in the fact that the equation $ax + by = c$ has one independent variable x and one dependent variable y, that is, it can be written as $y = -\frac{a}{b}x + \frac{c}{b}$, and the number of independent variables in an equation is the number of degrees of freedom in expressing the equation's solution points $(x, y) = \left(x, -\frac{a}{b}x + \frac{c}{b}\right)$. Similarly, the equation $ax + by + cz = d$ has two independent variables x and y, while only one dependent variable z, that is, it can be written as $z = -\frac{a}{c}x - \frac{b}{c}y + \frac{d}{c}$, and the number of independent variables in an equation is the number of degrees of freedom in expressing the equation's solution points $(x, y, z) = \left(x, y, -\frac{a}{c}x - \frac{b}{c}y + \frac{d}{c}\right)$. Since the dimension of a geometric object is really the number of degrees of freedom in expressing it, the equation $ax+by = c$ has dimension one and is a line while the equation $ax+by+cz = d$ has dimension two and is a plane.

Definition 2.1.3. The *dimension* of a geometric object is the minimum number of degrees of freedom needed to express the equation's solution points. Equivalently, the dimension of a geometric object is the minimum number of independent variables it takes to describe uniquely all of the points of the geometric object.

By independent variables, we mean variables that do not depend on any other variables. By dependent variables, we mean variables that do depend on other variables. If we see an expression, such as $z = f(x, y)$, z is a dependent variable, which depends on the two independent variables x and y. Now, it should be clear that a general linear equation

$$a_1 x_1 + a_2 x_2 + \cdots + a_n x_n = b \tag{2.4}$$

has dimension $n - 1$ as a geometric object in n-dimensional space $\mathbb{R}^n$. For example, $\mathbb{R}^2$ is two-dimensional space and often pictured as the xy-plane while $\mathbb{R}^3$ is three-dimensional space and pictured as xyz-space.

Definition 2.1.4. $\mathbb{R}^n$ is the collection of n-tuples of all real numbers:

$$\mathbb{R}^n = \{(x_1, x_2, \ldots, x_n) \,|\, x_k \in \mathbb{R} \text{ for } k = 1, 2, \ldots, n\}.$$

Definition 2.1.5. $\mathbb{C}^n$ is the collection of n-tuples of all complex numbers:

$$\mathbb{C}^n = \{(x_1, x_2, \ldots, x_n) \,|\, x_k \in \mathbb{C} \text{ for } k = 1, 2, \ldots, n\}$$

From the above two definitions, we can now see that points in $\mathbb{R}^n$ of the form

$$\left(x_1, x_2, \ldots, x_{n-1}, -\frac{a_1}{a_n}x_1 - \frac{a_2}{a_n}x_2 - \cdots - \frac{a_{n-1}}{a_n}x_{n-1} + \frac{b}{a_n}\right)$$

satisfy the general linear equation (2.4). If $a_n = 0$, then we simply solve for x_k, where $k \neq n$ and $a_k \neq 0$, with the resulting point still being a point of $\mathbb{R}^n$.

Now back to discussing linear systems of equations and their simultaneous solutions. From the information above, a line is one-dimensional and the intersection set of two lines is usually a point, which is zero-dimensional. We have already discussed the situations in which the intersection of two lines is not zero-dimensional, so we will not repeat ourselves here. If instead, we consider three random lines in the plane, then there is a high probability that the intersection set of these three random lines is empty. However, it is possible that the intersection set could be of dimension one, or zero, as well. The intersection set of a line and a plane is normally a single point, which is zero-dimensional. However, the simultaneous solution set could have dimension one, or even be empty. You should determine for yourself the required orientation between the plane and line for their resulting intersection to be one-dimensional or empty.

As well, a plane is two-dimensional and the intersection set of two planes is usually a line that is one-dimensional. The intersection set could be the plane itself, which is two-dimensional, or empty, which has no dimension. The intersection set of three planes is usually a point that is zero-dimensional. One can think of this as first intersecting two planes, which results in a line and then intersecting the resulting line with the third plane, which is usually a single point. The intersection set could also be the plane itself, which is two-dimensional, a line that is one-dimensional, or empty with no dimension. The intersection set of four planes is usually empty and has no dimension, although the intersection set could be of any dimension two or lower. Hopefully, you see a pattern developing here.

The general idea of this discussion is that in looking for the simultaneous solution or intersection set to a system of m linear equations, if each linear equation in the system is a geometric object of the same dimension n (i.e., $n+1$ variables), then the dimension of the solution set is usually $(n+1) - m = n - m + 1$, which is equivalent to stating that the dimension is simply given by taking the number of variables minus the number of equations, although it can have any dimension from n down to none. Thus, a square system of r linear equations, where we have as many equations as variables in each equation, typically has a solution set of dimension $r - r = 0$, and is a single point where the dimension of each equation is $r - 1$.

Let us do a few examples to illustrate what we have discovered. Since we cannot plot beyond three-dimensions, we stick to solving linear systems of equations in 3 variables so that each equation is a plane in $\mathbb{R}^3$.

Example 2.1.4. We begin by plotting and solving the linear system:

$$5x - 2y + 7z = 15$$
$$3x + 8y - z = -4 \tag{2.5}$$

This solution set should be one-dimensional, which is a line since these equations are a pair of two-dimensional planes, that is, the dimension of the solution set is the number of its variables minus the number of its equations, that is $3 - 2 = 1$. Figures 2.4 and 2.5 illustrate the intersection of these planes and the corresponding solution to system (2.5), respectively.

> f:= 5*x-2*y+7*z=15: h:= 3*x + 8*y - z = -4:
> g:= solve(f,z):
> k:= solve(h,z);

$$k := 3x + 8y + 4$$

> plot_g:= plot3d(g, x=-7..7, y=-7..7, axes = boxed, color = red):
> plot_k:= plot3d(k, x=-7..7, y=-7..7, axes = boxed, color = blue):
> display({plot_g, plot_k}, style=patchnogrid, orientation=[-9,48]);

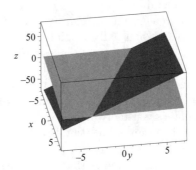

Figure 2.4: Plot of the planes f and g.

> solnline:= [op(solve({f,h}, {x,y}))];

$$solnline := \left[x = \frac{56}{23} - \frac{27}{23}z, y = -\frac{65}{46} + \frac{13}{23}z \right]$$

> op(2,solnline[1]);

$$\frac{56}{23} - \frac{27}{23}z$$

> plot_solnline:= spacecurve([op(2,solnline[1]), op(2,solnline[2]), z], z=-10..10, color = black, thickness=2):

> display({plot_g, plot_k, plot_solnline}, style = patchnogrid, orientation = [-67,78]);

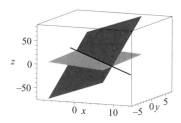

Figure 2.5: The intersection of f and g is the line *solnline*.

The solution *solnline* to system (2.5), provided by *Maple*, is one-dimensional and a line since it expresses the solution in terms of the single independent variable z for the two dependent variables x and y. This solution is clearly represented as such in Figure 2.5.

To solve system (2.5) by hand, we must first decide which variable we would like expressed in terms of the others. With our current system of two equations and three unknowns, we should be able to express two variables as a function of the third in the final solution. Considering that we had *Maple* express the x and y variables in terms of z, that is what we will do. A very simply way to do this, is to move the z variable, along with its coefficient, to the right-hand side (RHS) of each equation. If we focus on the resulting left-hand sides (LHS), then we can approach the problem just like we did in Example 2.1.1. To first solve for x in terms of z, we must cancel the y variables. To do this, we multiply the first equation of system (2.5) by 4, and add it to the second. To solve for y in terms of z, we multiply the first equation by 3, the second by -5, and add them.

$$
\begin{array}{ll}
4(5x - 2y = 15 - 7z) & 3(5x - 2y = 15 - 7z) \\
\underline{3x + 8y = -4 + z} & \underline{-5(3x + 8y = -4 + z)} \\
23x = 56 - 27z & -46y = 65 - 26z \\
x = \dfrac{56}{23} - \dfrac{27}{23}z & y = -\dfrac{65}{46} + \dfrac{13}{23}z
\end{array}
$$

From the work above, we see that the solution is a line L in $\mathbb{R}^3$ defined as follows:

$$
L = \left\{ \left(\frac{56}{23} - \frac{27}{23}z, \, -\frac{65}{46} + \frac{13}{23}z, \, z \right) \; \middle| \; z \in \mathbb{R} \right\}
$$

Clearly, the approach above to solving the system by hand can be generalized. If we have a system of m equations with n variables and $m < n$, then we keep only m variables on the LHS, moving the resulting $n - m$ variables to the RHS and then solving the resulting system in terms of the m variables left. This will normally yield a solution with $n - m$ independent variables, and m dependent variables.

Example 2.1.5. Now let us throw in another plane and see if we get a single point as solution. We will now attempt to solve and plot the linear system:

$$5x - 2y + 7z = 15$$
$$3x + 8y - z = -4$$
$$-9x + 6y + 10z = 7$$

> with(plottools):

> u:= -9*x + 6*y + 5*z = 7:

> solnpoint:= op(solve([f,h,u], [x,y,z]));

$$solnpoint := \left[x = \frac{13}{109}, y = -\frac{65}{218}, z = \frac{215}{109} \right]$$

> op(2,solnpoint[1]);

$$\frac{13}{109}$$

> plot_solnpoint:= sphere([op(2, solnpoint[1]), op(2, solnpoint[2]), op(2, solnpoint[3])], 1, color = black):

> v:= solve(u,z);

$$v := \frac{9}{5}x - \frac{6}{5}y + \frac{7}{5}$$

> plot_v:= plot3d(v, x=-7..7, y=-7..7, axes = boxed, color = tan):

> display({plot_g, plot_k, plot_v, plot_solnpoint}, view = [-10..10,-10..10,-10..10], orientation = [68,40]);

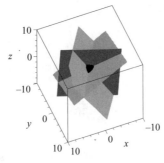

Figure 2.6: The intersection of the three planes is the single point *solnpoint*.

The point of intersection is difficult to see in Figure 2.6, however, if you enter these commands into *Maple*, you can rotate the figure any way you wish and will be able to see the point of intersection more clearly. At this point, we would once again like to remind you of the *Maple* worksheets located on the book's website, whose link is given in the preface. This allows you to easily download the worksheet containing all of the commands in this section so that you can rotate the above figure all you want. We do suggest that you type some of these commands in yourself in order to get used to *Maple* and its syntax. After all, watching a pianist perform may give you the idea of how to play the piano, but if you do not do any playing yourself you will still be a lousy pianist.

Now one last example where *Maple* can solve the system, but we cannot plot the linear equations and their simultaneous solution because the dimension of each linear equation's solution set is four since each equation has five variables.

Example 2.1.6. Let us solve the linear system given by

$$5x - 2y + 7z - 3w - t = 25$$
$$3x + 8y - z + 2w - 7t = -4$$
$$-9x + 6y + 10z + w + 5t = 7$$

Note that each of these linear equations represents a four-dimensional geometrical object. Since we have three linear equations, we expect the simultaneous solution to be a $5 - 3 = 2$-dimensional plane.

```
> solve({5*x - 2*y + 7*z - 3*w - t = 25, 3*x + 8*y - z + 2*w - 7*t = -4, -9*x
+ 6*y + 10*z + w + 5*t = 7}, {x, y, z});
```

$$\left\{ z = -\frac{112}{551}t + \frac{1310}{551} + \frac{124}{551}w, \ x = \frac{395}{551}t + \frac{762}{551} + \frac{94}{551}w, \ y = -\frac{795}{1102} - \frac{315}{1102}w + \frac{320}{551}t \right\}$$

The above solution has two independent variables w and t with three dependent variables x, y, and z. So the dimension of the solution to this system is two. If we solve the system consisting of only the first two linear equations, then we expect to get a simultaneous solution that is $5 - 2 = 3$-dimensional.

```
> solve({5*x-2*y+7*z-3*w-t=25, 3*x+8*y-z+2*w-7*t=-4 }, {x,y});
```

$$\left\{ x = \frac{11}{23}t + \frac{96}{23} - \frac{27}{23}z + \frac{10}{23}w, y = -\frac{95}{46} + \frac{13}{23}z - \frac{19}{46}w + \frac{16}{23}t \right\}$$

The above solution has three independent variables z, w, and t with two dependent variables x and y. So the dimension of the solution to this system is three. The two solutions we get above do agree with our dimensional analysis for each linear system.

All of the examples and equations that we have looked at thus far have been linear. In fact, most of the equations that we will look at will be linear. Linear equations are nice because they are easy to work with, many ideas in one or two dimensions can be generalized to higher dimensions. On the other hand, most equations that occur in science and mathematics are nonlinear, and the last couple of *Maple* problems illustrate just a small sample of the various types of nonlinear equations.

Homework Problems

1. Give conditions on two lines such that their intersection results in a simultaneous solution that is (a) of dimension zero, (b) of dimension one, (c) is empty.

2. Give conditions on a line and a plane such that the intersection of these two geometric objects results in a simultaneous solution that is (a) of dimension zero, (b) of dimension one, (c) is empty.

3. Give conditions on two planes such that their intersection results in a simultaneous solution that is (a) of dimension one, (b) of dimension two, (c) is empty.

4. Determine the dimension of the solutions to problem 2 in the *Maple* Problems section. You do not have to solve the systems by hand.

5. Solve by hand, for all of its intersection points (if any), each of the following linear systems of two lines:

(a) $\quad 3x - 2y = 9$ (b) $\quad x + 5y = 9$ (c) $\quad 7x + 14y = -21$
$\quad\quad 5x + 4y = -13$ $\quad\quad -2x - 10y = -2$ $\quad\quad x + 2y = -3$

6. Solve by hand, for all intersection points (if any), each of the following linear systems of three planes:

(a) $3x - 2y + z = 9$ (b) $x - 6y + 3z = 9$ (c) $-x + 3y + 8z = 9$
$\quad 5x + 4y - 7z = -13$ $\quad -2x + 4y - 7z = -2$ $\quad 5x + 4y - 3z = -2$
$\quad x - y - z = 2$ $\quad -x - 2y - 4z = 15$ $\quad 6x + y - 11z = -11$

7. Show that $ax + by = c$ and $dx + ey = f$, for a through f real constants, are two parallel lines exactly when $ae - bd = 0$. As a consequence of this, these two lines intersect with dimension zero exactly when $ae - bd \neq 0$.

8. Show that $ax + by = c$ and $dx + ey = f$, for a through f real con-

stants, are two parallel lines exactly when there is some real number k where $dx + ey = k(ax + by)$.

9. Show that $ax + by = c$ and $dx + ey = f$, for a through f real constants, has the single intersection point with coordinates $\left(\dfrac{ce - bf}{ae - bd}, \dfrac{af - cd}{ae - bd} \right)$, when $ae - bd \neq 0$.

10. Show that $ax + by = c$ and $dx + ey = f$, for a through f real constants, are two perpendicular lines exactly when $ad + be = 0$.

11. The definition of parallel planes is that they are either identical or they do not intersect. Show that $ax + by + cz = d$ and $ax + by + cz = e$, for a through e real constants, are two parallel planes.

12. Show that the two planes $ax + by + cz = d$ and $ex + fy + gz = h$, for a through h real constants, have the line of intersection given by

$$\left\{ x = \frac{bg - cf}{af - be} z + \frac{df - bh}{af - be}, y = \frac{ce - ag}{af - be} z + \frac{ah - ed}{af - be}, z \right\}$$

for the independent variable z when $af - be \neq 0$.

Maple Problems

1. Use *Maple* to graph the required lines and planes that illustrate your answers to *Homework* problems 1, 2, 3, 5, and 6.

2. Solve for the intersections of the following equations, and graph the simultaneous solutions where applicable.

(a) $3x + 4y = 4$
$\ 4x + 3y = 8$

(b) $x + 4y = 1$
$\ x + y = 1$
$\ -4x + y = -4$

(c) $x + 4y = 1$
$\ x + 2y = 1$
$\ -4x + y = -1$

(d) $3x - 5y + 6z = 1$
$\ 4x + 3y - 2z = 2$

(e) $3x - 5y + 6z = 1$
$\ 4x + 3y - 2z = 2$
$\ x - y + 2z = -1$

(f) $x + y + z = 1$
$\ x - y + z = 1$
$\ x - 2y + z = 2$

(g) $x + y + 3z = 1$
$\ x - 2y + z = 3$
$\ x - 2y + 2z = -2$

(h) $3x - 5y + 6z + 6t = 1$
$\ 4x + 3y - 2z - 3t = 2$
$\ x - y + 2z + 2t = -1$

(i) $x + y + z + t = 1$
$\ x - y + z + t = 1$
$\ x - 2y + z + t = 2$
$\ x - 2y + z + 2t = 2$

3. Use the *implicitplot* command in the *plots* package to plot the following nonlinear equations. You may need the plot option "grid = $[100,100]$" in order to get a good plot of these three curves, this option controls the number of points used in the coordinate directions for plotting.

   (a) $x^2 - 7xy + y^2 - 10x + 25y = 1$

   (b) $x^2 - 7xy + 15y^2 - 10x + 25y = 1$

   (c) $\cos(3x - y) - x - \sin(x + y) = \dfrac{1}{2}$

4. Use the *implicitplot3d* command in the *plots* package to plot the nonlinear equations. You will need the two plot options of "axes = boxed" and "scaling = constrained" in order to plot these surfaces accurately and to see exactly where they are located in space.

   (a) $(x - 9)^2 + (y - 2)^2 + (z + 5)^2 = 144$

   (b) $(x - 9)^2 + (y - 2)^2 - (z + 5)^2 = 144$

   (c) $(x - y + z)^2 - x^3 + \sin(z) = y$

5. Use the *solve* command to verify *Homework* problem 9.

6. Use the *solve* command to verify *Homework* problem 12.

2.2   Augmented Matrix of a Linear System and Row Operations

A general linear system of equations is always written so that each linear equation uses the same variables given in the same order. This makes it easier to write down the system, as well as to solve the system of equations simultaneously.

Definition 2.2.1. A *general linear system* of m equations in n variables $x_1, x_2, \ldots, x_n$ is of the form

$$
\begin{aligned}
a_{1,1}x_1 + a_{1,2}x_2 + \cdots + a_{1,n}x_n &= b_1 \\
a_{2,1}x_1 + a_{2,2}x_2 + \cdots + a_{2,n}x_n &= b_2 \\
&\vdots \\
a_{m,1}x_1 + a_{m,2}x_2 + \cdots + a_{m,n}x_n &= b_m
\end{aligned}
$$

The $a_{i,j}$'s above are called the coefficients of the linear system where the subscript i tells you the equation you are in and the subscript j tells you which variable it multiplies. The mathematical construct A, formed by the $a_{i,j}$'s is called the *matrix of coefficients* of this linear system, where the subscript i now tells you which row you are in and the subscript j tells you which column. The matrix A, of coefficients, is of size (dimension) $m \times n$ (read as "m by n") where m is the number of equations and n is the number of variables in the linear system. The array B formed by the b_i's is a column matrix of size $m \times 1$ and is called RHS matrix of the system. The array X formed by the x_j's is a column matrix of size $n \times 1$.

This system of m linear equations in n variables will generally have a simultaneous solution which has dimension $n - m$ as a geometric object in n-dimensional space $\mathbb{R}^n$ with the variables $x_1, x_2, \ldots, x_n$. In many practical situations where linear systems and their simultaneous solution set play a role, it is normal that the number of equations m is equal to the number of variables n. When $m = n$, the simultaneous solution set to the linear system usually consists of a single point, since its dimension is typically $n - m = 0$. When $m > n$, the number of simultaneous solutions is usually none while when $m < n$, the number of simultaneous solutions is usually infinite, which is backed up by our dimensional analysis.

Before we go any further, we will introduce the definition of a matrix, since we will be using the word quite frequently.

Definition 2.2.2. A *matrix*, or *two-dimensional array*, is a collection of real or complex numbers arranged in rows and columns. If the entries of a matrix A are real numbers, and there are m rows and n columns, we say that $A \in \mathbb{R}^{m \times n}$, likewise if the matrix B has complex entries, we say that $B \in \mathbb{C}^{m \times n}$.

A matrix with m rows and n columns is said to be of size $m \times n$. The following is the general structure of a matrix with m rows and n columns. Pay special attention to how the entries in the matrix are indexed.

$$A = \begin{bmatrix} a_{1,1} & a_{1,2} & a_{1,3} & \cdots & a_{1,n} \\ a_{2,1} & a_{2,2} & a_{2,3} & \cdots & a_{2,n} \\ & \vdots & & \vdots & \\ a_{m,1} & a_{m,2} & a_{m,3} & \cdots & a_{m,n} \end{bmatrix}$$

If there is only one column, then $n = 1$ and the matrix is called a *column matrix*. Similarly, if $m = 1$, the matrix is referred to as a *row matrix*. Column matrices will play an important role in vector algebra later on in the text.

The following definition of the augmented matrix will allow us to place into a single matrix all of the information contained in a linear system of equations. As such, we will be able to use this augmented matrix for a linear system to find all of the system's simultaneous solutions without the need to carry around the superfluous variable symbols of the system and its equal signs.

Definition 2.2.3. If A is an $m \times n$ matrix with entries $a_{i,j}$, and B is $m \times k$ matrix with entries $b_{i,j}$, the *augmented matrix* $(A|B)$ is the $m \times (n+k)$ matrix defined as:

$$(A|B) = \begin{bmatrix} a_{1,1} & a_{1,2} & a_{1,3} & \cdots & a_{1,n} & b_{1,1} & b_{1,2} & b_{1,3} & \cdots & b_{1,k} \\ a_{2,1} & a_{2,2} & a_{2,3} & \cdots & a_{2,n} & b_{2,1} & b_{2,2} & b_{2,3} & \cdots & b_{2,k} \\ & \vdots & & & \vdots & \vdots & & & & \vdots \\ a_{m,1} & a_{m,2} & a_{m,3} & \cdots & a_{m,n} & b_{m,1} & b_{m,2} & b_{m,3} & \cdots & b_{m,k} \end{bmatrix}$$

Let us do an example of forming the coefficient matrix A and the RHS matrix B for a linear system as well as the augmented matrix of the system $(A|B)$, which is the column B written in after the last column of A. The *Maple* commands *augment* or *concat* (short for concatenate) in the *linalg* package will join the matrix B to A.

Example 2.2.1. We will use the following linear system:

$$5x - 3y = 9$$
$$2x + 7y = -4$$

and find the augmented matrix C of this linear system.

> restart: with(linalg):
> A:= matrix([[5,-3],[2,7]]);

$$A := \begin{bmatrix} 5 & -3 \\ 2 & 7 \end{bmatrix}$$

> B:= matrix([[9],[-4]]);

$$B := \begin{bmatrix} 9 \\ -4 \end{bmatrix}$$

> C:= augment(A,B);

$$C := \begin{bmatrix} 5 & -3 & 9 \\ 2 & 7 & -4 \end{bmatrix}$$

This augmented matrix C for the system of linear equations represents all of the information of the original system and so can be used to solve the system without the need of the variables or equations. Each row of an augmented matrix for a linear system represents an equation of the system while each column but the last is a variable coefficient column. The last column of an augmented matrix always consists of the values from the RHS of the equations.

Example 2.2.2. Let C be the following augmented matrix of a system of linear equations:

$$C = \begin{bmatrix} 6 & 1 & 0 & 4 & -3 \\ -9 & 2 & 3 & -8 & 1 \\ 7 & 0 & -4 & 5 & 2 \end{bmatrix}$$

This system has three equations represented by the rows of C and four variables called x, y, z, and w represented by the first four columns of C. This augmented matrix C represents the linear system

$$6x + y + 4w = -3$$
$$-9x + 2y + 3z - 8w = 1 \tag{2.6}$$
$$7x - 4z + 5w = 2$$

Now, the question is: How can these augmented matrices be used to solve the underlying linear system of equations for their simultaneous solution? Let us take the augmented matrix C of the last example. We know that its solution set probably has dimension $4-3=1$, which says that we should be able to write the solution in terms of the single independent variable w with dependent variables x, y, and z. We defined w to be the independent variable since it corresponds to the last column of variables in the augmented matrix. You may find that for a particular problem it is easier to treat another variable as the independent variable. However, in a case like that, you can simply switch columns so that the correct variable is the second to last column in the matrix, just be sure to remember which column corresponds to each variable. Now, back to our problem. We want to manipulate the augmented matrix C for the original system into the new augmented matrix

$$D = \begin{bmatrix} 1 & 0 & 0 & d_{1,4} & d_{1,5} \\ 0 & 1 & 0 & d_{2,4} & d_{2,5} \\ 0 & 0 & 1 & d_{3,4} & d_{3,5} \end{bmatrix} \tag{2.7}$$

which is equivalent to the original augmented matrix C (same solution set) from which we can read off the solution

$$\{x = -d_{1,4}w + d_{1,5}, \; y = -d_{2,4}w + d_{2,5}, \; z = -d_{3,4}w + d_{3,5}\} \tag{2.8}$$

to the original system. Be sure to pay attention to the way in which the matrix D corresponds to the solution given above.

We can use the two row operations *mulrow* and *addrow* that are in the *linalg* package of *Maple* in order to manipulate the augmented matrix C into the new augmented matrix D. We need to solve the system without changing the solution to the system, and the *mulrow* and *addrow* commands correspond to multiplying a row by a nonzero scalar, and adding a multiple of one row to another row, respectively. The command "mulrow(C, r, α)" replaces the rth row of the matrix C by the nonzero scalar α times every entry of this row, while "addrow(C, r, k, α)" replaces the kth row of C by the sum of the nonzero scalar α times the rth row of C added to the kth row of C.

The main question we need to ask is: Why do these operations not change the equations? Given a linear equation of the form (2.4), notice that multiplying both sides of the equation by a nonzero constant α, resulting in

$$\alpha \left(a_1 x_1 + a_2 x_2 + \cdots + a_n x_n \right) = \alpha \, b \tag{2.9}$$

does not change the solution to the original equation, since we could later divide by α to retrieve the original equation. Similarly, given two equations

$$a_1 x_1 + a_2 x_2 + \cdots + a_n x_n = b$$
$$c_1 x_1 + c_2 x_2 + \cdots + c_n x_n = d$$

then the sum of the two equations is

$$(a_1 + c_1) x_1 + (a_2 + c_2) x_2 + \cdots + (a_n + c_n) x_n = b + d$$

We can retrieve either equation from this sum by subtracting either equation from the sum, and so replacing one of the two equations with this sum is reversible and does not change the simultaneous solution set of the original system.

Back to the problem at hand, when inspecting C, notice that in order to change the 6 at the top of the first column of C into a 1, we must multiply the first row of C by $\frac{1}{6}$. Then, we will multiply the new first row by 9 and add it to the second row in order to change -9 to 0. The rest of the row operations are given below to get the final augmented matrix of the system in the form of equation (2.7). Be sure to understand each step in the following sequence of commands:

> C:= matrix([[6, 1, 0, 4, -3], [-9, 2, 3, -8, 1], [7, 0, -4, 5, 2]]);

$$C := \begin{bmatrix} 6 & 1 & 0 & 4 & -3 \\ -9 & 2 & 3 & -8 & 1 \\ 7 & 0 & -4 & 5 & 2 \end{bmatrix}$$

> C1:= mulrow(C,1,1/6);

$$C1 := \begin{bmatrix} 1 & \frac{1}{6} & 0 & \frac{2}{3} & -\frac{1}{2} \\ -9 & 2 & 3 & -8 & 1 \\ 7 & 0 & -4 & 5 & 2 \end{bmatrix}$$

> C2:= addrow(C1,1,2,9);

$$C2 := \begin{bmatrix} 1 & \frac{1}{6} & 0 & \frac{2}{3} & -\frac{1}{2} \\ 0 & \frac{7}{2} & 3 & -2 & -\frac{7}{2} \\ 7 & 0 & -4 & 5 & 2 \end{bmatrix}$$

> C3:= addrow(C2,1,3,-7);

$$C3 := \begin{bmatrix} 1 & \frac{1}{6} & 0 & \frac{2}{3} & -\frac{1}{2} \\ 0 & \frac{7}{2} & 3 & -2 & -\frac{7}{2} \\ 0 & -\frac{7}{6} & -4 & \frac{1}{3} & \frac{11}{2} \end{bmatrix}$$

> C4:= mulrow(C3,2,2/7);

$$C4 := \begin{bmatrix} 1 & \frac{1}{6} & 0 & \frac{2}{3} & -\frac{1}{2} \\ 0 & 1 & \frac{6}{7} & -\frac{4}{7} & -1 \\ 0 & -\frac{7}{6} & -4 & \frac{1}{3} & \frac{11}{2} \end{bmatrix}$$

> C5:= addrow(C4,2,1,-1/6);

$$C5 := \begin{bmatrix} 1 & 0 & -\frac{1}{7} & \frac{16}{21} & -\frac{1}{3} \\ 0 & 1 & \frac{6}{7} & -\frac{4}{7} & -1 \\ 0 & -\frac{7}{6} & -4 & \frac{1}{3} & \frac{11}{2} \end{bmatrix}$$

> C6:= addrow(C5,2,3,7/6);

$$C6 := \begin{bmatrix} 1 & 0 & -\frac{1}{7} & \frac{16}{21} & -\frac{1}{3} \\ 0 & 1 & \frac{6}{7} & -\frac{4}{7} & -1 \\ 0 & 0 & -3 & -\frac{1}{3} & \frac{13}{3} \end{bmatrix}$$

> C7:= mulrow(C6,3,-1/3);

$$C7 := \begin{bmatrix} 1 & 0 & -\frac{1}{7} & \frac{16}{21} & -\frac{1}{3} \\ 0 & 1 & \frac{6}{7} & -\frac{4}{7} & -1 \\ 0 & 0 & 1 & \frac{1}{9} & -\frac{13}{9} \end{bmatrix}$$

> C8:= addrow(C7,3,1,1/7);

$$C8 := \begin{bmatrix} 1 & 0 & 0 & \frac{7}{9} & -\frac{34}{63} \\ 0 & 1 & \frac{6}{7} & -\frac{4}{7} & -1 \\ 0 & 0 & 1 & \frac{1}{9} & -\frac{13}{9} \end{bmatrix}$$

> C9:= addrow(C8,3,2,-6/7);

$$C9 := \begin{bmatrix} 1 & 0 & 0 & \frac{7}{9} & -\frac{34}{63} \\ 0 & 1 & 0 & -\frac{2}{3} & \frac{5}{21} \\ 0 & 0 & 1 & \frac{1}{9} & -\frac{13}{9} \end{bmatrix}$$

The matrix $C9$ given above is the augmented matrix of the new (but equivalent to (2.6)) linear system with equations:

$$\begin{aligned} x + \frac{7}{9}w &= -\frac{34}{63} \\ y - \frac{2}{3}w &= \frac{5}{21} \\ z + \frac{1}{9}w &= -\frac{13}{9} \end{aligned} \qquad (2.10)$$

If you followed the steps, you will notice that a value of 1 was gotten in the upper-left-hand entry $C_{1,1}$ by multiplying by the reciprocal of the entry. Notice the rest of the row changed accordingly, just like an equation would change if you multiplied everything by a scalar. The next step in the process was to make every entry in the matrix below $C_{1,1}$ equal to zero. This is done simply enough after the upper-left entry has been changed to a 1. (Make sure you understand why this is true.) Next, we move to entry $C_{2,2}$ and attempt to make it a 1. After that has been accomplished, every other entry in that column must be made into a zero. Finally, something similar is done with the third column and the entry $C_{3,3}$. The resulting matrix says that the solution to our original system is

$$\left\{ x = -\frac{7}{9}w - \frac{34}{63}, \ y = \frac{2}{3}w + \frac{5}{21}, \ z = -\frac{1}{9}w - \frac{13}{9} \right\}$$

This method of reducing the matrix to find the final answer will be explained in more detail in Chapter 3. Let us now check, using the *solve* command on the original system, to see if we get the same solution. This solution is clearly of dimension one as expected with one independent variable w expressing the solution using the three dependent variables x, y, and z.

> solve({6*x+y+4*w=-3, -9*x+2*y+3*z-8*w=1, 7*x-4*z+5*w=2}, {x,y,z});

$$\left\{ x = -\frac{34}{63} - \frac{7}{9}w, \ y = \frac{5}{21} + \frac{2}{3}w, \ z = -\frac{13}{9} - \frac{1}{9}w \right\}$$

The two solutions are determined to be the same using the row operations on the augmented matrix of the system or using *Maple's solve* command.

A similar approach can be used on complex valued matrices, as we show in the next example. The situation of a complex linear system is the same as the real case of the last example except that now our variables can take on complex values.

Example 2.2.3. Consider the following system of equations:

$$2x - 3iy - 4z = -1 + 2i$$
$$(2 - 2i)x + 3y + (1 - i)z = -2i \qquad (2.11)$$
$$(10 - 4i)x + (6 - 9i)y + (-10 - 2i)z = -3 + 2i$$

Using the same approach as that of the last example, we put the system of equations into a matrix of dimension 3×4 with complex entries. We will attempt to get values of 1 along the diagonal from the upper left to the lower right corner, and zeros off the diagonal of the left 3 columns of the matrix using only the *addrow* and *mulrow* commands. Note that in the *Maple* code to follow, we have combined two *addrow* commands in certain spots to save space. Also, you should note that *Maple* uses the letter I for the complex number written as i in the text of this book.

> B:= matrix(3,4,[2, -(3*I), -4, -1+2*I, (2-2*I), 3, (1-I), -2*I, (10-4*I), (6-9*I), (-10-2*I), -3+2*I]);

$$B := \begin{bmatrix} 2 & -3I & -4 & -1+2I \\ 2-2I & 3 & 1-I & -2I \\ 10-4I & 6-9I & -10-2I & -3+2I \end{bmatrix}$$

> B1:= mulrow(B,1,1/2);

$$B1 := \begin{bmatrix} 1 & -\frac{3}{2}I & -2 & -\frac{1}{2}+I \\ 2-2I & 3 & 1-I & -2I \\ 10-4I & 6-9I & -10-2I & -3+2I \end{bmatrix}$$

> B2:= addrow(addrow(B1,1,2,2*I-2),1,3,4*I-10);

$$B2 := \begin{bmatrix} 1 & -\frac{3}{2}I & -2 & -\frac{1}{2}+I \\ 0 & 6+3I & 5-5I & -1-5I \\ 0 & 12+6I & 10-10I & -2-10I \end{bmatrix}$$

> B3:= mulrow(B2,2,1/(6+3*I));

$$B3 := \begin{bmatrix} 1 & -\frac{3}{2}I & -2 & -\frac{1}{2}+I \\ 0 & 1 & \frac{1}{3}-I & -\frac{7}{15}-\frac{3}{5}I \\ 0 & 12+6I & 10-10I & -2-10I \end{bmatrix}$$

> B4:= addrow(addrow(B3,2,1,3/2*I),2,3,-12-6*I);

$$B4 := \begin{bmatrix} 1 & 0 & -\frac{1}{2} + \frac{1}{2}I & \frac{2}{5} + \frac{3}{10}I \\ 0 & 1 & \frac{1}{3} - I & -\frac{7}{15} - \frac{3}{5}I \\ 0 & 0 & 0 & 0 \end{bmatrix}$$

This new augmented matrix $B4$ gives the linear system:

$$x + \left(-\frac{1}{2} + \frac{1}{2}i\right) w = \frac{2}{5} + \frac{3}{10}i$$
$$y + \left(\frac{1}{3} - i\right) w = -\frac{7}{15} - \frac{3}{5}i \qquad (2.12)$$
$$0 = 0$$

which has the same solution set as the original augmented matrix and its linear system given in (2.11). The equation $0 = 0$ tells us that one equation of the original linear system is superfluous and does not help determine the simultaneous solution set of the system.

> solve({2*x-(3*I)*y-4*z = -1+2*I,(2-2*I)*x+3*y+(1-I)*z = -2*I, (10-4*I)*x + (6-9*I)*y+(-10-2*I)*z = -3+2*I}, {x,y});

$$\left\{x = -\frac{1}{2}Iz + \frac{2}{5} + \frac{3}{10}I + \frac{1}{2}z,\ y = \left(\frac{2}{15} - \frac{1}{15}I\right)(5I\,z - 1 - 5I - 5z)\right\}$$

> expand(%)[2];

$$y = Iz - \frac{1}{3}z - \frac{7}{15} - \frac{3}{5}I$$

From the final matrix, or the *solve* command output given above, we see that the solution to system of three equations with three unknowns given in equation (2.11) has an infinite number of solutions. The solution can be expressed with the x and y variables in terms of the z variable as follows:

$$\left\{x = \left(\frac{2}{5} + \frac{3}{10}i\right) + \left(\frac{1}{2} - \frac{1}{2}i\right)z,\ y = \left(-\frac{7}{15} - \frac{3}{5}i\right) + \left(-\frac{1}{3} + i\right)z\right\}$$

Homework Problems

1. Given an augmented matrix, describe how to determine if it is in the simplest form possible for finding the solution to the system of equations from which the original augmented matrix was constructed.

2. Given an augmented matrix, describe how to determine upon inspection whether or not the system of linear equations it represents has no solution, or an infinite number of solutions.

3. Explain how to find the equation of a line through two points (x_1, y_1), and (x_2, y_2) and a plane through three points (x_1, y_1, z_1), (x_2, y_2, z_2), and (x_3, y_3, z_3) using linear systems.

4. Convert each of the following systems to matrix form and determine to what set the resulting matrix belongs: $\mathbb{R}^{n \times m}$ or $\mathbb{C}^{n \times m}$, for specific values of m and n.

(a) $3x + 4y = 4$
    $4x + 3y = 8$

(b) $x + 4iy = 1$
    $x + y = i$
    $-4x + y = -4i$

(c) $x + 4y = 1$
    $x + 2y = 1$
    $-4x + y = -1$

(d) $3x - 5y + 6z = 1$
    $4x + 3y - 2z = 2$

(e) $3ix - 5y + 6z = 1$
    $4x + 3iy - 2z = 2$
    $x - y + 2iz = -i$

(f) $x + y + z = 1$
    $x - y + z = 1$
    $x - 2y + z = 2$

(g) $x + y + 3z = 1$
    $x - 2y + z = 3$
    $x - 2y + z = -2$

(h) $3x - 5y + 6z + 6t = 1$
    $4x + 3y - 2z - 3t = 2$
    $x - y + 2z + 2t = -1$

(i) $x + y + z + t = i$
    $x - y + z + t = 1$
    $x - 2y + z + t = 2$
    $x - 2y + z + 2t = 2$

5. Convert the following systems to matrix form and then reduce each to its final augmented form using the row operations discussed in this section. Explain each step and show the modified augmented matrix at each step.

(a) $2x - 3y = -5$
      $3x + 7y = 4$

(b) $x + y + z = 6$
      $-x + y - z = -2$
      $x + 2y - z = 2$

(c) $3ix - 4y = -26 - 21i$
      $7x + 5iy = -36 + 39i$

(d) $2x + 3y - 4z = -5$
      $4x - 3y + 9z = 13$
      $-6x + 9y + z = -8$

(e) $10x + 15y - 7z = -35$
      $-\dfrac{4}{3}x + 5y - 14z = -1$
      $35x - 35y + 7z = -96$

(f) $x - y + z = 0$
      $x + 3y - 2z = 4 - 2i$
      $-3x + y - 2iz = 2 + 4i$

6. Using the final augmented matrices from the previous exercise, express the solutions to the original systems in set notation using the given variables.

7. If the following augmented matrices represented a system of equations now

in reduced form, express the solutions in set notation.

a)
$$\begin{bmatrix} 1 & 0 & 0 & 3 \\ 0 & 1 & 0 & -6 \\ 0 & 0 & 1 & 2 \end{bmatrix}$$
b)
$$\begin{bmatrix} 1 & 0 & 0 & 3 & 1 \\ 0 & 1 & 0 & -6 & 0 \\ 0 & 0 & 1 & 2 & -2 \end{bmatrix}$$
c)
$$\begin{bmatrix} 1 & 0 & 0 & 3 \\ 0 & 1 & 0 & -6 \\ 0 & 0 & 0 & 1 \end{bmatrix}$$

Maple Problems

1. Verify your answers to *Homework* problem 3.

2. Solve the system of two lines given by $ax + by = c$ and $ex + fy = g$.

3. Solve the system of two planes given by $ax+by+cz = d$ and $ex+fy+gz = h$.

4. Solve the system of three planes given by $ax + by + cz = d$, $ex + fy + gz = h$ and $ix + jy + kz = l$.

5. Use the *solve* command to find the solutions to the following systems:

(a) $w - 2x + 3y - z = 3$
$\quad -2w + 3x + y + 5z = 2$
$\quad 6w - x - 6y - 3z = 0$
$\quad -3w - 3x + 2y + z = 0$

(b) $2w - x + 5z = 4$
$\quad -w + 4x + 8y + 9z = -3$
$\quad -3w + 2x + y - 3z = 2$
$\quad 7w + 3x - 6y + z = 2$

(c) $w - 5x + 3y + 7z = -8$
$\quad -9w + 3x + 4y - 6z = 3$
$\quad 5w - 8x + y - 5z = 3$

(d) $w - 5x + 3y + 7z = -8$
$\quad -9w + 3x + 4y - 6z = 3$
$\quad 5w - 8x + y - 5z = 3$
$\quad 6w + x + 9y + 13z = -16$

(e) $w - 5x + 3y + 7z = -8$
$\quad -9w + 3x + 4y - 6z = 3$
$\quad 5w - 8x + y - 5z = 3$
$\quad 6w + x + 9y + 13z = -15$

(f) $v - 5w + 3x + 7y - 8z = 9$
$\quad 3v + 4w - 6x + 3y + 5z = -8$
$\quad v - 5w + 3x + 6y + z = 9$
$\quad 13v - 16w + 2x - 3y + 6z = 2$
$\quad 2v - 3w + x - 6y + 7z = 3$

(g) $2w + 3x - y + 2z = 4$
   $2w + 7x - 2y - 9z = -5$
   $2ix + 5y - 13z = 5$
   $(2 - i)w + 2x - 3y + 5z = 6i$

(h) $2v - (5 + i)w + 3x + 6y + 7iz = 8$
   $3v + 7w + 2x + (6 + 3i)y = 6$
   $3v + (5 + 2i)w + 2x + 7y + 5z = -4i$
   $8v + 9w + (2 - 3i)y + 4z = 2$
   $v + w + (3 + i)x + 6y + 7z = 8$

6.   Redo *Maple* problem 5 using only the two row operations *mulrow* and *addrow* on the augmented matrix of each system. What is the dimension of the solution set to each system?

2.3   Some Matrix Arithmetic

In this section, we want to describe how to add, subtract, and multiply two matrices, as well as multiply matrices by scalars. Division of matrices will be postponed to a future section, since it is not obvious how to do division of matrices or even if it can be done at all. If we think of a matrix as analogous to a spreadsheet of data, then adding or subtracting two matrices should correspond to adding or subtracting the data in two spreadsheets covering the same topic. This tells us that we should add or subtract two matrices only if they have the same size and then by adding or subtracting corresponding entries. As well, if you want to multiply a matrix by a constant c, then multiply each entry of the matrix by c.

Example 2.3.1. Some examples would now be useful. Consider the following two matrices:

$$A = \begin{bmatrix} -2 & 5 & 1 & 9 \\ 3 & 0 & -4 & 7 \end{bmatrix}, B = \begin{bmatrix} 6 & -2 & 4 & 1 \\ -8 & -1 & 3 & 5 \end{bmatrix}$$

Let us find $A + B$, $A - B$, $-3A$ and $-3A + 5B$. In order to actually have *Maple* display the resulting matrices, we must use the *evalm* command. *Maple* is reluctant to actually print the resulting matrix otherwise, as can be seen in the following commands:

> with(linalg):
> A := matrix([[-2, 5, 1, 9], [3, 0, -4, 7]]):
> B := matrix([[6, -2, 4, 1], [-8, -1, 3, 5]]):
> A+B;

$$A + B$$

> evalm(A+B);

$$\begin{bmatrix} 4 & 3 & 5 & 10 \\ -5 & -1 & -1 & 12 \end{bmatrix}$$

> evalm(A-B);

$$\begin{bmatrix} -8 & 7 & -3 & 8 \\ 11 & 1 & -7 & 2 \end{bmatrix}$$

> evalm(-3*A);

$$\begin{bmatrix} 6 & -15 & -3 & -27 \\ -9 & 0 & 12 & -21 \end{bmatrix}$$

> evalm(-3*A+5*B);

$$\begin{bmatrix} 36 & -25 & 17 & -22 \\ -49 & -5 & 27 & 4 \end{bmatrix}$$

Definition 2.3.1. Let $A, B \in \mathbb{R}^{m \times n}$ and $c \in \mathbb{R}$, the *sum* $A + B \in \mathbb{R}^{m \times n}$ with $(A + B)_{i,j} = A_{i,j} + B_{i,j}$. Similarly, the *difference* $A - B \in \mathbb{R}^{m \times n}$ with $(A - B)_{i,j} = A_{i,j} - B_{i,j}$. Finally, $cA \in \mathbb{R}^{m \times n}$ with $(cA)_{i,j} = cA_{i,j}$.

Definition 2.3.2. Two matrices $A, B \in \mathbb{R}^{m \times n}$ are equal if and only if $A_{i,j} = B_{i,j}$ for all $1 \le i \le m$, $1 \le j \le n$. Symbolically, this is expressed as $A = B$.

Both of these definitions are true if $\mathbb{R}$ is replaced by $\mathbb{C}$, as is the case for all of our linear system and matrix discussions past and future.

Now we turn to matrix multiplication. The idea behind the definition of how to multiply two matrices is the desire to turn a system of linear equations into a single matrix equation of the form $AX = B$, where A is the matrix of coefficients, X is the column of variables, and B is the column of RHS values. Let us look at the square system case that is 2×2 (two equations in two variables), corresponding to the following two linear equations:

$$\begin{aligned} ax + by &= \alpha \\ cx + dy &= \beta \end{aligned} \tag{2.13}$$

The matrix of coefficients A, the column of variables X and the column of RHS values B are given by

$$A = \begin{bmatrix} a & b \\ c & d \end{bmatrix}, \quad X = \begin{bmatrix} x \\ y \end{bmatrix}, \quad \text{and } B = \begin{bmatrix} \alpha \\ \beta \end{bmatrix}$$

respectively. Now, the simplest way to write our system of linear equations as a matrix equation is

$$\begin{bmatrix} ax + by \\ cx + dy \end{bmatrix} = \begin{bmatrix} \alpha \\ \beta \end{bmatrix}$$

since equality of matrices should mean that they have the same size along with the same corresponding entries. Thus, if we want the matrix equation to be $AX = B$, we must have

$$\begin{bmatrix} a & b \\ c & d \end{bmatrix} \begin{bmatrix} x \\ y \end{bmatrix} = \begin{bmatrix} ax + by \\ cx + dy \end{bmatrix} = \begin{bmatrix} \alpha \\ \beta \end{bmatrix} \tag{2.14}$$

Now we have our definition of matrix multiplication on the most basic level. This tells us that in order to multiply two matrices together we must multiply a row of the left matrix times a column of the right matrix getting a single value that goes in their row, column location in the product matrix. When a row of the left matrix multiplies a column of the right matrix in a product, you multiply corresponding entries together of the two and add the results. In other words,

$$\begin{bmatrix} a & b \end{bmatrix} \begin{bmatrix} x \\ y \end{bmatrix} = [ax + by]$$

is the defining operation in matrix multiplication. Let us do a few examples of matrix multiplication, to see if we can find the correct pattern. *Maple* will perform the computations, and we will give the definition of matrix multiplication shortly. We start with a simple multiplication of two square matrices $A, B \in \mathbb{R}^{2 \times 2}$.

Example 2.3.2. If $A = \begin{bmatrix} 3 & 2 \\ -2 & 6 \end{bmatrix}$, $B = \begin{bmatrix} -1 & 5 \\ -2 & 1 \end{bmatrix}$, we wish to find both AB and BA. In *Maple*, matrix multiplication is written as &*.

```
> A:=matrix(2,2,[3,2, -2,6]):
> B:=matrix(2,2,[-1,5,-2,1]):
> evalm(A&*B);
```

$$\begin{bmatrix} -7 & 17 \\ -10 & -4 \end{bmatrix}$$

```
> evalm(B&*A);
```

$$\begin{bmatrix} -13 & 28 \\ -8 & 2 \end{bmatrix}$$

Notice that if we actually follow the rule given in equation (2.14), where we take a row of the left matrix, and multiply it by a column of the second matrix, we end up with the entries of the resulting matrix that *Maple* has given. For instance, if we take the first row of A, and multiply it by the first column of B, we get $3(-1) + 2(-2) = -7$. This happens to be the entry in first row, first column of the matrix AB. If we take the second row of A and multiply it by the first column of B, we get $(-2)(-1) + 6(-2) = -10$, which happens to be the second row, first column entry of the matrix AB. Do you see the pattern yet? Attempt to fill in the last column of AB, which corresponds to multiplying the first row and second rows of A by the second column of B.

From the two matrix multiplications, we also know that $AB \neq BA$. This may seem a bit strange at first, but it makes sense if we consider that we always multiply the rows of the left matrix by the columns of the second matrix. This will be discussed more after Example 2.3.3.

Example 2.3.3. Let us now try something a little more complicated. We want to find AB and BA, given

$$A = \begin{bmatrix} -2 & 5 & 1 & 9 \\ 3 & 0 & -4 & 7 \end{bmatrix}, \quad B = \begin{bmatrix} 6 & -8 & -3 \\ -2 & -1 & 6 \\ 3 & 5 & -9 \\ -7 & 4 & 1 \end{bmatrix}$$

> A := matrix([[-2, 5, 1, 9], [3, 0, -4, 7]]):
> B := matrix([[6, -8, -3], [-2, -1, 6], [3, 5, -9], [-7, 4, 1]]):
> evalm(A&*B);

$$\begin{bmatrix} -82 & 52 & 36 \\ -43 & -16 & 34 \end{bmatrix}$$

> evalm(B&*A);

Error, (in linalg:-multiply) non matching dimensions for vector/matrix product

Notice that we can perform the multiplication AB, but not BA. In order to multiply BA, the number of columns of B must equal the number of rows of A, which is not true. Since the rows of B have three entries (i.e., B has three columns) and the columns of A have only two entries (i.e., A has two rows) we cannot perform the multiplication BA since the number of columns of B is not equal to the number of row of A. However, since the matrix A is of size 2×4 while B is of size 4×3, we can perform the multiplication AB, which will result in a matrix of dimension 2×3.

In general, if the matrix A has size $m \times k$ and the matrix B has size $k \times n$, then we can multiply in the order AB with the matrix AB having size $m \times n$. The only matrices that we can multiply in either order are square matrices of the same size. If both A and B are matrices of size $n \times n$, then AB and BA both can be computed but are not normally equal, although both products are of size $n \times n$. Thus, matrix multiplication is not a commutative operation even when both orders of multiplication can be done.

Definition 2.3.3. Let A be a matrix of dimension $m \times k$ and B be a matrix of dimension $k \times n$. We define C to be the *product*f A and B, denoted $C = AB$, of dimension $m \times n$. Each entry of C can be computed by the following formula:

$$C_{i,j} = \sum_{r=1}^{k} A_{i,r} B_{r,j} \tag{2.15}$$

In simple terms, if we are to multiply two matrices A and B together and A is $m \times k$ and B is $k \times n$, the resulting matrix $C = AB$ will be a matrix of size $m \times n$. Furthermore, the entry in the ith row and jth column of C is found by taking the ith row of A and multiplying it by the jth column of B in the way previously described. This allows for a very systematic way of computing the product of two matrices.

Example 2.3.4. Let us do another example where both AB and BA exist, but are not equal. Let $A = \begin{bmatrix} -2 & 5 & 1 \\ 3 & 0 & -4 \\ 7 & -5 & -8 \end{bmatrix}$ and $B = \begin{bmatrix} 6 & -8 & -3 \\ -2 & -1 & 6 \\ 3 & 5 & -9 \end{bmatrix}$

```
> A := matrix([[-2, 5, 1], [3, 0, -4], [7, -5, -8]]):
> B := matrix([[6, -8, -3], [-2, -1, 6], [3, 5, -9]]):
> evalm(A&*B);
```

$$\begin{bmatrix} -19 & 16 & 27 \\ 6 & -44 & 27 \\ 28 & -91 & 21 \end{bmatrix}$$

```
> evalm(B&*A);
```

$$\begin{bmatrix} -57 & 45 & 62 \\ 43 & -40 & -46 \\ -54 & 60 & 55 \end{bmatrix}$$

Note how tedious matrix multiplication is to perform when computed by hand with matrices of dimension greater than 2×2.

Now, we know how to add matrices, subtract matrices, multiply a matrix by a scalar, and multiply two matrices together. All that is left is matrix division, or multiplication by matrix inverse. When we consider multiplication and division for real numbers, the unique multiplicative inverse of a nonzero number x is $\dfrac{1}{x} = x^{-1}$, with $x\,x^{-1} = 1$. If we consider a set of square matrices of size $n \times n$ (as example we could choose $\mathbb{R}^{n \times n}$ or $\mathbb{C}^{n \times n}$), then there is a matrix called the $n \times n$ identity matrix, denoted by I_n, which acts like the scalar multiplicative identity 1 under multiplication of these square matrices.

The reason we need to discuss identity matrices and matrix inverses is due to the fact that the matrix equation $AX = B$, which represents a system of linear equations, looks like the simple algebra equation $ax = b$ for constants a and b. Now the equation $ax = b$ can be solved for x if $a \neq 0$ and then we get the unique solution $x = \dfrac{b}{a}$. We would like to do the same for our matrix equation $AX = B$, but this requires we figure out how to divide by matrices like A. The identity matrix is the only square matrix with the property that if A is an arbitrary $n \times n$ matrix, then $AI_n = I_n A = A$. We will make use of identity matrices later when defining the multiplicative inverse A^{-1} of a square $n \times n$ matrix A, since if A^{-1} exists we will want it to satisfy $AA^{-1} = A^{-1}A = I_n$, hence multiplying by A^{-1} will be the same as dividing by A. The identity matrices are built into *Maple* in the *LinearAlgebra* package under the command *IdentityMatrix*.

Definition 2.3.4. The $n \times n$ *identity matrix*, denoted I_n, is the matrix that has all 1's on its main diagonal from upper left to lower right and all zeros elsewhere:

$$
I_n = \begin{bmatrix} 1 & & \cdots & & 0 \\ & 1 & & & \\ \vdots & & \ddots & & \vdots \\ & & & 1 & \\ 0 & & \cdots & & 1 \end{bmatrix}
$$

Example 2.3.5. Our last example for this section will show us that $AI_4 = I_4 A = A$ for the matrix defined below. Hence, we can think of the 4×4 identity matrix I_4 as the multiplicative identity for 4×4 matrices. Finally, we take an arbitrary 4×4 matrix B, and multiply it on the left and right by I_4.

```
> with(LinearAlgebra):
> id2:= IdentityMatrix(2);
```

$$
id2 := \begin{bmatrix} 1 & 0 \\ 0 & 1 \end{bmatrix}
$$

```
> id4:= IdentityMatrix(4);
```

$$
id4 := \begin{bmatrix} 1 & 0 & 0 & 0 \\ 0 & 1 & 0 & 0 \\ 0 & 0 & 1 & 0 \\ 0 & 0 & 0 & 1 \end{bmatrix}
$$

```
> A:= matrix([[1,3,2,4],[-1,-6,0,7],[3,-5,1,9],[0,8,2,1]]);
```

$$
A := \begin{bmatrix} 1 & 3 & 2 & 4 \\ -1 & -6 & 0 & 7 \\ 3 & -5 & 1 & 9 \\ 0 & 8 & 2 & 1 \end{bmatrix}
$$

> evalm(id4&*A);

$$
\begin{bmatrix}
1 & 3 & 2 & 4 \\
-1 & -6 & 0 & 7 \\
3 & -5 & 1 & 9 \\
0 & 8 & 2 & 1
\end{bmatrix}
$$

> evalm(A&*id4);

$$
\begin{bmatrix}
1 & 3 & 2 & 4 \\
-1 & -6 & 0 & 7 \\
3 & -5 & 1 & 9 \\
0 & 8 & 2 & 1
\end{bmatrix}
$$

> B:= matrix(4, 4, [ ]):
> evalm(B&*id4);

$$
\begin{bmatrix}
B_{1,1} & B_{1,2} & B_{1,3} & B_{1,4} \\
B_{2,1} & B_{2,2} & B_{2,3} & B_{2,4} \\
B_{3,1} & B_{3,2} & B_{3,3} & B_{3,4} \\
B_{4,1} & B_{4,2} & B_{4,3} & B_{4,4}
\end{bmatrix}
$$

> evalm(id4&*B);

$$
\begin{bmatrix}
B_{1,1} & B_{1,2} & B_{1,3} & B_{1,4} \\
B_{2,1} & B_{2,2} & B_{2,3} & B_{2,4} \\
B_{3,1} & B_{3,2} & B_{3,3} & B_{3,4} \\
B_{4,1} & B_{4,2} & B_{4,3} & B_{4,4}
\end{bmatrix}
$$

■

With the above definitions and discussion of linear systems, we are now in a position to prove the following theorem concerning the number of simultaneous solutions to a linear system.

Theorem 2.3.1. *A linear system of equations has either no solution, exactly one solution, or infinitely many solutions.*

Proof. Let $AX = B$ be the matrix form of a linear system of equations for A its coefficient matrix, X its column of variables, and B the column of its RHS values. We have already seen examples of linear systems with no solution and with exactly one solution using pairs of lines as our linear system. All that remains is for us to show that if the linear system has at least two distinct solutions X_1 and X_2 with $X_1 \neq X_2$, then it must have infinitely many solutions. Note that $AX_1 = B$ and $AX_2 = B$ since X_1 and X_2 are solutions to the linear system. Let t be any real variable. We claim that $W = X_1 + t(X_2 - X_1)$ is always a solution for any value of t. Note that when $t = 0$, $W = X_1$ while when $t = 1$, $W = X_2$. The variable W is actually the line through the two solutions X_1 and X_2, where you get one point on this line for each value of t. In order to see if the W's are solutions to $AX = B$, we need only replace X by W and see if AW is B or not. Now,

$$
\begin{aligned}
AW &= AX_1 + tA(X_2 - X_1) \\
&= B + t(AX_2 - AX_1) \\
&= B + t(B - B) \\
&= B + t0 \\
&= B
\end{aligned}
$$

So all the W's are solutions to $AX = B$, and there are an infinite number of W's, one for each value of t.   □

We conclude this section with a list of useful properties of matrix addition and multiplication. In the following table, a and b are scalars, while A, B, and C are arbitrary matrices. For each equation, remember that there are restrictions on the dimensions of the matrices that must be satisfied in order for the equation to make sense. The properties, however, will hold for any matrices for which the equation can be used. You will verify many of the following matrix arithmetic properties in the *Homework* and *Maple* problems at the end of this section.

<table>
<tr><td colspan="2" align="center">Basic Matrix Addition and
Multiplication Properties</td></tr>
<tr>
<td align="center">Matrix
Addition</td>
<td>$A + B = B + A$
$(A + B) + C = A + (B + C)$
$A + (-A) = 0$</td>
</tr>
<tr>
<td align="center">Scalar
Multiplication</td>
<td>$a(bA) = (ab)A$
$a(A + B) = aA + aB$
$(a + b)A = aA + bA$</td>
</tr>
<tr>
<td align="center">Matrix
Multiplication</td>
<td>$(AB)C = A(BC)$
$A(B + C) = AB + AC$
$(A + B)C = AC + BC$
$AI = IA = A$
$A0 = 0A = 0$
$a(AB) = (aA)B = A(aB)$
$AB \neq BA$ in general</td>
</tr>
</table>

Homework Problems

1. Consider the following matrices:

$$A = \begin{bmatrix} 1 & 3 \\ -2 & 1 \end{bmatrix} \qquad B = \begin{bmatrix} 5 & -3 \\ 2 & 7 \end{bmatrix} \qquad C = \begin{bmatrix} 9 & -6 \\ -8 & 2 \end{bmatrix}$$

$$D = \begin{bmatrix} 1 & -9 & 2 \\ -4 & 6 & 2 \\ -1 & 3 & 0 \end{bmatrix} \quad E = \begin{bmatrix} 4 & -9 & 2 \\ 6 & -1 & 3 \\ -7 & 2 & 5 \end{bmatrix} \quad F = \begin{bmatrix} 6 & -3 & 9 \\ -3 & 2 & -8 \\ 5 & 3 & 4 \end{bmatrix}$$

Perform the following matrix operations:

(a) $A - 2B$ (b) $5A - 3C$ (c) $2A + 3B - 4C$ (d) $3(A - B + C)$

(e) $2B + 5C$ (f) $D - 3F + 4E$ (g) $6D + 3F$ (h) $2E + 3F$

(i) $6D - 4E + 2F$ (j) $2(D - 3E) + 5F$ (k) $B - 4C + 3A$ (l) $2(6D - 5F)$

2. Consider the following matrices:

$$A = \begin{bmatrix} 1 & -4 \\ -2 & 2 \end{bmatrix} \qquad B = \begin{bmatrix} 1 & -7 & 8 \\ -2 & 2 & 3 \end{bmatrix} \qquad C = \begin{bmatrix} 4 & -1 \\ -2 & 9 \\ 6 & -2 \end{bmatrix}$$

$$D = \begin{bmatrix} 1 & -4 & 2 \\ 0 & 2 & 5 \\ -1 & 3 & 5 \end{bmatrix} \quad E = \begin{bmatrix} 1 & -7 & 8 \\ -2 & 2 & 3 \\ -2 & 1 & 6 \\ -2 & 8 & -7 \end{bmatrix} \quad F = \begin{bmatrix} 4 & -1 \\ -2 & 9 \\ 8 & -2 \\ 6 & -3 \end{bmatrix}$$

Determine which of the following matrix multiplications can be performed:

(a) AA (b) AB (c) AC (d) CA (e) BB

(f) FD (g) DA (h) BC (i) CF (j) DB

3. Find all 12 possible combinations of matrices from problem 2 that will allow a matrix multiplication to be performed.

4. Find a value of a for which the following two matrices satisfy $AB = BA$:

$$A = \begin{bmatrix} a & 3 \\ -2 & 1 \end{bmatrix}, \quad B = \begin{bmatrix} 1 & -3 \\ 2 & -2 \end{bmatrix}$$

5. Perform the following matrix multiplications:

(a) $\begin{bmatrix} 1 & 0 \\ 1 & -2 \end{bmatrix} \begin{bmatrix} -2 & -4 \\ 5 & 6 \end{bmatrix}$ (b) $\begin{bmatrix} 1 & -2 & 3 \\ 8 & -1 & -7 \end{bmatrix} \begin{bmatrix} -2 & 4 \\ 7 & 2 \\ 8 & -5 \end{bmatrix}$

(c) $\begin{bmatrix} 2 & -1 & 4 \end{bmatrix} \begin{bmatrix} -3 \\ 1 \\ -5 \end{bmatrix}$ (d) $\begin{bmatrix} -2 & 2 \\ 5 & 2 \\ 1 & -1 \end{bmatrix} \begin{bmatrix} -1 & 0 & 8 \\ -4 & 5 & 9 \end{bmatrix}$

(e) $\begin{bmatrix} 2 & -1 \\ 3 & 0 \\ -8 & 7 \\ -4 & 2 \end{bmatrix} \begin{bmatrix} -2 & 0 & -1 \\ 5 & 3 & 5 \end{bmatrix}$ (f) $\begin{bmatrix} -2 & 3 & 1 \\ -7 & -1 & 4 \\ 3 & 8 & 1 \end{bmatrix} \begin{bmatrix} 1 & -3 \\ 2 & 4 \\ 0 & 1 \end{bmatrix}$

6. Write the following systems of equations in the matrix form $AX = B$:

(a) $2x - 3y = 7$
          $4x + 5y = 2$

(b) $23x - 6y + 4z = 2$
          $14x + 6y - 5z = 4$
          $-5x + 4y = -1$

(c) $2w + 4y - 5z = 6$
          $-w + x - 4y + 3z = 7$
          $8w - 2x - 7z = 9$

(d) $w - 5x + 3y + 7z = -8$
          $-9w + 3x + 4y - 6z = 3$
          $5w - 8x + y - 5z = 3$
          $6w + x + 9y + 13z = -16$

7. Consider the following three matrices:

$$A = \begin{bmatrix} 2 & 0 & 4 \\ 5 & 1 & -2 \\ 1 & 6 & -7 \end{bmatrix} \quad B = \begin{bmatrix} -1 & 1 & 2 \\ -6 & 4 & 1 \\ 7 & 0 & 2 \end{bmatrix} \quad C = \begin{bmatrix} 4 & -2 & 9 \\ 5 & -8 & 0 \\ 2 & 1 & -1 \end{bmatrix}$$

Show the following hold (assuming $c \in \mathbb{R}$).

(a) $A(BC) = (AB)C$ (b) $A(B + C) = AB + AC$ (c) $(A + B)C = AC + BC$

(d) $c(AB) = (cA)B$ (e) $A(cB) = (Ac)B$ (f) $(AB)c = A(Bc)$

8. For part (a) of problem 7, what are the general dimensions of A, B, and C such that one can perform the matrix multiplication? Do the same for the matrices of parts (b) and (c).

9. Let c be a scalar and A be any 2×2 matrix, show that $cA = \begin{bmatrix} c & 0 \\ 0 & c \end{bmatrix} A$. Is a similar statement true for any size square matrix A? Generalize this to any size matrix A so that general scalar multiplication is turned into matrix

multiplication by a *diagonal matrix*?

10. Let A, B, C, and D be four 2×2 matrices. Let $K = \begin{bmatrix} A & 0 \\ 0 & B \end{bmatrix}$ and

$L = \begin{bmatrix} C & 0 \\ 0 & D \end{bmatrix}$ be two block diagonal 4×4 matrices with A, B, C, and D on
their diagonals, the 0's in these two matrices are the zero 2×2 matrices. Show
that $KL = \begin{bmatrix} AC & 0 \\ 0 & BD \end{bmatrix}$ and $K^n = \begin{bmatrix} A^n & 0 \\ 0 & B^n \end{bmatrix}$ for any positive integer n.
Generalize this problem where there is no restriction on the sizes of A, B, C,
and D as long as they are square matrices, although you may want some of
their sizes to be the same. Will this work for larger size block diagonal matrices?

11. Define the set $\mathbf{M}$ as follows:

$$\mathbf{M} = \left\{ \begin{bmatrix} a & b \\ -b & a \end{bmatrix} \middle| a, b \in \mathbb{R} \right\}$$

Show that the following are true.

(a) For any two $K, L \in \mathbf{M}$, $KL = LK$, that is, matrix multiplication in $\mathbf{M}$
is commutative.

(b) Show that

$$\begin{bmatrix} a & b \\ -b & a \end{bmatrix} \begin{bmatrix} a & -b \\ b & a \end{bmatrix} = \begin{bmatrix} a^2 + b^2 & 0 \\ 0 & a^2 + b^2 \end{bmatrix} = (a^2 + b^2)I_2$$

where I_2 is the 2×2 identity matrix.

(c) Show that

$$\begin{bmatrix} 0 & 1 \\ -1 & 0 \end{bmatrix}^2 = \begin{bmatrix} -1 & 0 \\ 0 & -1 \end{bmatrix} = -I_2$$

Hence $\begin{bmatrix} 0 & 1 \\ -1 & 0 \end{bmatrix}$ is the square root of $-I_2$.

(d) Do these properties of $\mathbf{M}$ remind you of any other set and its properties?

12. Let $A = \begin{bmatrix} a & b \\ c & d \end{bmatrix} \in \mathbb{R}^{2 \times 2}$. Let $C = \dfrac{1}{ad - bc} \begin{bmatrix} d & -b \\ -c & a \end{bmatrix}$, with $ad - bc \neq 0$.
Show that $AC = I_2$ and $CA = I_2$. What does this make C with respect to A?

13. Use the result from problem 12 to solve the linear system

$$5x - 7y = 11$$
$$9x + 2y = -4$$

Check your answer by another means.

14. Let your 2×2 linear system of equations be given as the matrix equation $AX = B$, where

$$A = \begin{bmatrix} a & b \\ c & d \end{bmatrix}, X = \begin{bmatrix} x \\ y \end{bmatrix}, \text{ and } B = \begin{bmatrix} \alpha \\ \beta \end{bmatrix}$$

where $ad - bc \neq 0$. Using the result of problem 12, what is the solution formula for X?

Maple Problems

1. Perform all twelve matrix multiplications that you found in *Homework* problem 3.

2. Verify your answers to *Homework* problem 5.

3. Compute $A^2, A^3, A^4, \ldots, A^n$ for each of the following matrices:

(a) $\begin{bmatrix} 0 & 0 & 0 & 0 \\ 1 & 0 & 0 & 0 \\ -1 & 2 & 0 & 0 \\ 2 & 4 & 7 & 0 \end{bmatrix}$
(b) $\begin{bmatrix} -2 & 0 & 0 \\ 0 & 5 & 0 \\ 0 & 0 & -7 \end{bmatrix}$
(c) $\begin{bmatrix} 0 & 1 & 3 \\ 0 & 0 & -8 \\ 0 & 0 & 0 \end{bmatrix}$

4. Find a pattern to the sequence $\{A^n\}$, $n = 1, 2, \ldots$ for each of the following matrices:

(a) $\begin{bmatrix} -1 & 1 & -1 \\ 0 & 1 & 0 \\ 1 & 0 & 1 \end{bmatrix}$
(b) $\begin{bmatrix} -1 & 1 & -1 \\ 0 & 1 & 1 \\ 0 & 0 & 0 \end{bmatrix}$
(c) $\begin{bmatrix} -1 & 1 & -1 \\ 1 & 1 & -1 \\ 1 & 1 & -1 \end{bmatrix}$

5. Show that for arbitrary $A, B, C \in \mathbb{R}^{4 \times 4}$ and $c \in \mathbb{R}$ that the following hold:

(a) $A(BC) = (AB)C$ (b) $A(B + C) = AB + AC$ (c) $(A + B)C = AC + BC$

(d) $c(AB) = (cA)B$ (e) $A(cB) = (Ac)B$ (f) $(AB)c = A(Bc)$

6. Use *Maple* to find the formula for A^{-1} given in *Homework* problem 12.

7. Use *Maple* to find the solution formula for X in *Homework* problem 14.

8. Use *Maple* to check *Homework* problem 10.

Chapter 3

Gauss-Jordan Elimination and Reduced Row Echelon Form

3.1 Gauss-Jordan Elimination and *rref*

Gauss-Jordan elimination is a popular method for solving systems of linear equations. Examples 2.2.2 and 2.2.3 of Section 2.2 were actually examples of Gauss-Jordan elimination, which we now wish to formalize. Gauss-Jordan elimination uses three types of row operations in order to change the original augmented matrix of the system into a new augmented matrix whose form allows us to solve the system by simple inspection. The three types of row operations allowed in Gauss-Jordan elimination are switching or swapping any two rows, multiplying a row by a nonzero number, and finally multiplying a row by a nonzero number and then adding it to another row. In *Maple*, these three row operations can be found in the *linalg* package under the names *swaprow*, *mulrow*, and *addrow*, respectively. The last two commands were introduced in Chapter 2 when we attempted to solve a system of three equations with four unknowns. These three row operations correspond to algebraic operations done on the linear system of equations that do not alter the solution set to the linear system, and they are the only operations that are needed to solve any linear system for its solution set.

The final augmented matrix of Gauss-Jordan elimination is called a *rref* matrix, which is short for *reduced row echelon form*. Note that the *rref* form is unique, that is, *rref* is independent of the order of the original rows of the augmented matrix, or equivalently, independent of the order of the original equations in the linear system. Every row of the *rref* matrix is required to be either all zeros or have a leading 1 in the row preceded by all zeros in its row. As you go down the rows of the *rref* matrix, the leading 1's of the matrix must move left to right with no two of them occurring in the same column. Furthermore, any row of all zeroes must be placed at the bottom of the final matrix in *rref* form. As well, each column of the *rref* matrix that contains a leading 1 must have all other entries in the column be 0. The following are all examples of matrices in *rref* form:

$$
\begin{bmatrix} 1 & 0 & 0 & a \\ 0 & 1 & 0 & b \\ 0 & 0 & 1 & c \end{bmatrix}
\qquad
\begin{bmatrix} 1 & a & 0 & b & c & d \\ 0 & 0 & 1 & e & f & g \end{bmatrix}
$$

$$
\begin{bmatrix} 1 & 0 & 0 & a & b & c \\ 0 & 1 & 0 & d & e & f \\ 0 & 0 & 1 & g & h & i \\ 0 & 0 & 0 & 0 & 0 & 0 \end{bmatrix}
\qquad
\begin{bmatrix} 1 & a & b & 0 & 0 & c \\ 0 & 0 & 0 & 1 & 0 & d \\ 0 & 0 & 0 & 0 & 1 & e \\ 0 & 0 & 0 & 0 & 0 & 0 \end{bmatrix}
\qquad (3.1)
$$

$$
\begin{bmatrix} 1 & 0 & a & b & c & d \\ 0 & 1 & e & f & g & h \\ 0 & 0 & 0 & 0 & 0 & 0 \\ 0 & 0 & 0 & 0 & 0 & 0 \end{bmatrix}
\qquad
\begin{bmatrix} 1 & a & b & c & 0 & 0 \\ 0 & 0 & 0 & 0 & 1 & 0 \\ 0 & 0 & 0 & 0 & 0 & 1 \\ 0 & 0 & 0 & 0 & 0 & 0 \end{bmatrix}
$$

The process of Gauss-Jordan elimination produces a *rref* matrix with the above described properties by working on one column at a time going left to right through the matrix and trying to obtain a leading 1 in place first as high up in each column as possible followed by zeros elsewhere in the column. The least used of the three types of row operation in Gauss-Jordan elimination is the switching of two rows. It is only used in order to get a nonzero value in a leading 1 location so that you can then turn it into a leading 1. The preferred final form of the *rref* matrix in Gauss-Jordan elimination is one where the upper-left-hand corner of the matrix is an identity matrix of some size. The command *rref* appears in the *linalg* package and performs Gauss-Jordan elimination.

Example 3.1.1. First, let us repeat a previous example, specifically Example 2.2.2. We want to solve the linear system

$$
6x + y + 4w = -3
$$
$$
-9x + 2y + 3z - 8w = 1
$$
$$
7x - 4z + 5w = 2
$$

The augmented matrix of this system is

$$
C = \begin{bmatrix} 6 & 1 & 0 & 4 & -3 \\ -9 & 2 & 3 & -8 & 1 \\ 7 & 0 & -4 & 5 & 2 \end{bmatrix}
$$

Our goal is to transform the above matrix to one of the form:

$$D = \begin{bmatrix} 1 & 0 & 0 & d_{1,4} & d_{1,5} \\ 0 & 1 & 0 & d_{2,4} & d_{2,5} \\ 0 & 0 & 1 & d_{3,4} & d_{3,5} \end{bmatrix}$$

Now we will use *Maple* to perform several operations used in Gauss-Jordan elimination. First, however, we will let *Maple* convert the system to *rref* form via the *rref* command.

> with(linalg):
> C:= matrix([[6, 1, 0, 4, -3], [-9, 2, 3, -8, 1], [7, 0, -4, 5, 2]]);

$$C := \begin{bmatrix} 6 & 1 & 0 & 4 & -3 \\ -9 & 2 & 3 & -8 & 1 \\ 7 & 0 & -4 & 5 & 2 \end{bmatrix}$$

> rref(C);

$$\begin{bmatrix} 1 & 0 & 0 & \frac{7}{9} & -\frac{34}{63} \\ 0 & 1 & 0 & -\frac{2}{3} & \frac{5}{21} \\ 0 & 0 & 1 & \frac{1}{9} & -\frac{13}{9} \end{bmatrix}$$

Next, we swap rows 1 and 3 of C using the *swaprow* command, and then use the *rref* command.

> Swap_C13:= swaprow(C,1,3);

$$Swap\_C13 := \begin{bmatrix} 7 & 0 & -4 & 5 & 2 \\ -9 & 2 & 3 & -8 & 1 \\ 6 & 1 & 0 & 4 & -3 \end{bmatrix}$$

> rref(Swap_C13);

$$\begin{bmatrix} 1 & 0 & 0 & \frac{7}{9} & -\frac{34}{63} \\ 0 & 1 & 0 & -\frac{2}{3} & \frac{5}{21} \\ 0 & 0 & 1 & \frac{1}{9} & -\frac{13}{9} \end{bmatrix}$$

> rref(swaprow(C,2,3));

$$\begin{bmatrix} 1 & 0 & 0 & \frac{7}{9} & -\frac{34}{63} \\ 0 & 1 & 0 & -\frac{2}{3} & \frac{5}{21} \\ 0 & 0 & 1 & \frac{1}{9} & -\frac{13}{9} \end{bmatrix}$$

It is clear that the *rref* matrix is independent of the order of the equations, or equivalently, the rows in the augmented matrix C of the linear system.

Notice that the 3×3 identity matrix appears on the left in the *rref* matrix above. This form of the final augmented matrix of a linear system is what Gauss-Jordan elimination tries to produce. This final augmented matrix says that the solution to our original system is

$$\left\{ x = -\frac{7}{9}w - \frac{34}{63}, y = \frac{2}{3}w + \frac{5}{21}, z = -\frac{1}{9}w - \frac{13}{9} \right\} \tag{3.2}$$

This solution is written in terms of one independent variable w, and so it is one-dimensional, which is a line in $\mathbb{R}^4$ since we have four variables x, y, z, and w. We can also express the solution given in (3.2) as a the sum of two column matrices:

$$X = \begin{bmatrix} x \\ y \\ z \\ w \end{bmatrix} = \begin{bmatrix} -\frac{34}{63} \\ \frac{5}{21} \\ -\frac{13}{9} \\ 0 \end{bmatrix} + \begin{bmatrix} -\frac{7}{9} \\ \frac{2}{3} \\ -\frac{1}{9} \\ 1 \end{bmatrix} w \tag{3.3}$$

This form will be used later when we discuss bases and dimensions of vector spaces. It should be clear that the solution expressed in (3.3) is one-dimensional, since there is a single parameter w multiplied by the second column matrix, and the first column matrix represents a single point, which is of dimension 0. Furthermore, if we express the solution given in (3.3) as $X = X_p + X_h w$, where

$$X_p = \begin{bmatrix} -\frac{34}{63} \\ \frac{5}{21} \\ -\frac{13}{9} \\ 0 \end{bmatrix} \text{ and } X_h = \begin{bmatrix} -\frac{7}{9} \\ \frac{2}{3} \\ -\frac{1}{9} \\ 1 \end{bmatrix}$$

then we get

$$\begin{bmatrix} 6 & 1 & 0 & 4 \\ -9 & 2 & 3 & -8 \\ 7 & 0 & -4 & 5 \end{bmatrix} X_p = \begin{bmatrix} -3 \\ 1 \\ 2 \end{bmatrix}, \quad \begin{bmatrix} 6 & 1 & 0 & 4 \\ -9 & 2 & 3 & -8 \\ 7 & 0 & -4 & 5 \end{bmatrix} X_h = \begin{bmatrix} 0 \\ 0 \\ 0 \end{bmatrix}$$

The only unfortunate aspect of using the command *rref* is that it does not actually give the solution to the original system. You must still write out the solution yourself from the *rref* matrix. On the other hand, the *solve* command gives the actual solution to the system where you can specify which variables to solve for in the system. In this case, you know in advance that the dimension of the solution is most likely one and that of the four variables x, y, z, and w, you can probably solve for any three of them in terms of the remaining fourth one.

> solve({6*x+y+4*w=-3, -9*x+2*y+3*z-8*w=1, 7*x-4*z+5*w=2 }, {x,y,z});

$$\left\{ x = -\frac{34}{63} - \frac{7}{9}w, y = \frac{5}{21} + \frac{2}{3}w, z = -\frac{13}{9} - \frac{1}{9}w \right\}$$

> solve({6*x+y+4*w=-3, -9*x+2*y+3*z-8*w=1, 7*x-4*z+5*w=2}, {y,z,w});

$$\left\{ w = -\frac{9}{7}x - \frac{34}{49}, y = -\frac{6}{7}x - \frac{11}{49}, z = \frac{1}{7}x - \frac{67}{49} \right\}$$

■

Example 3.1.2. Let us do another example, similar to the first. We want to solve the slightly altered linear system

$$6x + y + 4w = -3$$
$$-9x + 2y + 3z - 8w = 1$$
$$7x - z + 5w = 2$$

> C:= matrix([[6, 1, 0, 4, -3], [-9, 2, 3, -8, 1], [7, 0, -1, 5, 2]]);

$$C := \begin{bmatrix} 6 & 1 & 0 & 4 & -3 \\ -9 & 2 & 3 & -8 & 1 \\ 7 & 0 & -1 & 5 & 2 \end{bmatrix}$$

> rref(C);

$$\begin{bmatrix} 1 & 0 & -\frac{1}{7} & 0 & \frac{67}{7} \\ 0 & 1 & \frac{6}{7} & 0 & -\frac{59}{7} \\ 0 & 0 & 0 & 1 & -13 \end{bmatrix}$$

> solve({6*x+y+4*w=-3, -9*x+2*y+3*z-8*w=1, 7*x-z+5*w=2}, {x,y,w});

$$\left\{ w = -13, x = \frac{67}{7} + \frac{1}{7}z, y = -\frac{59}{7} - \frac{6}{7}z \right\}$$

From the *rref* matrix given above, we see that $w = -13$ and that x and y can be expressed as functions of the single variable z. Hence, the solution is one-dimensional, with the solution given in the result of the *solve* command above. In column matrix form, we would express this as

$$X = \begin{bmatrix} x \\ y \\ z \\ w \end{bmatrix} = \begin{bmatrix} \frac{67}{7} \\ -\frac{59}{7} \\ 0 \\ -13 \end{bmatrix} + \begin{bmatrix} \frac{1}{7} \\ -\frac{6}{7} \\ 1 \\ 0 \end{bmatrix} w$$

Something interesting happens when we try to solve the systems of equations in terms of the variables x, y, and z.

> solve({6*x+y+4*w=-3, -9*x+2*y+3*z-8*w=1, 7*x-z+5*w=2}, {x,y,z});

In this situation, since $w = -13$ is a fixed value, and as a result x, y, and z cannot be expressed in terms of w, nor can they be expressed in terms of each other making it impossible to solve this system for only the variables x, y, and z.

■

Now, let us look at an example of a linear system where there is no solution and how the *rref* matrix will reflect this fact.

Example 3.1.3. We want to solve the linear system

$$6x + y + 4w = -3$$
$$-9x + 2y + 3z - 8w = 1$$
$$-3x + 3y + 3z - 4w = 10$$

> C:= matrix([[6, 1, 0, 4, -3], [-9, 2, 3, -8, 1], [-3, 3, 3, -4, 10]]);

$$C := \begin{bmatrix} 6 & 1 & 0 & 4 & -3 \\ -9 & 2 & 3 & -8 & 1 \\ -3 & 3 & 3 & -4 & 10 \end{bmatrix}$$

> rref(C);

$$\begin{bmatrix} 1 & 0 & -\frac{1}{7} & \frac{16}{21} & 0 \\ 0 & 1 & \frac{6}{7} & -\frac{4}{7} & 0 \\ 0 & 0 & 0 & 0 & 1 \end{bmatrix}$$

> soln:= solve({6*x + y + 4*w = -3, -9*x + 2*y + 3*z - 8*w = 1, -3*x + 3*y + 3*z - 4*w = 10}, {x, y, z});

$$soln :=$$

The last row of the *rref* matrix gives the equation $0 = 1$ when converted back to an equation. Since this is clearly impossible, it indicates that this linear system has no solution and so is said to be inconsistent. This feature of the *rref* matrix will always appear when there is no solution. Also notice that when we use the *solve* command on the system, *Maple* returns no answer, thus the system has no solution.

■

There are also linear systems where the *rref* matrix contains rows that are all zeros. This indicates that at least one of the equations of the system is not needed in order to solve the system. The following is an example of this situation.

Example 3.1.4. We want to solve the linear system

$$6x + y + 4w = -3$$
$$-9x + 2y + 3z - 8w = 1$$
$$-3x + 3y + 3z - 4w = -2$$

> C:= matrix([[6, 1, 0, 4, -3], [-9, 2, 3, -8, 1], [-3, 3, 3, -4, -2]]);

$$C := \begin{bmatrix} 6 & 1 & 0 & 4 & -3 \\ -9 & 2 & 3 & -8 & 1 \\ -3 & 3 & 3 & -4 & -2 \end{bmatrix}$$

> rref(C);

$$\begin{bmatrix} 1 & 0 & -\frac{1}{7} & \frac{16}{21} & -\frac{1}{3} \\ 0 & 1 & \frac{6}{7} & -\frac{4}{7} & -1 \\ 0 & 0 & 0 & 0 & 0 \end{bmatrix}$$

> solve({6*x+y+4*w=-3, -9*x+2*y+3*z-8*w=1, -3*x+3*y+3*z-4*w=-2}, {x, y});

$$\left\{ x = -\frac{1}{3} - \frac{16}{21}w + \frac{1}{7}z, y = -1 + \frac{4}{7}w - \frac{6}{7}z \right\}$$

This linear system has a solution of dimension two in $\mathbb{R}^4$. The last row of all zeros in the *rref* matrix indicates that one of these three equations is superfluous to solving the system. Hence, any one of the equations can be written as a linear combination of the remaining two, as this is the only way that our three operations can reduce an entire row of an augmented matrix to all zeros. A linear equation is called a *linear combination* of other linear equations if it is a sum of scalar multiples of the other equations. Here, the third equation is the sum of the other two equations, and so it is a linear combination of them.

In the *linalg* package, there is the command, *pivot*, which allows one to select a nonzero entry $A_{i,j}$ of a matrix A and then use the third row operation of multiplying the ith row by a nonzero number and then adding it to another row to make every entry in the jth column, except $A_{i,j}$, into 0. This pivoting process along the diagonal entries of the augmented matrix of a linear system is the backbone of the command *rref* along with multiplying the rows with a nonzero diagonal entry by one over its value and an occasional use of swapping rows. Under the assumption that no entry on the diagonal is zero, one could use the *pivot* command first on $A_{1,1}$ to zero out all other entries in the first column, then move to $A_{2,2}$ and zero out all entries except for $A_{2,2}$ in the second column. This process is continued until you have reached and performed the *pivot* command on the last row in the matrix.

Example 3.1.5. Let us see *pivot* in action on a previous example along with the *mulrow* command to get diagonal 1's. This will produce the same result as *rref*. We want to solve the linear system

$$6x + y + 4w = -3$$
$$-9x + 2y + 3z - 8w = 1$$
$$7x - 4z + 5w = 2$$

whose augmented matrix of this system is

$$C = \begin{bmatrix} 6 & 1 & 0 & 4 & -3 \\ -9 & 2 & 3 & -8 & 1 \\ 7 & 0 & -4 & 5 & 2 \end{bmatrix}$$

> C:= matrix([[6, 1, 0, 4, -3], [-9, 2, 3, -8, 1], [7, 0, -4, 5, 2]]):
> Pivot_C11:= pivot(C, 1, 1);

$$Pivot\_C11 := \begin{bmatrix} 6 & 1 & 0 & 4 & -3 \\ 0 & \frac{7}{2} & 3 & -2 & -\frac{7}{2} \\ 0 & -\frac{7}{6} & -4 & \frac{1}{3} & \frac{11}{2} \end{bmatrix}$$

> Pivot_C22:= pivot(Pivot_C11, 2, 2);

$$Pivot\_C22 := \begin{bmatrix} 6 & 0 & -\frac{6}{7} & \frac{32}{7} & -2 \\ 0 & \frac{7}{2} & 3 & -2 & -\frac{7}{2} \\ 0 & 0 & -3 & -\frac{1}{3} & \frac{13}{3} \end{bmatrix}$$

> Pivot_C33:= pivot(Pivot_C22, 3, 3);

$$Pivot\_C33 := \begin{bmatrix} 6 & 0 & 0 & \frac{14}{3} & -\frac{68}{21} \\ 0 & \frac{7}{2} & 0 & -\frac{7}{3} & \frac{5}{6} \\ 0 & 0 & -3 & -\frac{1}{3} & \frac{13}{3} \end{bmatrix}$$

> mulrow(mulrow(mulrow(Pivot_C33, 1, 1/6), 2, 2/7), 3, -1/3);

$$
\begin{bmatrix}
1 & 0 & 0 & \frac{7}{9} & -\frac{34}{63} \\
0 & 1 & 0 & -\frac{2}{3} & \frac{5}{21} \\
0 & 0 & 1 & \frac{1}{9} & -\frac{13}{9}
\end{bmatrix}
$$

> rref(C);

$$
\begin{bmatrix}
1 & 0 & 0 & \frac{7}{9} & -\frac{34}{63} \\
0 & 1 & 0 & -\frac{2}{3} & \frac{5}{21} \\
0 & 0 & 1 & \frac{1}{9} & -\frac{13}{9}
\end{bmatrix}
$$

∎

All of the previous examples have been systems with more variables than equations that typically have an infinite number of solutions since their solution will have at least one arbitrary independent variable in it. A square system with as many variables as equations usually has only one solution.

Example 3.1.6. Let us solve the following square 4 by 4 system given by

$$
\begin{aligned}
-7x + 2y - 9z + 3w &= -5 \\
4x + y + 6z - 11w &= 15 \\
x - y + 3z + 8w &= 4 \\
9x + 12y - 7z + 5w &= -2
\end{aligned}
$$

The system is said to be 4×4 since there are four equations and four unknowns. In general, if there are m equations and n variables, the resulting augmented matrix will be an element of $\mathbb{R}^{m \times (n+1)}$. The augmented matrix of this system is

$$
C =
\begin{bmatrix}
-7 & 2 & -9 & 3 & -5 \\
4 & 1 & 6 & -11 & 15 \\
1 & -1 & 3 & 8 & 4 \\
9 & 12 & -7 & 5 & -2
\end{bmatrix}
$$

> C:= matrix([[-7,2,-9,3,-5], [4,1,6,-11,15], [1,-1,3,8,4], [9,12,-7,5,-2]]):
> R:= rref(C);

$$
R :=
\begin{bmatrix}
1 & 0 & 0 & 0 & -\frac{5881}{2056} \\
0 & 1 & 0 & 0 & \frac{8511}{2056} \\
0 & 0 & 1 & 0 & \frac{950}{257} \\
0 & 0 & 0 & 1 & -\frac{23}{2056}
\end{bmatrix}
$$

> soln:= delcols(R,1..4);

$$soln := \begin{bmatrix} -\dfrac{5881}{2056} \\[2mm] \dfrac{8511}{2056} \\[2mm] \dfrac{950}{257} \\[2mm] -\dfrac{23}{2056} \end{bmatrix}$$

> x:= soln[1,1]; y:= soln[2,1]; z:= soln[3,1]; w:= soln[4,1];

$$x := -\frac{5881}{2056}$$

$$y := \frac{8511}{2056}$$

$$z := \frac{950}{257}$$

$$w := -\frac{23}{2056}$$

This 4×4 square system has the unique solution

$$\{x = soln[1,1], y = soln[2,1], z = soln[3,1], w = soln[4,1]\}$$

■

Square systems typically have a single solution, which is why people prefer square systems whenever possible in solving a real-life problem.

Definition 3.1.1. A *determined*, or *square linear system*, is one in which the number of equations m equals the number of unknown variables n.

Definition 3.1.2. An *underdetermined linear system* is one in which the number of equations m is less than the number of unknown variables n.

Definition 3.1.3. An *overdetermined linear system* is one in which the number of equations m is greater than the number of unknown variables n.

Underdetermined systems typically have an infinite number of solutions, while overdetermined systems typically have no solution, as we shall see below.

Example 3.1.7. As an example, let us try to solve the overdetermined linear system

$$7x + 2y + 9z = 8$$
$$-3x + 5y + 6z = -2$$
$$11x - y + 4z = 3$$
$$8x + 13y - 10z = -11$$

The augmented matrix of this system is

$$C = \begin{bmatrix} 7 & 2 & 9 & 8 \\ -3 & 5 & 6 & -2 \\ 11 & -1 & 4 & 3 \\ 8 & 13 & -10 & -11 \end{bmatrix}$$

> C:= matrix([[7, 2, 9, 8], [-3, 5, 6, -2], [11, -1, 4, 3], [8, 13, -10, -11]]):
> rref(C);

$$\begin{bmatrix} 1 & 0 & 0 & 0 \\ 0 & 1 & 0 & 0 \\ 0 & 0 & 1 & 0 \\ 0 & 0 & 0 & 1 \end{bmatrix}$$

The last row of this *rref* matrix gives the equation $0 = 1$, which is impossible. This implies that our overdetermined system has no solution. However, if we are to remove the last equation from the original system, we would end up with the following augmented matrix that is determined, that is, three equations with three unknown variables:

$$C1 = \begin{bmatrix} 7 & 2 & 9 & 8 \\ -3 & 5 & 6 & -2 \\ 11 & -1 & 4 & 3 \end{bmatrix}$$

> C1:= matrix([[7, 2, 9, 8], [-3, 5, 6, -2], [11, -1, 4, 3]]):
> rref(C1);

$$\begin{bmatrix} 1 & 0 & 0 & -\frac{11}{10} \\ 0 & 1 & 0 & -\frac{43}{10} \\ 0 & 0 & 1 & \frac{27}{10} \end{bmatrix}$$

So by removing one equation, we get the unique solution:

$$\left\{ x = -\frac{11}{10}, \ y = -\frac{43}{10}, \ z = \frac{27}{10} \right\}$$

If we remove yet one more equation, we end up with two equations and three unknowns, which implies a solution of dimension one.

> C2:= matrix([[7, 2, 9, 8], [-3, 5, 6, -2]]):
> rref(C2);

$$\begin{bmatrix} 1 & 0 & \frac{33}{41} & \frac{44}{41} \\ 0 & 1 & \frac{69}{41} & \frac{10}{41} \end{bmatrix}$$

The solution to this system is clearly one-dimensional, given by

$$\left\{ x = -\frac{33}{41}z + \frac{44}{41}, \ y = -\frac{69}{41}z + \frac{10}{41} \right\}$$

Homework Problems

1. Give both an algebraic and geometric explanation of why solutions to underdetermined systems typically have an infinite number of solutions and why solutions to overdetermined systems typically have no solution.

2. Using your answer to problem 1, construct (if possible) underdetermined and overdetermined systems that have a single solution, that is, a solution of dimension zero.

3. Perform Gauss-Jordan elimination on the following systems of equations, but with one restriction: You are not allowed to swap row 1 with any other. Be sure to show each step in the process.

(a)
$$x - 3y + z = 6$$
$$-2x + 6y + 3z = 4$$
$$2x + 5y + 6z = 1$$

(b)
$$2x + 3y - 5z = 7$$
$$3x + 2y + 7z = 8$$
$$4x + 6y + 2z = 1$$

(c)
$$x + y + z = 1$$
$$2w + 2x + 3y + z = 4$$
$$2w + 3x + 4y + 2z = 5$$
$$4w + 6x + 8y + 4z = 10$$

(d)
$$w + x + y + z = 3$$
$$2w + 2x + 2y - 3z = 5$$
$$-w + x + y + z = 6$$
$$4w + 6x + y - 7z = 8$$

(e)
$$3x + 4y - 7z = 1$$
$$-2x + 4y - 8z = 2$$
$$5x + z = -1$$

(f)
$$3x + 4y - 7z = 1$$
$$-2x + 4y - 8z = 2$$
$$5x + z = -1$$
$$-3x + 4y + 3z = 2$$

4. Write out the solutions to each system from problem 3, and give the dimension of the solution and the space $\mathbb{R}^n$ it lies in.

5. Write out the solutions to each system from problem 3 in column matrix format using scalar multiplication by the independent variables.

6. Would Gauss-Jordan elimination be easier to implement in any of the systems of problem 3 if you were allowed to use the row swap operation? If so, explain and go through the process of Gauss-Jordan elimination again, using the row swap operation.

7. Perform Gauss-Jordan elimination on the following matrices.

(a) $\begin{bmatrix} 2 & -3 & 6 \\ 8 & -4 & 2 \end{bmatrix}$ (b) $\begin{bmatrix} 2 & -3 & 6 & 1 \\ 8 & -4 & 2 & 3 \\ 7 & -3 & 2 & 5 \end{bmatrix}$ (c) $\begin{bmatrix} 2 & -3 & 4 & -5 & -10 \\ 1 & 1 & -1 & 1 & 4 \\ 3 & 5 & -9 & 7 & 24 \\ 2 & 2 & -2 & 3 & 9 \end{bmatrix}$

8. Given the following two systems of equations:

(a) $3x + 4y = -1$ (b) $3x + 4y = 9$
         $4x - 2y = 6$ $4x - 2y = -10$

Explain how the following matrix can be used to solve both systems simultaneously, then do so:

$$\left[\begin{array}{cc|cc} 3 & 4 & -1 & 9 \\ 4 & -2 & 6 & -10 \end{array} \right]$$

9. Is it possible for a *rref* matrix to have more than one row whose corresponding linear equation is $0 = 1$? If no, explain why not. If yes, then give an example.

10. A square matrix is called *upper triangular* if all of its entries below the main diagonal from upper left to lower right are 0. If A is a square matrix, then must $rref(A)$ be upper triangular? If yes, explain why. If no, then give an example.

11. What would have to be true about a linear system so that the *rref* matrix of the augmented matrix of this system has all rows of all zeroes except for the first row? Give examples of the possibilities.

12. What would have to be true about a linear system so that the *rref* matrix of the augmented matrix of this system is an identity matrix? Give examples of the possibilities.

13. What would have to be true about a linear system so that the *rref* matrix of the augmented matrix of this system is an identity matrix to the left of the last column? Give examples of the possibilities.

Maple Problems

1. Solve *Homework* problems 3 and 5 using the *pivot*, *swaprow*, and *mulrow* commands.

2. Solve *Homework* problems 3 and 5 using the *addrow*, *mulrow*, and *swaprow* commands.

3. Solve *Homework* problems 3 and 5 using the *rref* command.

4. Perform Gauss-Jordan elimination using the *addrow*, *mulrow*, and *swaprow* commands on the following matrices.

(a) $\begin{bmatrix} 2 & -5 & 8 & 9 \\ 3 & 2 & -6 & 0 \\ -1 & -4 & 2 & 5 \end{bmatrix}$

(b) $\begin{bmatrix} 2 & -3 & 6 & 1 & -2 \\ -4 & -1 & 0 & 1 & 6 \\ 7 & 3 & -1 & 0 & 7 \\ -2 & -3 & 2 & 5 & 8 \end{bmatrix}$

(c) $\begin{bmatrix} 2 & -3 & 6 & 8 & 3 \\ 8 & -4 & -7 & 3 & 2 \\ 7 & -3 & 2 & -4 & -1 \\ -3 & 6 & -3 & -6 & -4 \end{bmatrix}$

(d) $\begin{bmatrix} 4 & 5 & -1 & -1 & 1 \\ 2 & 2 & -2 & 2 & 2 \\ 8 & 0 & 3 & -7 & 3 \\ -2 & -3 & 3 & 5 & -5 \end{bmatrix}$

(e) $\begin{bmatrix} 2 & -3 & 6 & 1 & -2 \\ -4 & -1 & 0 & 1 & 6 \\ 7 & 3 & -1 & 0 & 7 \\ -2 & -3 & 2 & 5 & 8 \end{bmatrix}$

(f) $\begin{bmatrix} -4 & -3 & 6 & -6 & 3 & -4 \\ 1 & -4 & -7 & 3 & 2 & 5 \\ 7 & 2 & 6 & -4 & -1 & 2 \\ 8 & 6 & -3 & -6 & -4 & -1 \\ -3 & 6 & -1 & 3 & 8 & 9 \end{bmatrix}$

5. For each *rref* matrix given below, find a corresponding original linear system of equations whose *rref* matrix yields the *rref* form. Your original system's augmented matrix should not look like the *rref* matrix you are attempting to reduce to.

(a) $\begin{bmatrix} 1 & 0 & a & b \\ 0 & 1 & 0 & c \\ 0 & 0 & 0 & 0 \end{bmatrix}$

(b) $\begin{bmatrix} 1 & a & 0 & b & c & d \\ 0 & 0 & 1 & e & f & g \end{bmatrix}$

(c) $\begin{bmatrix} 1 & 0 & 0 & a & b & c \\ 0 & 1 & 0 & d & e & f \\ 0 & 0 & 1 & g & h & i \\ 0 & 0 & 0 & 0 & 0 & 0 \end{bmatrix}$

(d) $\begin{bmatrix} 1 & a & b & 0 & 0 & c \\ 0 & 0 & 0 & 1 & 0 & d \\ 0 & 0 & 0 & 0 & 1 & e \\ 0 & 0 & 0 & 0 & 0 & 0 \end{bmatrix}$

(e) $\begin{bmatrix} 1 & 0 & a & b & c & d \\ 0 & 1 & e & f & g & h \\ 0 & 0 & 0 & 0 & 0 & 0 \\ 0 & 0 & 0 & 0 & 0 & 0 \end{bmatrix}$

(f) $\begin{bmatrix} 1 & a & b & c & 0 & 0 \\ 0 & 0 & 0 & 0 & 1 & 0 \\ 0 & 0 & 0 & 0 & 0 & 1 \\ 0 & 0 & 0 & 0 & 0 & 0 \end{bmatrix}$

6. Let $A = \begin{bmatrix} a & b \\ c & d \end{bmatrix}$. Perform *rref* on the augmented matrix $(A|I_2)$ of A with

the 2×2 identity matrix after it. You will get a 2×4 matrix $(I_2|B)$, with the 2×2 identity matrix followed by a 2×2 matrix B. Do you recognize B? If not, then find the two products AB and BA, do you now know what B is? Generalize this to larger square matrices A and see if it still works.

7. Let $A = \begin{bmatrix} a & b \\ c & d \end{bmatrix}$. The row operation of multiplying row 1 of A by k and then adding the result to row 2 of A gives addrow$(A, \text{row}1, \text{row}2, k) = \begin{bmatrix} a & b \\ ka+c & kb+d \end{bmatrix}$. If we perform this row operation on the 2×2 identity matrix I_2, we get $B = \begin{bmatrix} 1 & 0 \\ k & 1 \end{bmatrix}$. What is the product BA? Generalize this so that each of the three types of row operations applied to A can be done through multiplication by a matrix like B.

8. Solve in as many ways as possible the linear systems which have the augmented matrices in *Maple* problem 4 of this section. Write your solutions using column matrices.

3.2   Elementary Matrices

In Section 3.1, we discovered Gauss-Jordan elimination (*rref*) uses three different kinds of row operations to convert a matrix to the final augmented matrix, which in turn provides the solutions to the linear system. Each of these three row operations can be done through the use of matrix multiplication on the left by what are called *elementary matrices*. The multiplication by the elementary matrix E on the left must preserve the size of the matrix A.

Thus, if A is of size $l \times n$, then for E of size $m \times k$, we will only have EA possible if $k = l$, and of the same size as A if $m = l$. Now we see that all elementary matrices E must be square and of size $l \times l$ if they are to be applied, be left multiplication, to matrices A of size $l \times n$, and be size preserving.

Example 3.2.1. Let us do an example of the first row operation of swapping two rows and how it can be done using multiplication by an elementary matrix E. This type of elementary matrix is referred to as type I. First, we will define $A = \begin{bmatrix} 1 & 2 & 3 & 4 \\ 5 & 6 & 7 & 8 \\ 9 & 10 & 11 & 12 \end{bmatrix}$. We now ask, what matrix $E \in \mathbb{R}^{3 \times 3}$, will swap the first and third rows of A when you compute EA? So if $B = EA$, then $B = \begin{bmatrix} 9 & 10 & 11 & 12 \\ 5 & 6 & 7 & 8 \\ 1 & 2 & 3 & 4 \end{bmatrix}$. The answer, after some thought, is that we need

$$E = \begin{bmatrix} 0 & 0 & 1 \\ 0 & 1 & 0 \\ 1 & 0 & 0 \end{bmatrix}, \text{ which is the matrix gotten by swapping the first and third}$$

rows of the 3×3 identity matrix $I_3 = \begin{bmatrix} 1 & 0 & 0 \\ 0 & 1 & 0 \\ 0 & 0 & 1 \end{bmatrix}$. Multiplication of A by
I_3 on the left gives A back since $I_3 A = A$, so swapping I_3's rows should swap
A's rows, at least we hope it will. Notice that $EE = I_3$ (spend two minutes
performing the multiplication to convince yourself, it is time well spent) since
multiplication by E twice should bring us back to where we started as the second E swaps the rows back to their original positions.

> A:= matrix([[1, 2, 3, 4], [5, 6, 7, 8], [9, 10, 11, 12]]):
> E:= matrix([[0, 0, 1], [0, 1, 0], [1, 0, 0]]):
> evalm(E&*A);

$$\begin{bmatrix} 9 & 10 & 11 & 12 \\ 5 & 6 & 7 & 8 \\ 1 & 2 & 3 & 4 \end{bmatrix}$$

> evalm(E&*E);

$$\begin{bmatrix} 1 & 0 & 0 \\ 0 & 1 & 0 \\ 0 & 0 & 1 \end{bmatrix}$$

Example 3.2.2. Now let us do an example of the second row operation of
multiplying a row by a nonzero number c, and how it can be done in a similar
manner. We want to multiply the second row of the above matrix A by -5.
Then it seems the matrix E we need should be $E = \begin{bmatrix} 1 & 0 & 0 \\ 0 & -5 & 0 \\ 0 & 0 & 1 \end{bmatrix}$, which is
an elementary matrix of type II. In other words, the pattern seems to be that
to find the correct matrix E, do the row operation on the 3×3 identity matrix
I_3. The matrix that undoes this row operation (or undoes E as it may be said)
is $F = \begin{bmatrix} 1 & 0 & 0 \\ 0 & -\frac{1}{5} & 0 \\ 0 & 0 & 1 \end{bmatrix}$, since $FE = EF = I_3$.

> E:= matrix([[1, 0, 0], [0, -5, 0], [0, 0, 1]]):
> evalm(E&*A);

$$\begin{bmatrix} 1 & 2 & 3 & 4 \\ -25 & -30 & -35 & -40 \\ 9 & 10 & 11 & 12 \end{bmatrix}$$

> F:= matrix([[1, 0, 0], [0, -1/5, 0], [0, 0, 1]]):

> evalm(F&*E);

$$\begin{bmatrix} 1 & 0 & 0 \\ 0 & 1 & 0 \\ 0 & 0 & 1 \end{bmatrix}$$

■

Example 3.2.3. Now on to the final and third row operation of multiplying a row by a nonzero number c and adding it to another row. Let us now multiply row 1 of the matrix A above by -3 and add it to row 2. This should put the result in row 2 and leave the other rows the same. A type III elementary matrix is when this row operation is applied to I_3. For this instance, we have $E = \begin{bmatrix} 1 & 0 & 0 \\ -3 & 1 & 0 \\ 0 & 0 & 1 \end{bmatrix}$. The row operation that undoes this one is multiplying row 1 of the matrix A above by 3 and adding it to row 2 with matrix $F = \begin{bmatrix} 1 & 0 & 0 \\ 3 & 1 & 0 \\ 0 & 0 & 1 \end{bmatrix}$ so that $FE = EF = I_3$.

> E:= matrix([[1, 0, 0], [-3, 1, 0], [0, 0, 1]]):

> evalm(E&*A);

$$\begin{bmatrix} 1 & 2 & 3 & 4 \\ 2 & 0 & -2 & -4 \\ 9 & 10 & 11 & 12 \end{bmatrix}$$

> F:= matrix([[1, 0, 0], [3, 1, 0], [0, 0, 1]]):

> evalm(F&*E);

$$\begin{bmatrix} 1 & 0 & 0 \\ 0 & 1 & 0 \\ 0 & 0 & 1 \end{bmatrix}$$

■

In summary, we have introduced the important concept of elementary matrices, which has many theoretical uses in linear algebra. Gauss-Jordan elimination and *rref* can be done through left multiplication of the augmented matrix of the linear system by successive elementary matrices, one elementary matrix applied for each row operation done. Next, we give the formal definition of an elementary matrix, although you undoubtedly know what its going to be by now.

Definition 3.2.1. An *elementary matrix* E of size $k \times k$ is simply the modified $k \times k$ identity matrix I_k after exactly one elementary row operation has been applied to it.

Example 3.2.4. Now, let us do an example where we use multiple elementary matrices to find the *rref* matrix of the following matrix A. Consider the augmented matrix

$$A = \begin{bmatrix} 1 & 2 & -1 & 6 \\ 2 & -1 & 3 & 7 \\ 0 & 4 & 5 & -2 \end{bmatrix} \qquad (3.4)$$

We will perform Gauss-Jordan elimination using successive left multiplications by elementary matrices. Notice that we need to get rid of $A_{2,1} = 2$ first. To do this, we would take row 1, multiply it by -2, and add it to row 2. We would write it as $E_1 = \begin{bmatrix} 1 & 0 & 0 \\ -2 & 1 & 0 \\ 0 & 0 & 1 \end{bmatrix}$. If we perform the multiplication $E_1 A$, we get

$$E_1 A = \begin{bmatrix} 1 & 2 & -1 & 6 \\ 0 & -5 & 5 & -5 \\ 0 & 4 & 5 & -2 \end{bmatrix} \qquad (3.5)$$

Now, we want to get rid of $A_{1,2}$ and $A_{3,2}$, but before we do this, we will multiply row 2 by $-\frac{1}{5}$ to make our life easier. So $E_2 = \begin{bmatrix} 1 & 0 & 0 \\ 0 & -\frac{1}{5} & 0 \\ 0 & 0 & 1 \end{bmatrix}$ and we get

$$E_2 E_1 A = \begin{bmatrix} 1 & 2 & -1 & 6 \\ 0 & 1 & -1 & 1 \\ 0 & 4 & 5 & -2 \end{bmatrix} \qquad (3.6)$$

Now, we can zero out entry $A_{1,2}$ by taking -2 times row 2 and adding it to row 1. Similarly, for $A_{3,2}$ we take -4 times row 2 and add it to row 3. The elementary matrices are $E_3 = \begin{bmatrix} 1 & -2 & 0 \\ 0 & 1 & 0 \\ 0 & 0 & 1 \end{bmatrix}$ and $E_4 = \begin{bmatrix} 1 & 0 & 0 \\ 0 & 1 & 0 \\ 0 & -4 & 1 \end{bmatrix}$, respectively. Performing the multiplications gives

$$E_4 E_3 E_2 E_1 A = \begin{bmatrix} 1 & 0 & 1 & 4 \\ 0 & 1 & -1 & 1 \\ 0 & 0 & 9 & -6 \end{bmatrix} \qquad (3.7)$$

We are getting close to an answer now. Next step is to divide row 3 by 9. So

$$E_5 = \begin{bmatrix} 1 & 0 & 0 \\ 0 & 1 & 0 \\ 0 & 0 & \frac{1}{9} \end{bmatrix} \text{ with}$$

$$E_5 E_4 E_3 E_2 E_1 A = \begin{bmatrix} 1 & 0 & 1 & 4 \\ 0 & 1 & -1 & 1 \\ 0 & 0 & 1 & -\frac{2}{3} \end{bmatrix} \tag{3.8}$$

Last, we wish to zero out $A_{1,3}$ and $A_{2,3}$. To do this, we add minus row 3 to row 1, and row 3 to row 2. Thus $E_6 = \begin{bmatrix} 1 & 0 & -1 \\ 0 & 1 & 0 \\ 0 & 0 & 1 \end{bmatrix}$ and $E_7 = \begin{bmatrix} 1 & 0 & 0 \\ 0 & 1 & 1 \\ 0 & 0 & 1 \end{bmatrix}$, respectively. Finally, after multiplication by 7 elementary matrices, we have

$$E_7 E_6 E_5 E_4 E_3 E_2 E_1 A = \begin{bmatrix} 1 & 0 & 0 & \frac{14}{3} \\ 0 & 1 & 0 & \frac{1}{3} \\ 0 & 0 & 1 & -\frac{2}{3} \end{bmatrix} \tag{3.9}$$

One thing to notice and consider. If you want to speed up the process a little bit, consider the last two matrices, E_6 and E_7. We could combine these two together in a new matrix $E_{76} = E_7 E_6 = \begin{bmatrix} 1 & 0 & -1 \\ 0 & 1 & 1 \\ 0 & 0 & 1 \end{bmatrix}$ such that

$$E_{76} E_5 E_4 E_3 E_2 E_1 A = E_7 E_6 E_5 E_4 E_3 E_2 E_1 A \tag{3.10}$$

We will go through the same steps but with *Maple* this time.

> A:= matrix(3,4,[1,2,-1, 6, 2, -1,3, 7, 0, 4, 5, -2]);

$$A := \begin{bmatrix} 1 & 2 & -1 & 6 \\ 2 & -1 & 3 & 7 \\ 0 & 4 & 5 & -2 \end{bmatrix}$$

> E1:= matrix(3,3,[1,0,0,-2,1,0,0,0,1]);

$$E1 := \begin{bmatrix} 1 & 0 & 0 \\ -2 & 1 & 0 \\ 0 & 0 & 1 \end{bmatrix}$$

> evalm(E1&*A);

$$\begin{bmatrix} 1 & 2 & -1 & 6 \\ 0 & -5 & 5 & -5 \\ 0 & 4 & 5 & -2 \end{bmatrix}$$

> E2:= matrix(3,3,[1,0,0,0,-1/5,0,0,0,1]);

$$E2 := \begin{bmatrix} 1 & 0 & 0 \\ 0 & -\frac{1}{5} & 0 \\ 0 & 0 & 1 \end{bmatrix}$$

> evalm(E2&*E1&*A);

$$\begin{bmatrix} 1 & 2 & -1 & 6 \\ 0 & 1 & -1 & 1 \\ 0 & 4 & 5 & -2 \end{bmatrix}$$

> E3:= matrix(3,3,[1,-2,0,0,1,0,0,0,1]);

$$E3 := \begin{bmatrix} 1 & -2 & 0 \\ 0 & 1 & 0 \\ 0 & 0 & 1 \end{bmatrix}$$

> evalm(E3&*E2&*E1&*A);

$$\begin{bmatrix} 1 & 0 & 1 & 4 \\ 0 & 1 & -1 & 1 \\ 0 & 4 & 5 & -2 \end{bmatrix}$$

> E4:= matrix(3,3,[1,0,0,0,1,0,0,-4,1]);

$$E4 := \begin{bmatrix} 1 & 0 & 1 & 4 \\ 0 & 1 & -1 & 1 \\ 0 & 0 & 9 & -6 \end{bmatrix}$$

> evalm(E4&*E3&*E2&*E1&*A);

$$\begin{bmatrix} 1 & 0 & 1 & 4 \\ 0 & 1 & -1 & 1 \\ 0 & 0 & 9 & -6 \end{bmatrix}$$

> E5:= matrix(3,3,[1,0,0,0,1,0,0,0,1/9]);

$$E5 := \begin{bmatrix} 1 & 0 & 0 \\ 0 & 1 & 0 \\ 0 & 0 & \frac{1}{9} \end{bmatrix}$$

> evalm(E5&*E4&*E3&*E2&*E1&*A);

$$\begin{bmatrix} 1 & 0 & 1 & 4 \\ 0 & 1 & -1 & 1 \\ 0 & 0 & 1 & -\frac{2}{3} \end{bmatrix}$$

> E6:= matrix(3,3,[1,0,-1,0,1,0,0,0,1]);

$$E6 := \begin{bmatrix} 1 & 0 & -1 \\ 0 & 1 & 0 \\ 0 & 0 & 1 \end{bmatrix}$$

> evalm(E6&*E5&*E4&*E3&*E2&*E1&*A);

$$\begin{bmatrix} 1 & 0 & 0 & \frac{14}{3} \\ 0 & 1 & -1 & 1 \\ 0 & 0 & 1 & -\frac{2}{3} \end{bmatrix}$$

> E7:= matrix(3,3,[1,0,0,0,1,1,0,0,1]);

$$E7 := \begin{bmatrix} 1 & 0 & 0 \\ 0 & 1 & 1 \\ 0 & 0 & 1 \end{bmatrix}$$

> evalm(E7&*E6&*E5&*E4&*E3&*E2&*E1&*A);

$$\begin{bmatrix} 1 & 0 & 0 & \frac{14}{3} \\ 0 & 1 & 0 & \frac{1}{3} \\ 0 & 0 & 1 & -\frac{2}{3} \end{bmatrix}$$

Next, we will have *Maple* perform the *rref* command to verify that our application of elementary matrices was correct.

> rref(A);

$$\begin{bmatrix} 1 & 0 & 0 & \frac{14}{3} \\ 0 & 1 & 0 & \frac{1}{3} \\ 0 & 0 & 1 & -\frac{2}{3} \end{bmatrix}$$

Finally, in regards to the comment about replacing E_7 and E_6 with the single matrix E_{76} defined to be $E_{76} = \begin{bmatrix} 1 & 0 & -1 \\ 0 & 1 & 1 \\ 0 & 0 & 1 \end{bmatrix}$, if we replace the multiplication $E_7 E_6$ with the matrix E_{76} we will get the same result.

> E76:= matrix(3,3,[1,0,-1,0,1,1,0,0,1]);

$$E76 := \begin{bmatrix} 1 & 0 & -1 \\ 0 & 1 & 1 \\ 0 & 0 & 1 \end{bmatrix}$$

> evalm(E76&*E5&*E4&*E3&*E2&*E1&*A);

$$\begin{bmatrix} 1 & 0 & 0 & \frac{14}{3} \\ 0 & 1 & 0 & \frac{1}{3} \\ 0 & 0 & 1 & -\frac{2}{3} \end{bmatrix}$$

∎

Homework Problems

1. Given a system of n equations in n variables, what would the maximum number of left multiplications by elementary matrices be to convert the original augmented matrix representation of the system to *rref* form?

2. The example used in this section was a three equation, three variable system. Can elementary matrices be used on nonsquare systems? What are the restrictions?

3. Use left multiplication by elementary matrices to reduce the following systems of equations as far as possible. Also determine if the resulting matrix is in reduced *rref* form.

(a)
$$2x - 3y = 7$$
$$-2x + 5y = 1$$

(b)
$$-7x + 2y = 5$$
$$6x + 3y = 4$$

(c)
$$x - 3y + z = 6$$
$$-2x + 6y + 3z = 4$$
$$2x + 5y + 6z = 1$$

(d)
$$2x + 3y - 5z = 7$$
$$3x + 2y + 7z = 8$$
$$4x + 6y + 2z = 1$$

(e)
$$-3y + z = 6$$
$$-2x + 3z = 4$$
$$2x + 5y = 1$$

(f)
$$w - x - 3y + z = 6$$
$$w + 2x - y + 3z = 4$$
$$-w + 2x + 4z = 1$$
$$x - 2y + 5z = 1$$

4. Use left multiplication by elementary matrices to reduce the following systems of equations to *rref* form (or as close as possible). You may leave your answer in *rref* matrix form.

(a)
$$x - 3y + z = 6$$
$$-2x + 6y + 3z = 4$$

(b)
$$2x + 3y = 7$$
$$3x + 2y = 8$$
$$-2x - 5y = -2$$

(c) $\qquad$ $w + x + z = 1$ $\qquad$ (d) $\qquad$ $2x + 3y = 7$

$\qquad\qquad\qquad x + 3z = 4$ $\qquad\qquad\qquad\qquad 3x + 2y = 8$

$\qquad\qquad\qquad 6y + 2z = 1$ $\qquad\qquad\qquad\qquad x - y = 1$

5. Use the elementary matrices from part (a) of problem 3, on the following two corresponding systems. Explain what this implies.

(a) $\qquad$ $2x - 3y = 3$ $\qquad$ (b) $\qquad$ $2x - 3y = -2$

$\qquad\qquad\quad -2x + 5y = -4$ $\qquad\qquad\qquad -2x + 5y = -1$

6. (a) Give examples, E_1, E_2, and E_3, of each of the three types of 4×4 elementary matrices.

(b) Next, find for each of the elementary matrices E_1, E_2, and E_3, of part a another elementary matrix F_1, F_2, and F_3 of the same type and size as the corresponding E so that $F_k E_k = I_4$ and $E_k F_k = I_4$ for $k = 1, 2, 3$. Each $F_k = E_k^{-1}$, that is, each F_k is E_k's multiplicative inverse and vice versa.

(c) Compute $rref(E_k | I_4)$, for $k = 1, 2, 3$ where $(E_k | I_4)$ is the 4×8 augmented matrix. What is the result of these three $rref$'s?

7. (a) Give examples E_1, E_2, E_3, of each of the three types of 4×4 elementary matrices.

(b) Next, compute E_1^2, E_2^2, E_3^2 and E_1^3, E_2^3, E_3^3. Now give a general formula for E_1^m, E_2^m, E_3^m, where m is any positive integer.

(c) If you did problem 6, then find a general formula for E_1^m, E_2^m, E_3^m, where m is any negative integer.

(d) Can you put parts (b) and (c) of this problem together to get a general formula for E_1^m, E_2^m, E_3^m, where m is any integer?

8. Let E and F be two 3×3 elementary matrices. What must be true about E and F so that $EF = FE$, that is, E and F commute?

9. (a) Which type of elementary matrix is always a *diagonal matrix*? A diagonal matrix is a square matrix A, where all entries $A_{i,j} = 0$, when $i \neq j$.

(b) Which type of elementary matrix is always an *upper* or *lower triangular* matrix? An upper or lower triangular matrix is a square matrix A where in the upper triangular case all entries $A_{i,j} = 0$ when $i > j$ while in the lower triangular case all entries $A_{i,j} = 0$ when $i < j$.

Maple Problems

1. Verify the matrix multiplications used in *Homework* problem 3.

2. Verify the matrix multiplications used in *Homework* problem 4.

3. Use the *rref* command to row reduce each of the matrices created from the systems in *Homework* problem 4. Compare the row reduced matrices to your final answers from the original problem.

4. Use left multiplication by elementary matrices to solve the following systems of equations given their augmented matrices:

(a) $\begin{bmatrix} 2 & -5 & 8 & 9 \\ 3 & 2 & -6 & 0 \\ -1 & -4 & 2 & 5 \end{bmatrix}$
(b) $\begin{bmatrix} 2 & -3 & 6 & 1 \\ -4 & -1 & 0 & 1 \\ 7 & 3 & -1 & 0 \end{bmatrix}$

(c) $\begin{bmatrix} 2 & -3 & 6 & 8 & 3 \\ 8 & -4 & -7 & 3 & 2 \\ 7 & -3 & 2 & -4 & -1 \\ -3 & 6 & -3 & -6 & -4 \end{bmatrix}$
(d) $\begin{bmatrix} 4 & 5 & -1 & -1 & 1 \\ 2 & 2 & -2 & 2 & 2 \\ 8 & 0 & 3 & -7 & 3 \\ -2 & -3 & 3 & 5 & -5 \end{bmatrix}$

5. Use *Maple* to do *Homework* problem 6 for any size elementary matrix.

6. Use *Maple* to do *Homework* problem 7 for any size elementary matrix.

7. Use *Maple* to do *Homework* problem 8 for two elementary matrices, E and F, of the same size.

3.3   Sensitivity of Solutions to Error in the Linear System

We have spent a reasonable amount of time and effort attempting to solve linear systems of equations where we assume the values in the system are exact, thus without any error. One might expect that if the left-hand portion of an augmented matrix B is very close to another augmented matrix A, both of whom have the same last column, then both systems will have very similar solutions.

Example 3.3.1. As an example, consider the following two systems. The first we will call the *exact system*, and is given by

$$6x + 9y = 5$$
$$-2x + 7y = 3$$

We will denote the second system as the *approximate system*, and it is given by

$$6.1x + 9y = 5$$
$$-2x + 7y = 3$$

In matrix form, these two are given by $\begin{bmatrix} 6 & 9 & 5 \\ -2 & 7 & 3 \end{bmatrix}$ and $\begin{bmatrix} 6.1 & 9 & 5 \\ -2 & 7 & 3 \end{bmatrix}$, respectively. We can solve both of these systems of equations in *Maple* using the *solve* command (we will skip the matrix solution representation for now).

> with(linalg):
> evalf(solve({6*x + 9*y = 5, -2*x +7*y = 3}, {x,y}));

$$\{x = .1333333333, y = .4666666667\}$$

> evalf(solve({6.1*x + 9*y = 5, -2*x +7*y = 3},{x,y}));

$$\{x = .1317957166, y = .4662273476\}$$

It is apparent that the two solutions are very close. However, this is not always the case, so we will begin investigating how a small error in the variable coefficients or the RHS values of the equations can affect the solutions to a linear system.

Example 3.3.2. We will now solve the two almost identical linear systems with exact system being

$$187x + 790y = 5$$
$$201x + 850y = 37 \tag{3.11}$$

and approximate system

$$187x + 790y = 5$$
$$201.1x + 850y = 37 \tag{3.12}$$

Once again, we will use *Maple* to solve each system, however, in this example, we will work strictly with the matrices themselves and not the equations they were derived from.

> Ae:= matrix([[187,790.,5],[201,850,37]]);

$$Ae := \begin{bmatrix} 187 & 790. & 5 \\ 201 & 850 & 37 \end{bmatrix}$$

> rref(Ae);

$$\begin{bmatrix} 1 & 0 & -156.1250197 \\ 0 & 1 & 36.96250466 \end{bmatrix}$$

> Aa:= matrix([[187,790,5],[201.1,850,37]]);

$$Aa := \begin{bmatrix} 187 & 790 & 5 \\ 201.1 & 850 & 37 \end{bmatrix}$$

> rref(Aa);

$$\begin{bmatrix} 1 & 0 & -308.3951262 \\ 0 & 1 & 73.00618810 \end{bmatrix}$$

The solutions to the exact system (3.11) and approximate system (3.12), respectively, are

$$\{x = -156.1250197, y = 36.96250466\}, \{x = -308.3951262, y = 73.00618810\}$$

The solution to the approximate system has values almost double those of the solution to the exact system. This is clearly interesting, and the behavior is completely different than the first example. Since in real life there is almost always error in a linear system because its values are derived from measurements with inherent error, we must know if it is possible to avoid this sensitivity to error and also what causes it.

Since we used a two-dimensional system, it can be visualized easily in the plane. So let us solve for y in each equation of the exact system (3.11):

> solve(187.*x + 790*y = 5, y);

$$0.006329113924 - 0.2367088608x+$$

> solve(201.*x + 850*y = 37, y);

$$0.04352941176 - 0.2364705882x$$

Note that the solutions are close to parallel lines, since they have nearly identical slopes. This may be the cause of our sensitivity to slight amounts of error since a slight change in the slope of one line can move their intersection point a great distance from the original intersection point. If you find this hard to visualize, try the exercise with a pair of chopsticks.

Example 3.3.3. Now let us measure and plot the error in our intersection point for two almost identical parallel lines as a function of the change in slope of the second line. The error in our linear system's solution will be the distance in the xy-plane between the exact solution and the solution with an error in its slope from the first line.

Our system of linear, almost parallel, equations will be defined to be

$$
\begin{aligned}
0.23671x + y &= 0.00633 \\
(0.23647 + R)x + y &= 0.04353,
\end{aligned}
\tag{3.13}
$$

with the error in the slope being R. The exact system is the one for $R = 0$, and they become parallel lines when $R = 0.00024$ with no solution to the system. When $R = 0$, the solution to the system is given below.

> exact_soln:= delcols(rref(matrix([[0.23671, 1, 0.00633], [0.23647, 1, 0.04353]])), 1..2);

$$
exact\_soln := \begin{bmatrix} -154.9999944 \\ 36.69637868 \end{bmatrix}
$$

> exact_soln[1,1];

$$
-154.9999944
$$

> exact_soln[2,1];

$$
36.69637868
$$

The exact solution's x-coordinate is given by $exact\_soln[1, 1]$ and y-coordinate by $exact\_soln[2, 1]$. For $R \neq 0$, if we denote x_a and y_a as solutions to the approximate system, then to compute the distance from the exact solution we use the distance formula given by

$$
D = \sqrt{(x_a - exact\_soln[1, 1])^2 + (y_a - exact\_soln[2, 1])^2}
\tag{3.14}
$$

Instead of computing this by hand, we will have *Maple* do it for us and then graph the results.

> with(plots): with(plottools):
> approx_soln:= R -> delcols(rref(matrix([[.23671, 1, .00633], [.23647+R, 1, .04353]])) , 1..2):
> approx_soln(0);

$$
\begin{bmatrix} -154.9999944 \\ 36.69637868 \end{bmatrix}
$$

> approx_soln(0.1);

$$
\begin{bmatrix} 0.3728949480 \\ -0.8193796314 \end{bmatrix}
$$

> error_soln:= R -> sqrt((exact_soln[1,1]-approx_soln(R)[1,1]) ^2 + (exact_soln [2,1]-approx_soln(R)[2,1]) ^2):

> plot(error_soln, -1..1, view=[-0.3..0.3,120..180], axes=BOXED,numpoints = 5000);

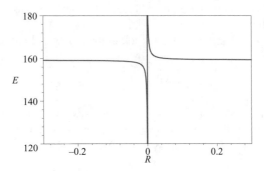

Figure 3.1: Plot of error for $-0.3 \le R \le 0.3$.

In Figure 3.1, the plot of the error (or distance) between the exact solution and the approximate system solutions has a vertical asymptote at $R = 0.00024$, which is the value of R that makes the lines parallel, and so system (3.13) is without a solution. The closer R is to this value, the larger the error. Note that this error is not a linear but an inverse relation. As R moves away from .00024 the error stabilizes at a value close to 160.

Let us also plot these solutions as a parametric plot in the xy-plane to see their behavior more clearly. The solutions form a line of points in our plot. Notice how quickly the dot moves at the beginning $(R = 0)$ versus the end of the animation near $R = 0.00024$. Figure 3.2 depicts one frame in the animation sequence.

> approx_soln1:= R -> approx_soln(R)[1,1]:

> approx_soln2:= R -> approx_soln(R)[2,1]:

> plot_Rstart:= textplot([approx_soln1(0), approx_soln2(0) + 5, "R = 0"], align = {ABOVE, RIGHT}, color=black):

> plot_Rstop:= textplot([approx_soln1(0.0002) + 20, approx_soln2(0.0002), "R = 0.0002"], align=RIGHT, color=black):

> plot_direction:= arrow([approx_soln1(0.00015), approx_soln2(0.00015) + 15], [approx_soln1(0.00018), approx_soln2(0.00018) + 15], 5, 15, .25, color=blue):

> plot_R:= textplot([approx_soln1(0.00017)+25, approx_soln2(0.00017) + 25, "R increasing"], align=RIGHT, color=black):

> plot_solns:= plot([approx_soln1, approx_soln2, 0..0.0002], color=red, thickness=3):

> animate_solns:= display([seq(ellipse([approx_soln1(.0000005*J),
approx_soln2(.0000005*J)], 15, 5, filled=true, color=red), J=0..400)],
insequence=true):

> display({plot_solns, plot_Rstart, plot_Rstop, plot_direction, plot_R,
animate_solns});

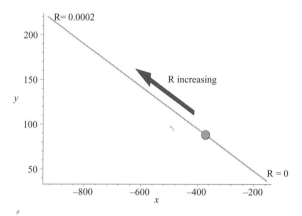

Figure 3.2: Parametrization of solutions in terms of R.

Example 3.3.4. In our next example, we look at a system of three equations
in three unknowns. The exact system is

$$
\begin{aligned}
5x - 2y + 8z &= 15 \\
-3x + 9y + 11z &= -4 \\
2x + 7y + 18.7z &= 20
\end{aligned}
\tag{3.15}
$$

while the approximate system is

$$
\begin{aligned}
5x - 2y + 8z &= 15 \\
-3x + 9y + 11z &= -4 \\
2x + 7y + 18.9z &= 20
\end{aligned}
\tag{3.16}
$$

As before, we will add a variable R into system (3.15) in order to look at error:

$$
\begin{aligned}
5x - 2y + 8z &= 15 \\
-3x + 9y + 11z &= -4 \\
2x + 7y + (18.7 + R)z &= 20
\end{aligned}
\tag{3.17}
$$

When $R = 0.3$ this system has no solution since, for this value of R, the last
equation is the sum of the first two equations with a different RHS value. We

will use the three-dimensional distance formula, similar to the two-dimensional one given in (3.14), and perform computations that mirror those done in the two-dimensional system of the previous example.

> exact_soln:= delcols(rref(matrix([[5.,-2,8,15], [-3,9,11,-4], [2,7,18.7,20]])), 1..3):

$$exact\_soln := \begin{bmatrix} 75.56410256 \\ 61.41025642 \\ -30.00000000 \end{bmatrix}$$

> rref(matrix([[5.,-2,8,15], [-3,9,11,-4], [2,7,19,20]]));

$$\begin{bmatrix} 1 & 0 & 2.410256410 & 0 \\ 0 & 1 & 2.025641026 & 0 \\ 0 & 0 & 0 & 1 \end{bmatrix}$$

> approx_soln:= delcols(rref(matrix([[5.,-2,8,15], [-3,9,11,-4], [2,7,18.9,20]])), 1..3);

$$approx\_soln := \begin{bmatrix} 220.1794872 \\ 182.9487179 \\ -90.00000000 \end{bmatrix}$$

> approx_soln:= R -> delcols(rref(matrix([[5.,-2,8,15], [-3,9,11,-4], [2,7,18.7+R, 20]])), 1..3):

> approx_soln(0);

$$\begin{bmatrix} 75.56410256 \\ 61.41025642 \\ -30.00000000 \end{bmatrix}$$

> approx_soln(0.2);

$$\begin{bmatrix} 220.1794872 \\ 182.9487179 \\ -90.00000000 \end{bmatrix}$$

> error_soln:= R -> sqrt((exact_soln[1,1] - approx_soln(R)[1,1]) ^2 + (exact_soln[2,1] - approx_soln(R)[2,1]) ^2 + (exact_soln[3,1] - approx_soln(R)[3,1]) ^2):

> plot(error_soln, -1..3, view=[-1..1,3..1000]);

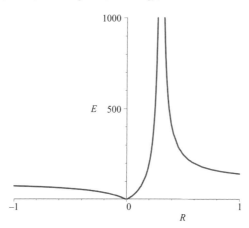

Figure 3.3: Plot of error for $-1 \leq R \leq 1$.

It is clear from the plot of the error, in Figure 3.3 above, that we have a vertical asymptote at $R = 0.3$. Let us also plot these solutions as a parametric plot in the xyz-space, $\mathbb{R}^3$, to see their behavior more clearly. The solutions form a line of points in our plot, similar to the previous two-dimensional problem. Notice the change in speed of the dot as the parameter R tends to the asymptotic value of $R = 0.3$. Figure 3.4 depicts one frame in this animation.

> approx_soln1:= R -> approx_soln(R)[1,1]:
> approx_soln2:= R -> approx_soln(R)[2,1]:
> approx_soln3:= R -> approx_soln(R)[3,1]:
> plot_Rstart:= textplot3d([approx_soln1(0), approx_soln2(0), approx_soln3(0) + 5, "R = 0"], align={ABOVE, RIGHT}, color=black):
> plot_Rstop:= textplot3d([approx_soln1(0.25) + 20, approx_soln2(0.25), approx_soln3(0.25),"R = 0.25"], align=RIGHT, color=black):
> plot_direction:= arrow([approx_soln1(0.15), approx_soln2(0.15), approx_soln3 (0.15) + 15], [approx_soln1(0.2), approx_soln2(0.2), approx_soln3(0.2) + 15], 5, 15, .25, cylindrical_arrow, color=blue):
> plot_R:= textplot3d([approx_soln1(0.17) + 25, approx_soln2(0.17), approx_soln3 (0.17) + 25, "R increasing"], align=RIGHT, color=black):
> plot_solns:= spacecurve([approx_soln1, approx_soln2, approx_soln3], 0..0.25, color=red, thickness=3):
> animate_solns:= display([seq(sphere([approx_soln1(.0025*J), approx_soln2 (.0025*J), approx_soln3(.0025*J)], 8, color=red), J=0..100)], insequence=true):

> display({plot_solns, plot_Rstart, plot_Rstop, plot_direction, plot_R, animate_solns }, axes = boxed);

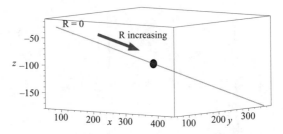

Figure 3.4: Parametrization of solutions in terms of R.

Our analysis above indicates that sensitivity to error in the variable coefficients occurs when at least one of the equations in the linear system can be "almost" formed from the other equations through a series of row operations made on the equations ignoring the RHS values of the equations. In a two-dimensional problem, this corresponds to having two lines that have nearly the same slope. In the three-dimensional setting, this corresponds to two planes having very similar planar slopes. However, in the three-dimensional setting, it is also possible that all three equations might have similar planar slopes.

Example 3.3.5. We now consider the following system:

$$3x + 2y - 4z = 2$$
$$-6x - 4y + 8.1z = -1 \qquad (3.18)$$
$$3x + 1.95y - 4z = 3$$

Solving this system gives

> solve({3*x+2*y-4*z=2, -6*x-4*y+(8.1)*z=-1, 3*x+(1.95)*y -4*z = 3}, {x, y, z});

$$\{x = 54., y = -20., z = 30.\}$$

Notice that the LHS of the second and third equations in system (3.18) are very close to multiples of the first equation. So we will rewrite the system of equations to reflect this.

$$3x + 2y - 4z = 2$$
$$3x + 2y - (4 + R)z = \frac{1}{2} \qquad (3.19)$$
$$3x + (2 + S)y - 4z = 3$$

The exact system is found when $S = -0.05$ and $R = 0.05$. We will now determine the distance from the exact solution in a slightly different manner, but which is equivalent to what we have done in the previous two examples. First, we will solve the above system of equations for arbitrary values of R and S. To do this, we will use the *solve* command.

> soln:= solve({3*x+2*y-4*z=2, 3*x+2*y-(4+R)*z=1/2, 3*x+(2+S)*y -4*z = 3}, {x,y,z});

$$soln := \left\{ x = \frac{2}{3} \frac{-R + SR + 3S}{SR}, y = \frac{1}{S}, z = \frac{3}{2R} \right\}$$

Notice that if $R = 0$ or $S = 0$, there will be no solution since the denominator in the solution for at least one variable would be zero. This situation corresponds to at least two of the equations in system (3.19) having the same LHS, but different RHS. We will now store the solutions to the actual solution in the variables $x0$, $y0$, and $z0$ and the approximate solutions in the variables xa, ya, and za.

> subs({R=0.05,S=-0.05}, soln);

$$\{x = 54.00000000, z = 30.00000000, y = -20.00000000\}$$

> x0:= 54: y0:=-20: z0:=30:
> xa:= 2*(-R+R*S+3*S)/(3*R*S); ya:=1/S; za:=3/(2*R);

$$xa := \frac{2}{3} \frac{-R + SR + 3S}{SR}$$

$$ya := \frac{1}{S}$$

$$za := \frac{3}{2R}$$

The distance (hence the error) from the exact solution to the approximate solution is given by

$$d = \sqrt{(x0 - xa)^2 + (y0 - ya)^2 + (z0 - za)^2} \tag{3.20}$$

The approximate solution depends on two variables, S and R, so instead of getting a one-dimensional curve, we get a two-dimensional surface, which can be seen in Figure 3.5.

> plot3d(sqrt((x0-xa)^2 + (y0-ya)^2+ (z0-za)^2), R=-0.3..0.3, S=-0.3..0.3, view = [-0.3..0.3, -0.3..0.3, 0..150], numpoints = 5000, orientation=[63,29], axes=BOXED, style=PATCHNOGRID);

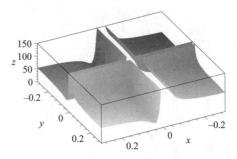

Figure 3.5: Graph of the error for $-0.3 \leq R, S \leq 0.3$.

The graph in Figure 3.5 can be rotated in many directions to get a better idea of the behavior of the error as a function of R and S.

There is another plot command, *countourplot*, which can help visualize the error, specially close to the exact system that occurs at $S = -0.05$ and $R = 0.05$. This graphing command allows you to pick values of the resulting function and graph the corresponding level surfaces. A level surface of the function $d(x_a, y_a, z_a)$ is the surface with equation $d(x_a, y_a, z_a) = c$ for some constant c. In the graph depicted in Figure 3.6, error values from 0 to 1 fall in the white circle, then from 1 to 2 in the next ring and so on until the darkest value corresponds to an error greater than 10. One can specify the desired contours and the corresponding colors as well.

> contourplot(sqrt((x0-xa)^2 + (y0-ya)^2 + (z0-za)^2), R=0.04..0.06, S=-0.06..-0.04, grid=[50,50], filled=true, contours = [0, 1, 2, 3, 4, 5, 6, 7, 8, 9, 10], coloring=[white, blue], view=[0.04..0.06, -0.06..-0.04]);

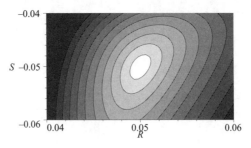

Figure 3.6: Graph of the error for $0.04 \leq R \leq 0.06$ and $-0.06 \leq S \leq -0.04$.

The countourplot shows how quickly the error builds up the further the values of S and R are away from the exact system values given by $S = -0.05$ and $R = 0.05$.

■

The moral of our story on sensitivity to error in a linear system is that it occurs mainly when at least one equation of the system is very close to being a linear combination of the other equations of the system. Geometrically, you should feel that sensitivity to error happens when at least one of the equations' plots is almost parallel to the plot of some linear combination of the other equations.

Homework Problems

Consider the following systems of equations:

(a) $-x + 0.048y = 6$ (b) $-x + 0.049y = 6$
          $2x - 0.1y = 24$ $2x - 0.1y = 24$

(c) $-x + 0.048y = 6.1$ (d) $-x + 0.048y = 6$
          $2x - 0.1y = 24$ $2.01x - 0.1y = 24$

We will denote system (a) as the exact system, while (b), (c), and (d) will be approximate systems. Answer the following questions:

1. Solve all four systems for the variables x and y.

2. Compute the Euclidean distance between the solution to the actual system and each of the three approximate systems.

3. The *Frobenius norm* of an $m \times n$ matrix A (with potentially complex entries) is one way to measure the magnitude, or length, of a matrix. It is given by the following formula:

$$\| A \|_F = \sqrt{\sum_{i=i}^{m} \sum_{j=1}^{n} |A_{i,j}|^2}$$

The Frobenius distance between two matrices A and B of the same size is $\| A - B \|_F$. Use this definition of the Frobenius norm to compute the distance between the augmented matrix corresponding to the exact system and the augmented matrices corresponding to approximate systems.

4. One might expect that the corresponding Frobenius norms from problem 3 should be arranged in the same order as the distances found in problem 2. Can you come up with a reason as to why this is not the case?

Consider the following systems of equations:

(a)
$$6w - 3x + 2y = 7$$
$$w - 2x + 4y = 5$$
$$4w + x - 7y = 0$$

(b)
$$6w - 3x + 2y = 7$$
$$w - 2x + 4y = 5$$
$$4w + x - 6.1y = 0$$

(c)
$$6w - 3x + 2y = 7$$
$$w - 2x + 4y = 5$$
$$5w + x - 7y = 0$$

(d)
$$6w - 3x + 2y = 7$$
$$w - 2x + 4y = 5$$
$$4w + x - 7y = -2$$

As before, we will denote system (a) as the exact system, while (b), (c), and (d) will be approximate systems. Answer the following questions:

5. Solve all four systems for the variables w, x, and y.

6. Compute the Euclidean distance between the solution to the exact system and each of the three approximate systems.

7. Compute the Frobenius norm of the distance between the augmented matrix corresponding to the exact system and the augmented matrices corresponding to approximate systems.

Maple Problems

Consider the following generalized two equation systems:

(a)
$$ax + by = c$$
$$max + (mb + \varepsilon)y = d$$

(b)
$$ax + by = c$$
$$max + \left(mb + \frac{1}{\varepsilon}\right)y = d$$

(c)
$$ax + by = c$$
$$max + m(b + \varepsilon)y = d$$

(d)
$$ax + by = c$$
$$max + m\left(b + \frac{1}{\varepsilon}\right)y = d$$

1. As an example, let $a = 5$, $b = -6$, $c = 1$, and $d = -1$. Explore how the solutions to each of these systems vary depending upon the values of m and ε. Included in your exploration should be graphs of the difference between solutions to these systems dependent on the parameters m and ε. Graphical depictions of the Frobenius norm defined in *Homework* problem 3 should also be included.

2. Repeat the process from problem 1, but this time, pick a couple of different values for d, some close to c and some further away from c. How does this affect your graphs from problem 1?

For the remaining questions for this section, we will focus on the following system of equations:

$$3x - 5y + 6z = -10$$
$$-2x + 4y - 7z = 1$$
$$5(1+R)x - \frac{28}{3}(1 - RS)y + \frac{29}{2}(R + S)z = -3$$

where the exact system corresponds to $R = 1$ and $S = 0$.

3. Solve the system of equations for the variables x, y, and z.

4. Determine all values of R and S that cause this system to have no solution.

5. Construct a three-dimensional graph of the Euclidean distance between the actual solution ($R = 1$ and $S = 0$) and approximate solutions in terms of R and S. Describe your resulting graph and how it corresponds to your answers to problems 3 and 4.

6. Construct a contour plot of the Euclidean distance between the actual solution ($R = 1$ and $S = 0$) and approximate solutions in terms of R and S. Describe your resulting graph and how it corresponds to your answers to problems 3 and 4.

Chapter 4

Applications of Linear Systems and Matrices

4.1 Applications of Linear Systems to Geometry

We begin this chapter with an application of linear systems to geometry. In this section, we will find the equations of *conic sections* that have a specified set of points on their graphs. Conic sections are ellipses, hyperbolas, and parabolas, which are all generated by intersecting planes with right circular cones. Their general equations are all of the form

$$Ax^2 + Bxy + Cy^2 + Dx + Ey + F = 0 \qquad (4.1)$$

for constants A through F with A, B, and C not all being zero at the same time. The quantity $B^2 - 4AC$ is called the *discriminant* of the conic section since the conic section is an ellipse when $B^2 - 4AC < 0$, a hyperbola when $B^2 - 4AC > 0$ and a parabola if $B^2 - 4AC = 0$.

The equation of a conic section (4.1) is said to be in standard form if $B = 0$, and the conic section itself is then said to be in standard position since it will have its axes parallel to the xy-coordinate axes. A standard form conic section is then an ellipse if $AC > 0$ (where you get a circle when $A = C$), a parabola if $AC = 0$, and a hyperbola if $AC < 0$. As well, the equation of a circle is always in standard form with $B = 0$ and $A = C$. There are also degenerate cases to equation (4.1). Consider the following examples:

$$x^2 + y^2 + 1 = 0, \ (x - 3)^2 + (y - 7)^2 = 0, \ (2x - 3y + 1)(x + y - 6) = 0 \quad (4.2)$$

the first of which has no solution, the second the single point $(3, 7)$, while the third is the product of two lines in $\mathbb{R}^2$, but is also in the form of (4.1) when expanded.

We will start with the most familiar of the conic sections, the circle. To arrive at the equation of a circle, we simply recognize that the values of A and C in (4.1) must be the same, and can thus divide through the entire equation by that quantity. This gives us the following equation:

$$x^2 + y^2 + Dx + Ey + F = 0 \tag{4.3}$$

We should remember that every circle, with center at the point (H, K) and radius R, can also be expressed by the equation

$$(x - H)^2 + (y - K)^2 = R^2$$

Since both of these equations for a circle contain three unknown constants D, E, and F, or H, K, and R, it should take three noncollinear points in the xy-plane to determine a unique circle passing through them. Notice that equation (4.3) can also be expressed as

$$Dx + Ey + F = - \left(x^2 + y^2\right)$$

which clearly indicates that the general equation of a circle is a linear equation in the three unknowns: D, E, and F. Since each of the three given points must satisfy the equation of the circle, we end up with the following linear system of equations in terms of the unknown variables D, E, and F:

$$
\begin{aligned}
Dx_1 + Ey_1 + F &= - \left(x_1^2 + y_1^2\right) \\
Dx_2 + Ey_2 + F &= - \left(x_2^2 + y_2^2\right) \\
Dx_3 + Ey_3 + F &= - \left(x_3^2 + y_3^2\right)
\end{aligned}
\tag{4.4}
$$

It is very easy to forget that the above system has only three unknowns, D, E, and F, just remember that the x_i's and y_i's are the coordinates of the three points our circle must pass through.

Example 4.1.1. By using system (4.4), we will find the circle passing through the three points $P(-7, 9)$, $Q(2, -13)$, and $T(8, 5)$, depicted in Figure 4.1. In addition, we will locate the coordinates (H, K) of the circle's center, and its radius R.

```
> with(plots): with(linalg):
> P:= [-7,9]: Q:= [2,-13]: T:= [8,5]:
> plotpts:= pointplot({P,Q,T}, symbol = diamond, symbolsize = 25, color =
blue):
> eqncircle:= D*x+E*y+F = -(x^2 + y^2):
> eqn1:= subs({x = P[1], y = P[2]}, eqncircle);
```

$$eqn1 := -7D + 9E + F = -130$$

> eqn2:= subs({x = Q[1], y = Q[2]}, eqncircle);

$$eqn2 := 2D - 13E + F = -173$$

> eqn3:= subs({x = T[1], y = T[2]}, eqncircle);

$$eqn3 := 8D + 5E + F = -89$$

> solns:= solve({eqn1, eqn2, eqn3}, {D,E,F});

$$solns := \left\{ D = \frac{179}{49}, E = \frac{169}{49}, F = -\frac{6638}{49} \right\}$$

> Circle:= subs(solns, eqncircle);

$$Circle := \frac{179}{49}x + \frac{169}{49}y - \frac{6638}{49} = -x^2 - y^2$$

> circ:= implicitplot(Circle, x = -20..12, y = -20..15, color = red):
> plotP:= textplot([-7,10, "P"], align = ABOVE):
> plotQ:= textplot([2,-12, "Q"], align = ABOVE):
> plotT:= textplot([8,6, "T"], align = ABOVE):
> display({plotpts, circ, plotP, plotQ, plotT}, scaling = constrained);

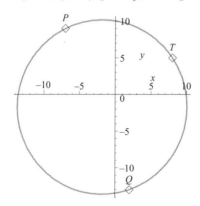

Figure 4.1: The unique circle passing through P, Q, and T.

> with(student):
> op(1,Circle);

$$\frac{179}{49}x + \frac{169}{49}y - \frac{6638}{49}$$

> op(2,Circle);

$$-x^2 - y^2$$

> completesquare(op(1,Circle)-op(2,Circle) = 0, y);

$$\left(y + \frac{169}{98}\right)^2 - \frac{1329609}{9604} + \frac{179}{49}x + x^2 = 0$$

> completesquare(%, x);

$$\left(y + \frac{169}{98}\right)^2 - \frac{680825}{4802} + \left(x + \frac{179}{98}\right)^2 = 0$$

Now, by using the *completesquare* command from the *student* package, we have found the center of the circle is located at coordinates $\left(-\frac{179}{98}, -\frac{169}{98}\right)$ with radius $\sqrt{\frac{680825}{4802}}$.

Maple had no problem solving this linear system for D, E, and F to get the equation of this circle. Now let us see if *Maple* can solve for D, E, and F if we use the standard form for the equation of a circle:

$$(x - H)^2 + (y - K)^2 = R^2$$

Notice that this form directly involves the center, but is also not linear in terms of H, K, and R (all three variables have highest degree two).

> eqncircle2:= (x-H)^2+(y-K)^2 = R^2:
> eqn4:= subs({x = P[1], y = P[2]}, eqncircle2);

$$eqn4 := (-7 - H)^2 + (9 - K)^2 = R^2$$

> eqn5:= subs({x = Q[1], y = Q[2]}, eqncircle2);

$$eqn5 := (2 - H)^2 + (-13 - K)^2 = R^2$$

> eqn6:= subs({x = T[1], y = T[2]}, eqncircle2);

$$eqn6 := (8 - H)^2 + (5 - K)^2 = R^2$$

> fsolve({eqn4, eqn5, eqn6}, {H,K,R}, R = 0.1..20);

$$\{H = -1.826530612, K = -1.724489796, R = 11.90711833\}$$

> evalf(sqrt(680825/4802));

$$11.90711833$$

Now we see that both of these methods find the same equation for a circle. The *solve* command undoubtedly used Gauss-Jordan elimination to solve the first system of equations since it was linear in the unknowns D, E, and F. For the second system, the *fsolve* command probably used the system of equations version of the Newton-Raphson method since the system of equations is not linear in the variables H, K, and R. For more information in regards to the Newton-Raphson algorithm for finding approximate solutions to nonlinear system of equations, visit the book's companion website.

Now we turn our attention to a somewhat different problem involving three points. A plane in space is uniquely determined by passing through three noncollinear points. On the other hand, the general equation of a plane in space is

$$Ax + By + Cz = D \tag{4.5}$$

for constants A through D, which is a linear equation in these four unknowns. Clearly, our three points will only generate three equations in these four unknowns which indicates we will get an infinite number of solutions to this linear system since the solution will have dimension one. We know that the plane is unique, so we should be able to reduce the problem to three equations in the three unknowns in order to get a single solution.

In order to resolve our dilemma, we must realize that the constant D, on the RHS of the planar equation, is 0 exactly when the plane goes through the origin $(0,0,0)$, and is otherwise nonzero. If we plot our three points and it is clear that the plane through them does not go through the origin (this is most likely to be the case), then $D \neq 0$ and we can divide equation (4.5) by D, yielding

$$Ex + Fy + Gz = 1$$

for constants E, F, and G. Now we are in business, and can solve this square linear system of three equations in the three unknowns E, F, and G, where each equation is determined by going through one of the three given points.

Example 4.1.2. As an example, we will now attempt to find and plot the equation of the plane through the three points $P(5, -8, 13)$, $Q(-7, 2, 4)$, and $R(16, 10, -9)$.

```
> P:= [5,-8,13]: Q := [-7,2,4]: R := [16,10,-9]:
> plotpts:= pointplot3d({[0,0,0], P, Q, R}, symbol = diamond, symbolsize = 50, color = blue):
> eqnplane:= E*x+F*y+G*z=1:
> eqn1:= subs({x = P[1], y = P[2], z = P[3]}, eqnplane);
```

$$eqn1 := 5E - 8F + 13G = 1$$

```
> eqn2:= subs({x = Q[1], y = Q[2], z = Q[3]}, eqnplane);
```

$$eqn2 := -7E + 2F + 4G = 1$$

> eqn3:= subs({x = R[1], y = R[2], z = R[3]}, eqnplane);

$$eqn3 := 16E + 10F - 9G = 1$$

> solns:= solve({eqn1, eqn2, eqn3}, {E, F, G});

$$solns := \left\{ E = \frac{1}{28}, F = \frac{363}{1624}, G = \frac{163}{812} \right\}$$

> Plane:= subs(solns, eqnplane);

$$Plane := \frac{1}{28}x + \frac{363}{1624}y + \frac{163}{812}z = 1$$

> f:= solve(Plane,z);

$$f := -\frac{29}{163}x - \frac{363}{326}y + \frac{812}{163}$$

> plotplane:= plot3d(f,x=-20..20,y=-20..20, color=red, style=patchnogrid):
> plotP:= textplot3d([5,-8,15,"P"], align = ABOVE, color = black):
> plotQ:= textplot3d([-7,2,6,"Q"], align = ABOVE, color = black):
> plotR:= textplot3d([16,10,-7,"R"], align = ABOVE, color = black):
> display({plotpts, plotplane, plotP, plotQ, plotR}, axes = boxed);

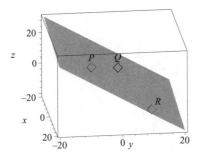

Figure 4.2: The plane fitting the points P, Q, and R.

The plane is plotted in Figure 4.2, along with the three points. The origin lies below the plane, and thus clearly the plane does not pass through the origin, but does pass through all three points P, Q, and R, as required.

Returning to our discussion of conic sections, it can be shown that any five points in the plane define a unique conic since two distinct conics can intersect in at most four points. Hence, all six constants, A through F, can somehow be solved for given five points. The five points should be plotted in order to determine which type of conic it is since if it is an ellipse, then it has an x^2 term, and we can divide the entire equation (4.1) by A. If the conic is a parabola it must have either an x^2 or a y^2 term, allowing us to divide by A or C. A hyperbola must have at least one of A, B, or C nonzero.

Example 4.1.3. Now we will have *Maple* find the equations and graphs of the conic sections through four fixed points, and a fifth variable point chosen from the unit circle with center at $(0, \frac{1}{2})$. As the fifth point rotates around the circle, the conic section changes from one type to another. For the four fixed points, we will pick the corners of the unit square: $\{[0,1],[1,1],[1,0],[0,0]\}$ and the fifth point will lie on the circle of radius 1, centered at $(0, \frac{1}{2})$. Figures 4.3 and 4.4 help to illustrate this example.

> Conic:= (x,y) -> x^2 + B*x*y + C*y^2 + D*x + E*y + F = 0;

$$Conic := (x,y) \rightarrow x^2 + Bxy + Cy^2 + Dx + Ey + F = 0$$

> pts:= [0,1], [1,0], [1,1], [0,0]:
> CircEqn:= t -> [cos(t), sin(t)+1/2]:
> eqns:= Conic(pts[1][1], pts[1][2]),Conic(pts[2][1], pts[2][2]),Conic(pts[3][1], pts[3][2]), Conic(pts[4][1], pts[4][2]), Conic(evalf(CircEqn(t))[1], evalf(CircEqn(t))[2]):
> soln:= solve({eqns}, {B,C,D,E,F });

$$soln := \left\{ B = 0, C = -\frac{4.\cos(t)(\cos(t) - 1.)}{4.\sin(t)^2 - 1.}, D = -1, E = \frac{4.\cos(t)(\cos(t) - 1.)}{4.\sin(t)^2 - 1.}, F = 0 \right\}$$

> par_soln:= subs(soln, Conic(x,y));

$$par\_soln := x^2 - \frac{4.\cos(t)(\cos(t) - 1.)y^2}{4.\sin(t)^2 - 1.} - 1.x + \frac{4.\cos(t)(\cos(t) - 1.)y}{4.\sin(t)^2 - 1.} = 0$$

> plot1:= pointplot({pts},symbol=DIAMOND, symbolsize=20,color=RED):
> plot2:= plot([cos(t),sin(t)+1/2,t=0..2*Pi], color=BLACK, linestyle=DASH, thickness=2):
> plotboth:= display({plot1,plot2}):
> opts_conic:= thickness = 3, numpoints = 3000, color = BLUE:
> an1:= animate(implicitplot, [par_soln, x = -3 .. 3, y = -3 .. 3, opts_conic], t = -Pi .. Pi, scaling = constrained, frames=100, background = plotboth):
> opts_ptplot:= symbol=CIRCLE,color=RED, symbolsize=22:
> an2:= animate(pointplot,[CircEqn(t),opts_ptplot],t=-Pi..Pi,frames=100):

> display({an1,an2});

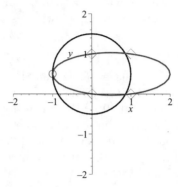

Figure 4.3: First frame in the animation, corresponding to $t = -\pi$. Conic section is depicted along with the five points that define it. The circle is the collection of points chosen as the fifth point to help define the conic section.

> plot1:= implicitplot(evalf(subs({t=-3*Pi/5},par_soln)), x=-3..3, y=-3..3, thickness = 2, numpoints = 3000, color = BLUE):

> plot2:= implicitplot(evalf(subs({t=-2*Pi/5},par_soln)), x=-3..3, y=-3..3, thickness = 2, numpoints = 3000, color = BLUE):

> plot3:= implicitplot(evalf(subs({t=2*Pi/3},par_soln)), x=-3..3, y=-3..3, thickness = 2, numpoints = 3000, color = BLUE):

> display({plot1, plot2, plot3, plotboth}, scaling = constrained);

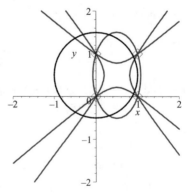

Figure 4.4: Several conic sections corresponding to $t = -\frac{3}{5}\pi, -\frac{2}{5}\pi$, and $\frac{2}{3}\pi$.

Homework Problems

1. Find equations of the circles that pass through the following sets of points:

    (a) $\left\{ (0, -1 - \sqrt{3}), (1, 1), (1 + \sqrt{3}, -2) \right\}$

    (b) $\{(3, 7), (3, 1), (6, 4)\}$

2. Find equations of the planes that pass through the following sets of points:

    (a) $\{(1, -2, -5), (-3, -2, 1), (-3, -1, 3)\}$

    (b) $\{(-4, 3, 8), (-6, -2, 1), (-3, 0, 3)\}$

3. Find the equation of the plane that passes through the following points:

$$\{(1, -2, 11), (-1, 1, -7), (2, 1, 2)\}$$

4. A sphere of radius r, centered at the point (a, b, c), can be expressed by the equation

$$(x - a)^2 + (y - b)^2 + (z - c)^2 = r^2$$

Construct a linear system of equations that can be solved to find the sphere that fits a set of data points in $\mathbb{R}^3$. How many points are required to determine a unique sphere?

5. Find the equation of the spheres that fit the following set of points:

    (a) $\left\{ (2, 3, 3 - 2\sqrt{3}), (4, 1 + 2\sqrt{3}, 3), (2, 1, -1), (6, 1, 3) \right\}$

    (b) $\{(-3, -1, -3), (0, -4, 1), (-2, 0, 1), (1, -1, 1)\}$

6. Explain what happens when you attempt to find the equation of a conic for which three of the given points are collinear.

7. (See Section 4.4 for more details about rotations in the plane.) Let (x', y') be a new coordinate system that is a rotation about the origin through the angle θ of the standard (x, y) Cartesian coordinate system. (This is actually a rotation about the origin of the x and y axes to produce the new x' and y' axes, respectively.) Then these two coordinate systems are related by the equations

$$x' = \cos(\theta)\, x - \sin(\theta)\, y$$
$$y' = \sin(\theta)\, x + \cos(\theta)\, y$$

or the single matrix equation

$$\begin{bmatrix} x' \\ y' \end{bmatrix} = \begin{bmatrix} \cos(\theta) & -\sin(\theta) \\ \sin(\theta) & \cos(\theta) \end{bmatrix} \begin{bmatrix} x \\ y \end{bmatrix}$$

Let

$$Ax^2 + Bxy + Cy^2 + Dx + Ey + F = 0$$

be a conic section in the xy-coordinate system. Find the equation of this conic section

$$A'x'^2 + B'x'y' + C'y'^2 + D'x' + E'y' + F' = 0$$

in the $x'y'$-coordinate system. In particular, find formulas for the coefficients A' through F' in terms of A through F and θ. Also, show that the discriminants of both equations are equal, that is,

$$B'^2 - 4A'C' = B^2 - 4AC$$

Also, find a formula in terms of A through F for the angle θ which makes $B' = 0$. Now find this angle θ and corresponding values of A' through F' for the conic given by

$$4x^2 + 6xy - 2y^2 + 7x + y - 1 = 0.$$

Maple Problems

1. Given the points $\{(-5, 0), (-2, 8), (0, 5)\}$, find the equations of the parabolas whose axes of symmetry are parallel to both the x- and y-axes, going through the three points. Also graph both parabolas with the points.

2. The points $\{(-1, -1), (0, 4), (0, -4), (2, 2), (2, -2)\}$ lie on an ellipse, find the equation of the ellipse.

3. Find the hyperbola that passes through the points

$$\{(-1, 2), (0, 3), (1, 1), (2, 4), (-2, 3)\}$$

and plot both the hyperbola and the points together on the same graph.

4. Using the points in problem 3, attempt to find the equation of the ellipse that passes through them.

5. Plug the points from problem 3 directly into the equation for the general conic section given in (4.1) and compare your result to that of problem 3. *Hint:*

assume $A \neq 0$.

6. Using the *Maple* code that generated the images in Figures 4.3 and 4.4, attempt to determine what happens when the fifth point that lies on the circle becomes collinear with a pair of points out of the four given.

7. Rerun the code that generated Figure 4.3, but this time, create a circle that lies completely inside the box generated by the four predetermined points. Points on this circle will be the location for the fifth point on the conic. Describe the changes, if any, in the overall behavior of the conic sections when the fifth point lies entirely inside the box.

8. Find the equation of the line through the two points $(3, -8)$ and $(-6, 11)$ using the solution to a linear system. Plot this line along with these two points.

9. Find the equation of the line in xyz-space $\mathbb{R}^3$ through the two points $(-5, 7, -11)$ and $(9, -2, 4)$. Plot this line in space along with these two points. *Hint: Think of this line as the line of intersection of two planes.*

10. Redo *Homework* problem 7 with as much of the work done by *Maple* as can be reasonably performed.

4.2   Applications of Linear Systems to Curve Fitting

In the Section 4.1, given a certain number of points, we were able to find a conic curve that passes through all the points. Next, we will attempt to generalize these ideas in an effort to fit a more generalized curve through a set of points. In particular, we will be interested in finding a linear combination of a certain type (e.g., trigonometric, exponential, or powers) of function whose graph will pass through all the points of a planar data set.

Definition 4.2.1. A function $F(x)$ is a *linear combination* of the functions

$$\{f_1(x), f_2(x), \ldots, f_n(x)\}$$

if $F(x)$ can be expressed as

$$F(x) = a_1 f_1(x) + a_2 f_2(x) + \cdots + a_n f_n(x) = \sum_{k=1}^{n} a_k f_k(x) \qquad (4.6)$$

for scalars $a_1, a_2, \ldots, a_n$.

As an example, it might be proper to use trigonometric functions, such as

$$\{\cos(0x) = 1, \cos(x), \sin(x), \cos(2x), \sin(2x), \ldots, \cos(nx), \sin(nx)\}$$

if the data set is known to be from a source that has a wave format, such as sound (music), light (fiber optics communication), or even stock prices. Notice that $\sin(0x) = 0$ and is thus omitted. Exponential functions, such as

$$\{e^{0x} = 1, e^{-x}, e^{x}, e^{-2x}, e^{2x}, \ldots, e^{-nx}, e^{nx}\}$$

would be used if the data set is from a source that can be very large and perhaps very small simultaneously, such as population (people or bacteria) or even the national debt and has horizontal asymptotes.

A polynomial is a linear combination of non-negative powers of x and so we begin with an example involving it. We want to find a quadratic polynomial $y = Ax^2 + Bx + C$ whose graph passes through a data set of three points, that is, we want the standard position parabola through three noncollinear points. This will require we solve a system of three linear equations (one for each point) in the three unknowns A, B, and C. Three points is the only number of points that give you a unique parabola. Consider, for instance, a line; a line is uniquely defined by two points. Given only one point, there are an infinite number of lines that go through it. If we have three points, the only way in which a line can go through all three is if they all lie on the same line, hence you would only need to use two points to determine the equation of the line. If the three points were not collinear, then no line passes through them. So if we have three points, we can find a unique parabola that goes through them. Given only one or two points, there are an infinite number of parabolas that pass through them. Given four points, the only way a parabola can pass through all four is if the points already lie on a predetermined parabola, in which case any three of the points will give enough information for us to find the equation of the parabola.

Example 4.2.1. As an example, we want to find the parabola through the three points $P(-21, 15)$, $Q(9, -4)$, and $T(37, 26)$.

> with(plots):
> P:= [-21,15]: Q := [9,-4]: T := [37,26]:
> plotpts:= pointplot({P,Q,T}, symbol = diamond, symbolsize = 25, color = blue):
> eqnparabola:= y = A*x^2 + B*x + C:
> eqn1:= subs({x = P[1], y = P[2]}, eqnparabola);

$$eqn1 := 15 = 441A - 21B + C$$

> eqn2:= subs({x = Q[1], y = Q[2]}, eqnparabola);

$$eqn2 := -4 = 81A + 9B + C$$

> eqn3:= subs({x = T[1], y = T[2]}, eqnparabola);

$$eqn3 := 26 = 1369A + 37B + C$$

> solns:= solve({eqn1, eqn2, eqn3}, {A,B,C});

$$solns := \left\{ A = \frac{179}{6090}, B = -\frac{1709}{6090}, C = -\frac{559}{145} \right\}$$

> Parabola:= subs(solns, eqnparabola);

$$Parabola := y = \frac{179}{6090}x^2 - \frac{1709}{6090}x - \frac{559}{145}$$

> quadfunc:= op(2,Parabola);

$$quadfunc := \frac{179}{6090}x^2 - \frac{1709}{6090}x - \frac{559}{145}$$

> parabla:= plot(quadfunc, x = -30..50, y = -20..50, color = red):
> plotP:= textplot([-21,17,"P"], align = ABOVE):
> plotQ:= textplot([9,-6,"Q"], align = BELOW):
> plotT:= textplot([37,28,"T"], align = ABOVE):
> display({plotpts, parabla, plotP, plotQ, plotT});

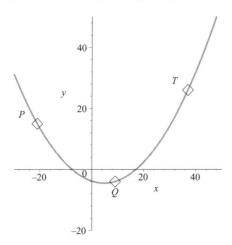

Figure 4.5: Parabola fitting the three data points.

In Figure 4.5, we clearly see that the parabola fits the data exactly. Upon inspecting the equations *eqn1*, *eqn2*, and *eqn3*, we can determine the general form for finding the coefficients A, B, and C to the parabola $y = Ax^2 + Bx + C$ given points $\{(x_1, y_1), (x_2, y_2), (x_3, y_3)\}$. Putting it in matrix form we have

$$
\begin{bmatrix}
x_1^2 & x_1 & 1 & y_1 \\
x_2^2 & x_2 & 1 & y_2 \\
x_3^2 & x_3 & 1 & y_3
\end{bmatrix}
$$

Performing Gauss-Jordan elimination on the above matrix will solve for A, B, and C that correspond to columns 1, 2, and 3 respectively.

■

Example 4.2.2. Now let us look at an example where the function desired is a linear combination of seven exponential functions. The data set we want the function to fit is the seven points

$$\{(0,2), (20,7), (40,15), (60,26), (80,47), (100,63), (120,72)\}$$

which is population data taken every 20 years in millions of people. The linear combination we need to fit the data to is

$$y = A + B e^{-x} + C e^{x} + D e^{-2x} + E e^{2x} + F e^{-3x} + G e^{3x}$$

so that we get a square 7×7 linear system of equations in the unknowns A through G. We will divide each x-coordinate by 100 for the actual data set used in order to prevent overflow in our arithmetic since otherwise our exponents become too large. It is quite normal to have to alter the data set in order to obtain better results, but you should remember that the resulting function corresponds to the modified data set, and not the original data.

> Dataset:= [[0,2],[.2,7],[.4,15],[.6,26],[.8,47],[1.0,63],[1.2,72]]:

> plotData:= pointplot(Dataset, symbol = diamond, symbolsize = 25, color = blue):

> eqnexponential:= y = A+B*exp(-x)+C*exp(x)+D*exp(-2*x)+E*exp(2*x) + F*exp(-3*x)+G*exp(3*x);

$eqnexponential := y = A + B e^{-x} + C e^{x} + D e^{-2x} + E e^{2x} + F e^{-3x} + G e^{3x}$

> eqns:= {seq(subs({x = Dataset[J][1], y = Dataset[J][2]}, eqnexponential), J=1..7)};

$eqns := \{2 = A + B e^{0} + C e^{0} + D e^{0} + E e^{0} + F e^{0} + G e^{0},$

$7 = A + B e^{-.2} + C e^{.2} + D e^{-.4} + E e^{.4} + F e^{-.6} + G e^{.6},$

$$15 = A + B\,e^{-.4} + C\,e^{.4} + D\,e^{-.8} + E\,e^{.8} + F\,e^{-1.2} + G\,e^{1.2},$$
$$26 = A + B\,e^{-.6} + C\,e^{.6} + D\,e^{-1.2} + E\,e^{1.2} + F\,e^{-1.8} + G\,e^{1.8},$$
$$47 = A + B\,e^{-.8} + C\,e^{.8} + D\,e^{-1.6} + E\,e^{1.6} + F\,e^{-2.4} + G\,e^{2.4},$$
$$63 = A + B\,e^{-1.0} + C\,e^{1.0} + D\,e^{-2.0} + E\,e^{2.0} + F\,e^{-3.0} + G\,e^{3.0},$$
$$72 = A + B\,e^{-1.2} + C\,e^{1.2} + D\,e^{-2.4} + E\,e^{2.4} + F\,e^{-3.6} + G\,e^{3.6}\}$$

> solns:= fsolve(eqns, {A, B, C, D, E, F, G});

$solns := \{A = -32050.17530, B = 42247.28669, C = 13136.24688,$
$D = -28641.39608, E = -2738.829082, F = 7820.471657, G = 228.3952254\}$

> LinearComb:= subs(solns, eqnexponential);

$LinearComb := y = -32050.17530 + 42247.28669\,e^{-x} + 13136.24688\,e^{x}$
$\quad - 28641.39608\,e^{-2x} - 2738.829082\,e^{2x} + 7820.471657\,e^{-3x} + 228.3952254\,e^{3x}$

> Expfunc:= op(2,LinearComb):
> plotLinearComb:= plot(Expfunc, x = 0..1.2, color = red):
> display({plotData, plotLinearComb});

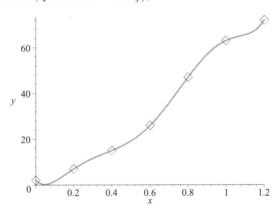

Figure 4.6: Exponential function fitting the modified data.

Now, we can get the linear combination that works for the original data set by replacing x by $\frac{x}{100}$.

> LinComb_OrigData:= subs(x = x/100, Expfunc);

$LinComb\_OrigData := -32050.17530 + 42247.28669\,e^{-\frac{1}{100}x}$
$\quad + 13136.24688\,e^{\frac{1}{100}x} - 28641.39608\,e^{-\frac{1}{50}x} - 2738.829082\,e^{\frac{1}{50}x}$
$\quad + 7820.471657\,e^{-\frac{3}{100}x} + 228.3952254\,e^{\frac{3}{100}x}$

Plotting the original data and the correctly adjusted function, as seen in Figure 4.7, gives the exact same graph as that depicted in Figure 4.6, but with the x-axis adjusted back to the correct scaling.

> Dataset2:= [[0, 2], [20, 7], [40, 15], [60, 26], [80, 47], [100, 63], [120, 72]]:

> plotData2:= pointplot(Dataset2, symbol = diamond, symbolsize = 25, color = blue):

> plotLinearComb2:= plot(LinComb_OrigData, x = 0 .. 120, color = red):

> display({plotData2, plotLinearComb2});

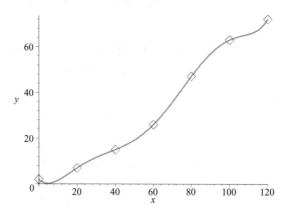

Figure 4.7: Exponential function fitting the original data.

Now let us attempt to generalize this process. If given a data set and function set

$$\{(x_1, y_1), (x_2, y_2), \ldots, (x_n, y_n)\}, \quad \{f_1(x), f_2(x), \ldots, f_n(x)\}$$

respectively, we can fit the data to a function expressed as the linear combination of functions, given in (4.6), by solving the system whose augmented matrix is given by

$$\begin{bmatrix} f_1(x_1) & f_2(x_1) & \cdots & f_n(x_1) & y_1 \\ f_1(x_2) & f_2(x_2) & \cdots & f_n(x_2) & y_2 \\ \vdots & \vdots & & \vdots & \vdots \\ f_1(x_n) & f_2(x_n) & \cdots & f_n(x_n) & y_n \end{bmatrix} \tag{4.7}$$

If there are more points in the data set than functions, then we end up with an overdetermined system that may not have a solution. In this case, the best that can be done is to get as "close" as possible to all the points with our resulting function. This can be accomplished by the *method of least squares*, which will be covered in Section 11.2.

Homework Problems

1. If a set of data has two points that have the same x-coordinate, but different y-coordinates, a problem occurs when attempting to perform Gauss-Jordan elimination. As an example, consider the function $y = Ax^2 + Bx + C$, and the set of points $\{(1,1),(2,3),(1,2)\}$. What is the problem and why does it occur?

2. Set up, but do not solve, the matrix required to find the constants to fit the following data points to the corresponding functions.

(a) $\{(0,0),(1,2),(-3,4)\}$, $\{1, x, x^2\}$

(b) $\{(0,0),(1,2),(-3,4),(-1,5)\}$, $\{1, x, x^2, x^3\}$

(c) $\{(2,3),(1,2),(-3,4),(-1,5)\}$, $\left\{1, x, x^2, \dfrac{1}{x}\right\}$

(d) $\{(0,1),(\pi,2),\left(-\frac{\pi}{4},-1\right)\}$, $\{1, \sin(x), \cos(x)\}$

3. So far, this section has been devoted to finding a one-dimensional curve of the form $y = a_1 f_1(x) + a_2 f_2(x) + \cdots + a_n f_n(x)$ given a set of n data points of the form (x_i, y_i). Discuss how this method can be extended to functions of two variables, given by $z = a_1 f_1(x,y) + a_2 f_2(x,y) + \cdots + a_n f_n(x,y)$, with n data points of the form (x_i, y_i, z_i).

4. Set up, but do not solve, the matrix required to find the constants to fit the following data points to the corresponding functions:

(a) $\{(0,1,2),(1,2,4),(-1,2,-1),(1,0,-3)\}$, $\{1, x, y, xy\}$

(b) $\{(0,1,2),(1,2,4),(-1,2,-1),(1,0,-3)\}$, $\{1, x, y, (x-y)^2\}$

5. The Lagrange polynomial $L(x)$ for a data set D_n of n points given by

$$\{(x_1, y_1),(x_2, y_2),\ldots,(x_n, y_n)\}$$

for distinct x-coordinates, is the smallest degree polynomial that passes through all of the points of the data set. What is the maximum degree of the Lagrange polynomial $L(x)$ passing through D_n?

Maple Problems

1. Solve the systems from *Homework* problems 2 and 4. Graph the points and the function on the same graph to verify that the solution curve does indeed fit the data.

2. This is a continuation of *Homework* problem 5.

(a) Find the equation of the Lagrange polynomial $L(x)$ which passes through the 6 points

$$\{(-5.258, 104.0773128), \ (0, 3.14159), \ (-3.1, 44.58859),$$
$$(-1.6, 18.05359), \ (4.9, 43.46859), \ (2.3, 5.92459)\}$$

(b) Now, plot together these six points and their Lagrange polynomial.

3. This is a continuation of *Homework* problem 5.

(a) Find the data set of 10 points equally spaced on the graph of $y = \sin(x)$ for $x \in \left[0, \frac{\pi}{2}\right]$.

(b) Find the equation of the Lagrange polynomial $L(x)$ that passes through these 10 points.

(c) Now plot together these 10 points, their Lagrange polynomial $L(x)$ and $y = \sin(x)$ for $x \in \left[0, \frac{\pi}{2}\right]$.

(d) The Lagrange polynomial $L(x)$ approximates $\sin(x)$ for angles in radians in the first quadrant. Compare the values of $\sin(x)$ and $L(x)$ for $x = \frac{\pi}{6}, \frac{\pi}{3}, \frac{\pi}{4}$. Do you believe that $L(x)$ is a very good approximation of $\sin(x)$ for angles in the first quadrant? Explain?

4. Let D_9 be the 9 point data set given by

$$D_9 = \{(-8, 3.9), \ (-6, -1.7), \ (-4, 5.5), \ (-2, 1.4),$$
$$(0, -3.2), \ (2, 4.2), \ (4, 0.3), \ (6, -2.8), \ (8, 5.1)\}$$

(a) Find the linear combination $F(x)$ of the trigonometric functions

$$\{1, \sin(x), \cos(x), \sin(2x), \cos(2x), \sin(3x), \cos(3x), \sin(4x), \cos(4x)\},$$

whose graph passes through the points of D_9.

(b) Now plot together D_9 and $F(x)$.

4.3 Applications of Linear Systems to Economics

Now we switch from purely mathematical or physical applications to ones that are financial. Our two financial applications will look at investing for retirement and taxes. The business application we will look at is the *Leontief input–output model*.

Example 4.3.1. Let us begin with taxes. We will call a tax system fair if there are no taxes paid on other taxes. In other words, you do not pay federal income tax on the amount you have paid on state, city and property taxes, and the same is true with regard to the other taxes. Let us now assume our tax system is fair and that you have a total taxable income of \$57,650. You also pay a 19.3% federal tax, a 4.25% state tax, a 0.55% city tax and a 2.35% property tax. What are the amounts of each tax paid, what percentage of your total taxable income is each tax and what is your overall tax rate as a percentage of your total taxable income? Let x, y, z, and w be your federal, state, city, and property tax amounts. Then we have the system of equations

$$x = 0.193(57650 - y - z - w)$$
$$y = 0.0425(57650 - x - z - w)$$
$$z = 0.0055(57650 - x - y - w)$$
$$w = 0.0235(57650 - x - y - z).$$

This system consists of four linear equations in four variables guaranteeing a unique solution, and now we can solve them.

> eqnfed:= x = 0.193*(57650 - y - z - w):
> eqnstate:= y = 0.0425*(57650 - x - z - w):
> eqncity:= z = 0.0055*(57650 - x - y - w):
> eqnprop:= w = 0.0235*(57650 - x - y - z):
> solve({eqnfed, eqnstate, eqncity, eqnprop}, {x, y, z, w});

$$\{w = 1056.535215, x = 10499.58507, y = 1948.670945, z = 242.7986482\}$$

> fedrate:= 10499.58507/57650;

$$fedrate := 0.1821263672$$

> staterate:= 1948.670945/57650;

$$staterate := 0.03380175100$$

> cityrate:= 0.004211598408;

$$cityrate := 0.004211598408$$

> proprate:= 1056.535215/57650;

$$proprate := 0.01832671665$$

> overallrate:= fedrate + staterate + cityrate + proprate;

$$overallrate := 0.2384664332$$

In this fair tax system, you end up paying roughly 24% of your total taxable income toward these four taxes.

◼

Example 4.3.2. As a second example, let us look at investing for retirement. When you retire you discover that you have a total of $248,000 saved for retirement. You decide to invest the total amount of this savings among three types of investments. One is a simple interest savings account at 4.35% per year, another is a certificate of deposit at simple interest of 6.15% per year and the third is a mutual fund at simple interest of 8.75% per year. You have decided that you want a yearly total yield of $19,500, and that three times as much should be invested in the mutual fund as the other two investments together. How much should you place in each type of investment?

Let x, y, and z be the amounts placed in each of the investments of savings, certificates and mutual fund, respectively. Then, we have the three linear equations:

$$x + y + z = 248000$$
$$0.0435x + 0.0615y + 0.0875z = 19500$$
$$z = 3(x + y)$$

Now we solve for x, y, and z.

> eqntotal:= x + y + z = 248000:
> eqnyield:= 0.0435*x + 0.0615*y + 0.0875*z = 19500:
> eqntriple:= z = 3*(x + y):
> solve({eqntotal, eqnyield, eqntriple}, {x, y, z});

$$\{x = 32666.66667, y = 29333.33333, z = 1.86000\,10^5\}$$

From the above calculations, $32,666.67 should go into savings, $29,333.33 should go into certificates, and $186,000 must go into the mutual fund.

◼

Now we get to the *Leontief open input–output business/economic model.* In this model, we have several industries or companies that are mutual suppliers to each other of their products or output. Each company needs as input a certain dollar amount of the other companies' outputs in order to meet their mutual production schedules and certain outside demands by customers other than these companies. This Leontief model is called closed if there are no outside demands for these companies' products. In order to best see this idea in action, we look at the following example.

Example 4.3.3. A town has four companies that produce coal, gas, steel, and electricity. Each company buys the others products, that is, each company needs as input a certain dollar amount of the other companies' outputs in order to meet its own production. Let us assume that the coal company, in order to produce \$1 of coal, needs \$0.01 of coal, \$0.05 of gas, \$0.08 of steel, and \$0.17 of electricity. As well, the gas company, in order to produce \$1 of gas needs \$0.03 of coal, \$0.01 of gas, \$0.06 of steel and \$0.14 of electricity. Also, the steel company, in order to produce \$1 of steel needs \$0.11 of coal, \$0.15 of gas, \$0.02 of steel, and \$0.09 of electricity. Finally, the electric company, in order to produce \$1 of electricity, needs \$0.10 of coal, \$0.23 of gas, \$0.05 of steel, and \$0.03 of electricity. Each month these four companies have outside demands from other companies and consumers of their products for \$18 million of coal, \$74 million of gas, \$51 million of steel, and \$106 million of electricity. What must the total monthly outputs be of these four companies to meet exactly these outside demands and their own mutual needs of each others products? How much of each companies total monthly output is used internally by the four companies?

Let c, g, s, and e be the total monthly dollar production of each of these companies. Then, we have the four production equations starting with coal, gas, steel, and electricity, respectively:

$$c - (0.01c + 0.03g + 0.11s + 0.10e) = 18,000,000$$
$$g - (0.05c + 0.01g + 0.15s + 0.23e) = 74,000,000$$
$$s - (0.08c + 0.06g + 0.02s + 0.05e) = 51,000,000$$
$$e - (0.17c + 0.14g + 0.09s + 0.03e) = 106,000,000.$$

This gives a square 4×4 linear system of equations for which we expect a unique solution.

> soln := solve({c - (.01*c + .03*g + .11*s + .10*e) = 18000000, g - (.05*c + .01*g + .15*s + .23*e) = 74000000, s - (.08*c + .06*g + .02*s + .05*e) = 51000000, e - (.17*c + .14*g + .09*s + .03*e) = 106000000}, {c, g, s, e});

$$soln := \{c = 4385084414\,10^7, e = 1.408393813\,10^8,$$
$$g = 1.203145999\,10^8, s = 7.017235979\,10^7\}$$

Now, this tells us that total monthly coal production must be \$43.8508 million, total monthly gas production must be \$120.3146 million, total monthly steel production must be \$70.1724 million and finally total monthly electrical production must be \$140.8494 million. Also, the internal use of coal, gas, steel, and electricity by these four companies is \$25.8508, \$46.3146, \$19.1724, and \$34.8394 million, respectively. All these internal costs can be computed by taking the difference of the total output and the outside demands.

> rhs(soln[1])-18000000;

$$2.585084414 \, 10^7$$

> rhs(soln[3])-74000000;

$$4.63145999 \, 10^7$$

> rhs(soln[4])-51000000;

$$1.917235979 \, 10^7$$

> rhs(soln[2])-106000000;

$$3.48393813 \, 10^7$$

■

Maple Problems

1. Solve Example 4.3.1 of this section without *solve*, instead use *rref* on the augmented matrix of this linear system.

2. Solve Example 4.3.2 of this section without *solve*, instead use *rref* on the augmented matrix of this linear system.

3. Solve Example 4.3.3 of this section without *solve*, instead use *rref* on the augmented matrix of this linear system.

4. Your great great aunt Maple has just left you an inheritance of \$8,278,325, you got this money because you are her only living relative and you persuaded her that you would take care of her 10 cats. Upon receiving this money, you immediately spent \$100,000 on yourself and then wisely decided to invest the rest. You have decided to invest the remaining amount in a combination of four different ways: a savings account earning simple interest of 3.55% annually, a certificate of deposit earning simple interest of 5.75% annually, a mutual fund earning simple interest of 7.25% annually, and a stock portfolio earning simple interest of 9.15% annually.

You want to earn \$525,000 annual interest on these investments, you want the mutual fund to be the sum of savings and the certificate of deposit and you also want the stock portfolio to have one-fifth the sum of the other three investments. How much money should be invested in each of these four ways? Use *rref* to solve the problem, and the verify your answer with the *solve* command.

5. In problem 4, increase the total annual earnings of $525,000 until the problem no longer has a positive solution for the four amounts of your investments. What is the maximum total amount of annual earnings you can receive to the nearest dollar if all other conditions of the problem are kept the same, and for this maximum annual earnings what are your four investment amounts to the nearest dollar?

6. A small business has a taxable annual income of $1,439,535. It pays taxes under a fair tax system where there are no taxes on taxes. The federal income tax rate is 13.25%, the state income tax rate is 4.15%, the county income tax rate is 0.35%, the city income tax rate is 0.15%, the property tax rate is 0.75%, and the school tax rate is 0.25%. What must this business pay on each of these six taxes, and what is their overall fair tax rate? Use *rref* to solve the problem checking your answer with *solve*.

7. In problem 6, replace all of the tax rates along with the taxable annual income of the business by different letters. Use *Maple* to find a formula in terms of these letters for the business' overall fair tax rate.

8. Use the Leontief economic model for this problem. A huge multinational corporation has seven industrial production divisions: natural gas, oil, coal, steel, electricity, plastics, and mineral mining. The corporation has determined that it takes the following amounts of each of these divisions to make $1 of a particular division's product.

	nat. gas	oil	coal	steel	elect	plast	minrls
one dollar of gas	0.01	0.03	0.02	0.07	0.15	0.13	0.01
one dollar of oil	0.02	0.01	0	0.03	0.05	0.11	0.04
one dollar of coal	0.03	0.01	0	0.05	0.08	0.02	0
one dollar of steel	0.04	0.01	0.07	0.02	0.09	0.01	0.02
one dollar of elect	0.12	0.01	0.04	0.02	0.03	0.01	0.01
one dollar of plast	0.05	0.07	0	0.01	0.04	0.01	0.01
one dollar of minrls	0.03	0.02	0.01	0.05	0.08	0.02	0

If the corporation has outside demands for $3B of natural gas, $5B of oil, $1B of coal, $4B of steel, $10B of electricity, $8B of plastics, and $6B of minerals per week, then how much must be each division's total production each week to meet all demands? How much of each division's total production goes to meet its internal demands per week? Use *rref* to solve the problem checking your answer with *solve*.

9. Find a general formula for the augmented matrix needed to solve a Leontief economic model in terms of the internal demand amounts as in the table for problem 8 and the outside demand amounts. Apply this formula to the Leontief example given in this section and then *rref* it to see if you get the same answer.

Research Projects

1. Leontief won the Nobel Prize in Economics for his work. Research what Leontief did and see if the concept of linear systems was useful in his work.

2. If you have some programming experience and/or are brave enough, write a *Maple* procedure (function) that inputs the list L of your different fair tax rates and your total annual taxable income I, and outputs the individual fair tax amounts in the same order as the rates were given as well as your overall fair tax rate.

3. If you have some programming experience and/or are brave enough, write a *Maple* procedure (function) which inputs the information for a Leontief economic model and outputs the total production amounts for each division of the corporation in the same order the information was input.

4.4   Applications of Matrix Multiplication to Geometry

In the plane and space, we can do rotations of objects about the origin using matrix multiplication. We will begin with rotating a point $P(x_0, y_0)$ in the xy-plane about the origin through an angle θ. Here, simple trigonometry will allow us to compute the coordinates of the new point $Q(x_1, y_1)$ after rotation by the angle θ. After this has been accomplished, we can turn the process of rotation into matrix multiplication. To begin, we recognize that the new point Q will be the same distance r from the origin as the original point P. So both points P and Q lie on the circle with center the origin and radius r. The point Q is just at an extra angle θ, as measured from the positive $x-$axis, than P is on this circle. If the point P is at angle ϕ, then the new point Q is at the angle $\phi + \theta$. Since all the points on the circle of radius r with center at the origin can be written as $(r\cos(\alpha), r\sin(\alpha))$ for position at angle α, measured from the positive x-axis, we now know that the point P has coordinates $(r\cos(\phi), r\sin(\phi))$ while the point Q has coordinates $(r\cos(\phi + \theta), r\sin(\phi + \theta))$. Simple trigonometry, right?

Example 4.4.1. We will now illustrate this concept by rotating, through an angle $\theta = \dfrac{4}{5}\pi$, the point $P\left(5\cos\left(\dfrac{\pi}{6}\right), 5\sin\left(\dfrac{\pi}{6}\right)\right)$. The angle ϕ for the point P is $\phi = \dfrac{\pi}{6}$. Then the new point Q is at $\left(5\cos\left(\dfrac{\pi}{6} + \dfrac{4}{5}\pi\right), 5\sin\left(\dfrac{\pi}{6} + \dfrac{4}{5}\pi\right)\right)$.

```
> with(plots): with(plottools): with(linalg):
> circ:= circle([0,0], 5, color = blue):
> P:= [5*cos(Pi/6), 5*sin(Pi/6)]:
> Q:= [5*cos(Pi/6 + 4*Pi/5), 5*sin(Pi/6 + 4*Pi/5)]:
> Arc:= arc([0,0], 5, Pi/6..Pi/6 + 4*Pi/5, color = red, thickness = 3):
> plotpts:= pointplot({P,Q}, symbol = diamond, symbolsize = 25, color =
red):
> plotP:= textplot([5*cos(Pi/6), 5*sin(Pi/6)+.3, "P"], align = ABOVE):
> plotQ:= textplot([5*cos(Pi/6 + 4*Pi/5), 5*sin(Pi/6 + 4*Pi/5)+.3, "Q"],
align = {ABOVE,LEFT}):
> line1:= line([0,0], P, color=black, thickness=2):
> line2:= line([0,0], Q, color=black, thickness=2 ):
> Angle1:= arc([0,0], 1.8, Pi/6..Pi/6 + 4*Pi/5, color=black, thickness=2):
> Angle2:= arc([0,0], 2.5, 0..Pi/6, color=black, thickness=2):
> LabelAngle1:= textplot([-.4,.6, q], align=ABOVE, font=[SYMBOL, 10]):
> LabelAngle2:= textplot([2,0.4, f], align =ABOVE, font=[SYMBOL, 10]):
> display({circ, Arc, plotpts, plotP, plotQ, line1, line2, Angle1, Angle2, La-
belAngle1, LabelAngle2}, scaling = constrained);
```

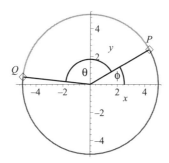

Figure 4.8: Original point P and the rotated point Q.

Now that we know from trigonometry how to go from the coordinates of the original point P to the coordinates of the new rotated point Q, as depicted in Figure 4.8, we can turn this process into a matrix multiplication. What we seek is a 2×2 rotation matrix A so that $Q = AP$ where P and Q are written as column matrices. The key to finding the rotation matrix A is again trigonometry, this time we need a trig identity. Recall that the point $P(x_0, y_0)$ also has coordinates $(r \cos(\phi), r \sin(\phi))$ while the point $Q(x_1, y_1)$ has coordinates $(r \cos(\phi + \theta), r \sin(\phi + \theta))$. We now need the trig identities for sine and cosine of a sum of two angles that *Maple* can give below.

> expand(cos(phi + theta), trig);

$$\cos(\phi) \cos(\theta) - \sin(\phi) \sin(\theta)$$

> expand(sin(phi+theta), trig);

$$\sin(\phi) \cos(\theta) + \cos(\phi) \sin(\theta)$$

Then, the point Q is given by

$$
\begin{aligned}
Q &= (r \cos(\phi + \theta), r \sin(\phi + \theta)) \\
&= (r \cos(\phi) \cos(\theta) - r \sin(\phi) \sin(\theta), r \sin(\phi) \cos(\theta) + r \cos(\phi) \sin(\theta)) \\
&= (\cos(\theta) x_0 - \sin(\theta) y_0, \cos(\theta) y_0 + \sin(\theta) x_0)
\end{aligned}
$$

Now, we have that Q, which has coordinates (x_1, y_1), can be expressed in terms of the original coordinates (x_0, y_0) of P and the angle θ. We have from the last equation that

$$(x_1, y_1) = (\cos(\theta) x_0 - \sin(\theta) y_0, \cos(\theta) y_0 + \sin(\theta) x_0) \tag{4.8}$$

If we convert these rows to columns, then we have

$$\begin{bmatrix} x_1 \\ y_1 \end{bmatrix} = \begin{bmatrix} \cos(\theta) x_0 - \sin(\theta) y_0 \\ \sin(\theta) x_0 + \cos(\theta) y_0 \end{bmatrix}$$

and by changing to matrix multiplication we have

$$\begin{bmatrix} x_1 \\ y_1 \end{bmatrix} = \begin{bmatrix} \cos(\theta) & -\sin(\theta) \\ \sin(\theta) & \cos(\theta) \end{bmatrix} \begin{bmatrix} x_0 \\ y_0 \end{bmatrix} \tag{4.9}$$

Now, we finally see that the 2×2 matrix

$$A_\theta = \begin{bmatrix} \cos(\theta) & -\sin(\theta) \\ \sin(\theta) & \cos(\theta) \end{bmatrix} \tag{4.10}$$

carries out the rotation about the origin through the angle θ and it does so by left multiplication, that is, $Q = A_\theta P$, where the initial point P and the new point Q are written as column matrices. We can now perform the rotation of P about the origin through the angle $\theta = \frac{4}{5}\pi$ using matrix multiplication.

> P:= matrix([[4.330127020], [2.5]]);

$$P := \left[\begin{array}{c} 4.330127020 \\ 2.5 \end{array} \right]$$

> A:= theta -> matrix([[cos(theta), -sin(theta)], [sin(theta), cos(theta)]]):
> Q:= evalf(evalm(A(4*Pi/5)&*P));

$$Q := \left[\begin{array}{c} -4.972609478 \\ .522642317 \end{array} \right]$$

We obtain the same result through matrix multiplication as we did above through trigonometry.

___

The origin is the simplest point to rotate about. We will now change the center of our rotation to the point $C(h, k)$. If we want to rotate the point $P(x_0, y_0)$ through the angle θ about the center C to get the point Q, then we can subtract C from P to get the new point $R = P - C$, which we can rotate about the origin and then translate it back to its correct location by adding the point C back on. In other words, the point $Q(x_1, y_1)$ can be gotten by

$$\left[\begin{array}{c} x_1 \\ y_1 \end{array} \right] = \left[\begin{array}{cc} \cos(\theta) & -\sin(\theta) \\ \sin(\theta) & \cos(\theta) \end{array} \right] \left[\begin{array}{c} x_0 - h \\ y_0 - k \end{array} \right] + \left[\begin{array}{c} h \\ k \end{array} \right] \tag{4.11}$$

___

Example 4.4.2. Instead of using a point for an example of this process, we will use this rotation formula to rotate the four pedaled flower parametric curve

$$(x(t), y(t)) = ((3\cos(4t) + 2)\cos(t) + 5, (3\cos(4t) + 2)\sin(t) + 12)$$

for $t \in [0, 2\pi]$, about the point $C(-4, -9)$ through the angle $\theta = \frac{7}{8}\pi$.

> f:= (3*cos(4*t) + 2)*cos(t) + 5:
> g:= (3*cos(4*t) + 2)*sin(t) + 12:
> plot_original:= plot([f,g, t=0..2*Pi], color = red):
> C:= matrix([[-4],[-9]]): P:= matrix([[f],[g]]):
> Q:= evalm(A(7*Pi/8)&*(P-C)+C);

$$Q := \left[\begin{array}{c} -\cos\left(\frac{1}{8}\pi\right)((3\cos(4t) + 2)\cos(t) + 9) - \sin\left(\frac{1}{8}\pi\right)((3\cos(4t) + 2)\sin(t) + 21) - 4 \\ \sin\left(\frac{1}{8}\pi\right)((3\cos(4t) + 2)\cos(t) + 9) - \cos\left(\frac{1}{8}\pi\right)((3\cos(4t) + 2)\sin(t) + 21) - 9 \end{array} \right]$$

> plot_rotated:= plot([Q[1,1],Q[2,1], t=0..2*Pi], color = blue):
> plot_C:= pointplot([C[1,1],C[2,1]], symbol = cross, symbolsize = 25):
> display({plot_original,plot_rotated,plot_C}, scaling = constrained);

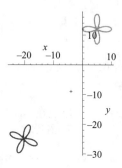

Figure 4.9: Original curve in the first quadrant, and rotated curve in the third.

Figure 4.9 clearly depicts the rotation of the parametric curve. Now let us animate this rotation so we can see some of the intermediate curves.
> Q:= theta -> evalm(A(theta)&*(P-C)+C):
> Q(7*Pi/8);

$$Q := \left[\begin{array}{c} -\cos\left(\tfrac{1}{8}\pi\right)\left((3\cos(4t)+2)\cos(t)+9\right) - \sin\left(\tfrac{1}{8}\pi\right)\left((3\cos(4t)+2)\sin(t)+21\right) - 4 \\ \sin\left(\tfrac{1}{8}\pi\right)\left((3\cos(4t)+2)\cos(t)+9\right) - \cos\left(\tfrac{1}{8}\pi\right)\left((3\cos(4t)+2)\sin(t)+21\right) - 9 \end{array} \right]$$

> Animation:= animate([Q(theta)[1,1], Q(theta)[2,1], t=0..2*Pi], theta = Pi/8..7*Pi/8, frames = 7, color = blue):
> display({Animation, plot_original, plot_C}, scaling = constrained);

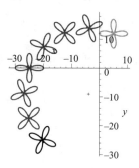

Figure 4.10: The original curve in the first quadrant is rotated counterclockwise around the "+" symbol through eight frames to the last curve in the third quadrant.

Since we cannot display the actual animated graphic in the book, Figure 4.10 is the next best thing. All the frames are depicted, from the original curve in the first quadrant, with all the intermediate frames, to the final rotated curve in the third quadrant. Each frame constitutes a rotation by an angle of $\frac{\pi}{8}$ in the counterclockwise direction starting with the original in the first quadrant, cumulating in a total rotation by $\frac{7}{8}\pi$ represented by the final curve in the third quadrant.

Homework Problems

1. Consider the point $P(3,0)$. Without using matrix multiplication, find the resulting points Q, R, and S after rotating P about the origin by angles $\frac{\pi}{4}$, $\frac{\pi}{2}$, and $\frac{3}{2}\pi$, respectively.

2. Use matrix multiplication to perform the rotations in problem 1.

3. Given a point P, let Q be the point corresponding to the rotation of P about the origin through an angle θ. Let R be the point corresponding to the rotation of Q about the origin through the angle ϕ. Verify that

$$A_\phi A_\theta = A_{\phi+\theta}$$

and thus that

$$R = A_{\phi+\theta}P = A_\phi A_\theta P = A_\theta A_\phi P$$

4. Geometrically, the same property discussed in problem 3 should hold for an arbitrary center of rotation. For instance, if we start with a point P, rotate it through an angle θ to the point Q, and the rotate Q through an angle ϕ to end up at R; this should be equivalent to starting at P and rotating through an angle of $\phi + \theta$, independent of the center. To show this, consider

$$Q = A_\theta(P - C) + C, \quad R = A_\theta(Q - C) + C$$

and prove that

$$A_{\phi+\theta}(P - C) + C = A_\phi [Q - C] + C$$

5. Find the coordinates of the point Q corresponding to the point $P(3,3)$ that has been rotated about the point $C(1,1)$ by an angle of $\theta = \frac{\pi}{4}$.

6. Given a point P, a point Q and a center of rotation C, how can one find the angle θ through which P was rotated to end up at Q?

7. Consider the points $P(4,5)$ and $Q(2, 2\sqrt{2}+3)$ and center of rotation $C(2,3)$. Determine the angle θ through which P was rotated about C to end up at point Q.

8. Find the coordinates of the point Q corresponding to the point $P(3,3)$ after it has been rotated about the point $C(1,3)$ by an angle of $\theta = \pi$. Consider this problem from a geometric point of view, explain how you could have known the answer without performing any matrix multiplication.

9. As discussed in this section, the matrix A_θ corresponds to a counter-clockwise rotation about the origin. How can you modify the matrix A_θ to perform clockwise rotations?

10. The process of rotation about a point can be generalized to three dimensions. Given a point $P(x_0, y_0, z_0)$, determine what rotations the following matrices perform upon the point P.

$$A_1 = \begin{bmatrix} 1 & 0 & 0 \\ 0 & \cos(\theta) & \sin(\theta) \\ 0 & -\sin(\theta) & \cos(\theta) \end{bmatrix}$$

$$A_2 = \begin{bmatrix} \cos(\theta) & 0 & -\sin(\theta) \\ 0 & 1 & 0 \\ \sin(\theta) & 0 & \cos(\theta) \end{bmatrix}$$

$$A_3 = \begin{bmatrix} \cos(\theta) & \sin(\theta) & 0 \\ -\sin(\theta) & \cos(\theta) & 0 \\ 0 & 0 & 1 \end{bmatrix}$$

11. (a) What 3×3 matrix R will carry out, by a single matrix multiplication by R, the following three consecutive rotations in space in the given order: first, rotate in space by the angle α about the x-axis followed by a rotation by the angle β about the y-axis followed by a rotation by the angle γ about the z-axis?

(b) Is it the same matrix R if we switch the order of these three consecutive rotations, explain?

12. How can you use matrix multiplication and addition/subtraction to rotate in space about a line parallel to one of the three coordinate axes?

13. Using the information learned in this section, do (or redo) *Homework* problem 7 of Section 4.1.

Maple Problems

1. Rotate the points P through the angles θ with centers of rotation C.

   (a) $P(1,2)$, $\theta = \dfrac{\pi}{6}$, $C(0,0)$ (b) $P(2,1)$, $\theta = \dfrac{\pi}{3}$, $C(1,0)$

   (c) $P(7,2)$, $\theta = -\dfrac{2}{3}\pi$, $C(9,2)$ (d) $P(0,0)$, $\theta = \dfrac{\pi}{2}$, $C(2,3)$

   (e) $P(1,1)$, $\theta = \dfrac{3}{4}\pi$, $C(-1,-1)$ (f) $P(3,-1)$, $\theta = -\dfrac{\pi}{4}$, $C(3,0)$

2. Using the results of *Homework* problem 10, perform the following rotations:

   (a) Rotate $P(1,1,1)$ about the x-axis by an angle of $\theta = \dfrac{\pi}{4}$.

   (b) Rotate $P(1,2,1)$ about the z-axis by an angle of $\theta = \dfrac{\pi}{3}$.

   (c) Rotate $P(1,1,2)$ about the y-axis by an angle of $\theta = \dfrac{2}{3}\pi$.

   (d) Rotate $P(1,1,1)$ about the z-axis by an angle of $\theta = \dfrac{\pi}{4}$, then take the resulting point and rotate it about the x-axis by an angle of $\phi = \dfrac{\pi}{3}$.

   (e) Rotate $P(1,1,1)$ about the x-axis by an angle of $\phi = \dfrac{\pi}{3}$, then take the resulting point and rotate it about the z-axis by an angle of $\theta = \dfrac{\pi}{4}$.

   (f) Rotate $P(2,3,4)$ about the x-axis by an angle of $\phi = \dfrac{\pi}{3}$, then take the resulting point and rotate it about that y-axis by an angle of $\theta = \dfrac{2}{3}\pi$ and finally take this second point and rotate it about the $z-$axis by an angle of $\rho = \dfrac{\pi}{4}$.

3. Construct a set of piecewise parametric functions that, when graphed, appear to be the first, middle, and last initials of your name. Rotate these initials about a point and animate the sequence.

4. A complex number z can be written as follows:

$$z = a + bi = |z|\, e^{i\phi} = |z|\cos(\phi) + |z|\sin(\phi)i$$

where $|z| = \sqrt{a^2 + b^2}$ is the modulus of z, the distance from z as the point (a,b) to the origin, and ϕ is the angle between the complex number z as the point (a,b) and the positive x-axis. Using complex numbers instead of matrices, find a formula for rotating the point $P(x_0, y_0)$ about the center point $C(h,k)$ counterclockwise through the angle ϕ to get the new point Q. Use this formula for Q on an example from this section to see that it is correct. *Hint: If you want to rotate a point $P(a, b)$ about the origin through the angle θ to get a new point Q, then rewrite P as a complex number z and think of what multiplying z by $e^{i\theta}$ will do to z.*

Research Projects

1. Research the real quaternions $\mathbb{H}$, which are a generalization of the complex numbers $\mathbb{C}$, and find out how they can be used to do rotations in space.

4.5   An Application of Matrix Multiplication to Economics

A modern company typically makes many kinds of similar items and has a production quota to meet. In the production of all of these items, the company uses several types of labor in varying amounts. It therefore knows how many man-hours of each type of labor (on average) it takes to make one type of item. The most convenient way to manipulate and change both production and man-hour data is to place the data in a matrix, which is our mathematical version of a spreadsheet. The type of example we will look at concerns the areas of business related to production management, quality control, and efficiency since we will look at production, man-hour usage, cost, and defective rate data for a quarter's production for one company.

Example 4.5.1. Our example concerns a large car manufacturer. It produces the five types of items: compact cars, sedans, sports cars, trucks, and SUVs using the four types of labor: upholstery work, metal work, electrical work, and general assembly. The following matrices $T1$ and $P1$ are the time matrix in man-hours and the production or order matrix for the first quarter of January, February, and March. We have

$$T1 = \begin{bmatrix} \textit{labor per vehicle} & \textit{compact car} & \textit{sedan} & \textit{sports car} & \textit{truck} & \textit{SUV} \\ \textit{upholstery} & 0.735 & 1.105 & 0.825 & 0.765 & 0.855 \\ \textit{metal} & 1.515 & 1.785 & 1.325 & 1.605 & 1.905 \\ \textit{electrical} & 1.105 & 1.325 & 0.905 & 1.535 & 1.725 \\ \textit{general assembly} & 2.465 & 2.535 & 2.015 & 2.935 & 3.265 \end{bmatrix}$$

and

$$
P1 = \begin{bmatrix}
vehicle\ per\ month & January & February & March \\
compact\ car & 7825 & 8635 & 8950 \\
sedan & 5085 & 5530 & 5815 \\
sports\ car & 1205 & 1365 & 1420 \\
truck & 4095 & 5140 & 5375 \\
SUV & 3820 & 4235 & 4680
\end{bmatrix}.
$$

> with(linalg):
> T1_headings:= matrix([[labor*per*vehicle, compact*car, sedan, sports*car, truck, SUV], [upholstery, .735, 1.105, .825, .765, .855], [metal, 1.515, 1.785, 1.325, 1.605, 1.905], [electrical, 1.105, 1.325, .905, 1.535, 1.725], [general*assembly, 2.465, 2.535, 2.015, 2.935, 3.265]]);

$$
T1 := \begin{bmatrix}
labor\ per\ vehicle & compact\ car & sedan & sports\ car & truck & SUV \\
upholstery & 0.735 & 1.105 & 0.825 & 0.765 & 0.855 \\
metal & 1.515 & 1.785 & 1.325 & 1.605 & 1.905 \\
electrical & 1.105 & 1.325 & 0.905 & 1.535 & 1.725 \\
general\ assembly & 2.465 & 2.535 & 2.015 & 2.935 & 3.265
\end{bmatrix}
$$

> P1_headings:= matrix([[vehicle*per*month, January, February, March], [compact*car, 7825, 8635, 8950], [sedan, 5085, 5530, 5815], [sports*car, 1205, 1365, 1420], [truck, 4095, 5140, 5375], [SUV, 3820, 4235, 4680]]);

$$
P1 := \begin{bmatrix}
vehicle\ per\ month & January & February & March \\
compact\ car & 7825 & 8635 & 8950 \\
sedan & 5085 & 5530 & 5815 \\
sports\ car & 1205 & 1365 & 1420 \\
truck & 4095 & 5140 & 5375 \\
SUV & 3820 & 4235 & 4680
\end{bmatrix}
$$

We now want to know how many total man-hours of each type of labor is needed for each of the three months of the first quarter in order to exactly meet the production quota, call this matrix the total manpower matrix $M1$. You can get the matrix $M1$ by multiplying $T1$ times $P1$, but first we must remove the headings. The commands *delrows* and *delcols* in the *linalg* package will take care of this removal. After we compute $M1$, its headings will be put in place.

> T1:= delcols(delrows(T1_headings,1..1),1..1);

$$
T1 := \begin{bmatrix}
0.735 & 1.105 & 0.825 & 0.765 & 0.855 \\
1.515 & 1.785 & 1.325 & 1.605 & 1.905 \\
1.105 & 1.325 & 0.905 & 1.535 & 1.725 \\
2.465 & 2.535 & 2.015 & 2.935 & 3.265
\end{bmatrix}
$$

> P1:= delcols(delrows(P1_headings,1..1),1..1);

$$P1 := \begin{bmatrix} 7825 & 8635 & 8950 \\ 5085 & 5530 & 5815 \\ 1205 & 1365 & 1420 \\ 4095 & 5140 & 5375 \\ 3820 & 4235 & 4680 \end{bmatrix}$$

> M1:= evalm(T1&*P1);

$$M1 := \begin{bmatrix} 18763.200 & 21136.525 & 22288.600 \\ 36377.800 & 41079.075 & 43362.800 \\ 29350.100 & 33299.525 & 35203.350 \\ 59098.300 & 66967.475 & 70719.900 \end{bmatrix}$$

> M1_headings:= matrix([[labor*per*month, January, February, March], [upholstery, 18763.200, 21136.525, 22288.600], [metal, 36377.800, 41079.075, 43362.800], [electrical, 29350.100, 33299.525, 35203.350], [general*assembly, 59098.300, 66967.475, 70719.900]]);

$$M1\_headings := \begin{bmatrix} labor\ per\ month & January & February & March \\ upholstery & 18763.200 & 21136.525 & 22288.600 \\ metal & 36377.800 & 41079.075 & 43362.800 \\ electrical & 29350.100 & 33299.525 & 35203.350 \\ general\ assembly & 59098.300 & 66967.475 & 70719.900 \end{bmatrix}$$

If we divide the total manpower matrix $M1$ by 8, then we know how many people must be scheduled per month for each type of labor in order to meet first quarter production assuming one person does an eight hour shift. The entries in this matrix should be rounded up to get integer values. The command *ceil* will round up to the next integer while *map* will send the *ceil* command to each entry of the matrix.

> people1:= map(ceil,evalm(1/8*M1));

$$people1 := \begin{bmatrix} 2346 & 2643 & 2787 \\ 4548 & 5135 & 5421 \\ 3669 & 4163 & 4401 \\ 7388 & 8371 & 8840 \end{bmatrix}$$

> people1_headings:= matrix([[people*per*month, January, February, March], [upholsterers, 2346, 2643, 2787], [metal*workers, 4548, 5135, 5421], [electrical*workers, 3669, 4163, 4401], [gen*assembly*workers, 7388, 8371, 8840]]);

$$people1\_headings := \begin{bmatrix} people\ per\ month & January & February & March \\ upholsterers & 2346 & 2643 & 2787 \\ metal\ workers & 4548 & 5135 & 5421 \\ electrical\ workers & 3669 & 4163 & 4401 \\ gen\ assembly\ workers & 7388 & 8371 & 8840 \end{bmatrix}$$

Let us assume that it is normal for the second quarter production orders to go up across the board by 4.15%, while in the second quarter increased efficiency of labor changes by 6.25%. Then our second quarter production matrix $P2 = (1 + .0415)P1 = 1.0415 \cdot P1$ while the second quarter time matrix is $T2 = (1 - 0.0625) \cdot T1 = 0.9375 \cdot T1$. So the second quarter total manpower matrix

$$M2 = T2 \cdot P2$$
$$= (0.9375 \cdot T1)(1.0415 \cdot P1)$$
$$= 0.9375 \cdot 1.0415 \cdot (T1 \cdot P1)$$
$$= 0.97640625 \cdot M1$$

This says that the total need for manpower has gone down by 2.36% from the first to second quarter. Let us now compute these matrices to see if this is correct.

> 0.9375*1.0415;

$$.97640625$$

> 1-0.97640625;

$$.02359375$$

> T2:= evalm(.9375*T1);

$$T2 := \begin{bmatrix} 0.6890625 & 1.0359375 & 0.7734375 & 0.7171875 & 0.8015625 \\ 1.4203125 & 1.6734375 & 1.2421875 & 1.5046875 & 1.7859375 \\ 1.0359375 & 1.2421875 & .8484375 & 1.4390625 & 1.6171875 \\ 2.3109375 & 2.3765625 & 1.8890625 & 2.7515625 & 3.0609375 \end{bmatrix}$$

> P2:= evalm(1.0415*P1);

$$P2 := \begin{bmatrix} 8149.7375 & 8993.3525 & 9321.4250 \\ 5296.0275 & 5759.4950 & 6056.3225 \\ 1255.0075 & 1421.6475 & 1478.9300 \\ 4264.9425 & 5353.3100 & 5598.0625 \\ 3978.5300 & 4410.7525 & 4874.2200 \end{bmatrix}$$

> M2:= evalm(T2&*P2);

$$M2 := \begin{bmatrix} 18320.50574 & 20637.83512 & 21762.72834 \\ 35519.51128 & 40109.86556 & 42339.70894 \\ 28657.62108 & 32513.86433 & 34372.77097 \\ 57703.94948 & 65387.46114 & 69051.35236 \end{bmatrix}$$

> evalm(.97640625*M1);

$$\begin{bmatrix} 18320.50575 & 20637.83511 & 21762.72834 \\ 35519.51128 & 40109.86557 & 42339.70894 \\ 28657.62108 & 32513.86433 & 34372.77096 \\ 57703.94948 & 65387.46114 & 69051.35236 \end{bmatrix}$$

> people2:= map(ceil,evalm(1/8*M2));

$$people2 := \begin{bmatrix} 2291 & 2580 & 2721 \\ 4440 & 5014 & 5293 \\ 3583 & 4065 & 4297 \\ 7213 & 8174 & 8632 \end{bmatrix}$$

> people2_headings:= matrix([[people*per*month, April, May, June], [upholsterers, 2291, 2580, 2721], [metal*workers, 4440, 5014, 5293], [electrical*workers, 3583, 4065, 4297], [gen*assembly*workers, 7213, 8174, 8632]]);

$$people2\_headings := \begin{bmatrix} people\ per\ month & April & May & June \\ upholsterers & 2291 & 2580 & 2721 \\ metal\ workers & 4440 & 5014 & 5293 \\ electrical\ workers & 3583 & 4065 & 4297 \\ gen\ assembly\ workers & 7213 & 8174 & 8632 \end{bmatrix}$$

Let us return to the first quarter. After each type of vehicle is produced, it is driven off the assembly line and given a thorough quality control test in order to find any defects before it is sent to the dealership. We have the defective rate matrix $D1$ for the first quarter as

$$D1 = \begin{bmatrix} labor/vehicle & compact\ car & sedan & sports\ car & truck & SUV \\ upholstery & 0.825\% & 1.15\% & 0.615\% & 0.905\% & 1.25\% \\ metal & 1.5\% & 1.785\% & 1.125\% & 1.905\% & 1.815\% \\ electrical & 0.735\% & 0.775\% & 0.615\% & 0.845\% & 0.875\% \\ gen.\ assembly & 1.65\% & 1.875\% & 1.415\% & 1.925\% & 1.965\% \end{bmatrix},$$

and the average cost of repair matrix $R1$ for the first quarter as

$$R1 = \begin{bmatrix} vehicle/labor\ type & upholstery & metal & electric & gen.\ assembly \\ compact\ car & \$7.85 & \$15.10 & \$9.15 & \$35.20 \\ sedan & \$10.25 & \$18.90 & \$11.25 & \$42.70 \\ sports\ car & \$8.15 & \$15.75 & \$9.45 & \$29.65 \\ truck & \$13.95 & \$21.05 & \$11.80 & \$51.75 \\ SUV & \$14.35 & \$20.60 & \$10.95 & \$48.15 \end{bmatrix}.$$

How much does it cost each of the months of the first quarter to make these repairs on these five types of vehicles? The answer is $R1 \cdot D1 \cdot P1$ since $D1 \cdot P1$ is the number of defective vehicles of each type per month of the first quarter. In order to have *Maple* compute this product for us, we need to delete all headings from these matrices, as well as the % and $ notation. We can also ask *Maple* to compute the total repair cost for the first quarter by summing the entries in the matrix $R1 \cdot D1 \cdot P1$.

> D1:= matrix([[.00825, .0115, .00615, .00905, .0125], [.015, .01785, .01125, .01905, .01815], [.00735, .00775, .00615, .00845, .00875], [.0165, .01875, .01415, .01925, .01965]]);

$$D1 := \begin{bmatrix} 0.00825 & 0.0115 & 0.00615 & 0.00905 & 0.0125 \\ 0.015 & 0.01785 & 0.01125 & 0.01905 & 0.01815 \\ 0.00735 & 0.00775 & 0.00615 & 0.00845 & 0.00875 \\ 0.0165 & 0.01875 & 0.01415 & 0.01925 & 0.01965 \end{bmatrix}$$

> map(ceil, evalm(D1&*P1));

$$\begin{bmatrix} 216 & 243 & 257 \\ 370 & 419 & 442 \\ 173 & 196 & 206 \\ 396 & 448 & 473 \end{bmatrix}$$

> R1:= matrix([[7.85, 15.10, 9.15, 35.20], [10.25, 18.90, 11.25, 42.70], [8.15, 15.75, 9.45, 29.65], [13.95, 21.05, 11.80, 51.75], [14.35, 20.60, 10.95, 48.15]]);

$$R1 := \begin{bmatrix} 7.85 & 15.10 & 9.15 & 35.20 \\ 10.25 & 18.90 & 11.25 & 42.70 \\ 8.15 & 15.75 & 9.45 & 29.65 \\ 13.95 & 21.05 & 11.80 & 51.75 \\ 14.35 & 20.60 & 10.95 & 48.15 \end{bmatrix}$$

> Costs1:= map(ceil, evalm(R1&*(D1&*P1)));

$$Costs1 := \begin{bmatrix} 22758 & 25766 & 27186 \\ 28004 & 31706 & 33453 \\ 20920 & 23685 & 24991 \\ 33267 & 37662 & 39739 \\ 31617 & 35793 & 37768 \end{bmatrix}$$

> Costs1_headings:= matrix([[type*of*vehicle*per*month, January, February, March], [compact*car, 22758, 25766, 27186], [sedan, 28004, 31706, 33453], [sports*car, 20920, 23685, 24991], [truck, 33267, 37662, 39739], [SUV, 31617, 35793, 37768]]);

$$Costs1\_headings := \begin{bmatrix} \textit{type of vehicle per month} & \textit{January} & \textit{February} & \textit{March} \\ \textit{compact car} & 22758 & 25766 & 27186 \\ \textit{sedan} & 28004 & 31706 & 33453 \\ \textit{sports car} & 20920 & 23685 & 24991 \\ \textit{truck} & 33267 & 37662 & 39739 \\ \textit{SUV} & 31617 & 35793 & 37768 \end{bmatrix}$$

The total cost of making repairs under the quality control to the five types of vehicles from the four kinds of work for the first quarter is the sum of the entries to the *Costs1* matrix. It is found below to be $454,315.

> Total_Repair_Cost1:= add(add(Costs1[i,j],i=1..5),j=1..3);

$$Total\_Repair\_Cost1 := 454315$$

■

Maple Problems

All of the following *Maple* problems will use the information given in the example from this section as this year's data, where each one is a continuation of the previous problem. In each of these *Maple* problems compute the required matrices with and without their appropriate headings.

1. In next year's first quarter, it is anticipated that all of the following will occur: compact car orders will go up by an average of 3.75% each month, sedan orders will go down by an average of 2.95% each month, sports car orders will go down by an average of 4.30% each month, truck orders will go up by an average of 1.65% each month and SUV orders will go up by an average of 2.15% each month. What is next year's production matrix $P1$ for the first quarter?

2. In next year's first quarter, it is anticipated that all of the following will occur due to changes in practices and automation: next year's first quarter man-hours for upholstery will go down by an average of 2.55% for each type of vehicle, man-hours for metal work will go up by an average of 1.35% for each type of vehicle, man-hours for electrical work will go down by an average of 3.75% for each type of vehicle and man-hours for general assembly will go down by an average of 0.85% for each type of vehicle. What is next year's time matrix $T1$ for the first quarter?

3. (a) In next year's first quarter, what is the total manpower matrix $M1$?

(b) What is the total amount of people (*people1* matrix) needed for next year's first quarter by month and by the type of labor done if one person is needed to fill one eight hour shift per day?

(c) If upholsterers make an average of $21.25 per hour including benefits, metal workers make an average of $24.50 per hour including benefits, electrical workers make an average of $26.75 per hour including benefits and general assembly workers make an average of $27.85 per hour including benefits, then what is the cost matrix $C1$ for next year's first quarter by month and type of labor? What is the total labor cost for next year's first quarter for these four types of assembly line workers? What is the average salary including benefits for next year's first quarter for all of these four types of assembly line workers?

4.   (a) In next year's second quarter, what is the production matrix $P2$ if it reflects an across the board average drop in orders by 2.65% from the first quarter of next year?

(b) In next year's second quarter, what is the time matrix $T2$ if it reflects an across the board average rise in times by 1.35% from the first quarter of next year due to new safety regulations?

(c) In next year's second quarter, what is the total man power matrix $M2$, and by what percentage has the need for man power changed from the first quarter to the second quarter of next year?

(d) What is the total amount of people (people2 matrix) needed for next year's second quarter by month and by the type of labor done if one person is needed to fill one eight hour shift per day?

(e) If all labor costs have gone up from the first quarter of next year to the second quarter of next year by an average of 2.25%, then what is the cost matrix $C2$ for next year's second quarter by month and type of labor? What is the total labor cost for next year's second quarter for these four types of assembly line workers? What is the average salary including benefits for next year's second quarter for all of these four types of assembly line workers?

5. (a) Find the defective rate matrix $D2$ for the second quarter of next year if it is anticipated that these defective rates will have gone down by an average of 1.75% from the first quarter of this year.

(b) Find the average cost of repair matrix $R2$ for the second quarter of next year if it is anticipated that these repair costs will have gone up by an average of 2.15% from the first quarter of this year.

(c) Find the repair costs matrix $Costs2$ for the second quarter of next year.

d) Find the total repair costs for the second quarter of next year. What is the average repair cost per vehicle produced for the second quarter of next year?

Chapter 5

Determinants, Inverses, and Cramer's Rule

5.1   Determinants and Inverses from the Adjoint Formula

We introduced matrix arithmetic in Chapter 2, however, we left out matrix division. We now concern ourselves with extending matrix arithmetic to matrix division, or more precisely: multiplication by the multiplicative inverse. The reason for matrix multiplicative inverses arises in the following situation: If we are given a linear system expressed as the matrix equation $AX = B$, where A is the coefficient matrix, X the variable column, and B the column of the RHS values, then we want to know when we can divide by A, or equivalently, multiply the entire equation by A's multiplicative inverse A^{-1}, to get the unique solution, $X = A^{-1}B$, to the system.

When might our matrix equation $AX = B$ have a unique solution of the form $X = A^{-1}B$? From our past experience, we know that it is usual to get a single solution to a linear system only when it is square, which occurs only when the matrix of coefficients A is square. This tells us that we should only look for multiplicative inverses A^{-1} for square matrices A.

Remember that in the set of square matrices $\mathbb{R}^{n \times n}$, the (multiplicative) identity matrix I_n was defined to be the $n \times n$ matrix with ones on its main diagonal from upper left to lower right and zeros everywhere else. For any $A \in \mathbb{R}^{n \times n}$, we have both $AI_n = A$ and $I_n A = A$. Now if A has a multiplicative inverse K with $K \in \mathbb{R}^{n \times n}$, then K must satisfy both $AK = I_n$ and $KA = I_n$. These last two equations define what we mean by the multiplicative inverse, A^{-1}, of a square matrix A.

Definition 5.1.1. The *inverse* of an $n \times n$ matrix A, denoted A^{-1}, is the unique matrix of dimension $n \times n$ that satisfies $A A^{-1} = A^{-1} A = I_n$.

The simplest case to study is a 1×1 matrix $[a]$ with only one entry, a. The real numbers $\mathbb{R}$ are the same as the 1×1 real matrices $[a]$, with the same multiplication, and so if $a \neq 0$,

$$[a]^{-1} = [a^{-1}] = \left[\frac{1}{a}\right] \tag{5.1}$$

We now move on to the case of 2×2 matrices and ask the questions: When, and how, does a matrix $A \in \mathbb{R}^{2 \times 2}$ have a multiplicative inverse? It should be clear that not all 2×2 matrices will have multiplicative inverses, since there should be matrices that have properties similar to those of the number zero in regards to multiplicative inverses in the scalar case.

Example 5.1.1. Let us see an example of a 2×2 matrix that has no multiplicative inverse. If $A = \begin{bmatrix} 0 & 0 \\ 0 & d \end{bmatrix}$ for $d \in \mathbb{R}$, then A has no inverse K. To see this, if we arbitrarily define $K = \begin{bmatrix} \alpha & \beta \\ \theta & \delta \end{bmatrix}$, then no choice of α through δ will let $AK = I_2 = \begin{bmatrix} 1 & 0 \\ 0 & 1 \end{bmatrix}$ since the multiplication $AK = \begin{bmatrix} 0 & 0 \\ d\theta & d\delta \end{bmatrix}$. We will have *Maple* verify our above calculations quickly.

```
> with(linalg):
> A:= matrix([[0, 0], [0, d]]):
> K:= matrix([[alpha, beta], [theta, delta]]):
> evalm(A&*K);
```

$$\begin{bmatrix} 0 & 0 \\ d\theta & d\delta \end{bmatrix}$$

From this one simple example, it becomes immediately apparent that there are an infinite number of 2×2 matrices that have no multiplicative inverse. This is much different than the scalar case, where zero was the only number with no multiplicative inverse. Now, we must strive to find a generalized method to determine whether or not an arbitrary matrix has an inverse, and if an inverse exists, we must also find a method of computing it. Our next step is to generalize our matrix $A \in \mathbb{R}^{2 \times 2}$ and attempt to solve our multiplicative inverse problem. We will treat this problem as a system of four equations with four unknowns. Let $A = \begin{bmatrix} a & b \\ c & d \end{bmatrix}$ for real constants a through d. We want to find the matrix $K = \begin{bmatrix} \alpha & \beta \\ \theta & \delta \end{bmatrix}$ so that $AK = I_2$. The matrix equation $AK = I_2$

will give us a linear system of four equations in the unknowns α through δ. Although it is not difficult to perform the algebra, we will have *Maple* find the formulas for the unknown variables α through δ in terms of known constants a through d.

> A:= matrix([[a, b], [c, d]]);

$$\begin{bmatrix} a & b \\ c & d \end{bmatrix}$$

> P:= evalm(A&*K);

$$P := \begin{bmatrix} a\alpha + b\theta & a\beta + b\delta \\ c\alpha + d\theta & c\beta + d\delta \end{bmatrix}$$

> eqn1:= P[1,1] = 1;

$$eqn1 := a\alpha + b\theta = 1$$

> eqn2:= P[1,2] = 0;

$$eqn2 := a\beta + b\delta = 0$$

> eqn3:= P[2,1] = 0;

$$eqn3 := c\alpha + d\theta = 0$$

> eqn4:= P[2,2] = 1;

$$eqn4 := c\beta + d\delta = 1$$

> solve({eqn1,eqn2,eqn3,eqn4}, {alpha,beta,theta,delta});

$$\left\{ \alpha = \frac{d}{ad - cb}, \beta = -\frac{b}{ad - cb}, \delta = \frac{a}{ad - cb}, \theta = -\frac{c}{ad - cb} \right\}$$

This tells us that the multiplicative inverse of A has the form

$$K = \frac{1}{ad - bc} \begin{bmatrix} d & -b \\ -c & a \end{bmatrix}$$

as long as $ad - bc \neq 0$. Thus, if A is defined as before, then we say $A^{-1} = K$ provided that $ad - bc \neq 0$. The value $ad - bc$ for the 2×2 matrix A is called its *determinant* since its value being zero or not determines if A has an inverse or not. The *Maple* command *det*, in the *linalg* package, will compute the determinant of a matrix, furthermore, the command *inverse* will find the inverse to a square matrix, assuming it exists.

> K:= evalm(1/(a*d-b*c)*matrix([[d, -b], [-c, a]]));

$$K := \begin{bmatrix} \dfrac{d}{ad - cb} & -\dfrac{b}{ad - cb} \\ -\dfrac{c}{ad - cb} & \dfrac{a}{ad - cb} \end{bmatrix}$$

> simplify(evalm(A&*K));

$$\begin{bmatrix} 1 & 0 \\ 0 & 1 \end{bmatrix}$$

> simplify(evalm(K&*A));

$$\begin{bmatrix} 1 & 0 \\ 0 & 1 \end{bmatrix}$$

> det(A);

$$ad - cb$$

> inverse(A);

$$\begin{bmatrix} \dfrac{d}{ad-cb} & -\dfrac{b}{ad-cb} \\ -\dfrac{c}{ad-cb} & \dfrac{a}{ad-cb} \end{bmatrix}$$

Using the above commands, *Maple* has checked the formula for the inverse K of A and shown that both $AK = I_2$ and $KA = I_2$ are true. In the above formula for A^{-1}, we see that each entry looks like the determinant of a 1×1 matrix (which is the number itself), divided by the determinant of A. In fact,

$$A_{1,1}^{-1} = \frac{d}{\det(A)} = \frac{1}{\det(A)} \det([d])$$

$$= \frac{1}{\det(A)} \det(\text{matrix } A \text{ with the first row and column removed})$$

$$A_{2,2}^{-1} = \frac{a}{\det(A)} = \frac{1}{\det(A)} \det([a])$$

$$= \frac{1}{\det(A)} \det(\text{matrix } A \text{ with the second row and column removed})$$

$$A_{1,2}^{-1} = -\frac{b}{\det(A)} = -\frac{1}{\det(A)} \det([b])$$

$$= -\frac{1}{\det(A)} \det(\text{matrix } A \text{ with the second row and first column removed})$$

$$A_{2,1}^{-1} = -\frac{c}{\det(A)} = -\frac{1}{\det(A)} \det([c])$$

$$= -\frac{1}{\det(A)} \det(\text{matrix } A \text{ with the first row and second column removed})$$

Now that the 2×2 case has been solved, we want to find a pattern for computing the entries of A^{-1}, where A is a square matrix of arbitrary dimension $n \times n$. To do this, we will need to introduce some new terminology.

Definition 5.1.2. Given $A \in \mathbb{R}^{m \times n}$, the *transpose* of A, denoted A^T, is an element of $\mathbb{R}^{n \times m}$ defined by $A_{i,j}^T = A_{j,i}$.

Simply put, the transpose of a matrix is computed by swapping the rows and columns of the matrix.

Example 5.1.2. If $A \in \mathbb{R}^{3 \times 3}$ and $B \in \mathbb{R}^{2 \times 4}$ are given by

$$A = \begin{bmatrix} 1 & -2 & 2 \\ 4 & 5 & 1 \\ 0 & 3 & -1 \end{bmatrix}, \quad B = \begin{bmatrix} -1 & 3 & 0 & 4 \\ 6 & 7 & -2 & 1 \end{bmatrix} \qquad (5.2)$$

then $A^T \in \mathbb{R}^{3 \times 3}$ and $B^T \in \mathbb{R}^{4 \times 2}$ are given, respectively, by

$$A^T = \begin{bmatrix} 1 & 4 & 0 \\ -2 & 5 & 3 \\ 2 & 1 & -1 \end{bmatrix}, \quad B^T = \begin{bmatrix} -1 & 6 \\ 3 & 7 \\ 0 & -2 \\ 4 & 1 \end{bmatrix}$$

The following is a table of properties for matrix transpose. The formulas given hold for arbitrary matrices A and B, and scalar c, assuming the standard matrix operations of addition and multiplication can be performed on A and B.

<table>
<tr><th colspan="2">Matrix Transpose Properties</th></tr>
<tr><td>$(A^T)^T = A$</td><td>$(A + B)^T = A^T + B^T$</td></tr>
<tr><td>$(AB)^T = B^T A^T$</td><td>$(cA)^T = cA^T$</td></tr>
</table>

Definition 5.1.3. Given $A \in \mathbb{R}^{n \times n}$, the *minor* $M_{i,j}$ of the element $A_{i,j}$ is the determinant of the resulting $(n-1) \times (n-1)$ matrix found by removing row i and column j from A.

Note that the minor of element $A_{i,j}$ is a scalar, found by taking the determinant of a matrix of dimension $(n-1) \times (n-1)$.

Example 5.1.3. If we consider the matrix A from equation (5.2), which was used as an example for transpose operation, the minor $M_{1,1}$ of $A_{1,1}$ is given by

$$M_{1,1} = \det \left(\begin{bmatrix} 5 & 1 \\ 3 & -1 \end{bmatrix} \right) = -8$$

On the other hand, the minor $M_{2,3}$ of $A_{2,3}$ is calculated as

$$M_{2,3} = \det\left(\begin{bmatrix} 1 & -2 \\ 0 & 3 \end{bmatrix}\right) = 3$$

We now turn to the arbitrary 3×3 matrix inverse case. In the process, we will find the determinant of a 3×3 matrix. We define the matrix A, and its potential inverse K as follows:

$$A = \begin{bmatrix} a & b & c \\ d & e & f \\ g & h & i \end{bmatrix}, \quad K = \begin{bmatrix} p & q & r \\ u & v & w \\ x & y & z \end{bmatrix}$$

for real constants a through i. We require that $KA = I_3$ and $AK = I_3$. We choose the latter equation, $AK = I_3$, which will give us a linear system of nine equations in the unknowns p through z. We will have *Maple* do the work for us and find the formulas for p through z in terms of a through i.

```
> with(LinearAlgebra):
> A:= matrix([[a, b, c], [d, e, f], [g, h, i]]);
```

$$A := \begin{bmatrix} a & b & c \\ d & e & f \\ g & h & i \end{bmatrix}$$

```
> K:= matrix([[p, q, r], [u, v, w], [x, y, z]]);
```

$$K := \begin{bmatrix} p & q & r \\ u & v & w \\ x & y & z \end{bmatrix}$$

```
> P:= evalm(A&*K);
```

$$P := \begin{bmatrix} ap + bu + cx & aq + bv + cy & ar + bw + cz \\ dp + eu + fx & dq + ev + fy & gp + hu + ix \\ gp + hu + ix & gq + hv + iy & gr + hw + iz \end{bmatrix}$$

```
> Eqns:= {seq(seq(evalm(P-IdentityMatrix(3))[k,j] = 0,k=1..3), j=1..3)};
```

$$Eqns := \{aq + bv + cy = 0, ar + bw + cz = 0, dp + eu + fx = 0,$$
$$dr + ew + fz = 0, gp + hu + ix = 0, gq + hv + iy = 0,$$
$$ap + bu + cx - 1 = 0, dq + ev + fy - 1 = 0, gr + hw + iz - 1 = 0\}$$

> solns:= solve(Eqns,{p,q,r,u,v,w,x,y,z});

$$solns := \left\{ p = \frac{ei - hf}{cdh - cge - dib - fah + gfb + iae}, \right.$$

$$q = -\frac{bi - hc}{aei - ahf - dbi - egc + hdc + gbf}, r = \frac{-ec + bf}{aei - ahf - dbi - egc + hdc + gbf},$$

$$u = -\frac{di - gf}{aei - ahf - dbi - egc + hdc + gbf}, v = \frac{ai - gc}{aei - ahf - dbi - egc + hdc + gbf},$$

$$w = -\frac{af - dc}{aei - ahf - dbi - egc + hdc + gbf}, x = \frac{dh - ge}{aei - ahf - dbi - egc + hdc + gbf},$$

$$\left. y = -\frac{ah - gb}{aei - ahf - dbi - egc + hdc + gbf}, z = \frac{ae - db}{aei - ahf - dbi - egc + hdc + gbf} \right\}$$

This tells us that the multiplicative inverse K, to A, has the formula

$$K = \frac{1}{aei - ahf - dbi - egc + hdc + gbf} \begin{bmatrix} ei - hf & -bi + hc & -ec + bf \\ -di + gf & ai - gc & -af + dc \\ dh - ge & -ah + gb & ae - db \end{bmatrix} \tag{5.3}$$

as long as $aei - ahf - dbi - egc + hdc + gbf \neq 0$. Similar to the 2×2 case, the term required to be nonzero is a determinant for 3×3 matrices. Once again, we denote $K = A^{-1}$.

> for m from 1 to 3 do
    for n from 1 to 3 do
        K[m,n]:= eval(K[m,n],solns);
    end do:
  end do:

> K[3,1];

$$\frac{dh - ge}{aei - ahf - dbi - egc + hdc + gbf}$$

> simplify(evalm(A&*K));

$$\begin{bmatrix} 1 & 0 & 0 \\ 0 & 1 & 0 \\ 0 & 0 & 1 \end{bmatrix}$$

> simplify(evalm(K&*A));

$$\begin{bmatrix} 1 & 0 & 0 \\ 0 & 1 & 0 \\ 0 & 0 & 1 \end{bmatrix}$$

> det(A);

$$aei - ahf - dbi - egc + hdc + gbf$$

In the formula for A^{-1} given in equation (5.3), we see that each entry looks like the determinant of a 2×2 matrix divided by the determinant of A. In fact,

$$A_{1,1}^{-1} = \frac{ei - hf}{\det(A)} = \frac{1}{\det(A)} \det \left(\begin{bmatrix} e & f \\ h & i \end{bmatrix} \right)$$

$$= \frac{1}{\det(A)} \det(\text{matrix } A \text{ with the first row and first column removed}).$$

Let us check one more entry of A^{-1}, but off the main diagonal. We have

$$A_{3,2}^{-1} = \frac{-ah + gb}{\det(A)} = -\frac{1}{\det(A)} \det \left(\begin{bmatrix} a & b \\ g & h \end{bmatrix} \right)$$

$$= \frac{(-1)^{3+2}}{\det(A)} \det(\text{matrix } A \text{ with the second row, third column removed}).$$

Furthermore, if we look at the term we defined to be the determinant of the 3×3 matrix, notice that we can factor it as

$$aei - ahf - dbi - egc + hdc + gbf = a(ei - fh) - b(di - fg) + c(dh - eg),$$

or

$$aei - ahf - dbi - egc + hdc + gbf = a(ei - fh) - d(bi - ch) + g(bf - ce).$$

The terms in parentheses are minors of the original matrix by using the first row and column, respectively. We leave it as an exercise to rewrite the LHS in terms of minors using second and third rows and columns. It can be done!

We can now define the determinant of a square $n \times n$ matrix. We will define the determinant as the expansion along the first row of the matrix using the determinants (minors) of one size smaller matrices based on it being true for $n \times n$ matrices of sizes $n = 1, 2, 3$.

Definition 5.1.4. The *determinant* of an $n \times n$ matrix A, denoted $\det(A)$, is the scalar given by the formula

$$\det(A) = \sum_{j=1}^{n} (-1)^{1+j} A_{1,j} M_{1,j} \tag{5.4}$$

Definition 5.1.5. The *cofactor* $C_{i,j}$ of $A_{i,j}$ is defined to be $C_{i,j} = (-1)^{i+j} M_{i,j}$.

Using the two minors calculated previously, notice that $C_{1,1} = (-1)^{1+1} \cdot (-8) = -8$, and $C_{2,3} = (-1)^{2+3} \cdot 3 = -3$.

Definition 5.1.6. The *cofactor matrix* C of the same square size as A consists of the cofactors of A, that is,

$$C_{i,j} = (-1)^{i+j} M_{i,j} \tag{5.5}$$

Closely related to the cofactor matrix is the *adjoint matrix*, which we define next.

Definition 5.1.7. The *adjoint* of a square matrix A, denoted $\mathrm{adj}(A)$, is the transpose of its cofactor matrix C:

$$\mathrm{adj}(A) = C^T \tag{5.6}$$

Example 5.1.4. We have already computed two entries of the cofactor matrix of A given in (5.2), we leave it as an exercise for you to find the rest. The following is the cofactor matrix of A:

$$C = \begin{bmatrix} -8 & 4 & 12 \\ 4 & -1 & -3 \\ -12 & 7 & 13 \end{bmatrix}$$

The adjoint matrix is simple to calculate once we have the cofactor matrix. With the definitions of the cofactor and adjoint matrices, we now have enough information to define the *inverse* of a square matrix.

Definition 5.1.8. Given a matrix $A \in \mathbb{R}^{n \times n}$, the *inverse matrix*, denoted A^{-1}, is defined by the following formula, where C is the cofactor matrix of A:

$$A^{-1} = \frac{1}{\det(A)} C^T = \frac{1}{\det(A)} \mathrm{adj}(A) \tag{5.7}$$

If $\det(A) = 0$, then the matrix A has no inverse, and is said to be *singular*, else we call the matrix A *nonsingular*, or *invertible*.

The formula given above is referred to as the *adjoint formula*. The command "adjoint(A)" in *Maple* will compute the adjoint matrix of A.

We now introduce some properties of the matrix inverse in the following table. Notice that there is no formula to relate $(A + B)^{-1}$ to the sum of inverses.

Matrix Inverse Properties	
$(A^{-1})^{-1} = A$	$(AB)^{-1} = B^{-1}A^{-1}$
$(cA)^{-1} = \frac{1}{c}A^{-1}$	$(A^T)^{-1} = (A^{-1})^T$
$AA^{-1} = I_n = A^{-1}A$	

Example 5.1.5. Putting all of the previous definitions together, we can continue our work on the inverse of the matrix A from (5.2). The adjoint of A is given by:

$$\text{adj}(A) = \begin{bmatrix} -8 & 4 & -12 \\ 4 & -1 & 7 \\ 12 & -3 & 13 \end{bmatrix}$$

To compute A^{-1} using the adjoint formula (5.7), we need one last piece of information, $\det(A)$. We have previously defined this scalar for arbitrary 3×3 matrices, and applying the formula to the matrix A yields $\det(A) = 8$. Now we have enough information to compute the inverse of A:

$$A^{-1} = \frac{1}{8} \begin{bmatrix} -8 & 4 & -12 \\ 4 & -1 & 7 \\ 12 & -3 & 13 \end{bmatrix} = \begin{bmatrix} -1 & \frac{1}{2} & -\frac{3}{2} \\ \frac{1}{2} & -\frac{1}{8} & \frac{7}{8} \\ \frac{3}{2} & -\frac{3}{8} & \frac{13}{8} \end{bmatrix}$$

We can verify through matrix multiplication that we really have found the inverse to A. Remember that we should end up with $AA^{-1} = I_3 = A^{-1}A$.

$$\begin{bmatrix} 1 & -2 & 2 \\ 4 & 5 & 1 \\ 0 & 3 & -1 \end{bmatrix} \begin{bmatrix} -1 & \frac{1}{2} & -\frac{3}{2} \\ \frac{1}{2} & -\frac{1}{8} & \frac{7}{8} \\ \frac{3}{2} & -\frac{3}{8} & \frac{13}{8} \end{bmatrix} = \begin{bmatrix} 1 & 0 & 0 \\ 0 & 1 & 0 \\ 0 & 0 & 1 \end{bmatrix}$$

and

$$\begin{bmatrix} -1 & \frac{1}{2} & -\frac{3}{2} \\ \frac{1}{2} & -\frac{1}{8} & \frac{7}{8} \\ \frac{3}{2} & -\frac{3}{8} & \frac{13}{8} \end{bmatrix} \begin{bmatrix} 1 & -2 & 2 \\ 4 & 5 & 1 \\ 0 & 3 & -1 \end{bmatrix} = \begin{bmatrix} 1 & 0 & 0 \\ 0 & 1 & 0 \\ 0 & 0 & 1 \end{bmatrix}$$

Example 5.1.6. Now, let us use the inverse of a square matrix to solve a square linear system. We wish to solve the system

$$\begin{aligned} 5x - 7y + 2z - w &= -3 \\ 9x + 3y - 5z + 8w &= -11 \\ -6x + y - z + 7w &= 0 \\ x - 4y - 3z + 5w &= 6 \end{aligned} \tag{5.8}$$

for the unique solution $X = A^{-1}B$.

> System:= {5*x - 7*y + 2*z - w = -3, 9*x + 3*y - 5*z + 8*w = - 11, -6*x + y - z + 7*w = 0, x - 4*y - 3*z + 5*w = 6}:

> solve(System, {x,y,z,w});

$$\left\{ w = -\frac{1181}{837}, x = -\frac{1106}{837}, y = -\frac{340}{279}, z = -\frac{2651}{837} \right\}$$

> A:= matrix([[5,-7,2,-1], [9,3,-5,8], [-6,1,-1,7], [1,-4,-3,5]]);

$$A := \begin{bmatrix} 5 & -7 & 2 & -1 \\ 9 & 3 & -5 & 8 \\ -6 & 1 & -1 & 7 \\ 1 & -4 & -3 & 5 \end{bmatrix}$$

> Ainv:= inverse(A);

$$Ainv := \begin{bmatrix} \dfrac{157}{2511} & \dfrac{19}{279} & -\dfrac{58}{2511} & -\dfrac{161}{2511} \\ -\dfrac{37}{837} & \dfrac{5}{93} & \dfrac{19}{837} & -\dfrac{106}{837} \\ \dfrac{592}{2511} & \dfrac{13}{279} & \dfrac{533}{2511} & -\dfrac{815}{2511} \\ \dfrac{235}{2511} & \dfrac{16}{279} & \dfrac{377}{2511} & -\dfrac{209}{2511} \end{bmatrix}$$

> B:= matrix([[-3], [-11], [0], [6]]);

$$B := \begin{bmatrix} -3 \\ -11 \\ 0 \\ 6 \end{bmatrix}$$

> evalm(Ainv&*B);

$$\begin{bmatrix} -\dfrac{1106}{837} \\ -\dfrac{340}{279} \\ -\dfrac{2651}{837} \\ -\dfrac{1181}{837} \end{bmatrix}$$

> rref(augment(A,B));

$$\begin{bmatrix} 1 & 0 & 0 & 0 & -\dfrac{1106}{837} \\ 0 & 1 & 0 & 0 & -\dfrac{340}{279} \\ 0 & 0 & 1 & 0 & -\dfrac{2651}{837} \\ 0 & 0 & 0 & 1 & -\dfrac{1181}{837} \end{bmatrix}$$

■

All three methods (*solve*, *inverse*, and *rref*) used above to solve this system agree on the final answer. One may wonder what the advantage would be to solving a system of equations using matrix inverses, as opposed to performing Gauss-Jordan elimination. Consider the situation in which you were asked to solve two linear systems of equations, which in matrix form can be written as $AX = B_1$ and $AX = B_2$, where the matrix A on the left side of each equation is the same. The solution to the two systems are given, respectively, by

$$X = A^{-1}B_1, \ X = A^{-1}B_2$$

So if we compute A^{-1} once, we can solve two separate linear systems by performing simple matrix multiplication.

Example 5.1.7. Consider the following system of equations:

$$\begin{aligned}
5x - 7y + 2z - w &= 3 \\
9x + 3y - 5z + 8w &= 11 \\
-6x + y - z + 7w &= 6 \\
x - 4y - 3z + 5w &= -1
\end{aligned} \tag{5.9}$$

Notice that the coefficients on the LHS of systems (5.8) and (5.9) are the same. Thus in matrix form $AX = B$, the matrix A will be the same.

> System2:= {5*x - 7*y + 2*z - w = 3, 9*x + 3*y - 5*z + 8*w = 11, -6*x + y - z + 7*w = 6, x - 4*y - 3*z + 5*w = -1}:
> solve(System2, {x,y,z,w});

$$\left\{ w = \frac{4760}{2511}, x = \frac{2165}{2511}, y = \frac{604}{837}, z = \frac{7076}{2511} \right\}$$

> Bhat:= matrix([[3], [11], [6], [-1]]);

$$Bhat := \begin{bmatrix} 3 \\ 11 \\ 6 \\ -1 \end{bmatrix}$$

> evalm(Ainv&*Bhat);

$$\begin{bmatrix} \frac{2165}{2511} \\ \frac{604}{837} \\ \frac{7076}{2511} \\ \frac{4760}{2511} \end{bmatrix}$$

> rref(augment(A,Bhat));

$$\begin{bmatrix} 1 & 0 & 0 & 0 & \frac{2165}{2511} \\ 0 & 1 & 0 & 0 & \frac{604}{837} \\ 0 & 0 & 1 & 0 & \frac{7076}{2511} \\ 0 & 0 & 0 & 1 & \frac{4760}{2511} \end{bmatrix}$$

■

One last interesting comment before we prove the matrix inverse properties tabled earlier in this section. In the case of the 3×3 matrix A as defined previous, the determinant can be found by augmenting to A its first two columns and then summing the three products down the diagonals from upper left to lower right followed by subtracting the three products up the three diagonals from lower left to upper right. Unfortunately, this method does not generalize to larger matrices.

> A:= matrix([[a, b, c], [d, e, f], [g, h, i]]):
> delcols(augment(A,A),6..6);

$$\begin{bmatrix} a & b & c & a & b \\ d & e & f & d & e \\ g & h & i & g & h \end{bmatrix}$$

> det(A);

$$aei - ahf - dbi - egc + hdc + gbf$$

Theorem 5.1.1. *Let A and B be two square $n \times n$ matrices. Then, we have the following useful and interesting matrix inverse properties.*

(a) The inverse matrix A^{-1} is unique, if it exists.

(b) $(AB)^{-1} = B^{-1}A^{-1}$ if both A^{-1} and B^{-1} exist.

(c) $(A^T)^{-1} = (A^{-1})^T$ if A^{-1} exists.

Proof. (a) Assume that A has two different inverses K and L, so $K - L \neq 0$. From Definition 5.1.1, we must have both $KA = AK = I_n$ and $LA = AL = I_n$. Then $KA = LA$ or $(K - L)A = 0$. If we multiply this last equation on the right by L, we get $(K - L)AL = 0L$, or $(K - L)I_n = 0$. This last equation is of course the same as $K - L = 0$, which is a contradiction to $K - L \neq 0$, and so our assumption is false, which forces $K = L$ and the inverse to A is unique.

(b) Let both A^{-1} and B^{-1} exist. Then we have

$$
\begin{aligned}
(AB)\left(B^{-1}A^{-1}\right) &= A\left(BB^{-1}\right)A^{-1} \\
&= A\,I_n\,A^{-1} \\
&= A\,A^{-1} \\
&= I_n
\end{aligned}
$$

Similarly, we also have

$$
\begin{aligned}
\left(B^{-1}A^{-1}\right)(AB) &= B\left(AA^{-1}\right)B^{-1} \\
&= B\,I_n\,B^{-1} \\
&= B^{-1}B \\
&= I_n
\end{aligned}
$$

So, by Definition 5.1.1 and part (a) of this theorem, $(AB)^{-1} = B^{-1}A^{-1}$.

(c) Let A^{-1} exist. Then using the property of the transpose that $(CD)^T = D^T C^T$, for any two matrices C and D, where CD exists, we have

$$
\begin{aligned}
\left[\left(A^{-1}\right)^T A^T\right]^T &= \left(A^T\right)^T \left[\left(A^{-1}\right)^T\right]^T \\
&= A\,A^{-1} \\
&= I_n
\end{aligned}
$$

Therefore, we have $(A^{-1})^T A^T = I_n$ by taking the transpose of the above result and using the facts that $\left(C^T\right)^T = C$ and $I_n^T = I_n$. In a very similar fashion, we can show that $A^T (A^{-1})^T = I_n$. So, by Definition 5.1.1 and part (a) of this theorem, $(A^T)^{-1} = (A^{-1})^T$. $\qquad\square$

Homework Problems

1. Compute the transpose of the following matrices.

(a) $\begin{bmatrix} 1 & 3 \\ -2 & -5 \end{bmatrix}$ (b) $\begin{bmatrix} -2 \\ 0 \\ 3 \end{bmatrix}$ (c) $\begin{bmatrix} 2 & 0 & -1 \\ 2 & -4 & 4 \end{bmatrix}$

(d) $\begin{bmatrix} 2 & -4 \\ -1 & 0 \\ 8 & 5 \end{bmatrix}$ (e) $\begin{bmatrix} -1 & 0 & 1 \\ 0 & 1 & 2 \\ 1 & 2 & 1 \end{bmatrix}$ (f) $\begin{bmatrix} -2 & 1 & 9 \\ -4 & 2 & 8 \\ -1 & 4 & -5 \end{bmatrix}$

$$
\text{(g)} \begin{bmatrix} 0 & -4 & -8 \\ 4 & 0 & -5 \\ 8 & 5 & 0 \end{bmatrix} \quad
\text{(h)} \begin{bmatrix} -1 & 0 & 1 & 2 \\ 7 & -1 & 2 & -4 \\ 1 & 4 & 1 & 2 \end{bmatrix} \quad
\text{(i)} \begin{bmatrix} 0 & 0 & 0 & 1 \\ 0 & 0 & 1 & 0 \\ 1 & 0 & 0 & 0 \\ 0 & 1 & 0 & 0 \end{bmatrix}
$$

2. A matrix is *symmetric* if $A^T = A$. Which of the matrices from problem 1 are symmetric?

3. A matrix is *antisymmetric* if $A^T = -A$. Which of the matrices from problem 1 are anti-symmetric?

4. Compute the determinants of the following matrices:

$$
\text{(a)} \begin{bmatrix} 2 & -3 \\ 8 & -4 \end{bmatrix} \quad
\text{(b)} \begin{bmatrix} 2 & -2 \\ 5 & 1 \end{bmatrix} \quad
\text{(c)} \begin{bmatrix} 2 & -8 \\ -4 & 16 \end{bmatrix}
$$

$$
\text{(d)} \begin{bmatrix} \dfrac{3}{5} & -\dfrac{1}{10} \\ 8 & -\dfrac{4}{3} \end{bmatrix} \quad
\text{(e)} \begin{bmatrix} 0 & -\dfrac{2}{3} \\ 5 & \dfrac{1}{10} \end{bmatrix} \quad
\text{(f)} \begin{bmatrix} 1 & 3 & 0 \\ -4 & 16 & 3 \\ 0 & -3 & -5 \end{bmatrix}
$$

$$
\text{(g)} \begin{bmatrix} 1 & -3 & 1 \\ 9 & -4 & 0 \\ -3 & 5 & 2 \end{bmatrix} \quad
\text{(h)} \begin{bmatrix} 2 & -3 & 8 \\ -4 & 0 & 1 \\ 5 & -2 & 4 \end{bmatrix} \quad
\text{(i)} \begin{bmatrix} 1 & -3 & -5 \\ 5 & 4 & 5 \\ -1 & 3 & 5 \end{bmatrix}
$$

5. Compute the cofactor matrix to each of the matrices from problem 4.

6. Compute the inverse matrix to each of the matrices from problem 4, using the cofactor matrices from problem 5.

7. Use your answers (if possible) to problem 6 to help solve the following systems:

(a) $2x - 3y = 6$
$8x - 4y = 4$

(b) $2x - 2y = 7$
$5x + y = 8$

(c) $2x - 3y = -1$
$8x - 4y = 3$

(d) $2x - 2y = 6$
$5x + y = -5$

(e) $x - 3y + z = 1$
$9x - 4y = 4$
$-3x + 5y + 2z = 1$

(f) $2x - 3y + 8z = 3$
$-4x + z = 5$
$5x - 2y + 4z = 6$

(g) $x - 3y + z = 8$
$$9x - 4y = -2$$
$$-3x + 5y + 2z = 3$$

(h) $x - 3y - 5z = 2$
$$5x + 4y + 5z = -1$$
$$-x + 3y + 5z = 0$$

8. Determine values of λ such that the following matrices are not invertible. The values of λ that make each of the following matrices singular are called *eigenvalues*. In general, eigenvalues are found by solving for λ the equation $\det (A - \lambda I_n) = 0$, for $A \in \mathbb{R}^{n \times n}$.

(a) $\begin{bmatrix} 3 - \lambda & 1 \\ -1 & 1 - \lambda \end{bmatrix}$ (b) $\begin{bmatrix} 3 - \lambda & 1 \\ 1 & 3 - \lambda \end{bmatrix}$ (c) $\begin{bmatrix} -\lambda & 3 & 4 \\ 4 & -4 - \lambda & -8 \\ 6 & -9 & -10 - \lambda \end{bmatrix}$

9. A matrix A is *diagonal* if $A_{i,j} = 0$ for $i \neq j$. Entries on the diagonal are not required to be nonzero, however, for this problem, assume that $A_{i,i} \neq 0$ for $1 \leq i \leq n$. Show that the inverse matrix to A is a diagonal matrix with entries $\frac{1}{A_{i,i}}$.

10. A matrix A is *upper triangular* if $A_{i,j} = 0$ for $i > j$, and is *lower triangular* of $A_{i,j} = 0$ for $i < j$. Is the inverse of a lower/upper triangular matrix D also a lower/upper triangular matrix?

11. Compute the inverses of the following matrices:

(a) $\begin{bmatrix} 3 & 0 & 0 \\ 0 & -4 & 0 \\ 0 & 0 & -1 \end{bmatrix}$ (b) $\begin{bmatrix} -2 & 0 & 0 \\ 0 & 6 & 0 \\ 0 & 0 & 9 \end{bmatrix}$ (c) $\begin{bmatrix} \frac{5}{7} & 0 & 0 \\ 0 & -\frac{1}{4} & 0 \\ 0 & 0 & \frac{2}{3} \end{bmatrix}$

12. A matrix A is *orthogonal* if its transpose is equal to its inverse, that is, $A^{-1} = A^T$. Explain why a symmetric or anti-symmetric or orthogonal matrix must be square.

13. Let A be a square matrix. Show that $A + A^T$ is symmetric while $A - A^T$ is antisymmetric.

14. Let A be a square matrix. Show that A can be written as the sum of a symmetric and an antisymmetric matrix.

15. Let A be any matrix. Show that both AA^T and $A^T A$ are symmetric matrices.

16. Explain why $(AB)^T = B^T A^T$.

17. Let n be any positive integer and A be any invertible square matrix. Show that $(A^n)^{-1} = (A^{-1})^n$.

18. Let E be an elementary matrix. Does E always have an inverse, and if so, is E^{-1} also an elementary matrix?

Maple Problems

1. Solve the following systems by first converting them to the form $AX = B$, and then computing $X = A^{-1}B$.

(a) $\begin{aligned} 5x + 5y &= 1 \\ 2x + 3y &= 7 \end{aligned}$
  (b) $\begin{aligned} 5x + 5y &= 7 \\ 2x + 3y &= 1 \end{aligned}$

(c) $\begin{aligned} -4x + 3y &= -1 \\ 7x - 9y &= -13 \end{aligned}$
  (d) $\begin{aligned} -4x + 3y &= -13 \\ 7x - 9y &= -1 \end{aligned}$

(e) $\begin{aligned} 5x + 2y + 3z &= 0 \\ 4x - 7y - z &= 4 \\ 3x + 2z &= -1 \end{aligned}$
  (f) $\begin{aligned} 5x + 2y + 3z &= 1 \\ 4x - 7y - z &= -4 \\ 3x + 2z &= 2 \end{aligned}$

(g) $\begin{aligned} 7x + 5y - 3z &= 1 \\ 2x + 6y + 8z &= -2 \\ -4x - 5y + 7z &= 1 \end{aligned}$
  (h) $\begin{aligned} 7x + 5y - 3z &= -1 \\ 2x + 6y + 8z &= 2 \\ -4x - 5y + 7z &= -1 \end{aligned}$

(i) $\begin{aligned} 2w + 3x + 7y - 3z &= 1 \\ 8w - 2x + 5y + 8z &= -2 \\ -3w - 2x - y + 7z &= 1 \\ 13w - 4x - 5y + z &= 1 \end{aligned}$
  (j) $\begin{aligned} 2w + 3x + 7y - 3z &= 2 \\ 8w - 2x + 5y + 8z &= -4 \\ -4x - 5y + 7z &= 2 \\ 13w - 4x - 5y + z &= 2 \end{aligned}$

2. Find the values of λ so that the following matrices are singular:

(a) $\begin{bmatrix} 3 - \lambda & 0 & 0 \\ 0 & -4 - \lambda & 0 \\ 0 & 0 & -1 - \lambda \end{bmatrix}$
  (b) $\begin{bmatrix} 3 - \lambda & 2 & -5 \\ 0 & -4 - \lambda & 1 \\ 0 & 0 & -1 - \lambda \end{bmatrix}$

(c) $\begin{bmatrix} 3-\lambda & 1 & -1 \\ 1 & 5-\lambda & 0 \\ 0 & -14 & -1-\lambda \end{bmatrix}$
(d) $\begin{bmatrix} 3-\lambda & 1 & 0 \\ 1 & 5-\lambda & -14 \\ -1 & 0 & -1-\lambda \end{bmatrix}$

(e) $\begin{bmatrix} 1-\lambda & 1 & 4 & -7 \\ 0 & -2-\lambda & 3 & 1 \\ 0 & 0 & 1-\lambda & -2 \\ 0 & 0 & 0 & 2-\lambda \end{bmatrix}$
(f) $\begin{bmatrix} -\lambda & 1 & -1 & 1 \\ 1 & -\lambda & -1 & 1 \\ 1 & -1 & -\lambda & 1 \\ 1 & -1 & 1 & -\lambda \end{bmatrix}$

3. Compute the determinant of the matrix

$$A = \begin{bmatrix} 5 & 1 & 2 & -9 & 8 \\ 0 & 1 & 0 & 1 & 0 \\ 1 & -1 & 0 & 1 & 1 \\ 1 & 0 & 1 & 0 & 1 \\ 3 & 1 & -6 & -2 & 4 \end{bmatrix}$$

using expansions along the first row for all successive determinants of the minors. Check that your work is correct.

4. Compute the adjoint of the matrix A from problem 3.

5. Compute the inverse of the matrix A from problem 3.

6. For the two general 2×2 matrices $A = \begin{bmatrix} a & b \\ c & d \end{bmatrix}$ and $B = \begin{bmatrix} \alpha & \beta \\ \delta & \gamma \end{bmatrix}$, show that $\det(AB) = \det(A)\det(B)$.

7. Prove that $\det(AB) = \det(A)\det(B)$ for two arbitrary 3×3 matrices.

8. Using the generalization of the previous two problems, explain why $C_{AB} = C_A C_B$, where these three matrices are the cofactor matrices of the square $n \times n$ matrices A, B and AB. Do not use *Maple* for this problem.

9. Find the determinant and inverse for the matrix

$$A = \begin{bmatrix} 5-i & 2+7i & -i \\ 1+i & 3i & 1-i \\ 2 & -9+i & 4i \end{bmatrix}$$

verifying in as many ways as possible your results.

10. Using the matrix A from problem 9, solve in as many ways as possible the matrix equation $AX = B$, where

$$B = \begin{bmatrix} 2+3i \\ 6-i \\ -8+i \end{bmatrix}$$

11. Using the matrix A from problem 9, verify *Homework* problem 17 for $n = 5$.

5.2   Determinants by Expanding Along Any Row or Column

We spent quite a bit of time in the last section talking about determinants while trying to find inverses of square matrices. We now focus our attention solely on determinants. In this section, we will find a general method for finding the determinant of any square $n \times n$ matrix A based on the adjoint formula for its inverse A^{-1}.

Remember that the adjoint formula is given by (5.7), which we rewrite here for completeness:

$$A^{-1} = \frac{1}{\det(A)} C^T \tag{5.10}$$

Here C is the cofactor matrix of A, and C^T is the transpose matrix of C, which means the columns of C^T are actually the rows of C. Now, if we multiply the cofactor formula by A on the left, then we have

$$AA^{-1} = \frac{1}{\det(A)} AC^T$$

which means that

$$AC^T = \det(A)I_n$$

The $n \times n$ matrix $D = \det(A)I_n$ is simply the diagonal matrix whose diagonal entries are all the scalar value $\det(A)$:

$$D = \det(A)I_n = \begin{bmatrix} \det(A) & & \cdots & & 0 \\ & \det(A) & & \vdots \\ \vdots & & \ddots & \\ 0 & \cdots & & \det(A) \end{bmatrix} \tag{5.11}$$

This fact now gives us n different ways to compute the determinant of A as we shall now see.

First, we notice that D has diagonal entries $D_{i,i} = \det(A)$ for $1 \le i \le n$.

From above, we also know that $D = AC^T$. So for $i = 1$, we have

$$\det(A) = D_{1,1} = \sum_{j=1}^{n} A_{1,j} C_{j,1}^T$$

$$= \text{(the first row of } A\text{)(the first column of } C^T)$$
$$= \text{(the first row of } A\text{)(the first row of } C)$$
$$= A_{1,1} C_{1,1} + A_{1,2} C_{1,2} + \cdots + A_{1,n} C_{1,n}$$
$$= \sum_{j=1}^{n} A_{1,j} C_{1,j}$$
$$= \sum_{j=1}^{n} (-1)^{1+j} A_{1,j} M_{1,j}$$

Notice that the last expression is simply the sum of the products of the first row entries of A times their respective signed minors. This method of computing the determinant of A is called *expanding along its first row*. Since $\det(A) = D_{i,i}$ for any $1 \leq i \leq n$, we can expand along any row of A to find its determinant. If we expand along the ith row of A, very little changes:

$$\det(A) = \sum_{j=1}^{n} (-1)^{i+j} A_{i,j} M_{i,j} \qquad (5.12)$$

In order to efficiently use formula (5.12) to compute the determinant of A, we should find $\det(A)$ by expanding along the row which contains the largest number of zero entries. Remember that $D = \det(A)I_n$ has off diagonal entries $D_{i,j} = 0$ for all $i \neq j$ from 1 to n. Then, for example, with $i = 1$ and $j = 2$, we have

$$D_{1,2} = \sum_{j=1}^{n} A_{1,j} C_{j,2}^T$$

$$= \text{(the first row of } A\text{)(the second column of } C^T)$$
$$= \text{(the first row of } A\text{)(the second row of } C)$$
$$= A_{1,1} C_{2,1} + A_{1,2} C_{2,2} + \cdots + A_{1,n} C_{2,n}$$
$$= \sum_{j=1}^{n} A_{1,j} C_{2,j} = 0$$

The above formula tells us that the sum of the products of the first-row entries of A times the corresponding cofactors for the second row of A will result in a value of 0. That is,

$$\sum_{j=1}^{n} (-1)^{2+j} A_{1,j} \det(\text{matrix } A \text{ with second row and } j\text{th column removed}) = 0$$

This generalizes to any off diagonal location with

$$\sum_{j=1}^{n}(-1)^{k+j}A_{i,j} \det(\text{matrix } A \text{ with } k\text{th row and } j\text{th column removed}) = 0,$$

for $1 \leq i \neq k \leq n$. However, note that the RHS of the above equation is 0, not $\det(A)$. Therefore, we cannot derive the value of $\det(A)$ from it.

In a similar fashion, we can start with the adjoint formula and multiply both sides by A on the right instead of the left. This gives

$$A^{-1}A = \frac{1}{\det(A)} C^T A$$

or

$$C^T A = \det(A)I_n$$

We once again define $D = \det(A)I_n$ as before. Then for $i = 1$, we have

$$\det(A) = D_{1,1} = \sum_{j=1}^{n} C_{1,j}^T A_{j,1}$$

$$= (\text{the first row of } C^T)(\text{the first column of } A)$$
$$= (\text{the first column of } C)(\text{the first column of } A)$$
$$= C_{1,1}A_{1,1} + C_{2,1}A_{2,1} + \cdots + C_{n,1}A_{n,1}$$
$$= \sum_{j=1}^{n} C_{j,1}A_{j,1}$$

This is the sum of the products of the first column entries of A times their respective cofactors. This method of computing the determinant of A is called *expanding along its first column*. Similarly to using rows, you can now expand along any column of A to find its determinant, not just the first column. If we expand along the jth column of A, then

$$\det(A) = \sum_{i=1}^{n}(-1)^{i+j}A_{i,j}M_{i,j} \tag{5.13}$$

You should compare this definition of $\det(A)$ to the definition given in (5.12), and notice that only the index has changed. Again, examining off diagonal entries we have

$$\sum_{i=1}^{n}(-1)^{i+k}A_{i,j} \det(\text{matrix } A \text{ with } i\text{th row and } k\text{th column removed}) = 0$$

for $1 \leq j \neq k \leq n$. This method for finding the determinant of a square $n \times n$ matrix A is only practical for small values of n or if A is a *sparse matrix* meaning that it has a very large number of zero entries. These two expansion methods also tell us that if the matrix A has a row or column of all zero entries, then its determinant is zero and A has no inverse. So now we have $2n$ ways of computing the determinant of an $n \times n$ matrix, none of which involve computing an inverse.

Example 5.2.1. Let us find the determinant of the matrix

$$A = \begin{bmatrix} -9 & 5 & 0 & 2 \\ 4 & -1 & 3 & 7 \\ 6 & -2 & 0 & -8 \\ 1 & 10 & -4 & 11 \end{bmatrix}$$

by expanding along the third column since it has two zeros in it, and also along the first row. First, we expand along the third column:

$$\det(A) = (-1)^{1+3} \cdot 0 \cdot \det \left(\begin{bmatrix} 4 & -1 & 7 \\ 6 & -2 & -8 \\ 1 & 10 & 11 \end{bmatrix} \right)$$

$$+ (-1)^{2+3} \cdot 3 \cdot \det \left(\begin{bmatrix} -9 & 5 & 2 \\ 6 & -2 & -8 \\ 1 & 10 & 11 \end{bmatrix} \right)$$

$$+ (-1)^{3+3} \cdot 0 \cdot \det \left(\begin{bmatrix} -9 & 5 & 2 \\ 4 & -1 & 7 \\ 1 & 10 & 11 \end{bmatrix} \right)$$

$$+ (-1)^{4+3} \cdot (-4) \cdot \det \left(\begin{bmatrix} -9 & 5 & 2 \\ 4 & -1 & 7 \\ 6 & -2 & -8 \end{bmatrix} \right)$$

Notice that two of the terms are zero, hence we have

$$\det(A) = (-1)^{2+3} \cdot 3 \cdot \det \left(\begin{bmatrix} -9 & 5 & 2 \\ 6 & -2 & -8 \\ 1 & 10 & 11 \end{bmatrix} \right)$$

$$+ (-1)^{4+3} \cdot (-4) \cdot \det \left(\begin{bmatrix} -9 & 5 & 2 \\ 4 & -1 & 7 \\ 6 & -2 & -8 \end{bmatrix} \right)$$

In a similar fashion, we compute the determinant by expanding across the first row:

$$\det(A) = (-1)^{1+1} \cdot (-9) \cdot \det \left(\begin{bmatrix} -1 & 3 & 7 \\ -2 & 0 & -8 \\ 10 & -4 & 11 \end{bmatrix} \right)$$

$$+ (-1)^{1+2} \cdot 5 \cdot \det \left(\begin{bmatrix} 4 & 3 & 7 \\ 6 & 0 & -8 \\ 1 & -4 & 11 \end{bmatrix} \right)$$

$$+ (-1)^{1+4} \cdot (2) \cdot \det \left(\begin{bmatrix} 4 & -1 & 3 \\ 6 & -2 & 0 \\ 1 & 10 & -4 \end{bmatrix} \right)$$

where we did not bother to include the term corresponding to the 0 entry in row 1 column 3. In both expansions, we still have to compute determinants of 3×3 matrices to find the determinant of A. Instead of finishing this process by hand, we will have *Maple* do it for us. The command *minor* from the *linalg* package will help us in these two expansions.

> with(linalg):
> A:= matrix([[-9, 5, 0, 2], [4, -1, 3, 7], [6, -2, 0, -8], [1, 10, -4, 11]]):
> minor(A,2,3);

$$\begin{bmatrix} -9 & 5 & 2 \\ 6 & -2 & -8 \\ 1 & 10 & 11 \end{bmatrix}$$

In the following *Maple* commands, we will compute the determinant of A by expansion along the first row and then also the third column, using the *det* and *minor* commands.

> Expand_column3:= add((-1)^(i+3)*A[i,3]*det(minor(A,i,3)), i=1..4);

$$Expand\_column3 := 2976$$

> Expand_row1:= add((-1)^(1+j)*A[1,j]*det(minor(A,1,j)), j=1..4);

$$Expand\_row1 := 2976$$

> det(A);

$$2976$$

> Expand_row2androw3:= add((-1)^(3+j)*A[2,j]*det(minor(A,3,j)), j=1..4);

$$Expand\_row2androw3 := 0$$

> Expand_column1andcolumn3:= add((-1)^(i + 3)*A[i,1]*det(minor(A, i, 3)), i = 1..4);

$$Expand\_column1andcolumn3 := 0$$

Example 5.2.2. Let us use the expansion method to check our formula for the determinant of a 3×3 matrix. Let $A = \begin{bmatrix} a & b & c \\ d & e & f \\ g & h & k \end{bmatrix}$ for constants a through k.

We will find the 3×3 determinant formula by expanding along the first column of A and then also along the third row, verifying that they give the same result.

> A:= matrix([[a, b, c], [d, e, f], [g, h, k]]);

$$A := \begin{bmatrix} a & b & c \\ d & e & f \\ g & h & k \end{bmatrix}$$

> Expand_column1:= add((-1)^(i+1)*A[i,1]*det(minor(A,i,1)), i=1..3);

$$Expand\_column1 := a(ek - fh) - d(bk - ch) + g(bf - ce)$$

> simplify(%);

$$aek - afh - dbk + dch + gbf - gce$$

> det(A);

$$aek - afh - dbk + dch + gbf - gce$$

> Expand_row3:= add((-1)^(3+j)*A[3,j]*det(minor(A,3,j)), j=1..3);

$$Expand\_row3 := g(bf - ce) - h(af - cd) + k(ae - bd)$$

> simplify(%);

$$aek - afh - dbk + dch + gbf - gce$$

It is important to remember that taking determinants requires many computations. To compute the determinant of a 3×3 matrix, three 2×2 matrix determinants must be found. Similarly, for a 4×4 matrix, four 3×3 determinants must be found, but for each of the four 3×3 matrices, three determinants 2×2 must be computed. This would entail computing 12 determinants total for a 4×4 matrix, assuming that there were no zero-valued entries. Clearly, if there are any zero entries in the matrix, one should attempt to compute the determinant along a row or column in which the zero entry resides.

We end this section with a very important theorem relating determinants to the number of solutions to a square linear system.

Theorem 5.2.1. *Given the system $AX = B$, where $A \in R^{n \times n}$, if $\det(A) \neq 0$, then there exists a unique solution, $X = A^{-1}B$, to the system. As a consequence, if the system has either no solution or an infinite number of solutions, then $\det(A) = 0$.*

Proof. Clearly, if $\det(A) \neq 0$, then an inverse exists due to the adjoint formula

$$A^{-1} = \frac{1}{\det(A)} C^T$$

Therefore, one can perform the multiplication $A^{-1}B$, which is the solution to the system. We will prove a more general form of this theorem in Section 5.5 □

Example 5.2.3. Consider the following two systems:

(a) $2x - 4y = 4,$ (b) $2x - 4y = 4$

     $-4x + 8y = -8$ $-4y + 8y = -5$

Notice that system (a) has an infinite number of solutions, since any solution can be expressed in one of the following two forms, depending on which variable you wish to solve for

$$\begin{bmatrix} x \\ y \end{bmatrix} = \begin{bmatrix} 2y + 2 \\ y \end{bmatrix}, \quad \begin{bmatrix} x \\ y \end{bmatrix} = \begin{bmatrix} x \\ \frac{1}{2}x - 1 \end{bmatrix}$$

On the other hand, system (b) has no solution at all, even though the coefficient matrix is the same as that from system (a). The difference is that in system (a), the second equation is simply a multiple of the first, which is not the same for system (b).

Finally, we give a table of four important properties of determinants for square $n \times n$ matrices, and after the table, we show that the first property is true for arbitrary 2×2 matrices with the help *Maple*.

Matrix Determinant Properties	
$\det(AB) = \det(A)\det(B)$	$\det(A^{-1}) = \dfrac{1}{\det(A)}$
$\det(A^T) = \det(A)$	$\det(cA) = c^n \ \det(A)$

If A has a row or column of all zeroes, then $\det(A) = 0$

If two rows (or columns) of A are identical, then $\det(A) = 0$.

> A:= matrix(2,2,[a,b,c,d]);

$$A := \begin{bmatrix} a & b \\ c & d \end{bmatrix}$$

> B:= matrix(2,2,[e,f,g,h]);

$$B := \begin{bmatrix} e & f \\ g & h \end{bmatrix}$$

> det(A&*B);

$$aedh + bgcf - afdg - bhce$$

> expand(det(A)*det(B));

$$aedh + bgcf - afdg - bhce$$

Homework Problems

1. Compute the determinants of the following matrices by expanding along the first row.

(a) $\begin{bmatrix} 1 & -1 & 1 \\ -1 & -1 & 0 \\ 1 & 0 & 0 \end{bmatrix}$
(b) $\begin{bmatrix} 2 & -2 & 2 \\ -2 & -2 & 0 \\ 2 & 0 & 0 \end{bmatrix}$
(c) $\begin{bmatrix} 1 & -1 & 1 & 0 \\ 0 & 1 & -1 & 1 \\ 1 & -1 & -1 & 0 \\ -1 & 1 & 0 & 1 \end{bmatrix}$

(d) $\begin{bmatrix} 3 & -2 & 2 & 1 \\ 1 & -1 & 6 & 2 \\ 2 & -1 & 0 & 0 \\ -2 & 1 & 4 & 1 \end{bmatrix}$
(e) $\begin{bmatrix} 1 & 2 & 3 & 4 \\ 5 & 6 & 7 & 8 \\ 9 & 10 & 11 & 12 \\ 13 & 14 & 15 & 16 \end{bmatrix}$
(f) $\begin{bmatrix} 3 & 6 & -1 & 3 \\ 0 & -1 & 6 & 7 \\ 0 & 0 & 4 & 8 \\ 0 & 0 & 0 & 1 \end{bmatrix}$

2. Compute the determinants of the matrices from problem 1 by expanding along the second column.

3. For each of the matrices in problem 1, which row or column would be the best choice to expand upon in computing the determinant?

4. Compute the determinants of the matrices from problem 1 using the row or column that you found in problem 3.

5. Which of the matrices from problem 1 are singular?

6. Prove that the determinant of any upper triangular matrix U or lower triangular matrix L is simply the product of the diagonal elements: If

$$U_{i,j} = 0 \text{ for } 1 \leq j < i < n \text{ and } L_{i,j} = 0 \text{ for } 1 \leq i < j < n$$

then

$$\det(U) = \prod_{i=1}^{n} U_{i,i} \text{ and } \det(L) = \prod_{i=1}^{n} L_{i,i}$$

7. Suppose that a matrix A can be written as $A = LU$, where L is lower triangular and U is upper triangular, specifically given

$$A = \begin{bmatrix} 1 & 0 & 0 \\ g & 1 & 0 \\ h & i & 1 \end{bmatrix} \begin{bmatrix} a & b & c \\ 0 & d & e \\ 0 & 0 & f \end{bmatrix}$$

What is $\det(A)$? This process of decomposing a matrix into the product of an upper and lower triangular matrix is known as LU *factorization*.

8. Find values of λ such that the following systems of equations have a non-trivial (nonzero) solution.

(a) $\begin{aligned} (6 - \lambda)x - 4y &= 0 \\ -2x + (4 - \lambda)y &= 0 \end{aligned}$ (b) $\begin{aligned} (-3 - \lambda)x + 5y &= 0 \\ 7x + (-1 - \lambda)y &= 0 \end{aligned}$

(c) $\begin{aligned} (12 - \lambda)x + y &= 0 \\ -6x + (5 - \lambda)y &= 0 \end{aligned}$ (d) $\begin{aligned} (2 - \lambda)x + 8y &= 0 \\ 6x + (4 - \lambda)y &= 0 \end{aligned}$

9. For each system from problem 8, find the corresponding nontrivial solutions for each value of λ found.

10. Let c be a scalar and A be a $k \times l$ matrix. Explain why $cA = D_c A$, where D_c is the $k \times k$ diagonal matrix with all c's on its diagonal.

11. Use your argument from problem 10 to show that $\det(cA) = c^n \det(A)$ if A is any $n \times n$ matrix and c is any scalar.

12. Show that $\det(A^{-1}) = \dfrac{1}{\det(A)}$.

13. Explain why $\det(A^T) = \det(A)$.

14. Let A be an $n \times n$ matrix with $\det(A) \neq 0$.

(a) Find a formula for $\det(C)$ in terms of n and $\det(A)$, where C is the cofactor matrix of A.

(b) Find a formula for C^{-1} in terms of A and n if C is A's cofactor matrix.

15. Compute the determinants of each of the three types of $n \times n$ elementary matrices.

16. Let $P(x_0, y_0)$ and $Q(x_1, y_1)$ be two distinct points of $\mathbb{R}^2$.

(a) Show that the line through these two points has the equation

$$\det\left(\begin{bmatrix} x_0 & y_0 & 1 \\ x_1 & y_1 & 1 \\ x & y & 1 \end{bmatrix}\right) = 0$$

(b) Use the formula in part (a) to find the equation of the line through the two points $P(-7, 4)$ and $Q(9, -5)$.

17. Let $P(x_0, y_0, z_0)$, $Q(x_1, y_1, z_1)$, and $R(x_2, y_2, z_2)$ be three noncollinear points of $\mathbb{R}^3$.

(a) Show that the plane through these three points has the equation

$$\det\left(\begin{bmatrix} x_0 & y_0 & z_0 & 1 \\ x_1 & y_1 & z_1 & 1 \\ x_2 & y_2 & z_2 & 1 \\ x & y & z & 1 \end{bmatrix}\right) = 0$$

(b) Use the formula in part (a) to find the equation of the plane through the three points $P(-7, 4, 2)$, $Q(9, -5, 8)$, and $R(6, 11, -3)$.

18. Let $P(x_0, y_0)$, $Q(x_1, y_1)$, and $R(x_2, y_2)$ be three noncollinear points of $\mathbb{R}^2$.

(a) Show that the circle through these three points has the equation

$$\det\left(\begin{bmatrix} x_0^2 + y_0^2 & x_0 & y_0 & 1 \\ x_1^2 + y_1^2 & x_1 & y_1 & 1 \\ x_2^2 + y_2^2 & x_2 & y_2 & 1 \\ x^2 + y^2 & x & y & 1 \end{bmatrix}\right) = 0$$

(b) Use the formula in part (a) to find the equation of the circle through the three points $P(-7, 4)$, $Q(9, -5)$, and $R(6, 11)$.

19. Can you revise problem 18 in order to find a determinant equation for a general conic section passing through a certain number of points in the xy-plane? If yes, then test your formula on a set of points. If no, then explain why this is impossible.

20. Can you revise problem 18 in order to find a determinant equation for a sphere passing through a certain number of points in space? If yes, then test your formula on a set of points. If no, then explain why this is impossible.

21. Verify that any 2×2 or 3×3 matrix A that has two identical rows (or columns) must have $\det(A) = 0$.

Maple Problems

1. Use the *minor* and *add* commands to compute the following determinants.

(a) $\begin{bmatrix} 1 & -1 & 1 & 2 \\ -1 & -1 & 0 & 2 \\ 1 & 0 & 0 & 5 \\ 1 & -3 & -1 & 4 \end{bmatrix}$

(b) $\begin{bmatrix} 1 & -1 & 1 & 3 \\ 9 & -1 & -4 & 9 \\ 9 & 6 & 3 & 5 \\ 12 & -1 & -1 & 0 \end{bmatrix}$

(c) $\begin{bmatrix} 7 & -13 & 8 & -5 \\ -12 & -4 & 17 & -3 \\ -12 & 8 & 5 & -7 \\ -7 & 1 & 4 & 9 \end{bmatrix}$

(d) $\begin{bmatrix} 3 & -2 & 2 & 1 & 4 \\ 1 & -1 & 6 & 2 & 8 \\ 2 & -1 & 0 & 0 & 1 \\ -2 & 1 & 4 & 1 & 4 \\ 6 & -2 & -5 & 2 & 1 \end{bmatrix}$

(e) $\begin{bmatrix} 3 & 0 & 2 & 7 & -3 \\ 1 & 5 & 2 & -2 & 2 \\ 2 & 8 & -1 & 3 & 0 \\ 0 & 5 & 4 & -1 & 4 \\ 6 & -2 & -5 & 2 & 0 \end{bmatrix}$

(f) $\begin{bmatrix} 3 & 0 & 6 & 0 & 1 \\ 0 & 7 & 2 & 7 & 2 \\ 0 & 3 & 1 & 1 & 0 \\ 5 & 10 & 0 & -3 & 4 \\ -6 & 0 & 0 & 3 & 6 \end{bmatrix}$

2. Use the *det* command to compute the determinants from problem 1.

3. Compute the cofactor matrix for each of the matrices in problem 1.

4. Compute the adjoint matrix for each of the matrices in problem 1.

5. Compute the inverse matrix of each invertible matrix from problem 1.

6. Find values of λ such that the following systems of equations have a non-trivial (nonzero) solution.

(a) $\quad (2 - \lambda)x + 4y = 0$
$\qquad -2x + (-4 - \lambda)y = 0$

(b) $\quad (1 - \lambda)x + 4y = 0$
$\qquad -3x + (-7 - \lambda)y = 0$

(c) $\quad (2 - \lambda)x + y + z = 0$
$\qquad 2x + (3 - \lambda)y + 2z = 0$
$\qquad -3x - 3y + (-2 - \lambda)z = 0$

(d) $\quad (1 - \lambda)x - y + z = 0$
$\qquad 2x + (-1 - \lambda)y - 2z = 0$
$\qquad 2x - 2y + (2 - \lambda)z = 0$

7. For each system from problem 6, find the corresponding nontrivial solutions for each value of λ found.

8. Using some of the 5×5 matrices in problem 1, and with $c = -\frac{1}{3}$, verify that the four determinant properties given in this section are true.

9. In as many ways as possible, solve the system of linear equations with the augmented matrix

$$
\begin{bmatrix}
5 & -1 & 3 & 9 & 0 & -7 & -9 \\
-4 & 2 & 7 & 1 & 5 & 0 & 2 \\
0 & -8 & 2 & -5 & 1 & 4 & -5 \\
1 & 1 & 1 & 2 & 3 & 1 & 7 \\
-2 & 11 & -4 & 1 & 2 & 5 & 0 \\
4 & 0 & 1 & -13 & 1 & 9 & 1
\end{bmatrix}
$$

10. Do *Homework* problem 16 part (b), plotting the points and the line together.

11. Do *Homework* problem 17 part (b), plotting the points and the plane together.

12. Do *Homework* problem 18 part (b), plotting the points and the circle together.

13. Do *Homework* problem 19, plotting the points and the conic section together.

14. Do *Homework* problem 20 (if you can find a formula), plotting the points and the sphere together.

15. (a) For two arbitrary 3×3 matrices A and B, verify that $\det(AB) = \det(A)\det(B)$.

(b) Repeat this verification for two "randomly chosen" 4×4 matrices A and B with complex entries.

16. Verify that any 4×4 matrix A which has two identical rows (or columns) must have $\det(A) = 0$.

5.3 Determinants Found by Triangularizing Matrices

In the previous section, we looked at how to compute determinants by expanding along any row or column. This expansion method allows us to see that if a matrix has a row or column with only one nonzero entry, then computing its determinant is significantly easier. We begin this section by introducing some terminology that you may have noticed in some of *Homework* problems from Sections 5.1 and 5.2.

Definition 5.3.1. A square matrix U is called *upper triangular* if, below its main diagonal, all of its entries are zero (i.e. $U_{i,j} = 0$ for all $i > j$).

The following are all examples of upper triangular matrices:

$$
\begin{bmatrix} 1 & -2 & 2 \\ 0 & 3 & 4 \\ 0 & 0 & -2 \end{bmatrix}, \quad
\begin{bmatrix} -1 & 5 & 2 & 3 \\ 0 & 1 & -6 & 7 \\ 0 & 0 & -2 & -2 \\ 0 & 0 & 0 & 1 \end{bmatrix}, \quad
\begin{bmatrix} 7 & -3 & 2 & 3 & 9 \\ 0 & 2 & 8 & 7 & -6 \\ 0 & 0 & -2 & -2 & 2 \\ 0 & 0 & 0 & 1 & 4 \\ 0 & 0 & 0 & 0 & -2 \end{bmatrix}
$$

Definition 5.3.2. A square matrix L is called *lower triangular* if, above its main diagonal, all of its entries are zero (i.e. $L_{i,j} = 0$ for all $i < j$).

The following are all examples of lower triangular matrices, which are simply the transposed matrices of the upper triangular examples:

$$
\begin{bmatrix} 1 & 0 & 0 \\ -2 & 3 & 0 \\ 2 & 4 & -2 \end{bmatrix}, \quad
\begin{bmatrix} -1 & 0 & 0 & 0 \\ 5 & 1 & 0 & 0 \\ 2 & -6 & -2 & 0 \\ 3 & 7 & -2 & 1 \end{bmatrix}, \quad
\begin{bmatrix} 7 & 0 & 0 & 0 & 0 \\ -3 & 2 & 0 & 0 & 0 \\ 2 & 8 & -2 & 0 & 0 \\ 3 & 7 & -2 & 1 & 0 \\ 9 & -6 & 2 & 4 & -2 \end{bmatrix}
$$

Definition 5.3.3. A square matrix T is called *triangular* if it is either upper or lower triangular.

If we expand along the first column of an upper triangular matrix U, and along first columns of the successive minors, then its determinant is the product of its main diagonal entries. This was left for you to prove in *Homework* problem 8 of the Section 5.2. If we expand along the first row of a lower triangular matrix L and along first rows of the successive minors, then its determinant is also the product of its main diagonal entries. The result of these two statements is the following theorem.

Theorem 5.3.1. *If T is an $n \times n$ triangular matrix, then its determinant is the product of its main diagonal entries or*

$$\det(T) = \prod_{i=1}^{n} T_{i,i} \tag{5.14}$$

Let us have *Maple* check this for us in the general 3×3 case, as well as in a 6×6 case with numerical entries.

```
> with(linalg):
> U:= matrix([[a,b,c], [0,d,e], [0,0,f]]);
```

$$U := \begin{bmatrix} a & b & c \\ 0 & d & e \\ 0 & 0 & f \end{bmatrix}$$

```
> Expand_column1:= add((-1)^(i+1)*U[i,1]*det(minor(U,i,1)), i=1..3);
```

$$Expand\_column1 := adf$$

```
> L:= matrix([[a,0,0], [b,c,0], [d,e,f]]);
```

$$L := \begin{bmatrix} a & 0 & 0 \\ b & c & 0 \\ d & e & f \end{bmatrix}$$

```
> Expand_row1:= add((-1)^(1+j)*L[1,j]*det(minor(L,1,j)), j=1..3);
```

$$Expand\_row1 := acf$$

```
> T:= matrix([[-7, 2, 1, -5, 3, 8], [0, 4, -1, 6, 2, 1], [0, 0, 9, -1, 7, 2], [0, 0, 0, -4, 10, 1], [0, 0, 0, 0, 6, -7], [0, 0, 0, 0, 0, 11]]);
```

$$T := \begin{bmatrix} -7 & 2 & 1 & -5 & 3 & 8 \\ 0 & 4 & -1 & 6 & 2 & 1 \\ 0 & 0 & 9 & -1 & 7 & 2 \\ 0 & 0 & 0 & -4 & 10 & 1 \\ 0 & 0 & 0 & 0 & 6 & -7 \\ 0 & 0 & 0 & 0 & 0 & 11 \end{bmatrix}$$

> mul(T[i,i], i=1..6);

$$66528$$

> det(T);

$$66528$$

If we can quickly transform a general square matrix A into a new triangular matrix T without changing the value of its determinant, then we would have an easy method of computing its determinant. The way to change a general square matrix into a triangular matrix is through the three elementary row operations and their corresponding matrix multiplication counterparts. Recall that these three row operations are (1) swap two rows, (2) multiply a row by a nonzero number, and lastly (3) multiply a row by a nonzero number and add it to another row. Let us see how these three row operations affect the determinant of a square matrix by examining some examples.

Example 5.3.1. We first focus on the swapping of rows elementary operations.

> A:= matrix([[a,b,c], [d,e,f], [g,h,i]]);

$$A := \begin{bmatrix} a & b & c \\ d & e & f \\ g & h & i \end{bmatrix}$$

> det(A);

$$aei - afh - dbi + dch + gbf - gce$$

> Swap_1with3:= swaprow(A,1,3);

$$Swap\_1with3 := \begin{bmatrix} g & h & i \\ d & e & f \\ a & b & c \end{bmatrix}$$

> det(Swap_1with3);

$$gce - gbf - dch + dbi + afh - aei$$

> Swap_2with3:= swaprow(A,2,3);

$$Swap\_2with3 := \begin{bmatrix} a & b & c \\ g & h & i \\ d & e & f \end{bmatrix}$$

> det(Swap_2with3);

$$gce - gbf - dch + dbi + afh - aei$$

The swap row operation has changed the sign of the determinant of A. If we do an even number of row swaps on A, then its determinant is unchanged, while if we do an odd number, then its determinant changes sign.

Example 5.3.2. Next up is the operation of multiplying a row of the matrix A by a scalar k.

> Mul_row2byk:= mulrow(A,2,k);

$$Mul\_row2byk := \begin{bmatrix} a & b & c \\ kd & ke & kf \\ g & h & i \end{bmatrix}$$

> det(Mul_row2byk);

$$akei - akfh - kdbi + kdch + gbkf - gcke$$

> factor(%);

$$k(aei - afh - dbi + dch + gbf - gce)$$

From this example, we see that if we multiply a row of A by a nonzero number k, then the determinant of the new matrix is k times the determinant of A. In general, the easiest way to see this, is to expand along the row that was multiplied by k, and then notice that a common factor of k can be pulled out of every term.

Example 5.3.3. Now, we focus on the last elementary row operation.

> Add_ktimesrow1torow2:= addrow(A,1,2,k);

$$Add\_ktimesrow1torow2 := \begin{bmatrix} a & b & c \\ ak+d & bk+e & kc+f \\ g & h & i \end{bmatrix}$$

> det(Add_ktimesrow1torow2);

$$aei - afh - dbi + dch + gbf - gce$$

> % - det(A);

$$0$$

The final row operation, multiplying a row by a nonzero number k and adding it to another row, has no effect on the determinant of the original matrix A. If you expand in the new matrix along the row that has been added to, then you can see (using the expansion results of the Section 5.2) that the new matrix has the same determinant as the original matrix A. ∎

Determinant Properties with Elementary Matrices

Elementary Matrix of Type I - swapping of two rows	$\det(E_1 A) = -\det(A)$
Elementary Matrix of Type II - multiply a row by k	$\det(E_2 A) = k \det(A)$
Elementary Matrix of Type III - add k times one row to a second row	$\det(E_3 A) = \det(A)$

It is time to put this information together to find a way to compute general determinants for any square matrix A. Our approach will be to only use the third row operation of multiplying a row by a nonzero number k and adding it to another row, to change A into an upper triangular matrix U. Then both matrices A and U have the same determinant. We know that the determinant of U will be the product of its main diagonal entries. This gives a very practical way of computing determinants, even for large matrices A.

Example 5.3.4. Now it is time for an example using a 7×7 matrix A. The *gausselim* command from the *linalg* package will use the swap row operation and the row operation of multiplying a row by a nonzero number k and adding it to another row to change the matrix A into an upper triangular matrix U. The only drawback of using the *gausselim* command to upper triangularize A, is that the swap row operation will change the sign of the determinant, and so we might get a negative determinant when it is actually positive, or vice versa.

```
> A:= matrix([[-9,3,0,2,-4,5,1], [6,8,-7,-1,2,10,3], [2,5,-1,-15,6,2,-4], [11,7,5,4,-8,-1,6], [-7,1,1,2,5,-3,8], [1,-12,9,0,4,-2,13], [0,-9,7,2,6,3,-4]]);
```

$$
A := \begin{bmatrix}
-9 & 3 & 0 & 2 & -4 & 5 & 1 \\
6 & 8 & -7 & -1 & 2 & 10 & 3 \\
2 & 5 & -1 & -15 & 6 & 2 & -4 \\
11 & 7 & 5 & 4 & -8 & -1 & 6 \\
-7 & 1 & 1 & 2 & 5 & -3 & 8 \\
1 & -12 & 9 & 0 & 4 & -2 & 13 \\
0 & -9 & 7 & 2 & 6 & 3 & -4
\end{bmatrix}
$$

> U:= gausselim(A);

$$U := \begin{bmatrix} -9 & 3 & 0 & 2 & -4 & 5 & 1 \\ 0 & -9 & 7 & 2 & 6 & 3 & -4 \\ 0 & 0 & \frac{7}{9} & \frac{23}{9} & 6 & \frac{50}{3} & -\frac{7}{9} \\ 0 & 0 & 0 & \frac{17}{63} & \frac{473}{63} & -\frac{412}{63} & \frac{70}{9} \\ 0 & 0 & 0 & 0 & \frac{944}{17} & -\frac{940}{17} & \frac{1352}{17} \\ 0 & 0 & 0 & 0 & 0 & -\frac{81}{236} & -\frac{29265}{118} \\ 0 & 0 & 0 & 0 & 0 & 0 & \frac{1709254}{9} \end{bmatrix}$$

> mul(U[i,i], i=1..7);

$$-61533144$$

> det(U);

$$-61533144$$

> det(A);

$$61533144$$

The determinant of the matrix U above is negative, while the determinant of A is positive. This shows that the command *gausselim* performed an odd number of row swaps on the matrix A to get U.

■

In order to determine the correct sign of the determinant of the matrix $A \in \mathbb{R}^{n \times n}$, we need to keep track of the number of row swaps performed by *gausselim* on A to get U. The *gausselim* command has two extra options. The first option is "r", which returns the *rank* of the matrix A.

Definition 5.3.4. The *rank* of a matrix A is the number of nonzero rows of the matrix $rref(A)$.

The second option, "d", will return the determinant of A if the rank of A is n. The command *gausselim* computes the determinant by computing the product of the diagonal entries of U, and then correcting the sign if necessary from the number of row swaps used in finding U.

> U:= gausselim(A,'r','d');

$$
U := \begin{bmatrix}
-9 & 3 & 0 & 2 & -4 & 5 & 1 \\
0 & -9 & 7 & 2 & 6 & 3 & -4 \\
0 & 0 & \frac{7}{9} & \frac{23}{9} & 6 & \frac{50}{3} & -\frac{7}{9} \\
0 & 0 & 0 & \frac{17}{63} & \frac{473}{63} & -\frac{412}{63} & \frac{70}{9} \\
0 & 0 & 0 & 0 & \frac{944}{17} & -\frac{940}{17} & \frac{1352}{17} \\
0 & 0 & 0 & 0 & 0 & -\frac{81}{236} & -\frac{29265}{118} \\
0 & 0 & 0 & 0 & 0 & 0 & \frac{1709254}{9}
\end{bmatrix}
$$

> r;

$$7$$

> d;

$$61533144$$

Remember that we could also have *Maple* perform the *pivot* command, which allow us to carry out the process of the *gausselim* command ourselves. Combined with the *swaprow* command, we could then keep track of the number of times a row swap was performed to ensure that our determinant had the correct sign. As discussed previously, *pivot* allows us to pivot about any non-zero entry $A_{i,j}$, which means *pivot* will multiply the ith row by an appropriate value and add it to other rows to place zeros in the jth column, aside from the entry about which we are pivoting. For our current situation, we only want entries below the diagonal in the jth column to be zeroed out. One detail we did not include in our previous discussion of the *pivot* command, is that you can also specify a range of columns where you want your zeros to go: either above or below the ith row. This will allow us to construct a triangular matrix instead of a *rref* matrix.

> Pivot_A11:= pivot(A, 1, 1);

$$
Pivot\_A11 := \begin{bmatrix}
-9 & 3 & 0 & 2 & -4 & 5 & 1 \\
0 & 10 & -7 & \frac{1}{3} & -\frac{2}{3} & \frac{40}{3} & \frac{11}{3} \\
0 & \frac{17}{3} & -1 & -\frac{131}{9} & \frac{46}{9} & \frac{28}{9} & -\frac{34}{9} \\
0 & \frac{32}{3} & 5 & \frac{58}{9} & -\frac{116}{9} & \frac{46}{9} & \frac{65}{9} \\
0 & -\frac{4}{3} & 1 & \frac{4}{9} & \frac{73}{9} & -\frac{62}{9} & \frac{65}{9} \\
0 & -\frac{35}{3} & 9 & \frac{2}{9} & \frac{32}{9} & -\frac{13}{9} & \frac{118}{9} \\
0 & -9 & 7 & 2 & 6 & 3 & -4
\end{bmatrix}
$$

> Pivot_A22:= pivot(Pivot_A11, 2, 2, 3..7);

$$Pivot\_A22 := \begin{bmatrix} -9 & 3 & 0 & 2 & -4 & 5 & 1 \\ 0 & 10 & -7 & \frac{1}{3} & -\frac{2}{3} & \frac{40}{3} & \frac{11}{3} \\ 0 & 0 & \frac{89}{30} & -\frac{1327}{90} & \frac{247}{45} & -\frac{40}{9} & -\frac{527}{90} \\ 0 & 0 & \frac{187}{15} & \frac{274}{45} & -\frac{548}{45} & -\frac{82}{9} & \frac{149}{45} \\ 0 & 0 & \frac{1}{15} & \frac{22}{45} & \frac{361}{45} & -\frac{46}{9} & \frac{347}{45} \\ 0 & 0 & \frac{5}{6} & \frac{11}{18} & \frac{25}{9} & \frac{127}{9} & \frac{313}{18} \\ 0 & 0 & \frac{7}{10} & \frac{23}{10} & \frac{27}{5} & 15 & -\frac{7}{10} \end{bmatrix}$$

> Pivot_A33:= pivot(Pivot_A22, 3, 3, 4..7);

$$Pivot\_A33 := \begin{bmatrix} -9 & 3 & 0 & 2 & -4 & 5 & 1 \\ 0 & 10 & -7 & \frac{1}{3} & -\frac{2}{3} & \frac{40}{3} & \frac{11}{3} \\ 0 & 0 & \frac{89}{30} & -\frac{1327}{90} & \frac{247}{45} & -\frac{40}{9} & -\frac{527}{90} \\ 0 & 0 & 0 & \frac{18169}{267} & -\frac{9410}{267} & \frac{2554}{267} & \frac{7454}{267} \\ 0 & 0 & 0 & \frac{73}{89} & \frac{703}{89} & -\frac{446}{89} & \frac{698}{89} \\ 0 & 0 & 0 & \frac{423}{89} & \frac{110}{89} & \frac{1367}{89} & \frac{1694}{89} \\ 0 & 0 & 0 & \frac{1543}{267} & \frac{1096}{267} & \frac{4285}{267} & \frac{182}{267} \end{bmatrix}$$

> Pivot_A44:= pivot(Pivot_A33, 4, 4, 5..7);

$$Pivot\_A44 := \begin{bmatrix} -9 & 3 & 0 & 2 & -4 & 5 & 1 \\ 0 & 10 & -7 & \frac{1}{3} & -\frac{2}{3} & \frac{40}{3} & \frac{11}{3} \\ 0 & 0 & \frac{89}{30} & -\frac{1327}{90} & \frac{247}{45} & -\frac{40}{9} & -\frac{527}{90} \\ 0 & 0 & 0 & \frac{18169}{267} & -\frac{9410}{267} & \frac{2554}{267} & \frac{7454}{267} \\ 0 & 0 & 0 & 0 & \frac{151233}{18169} & -\frac{93144}{18169} & \frac{136380}{18169} \\ 0 & 0 & 0 & 0 & \frac{67180}{18169} & \frac{266929}{18169} & \frac{310396}{18169} \\ 0 & 0 & 0 & 0 & \frac{128962}{18169} & \frac{276829}{18169} & -\frac{30692}{18169} \end{bmatrix}$$

> Pivot_A55:= pivot(Pivot_A44, 5, 5, 6..7);

$$
Pivot\_A55 := \begin{bmatrix}
-9 & 3 & 0 & 2 & -4 & 5 & 1 \\
0 & 10 & -7 & \frac{1}{3} & -\frac{2}{3} & \frac{40}{3} & \frac{11}{3} \\
0 & 0 & \frac{89}{30} & -\frac{1327}{90} & \frac{247}{45} & -\frac{40}{9} & -\frac{527}{90} \\
0 & 0 & 0 & \frac{18169}{267} & -\frac{9410}{267} & \frac{2554}{267} & \frac{7454}{267} \\
0 & 0 & 0 & 0 & \frac{151233}{18169} & -\frac{93144}{18169} & \frac{136380}{18169} \\
0 & 0 & 0 & 0 & 0 & \frac{855411}{50411} & \frac{693124}{50411} \\
0 & 0 & 0 & 0 & 0 & \frac{988455}{50411} & -\frac{407828}{50411}
\end{bmatrix}
$$

> Pivot_A66:= pivot(Pivot_A55, 6, 6, 7..7);

$$
Pivot\_A66 := \begin{bmatrix}
-9 & 3 & 0 & 2 & -4 & 5 & 1 \\
0 & 10 & -7 & \frac{1}{3} & -\frac{2}{3} & \frac{40}{3} & \frac{11}{3} \\
0 & 0 & \frac{89}{30} & -\frac{1327}{90} & \frac{247}{45} & -\frac{40}{9} & -\frac{527}{90} \\
0 & 0 & 0 & \frac{18169}{267} & -\frac{9410}{267} & \frac{2554}{267} & \frac{7454}{267} \\
0 & 0 & 0 & 0 & \frac{151233}{18169} & -\frac{93144}{18169} & \frac{136380}{18169} \\
0 & 0 & 0 & 0 & 0 & \frac{855411}{50411} & \frac{693124}{50411} \\
0 & 0 & 0 & 0 & 0 & 0 & -\frac{6837016}{285137}
\end{bmatrix}
$$

> mul(Pivot_A66[i,i], i=1..7);

$$61533144$$

> det(A);

$$61533144$$

In doing these six partial pivots about the first six diagonal entries of A, we have found an upper triangular matrix, $Pivot\_A66$, whose determinant is the same as that of A, since we did not use *swaprow* at all. Notice that the upper triangular matrix, $Pivot\_A66$, which we have computed above, is different than the one given by using the *gausselim* command. The command *gausselim* is not doing what is expected by its description of using *pivot* and only using *swaprow* when necessary. However, the product of the diagonals for U and $Pivot\_A66$ still agree.

We now know that a square matrix A can be converted to an upper (or lower if need be) triangular matrix T through multiplication by elementary matrices $E_1, E_2, \ldots, E_r$, where E_1 is the first row operation applied to A and E_r, is the last. That is,

$$T = E_r E_{r-1} \cdots E_2 E_1 A$$

In a similar manner, instead of triangularizing A, we can use row operations through their elementary matrices $E_1, E_2, \ldots, E_k$ to convert A to its *rref* matrix rref(A), with

$$\text{rref}(A) = E_k E_{k-1} \cdots E_2 E_1 A$$

Since we have the above equation, we can take a determinant of both sides:

$$\det(\text{rref}(A)) = \det(E_k E_{k-1} \cdots E_2 E_1 A)$$

Using the table of matrix determinant properties in Section 5.2, specifically $\det(AB) = \det(A)\det(B)$, we know that the determinant of a product of square matrices is the product of the determinants of the matrices. We now have

$$\det(\text{rref}(A)) = \det(E_k)\det(E_{k-1}) \cdots \det(E_2)\det(E_1 A) \qquad (5.15)$$

Also, as we have just discussed, the determinant of a type III elementary matrix is 1. Similarly, the determinant of a type II elementary matrix E, for a nonzero scalar c, is $\det(E) = c \neq 0$. Finally, the determinant of a type I elementary matrix is -1. If we now combine this information with equation (5.15), we have that $\det(A) = 0$ exactly when $\det(\text{rref}(A)) = 0$, since $\det(E_j) \neq 0$ for all j. As a result of this, we can conclude with the following theorem, since for a square matrix A, rref(A) will be the identity matrix unless rref(A) contains at least one row of all zeroes (why?). A row of all zeros in rref(A) indicates that at least one row of A is a linear combination of the other rows of A (why?).

Theorem 5.3.2. *Given a square matrix A, $\det(A) = 0$ if and only if at least one row of A is a linear combination of the other rows of A. As well, since $\det(A^T) = \det(A)$, $\det(A) = 0$ if and only if at least one column of A is a linear combination of the other columns of A.*

Homework Problems

1. Compute the determinants of the following matrices.

(a) $\begin{bmatrix} 1 & -1 & 1 \\ 0 & -1 & 0 \\ 0 & 0 & 1 \end{bmatrix}$

(b) $\begin{bmatrix} 1 & 0 & 0 & 0 \\ 9 & 1 & 0 & 0 \\ 3 & 6 & 1 & 0 \\ 12 & -1 & -1 & 1 \end{bmatrix}$

(c) $\begin{bmatrix} 1 & -13 & -5 & -3 \\ 0 & -4 & 3 & 1 \\ 0 & 0 & 5 & 7 \\ 0 & 0 & 0 & 9 \end{bmatrix}$

(d) $\begin{bmatrix} 3 & 0 & 0 & 0 & 0 \\ 1 & 1 & 0 & 0 & 0 \\ 2 & 9 & 7 & 0 & 0 \\ -2 & 8 & 4 & 4 & 0 \\ 6 & -2 & -5 & 2 & 1 \end{bmatrix}$

(e) $\begin{bmatrix} 3 & 3 & 0 & 0 \\ 1 & 5 & 0 & 0 \\ 0 & 0 & -7 & 3 \\ 0 & 0 & 4 & 8 \end{bmatrix}$
(f) $\begin{bmatrix} 3 & 8 & 0 & 0 & 0 \\ -2 & 7 & 0 & 0 & 0 \\ 0 & 0 & 1 & 0 & 0 \\ 0 & 0 & 0 & 5 & -8 \\ 0 & 0 & 0 & 3 & 6 \end{bmatrix}$

2. Compute the determinants of the following matrices by converting each to upper triangular form using only type III elementary matrices.

(a) $\begin{bmatrix} 1 & -1 & 3 \\ 2 & 7 & 7 \\ -4 & 8 & 1 \end{bmatrix}$
(b) $\begin{bmatrix} 1 & 1 & -2 & 3 \\ 9 & -5 & 7 & 8 \\ 3 & 6 & 2 & -3 \\ 12 & -1 & -1 & 1 \end{bmatrix}$
(c) $\begin{bmatrix} 1 & 4 & 9 & 5 \\ 3 & 7 & -7 & 3 \\ 2 & -11 & 5 & -1 \\ 8 & 9 & 0 & 2 \end{bmatrix}$

3. Compute the determinants of the matrices from problem 2 by converting each to lower triangular form using only type III elementary matrices.

4. Compute the determinants of the following matrices by converting each to upper triangular form:

(a) $\begin{bmatrix} 2 & -1 & 3 & 1 \\ 4 & -2 & 5 & 5 \\ 3 & 7 & -1 & 4 \\ 6 & 2 & 2 & -8 \end{bmatrix}$
(b) $\begin{bmatrix} 1 & -2 & 0 & 4 \\ -1 & 2 & 1 & 3 \\ 2 & -4 & 0 & 5 \\ 5 & -3 & 2 & 1 \end{bmatrix}$

5. Verify your answers to problem 4 by computing the determinant via the method of expanding along a row or column.

6. Let A and B be two square matrices of any two sizes.

(a) Show that the square diagonal block matrix $C = \begin{bmatrix} A & 0 \\ 0 & B \end{bmatrix}$ for appropriate size 0 matrices has $\det(C) = \det(A)\det(B)$.

(b) Explain why
$$C^{-1} = \begin{bmatrix} A^{-1} & 0 \\ 0 & B^{-1} \end{bmatrix}$$
As a consequence, what is the inverse of a diagonal matrix D?

7. Let A be a square $n \times n$ matrix. If A can be lower/upper triangularized using only type III elementary row operations, then how many of these operations do you expect to need? Give you answer as simply as possible.

8. Give an example of each type of elementary matrix for size 3×3 and give their inverses.

9. If A is a product of elementary matrices of size $n \times n$, then is A invertible or not, and why?

10. Let A be a square matrix with $\det(A) \neq 0$. Can you find A^{-1} by successively applying elementary matrices to A in order to produce the $n \times n$ identity matrix I_n? If yes, explain how. If no, explain why not.

11. For a square matrix A, explain why $\text{rref}(A)$ will be the identity matrix unless $\text{rref}(A)$ contains at least one row of all zeroes.

Maple Problems

1. Use the *gausselim* command to convert the following matrices to upper triangular form.

(a) $\begin{bmatrix} 1 & -1 & 6 & 9 \\ -1 & -1 & 3 & -8 \\ 3 & 2 & -1 & 7 \\ 8 & -4 & 1 & 5 \end{bmatrix}$
(b) $\begin{bmatrix} 1 & 12 & -3 & 8 \\ 9 & 1 & -2 & 6 \\ 3 & 6 & 1 & -1 \\ 0 & -1 & -1 & 1 \end{bmatrix}$

(c) $\begin{bmatrix} 1 & -13 & -5 & -3 \\ -6 & -4 & 3 & 5 \\ 8 & 8 & 5 & 7 \\ 4 & -3 & 9 & -1 \end{bmatrix}$
(d) $\begin{bmatrix} 3 & -1 & 5 & 7 & -8 \\ 1 & 2 & 3 & -9 & 7 \\ 2 & 9 & 7 & 3 & -1 \\ -2 & 5 & 4 & 4 & 3 \\ 6 & 7 & -5 & 2 & 8 \end{bmatrix}$

(e) $\begin{bmatrix} 3 & 3 & 1 & 2 & 5 \\ 1 & 5 & 8 & 0 & 8 \\ 9 & 6 & -7 & 3 & 2 \\ -2 & -2 & 4 & 8 & 3 \\ 8 & 6 & 4 & 8 & 3 \end{bmatrix}$
(f) $\begin{bmatrix} 3 & 8 & -2 & -1 & -3 \\ -2 & 7 & 0 & 2 & 0 \\ 0 & 6 & 1 & 0 & 4 \\ 3 & 3 & 0 & 5 & -8 \\ 5 & -1 & 0 & 3 & 6 \end{bmatrix}$

2. Use the *pivot* command to convert the matrices from problem 1 to upper triangular form.

3. Compare your answers from problems 1 and 2.

4. Compute the determinants of the matrices from problem 1.

5. Let

$$
A = \begin{bmatrix}
\frac{1}{2} & -\frac{3}{7} & \frac{5}{11} & -\frac{4}{9} \\
-\frac{6}{5} & \frac{1}{10} & \frac{11}{9} & -\frac{3}{2} \\
\frac{8}{11} & \frac{1}{3} & -\frac{1}{4} & \frac{1}{5} \\
\frac{2}{3} & \frac{3}{5} & \frac{5}{12} & \frac{9}{2}
\end{bmatrix}
$$

Notice that this matrix has all rational entries instead of just integers that we have been used to using.

(a) Find the smallest positive scalar c so that $A = cB$ where B is 4×4 also, but B has all integer entries.

(b) Now triangularize both A and B in order to compute their determinants. Is $\det(A) = c^4 \det(B)$?

(c) Using the two 4×4 matrices in part (b), triangularize their product AB and see if $\det(AB) = \det(A)\det(B)$.

(d) Compute A^3 and triangularize it. Is $\det(A^3) = \det(A)^3$?

5.4   *LU* Factorization

We now take a slight detour from our study of determinants and focus our attention on the process of triangularization that we introduced in the previous section. If we focus our attention on how we obtained the upper triangular matrix U from A, we will notice that we "threw away" some information. More specifically, if we look at the 7×7 matrix A from Section 5.3, the information we did not keep was the multiplicative factors for the pivot row to cancel out all entries below the diagonal.

We can construct a second matrix L that contains these multiplicative factors, with a sign change, on the lower diagonal region located in the same position as the entry that was zeroed out. This will result in a lower triangular matrix that will be the L in the LU factorization of A, where U is our upper triangular matrix conversion of A through pivoting. The two triangular matrices L and U will now factor A as $A = LU$.

Example 5.4.1. We start with a smaller (than 7×7) example, which will be more tractable to help illustrate this process, with the use of *Maple*.

$$A = \begin{bmatrix} 3 & 4 & -2 & 5 \\ 2 & -7 & -8 & 9 \\ 1 & 4 & 2 & -5 \\ -2 & 3 & 4 & -1 \end{bmatrix}$$

```
> with(linalg):
> A:= matrix(4,4, [3,4,-2,5,2,-7,-8,9,1,4,2,-5,-2,3,4,-1] ):
> L:= matrix(4,4, [1,0,0,0,0,1,0,0,0,0,1,0,0,0,0,1] );
```

$$L := \begin{bmatrix} 1 & 0 & 0 & 0 \\ 0 & 1 & 0 & 0 \\ 0 & 0 & 1 & 0 \\ 0 & 0 & 0 & 1 \end{bmatrix}$$

```
> A1:= addrow(A, 1, 2, -2/3);
```

$$A1 := \begin{bmatrix} 3 & 4 & -2 & 5 \\ 0 & -\frac{29}{3} & -\frac{20}{3} & \frac{17}{3} \\ 1 & 4 & 2 & -5 \\ -2 & 3 & 4 & -1 \end{bmatrix}$$

```
> L[2,1]:= 2/3:
> A2:= addrow(A1, 1, 3, -1/3);
```

$$A2 := \begin{bmatrix} 3 & 4 & -2 & 5 \\ 0 & -\frac{29}{3} & -\frac{20}{3} & \frac{17}{3} \\ 0 & \frac{8}{3} & \frac{8}{3} & -\frac{20}{3} \\ -2 & 3 & 4 & -1 \end{bmatrix}$$

```
> L[3,1]:= 1/3:
> A3:= addrow(A2, 1, 4, 2/3);
```

$$A3 := \begin{bmatrix} 3 & 4 & -2 & 5 \\ 0 & -\frac{29}{3} & -\frac{20}{3} & \frac{17}{3} \\ 0 & \frac{8}{3} & \frac{8}{3} & -\frac{20}{3} \\ 0 & \frac{17}{3} & \frac{8}{3} & \frac{7}{3} \end{bmatrix}$$

```
> L[4,1]:= -2/3:
```

> A4:= addrow(A3, 2, 3, 8/29);

$$A4 := \begin{bmatrix} 3 & 4 & -2 & 5 \\ 0 & -\frac{29}{3} & -\frac{20}{3} & \frac{17}{3} \\ 0 & 0 & \frac{24}{29} & -\frac{148}{29} \\ 0 & \frac{17}{3} & \frac{8}{3} & \frac{7}{3} \end{bmatrix}$$

> L[3,2]:= -8/29:

> A5:= addrow(A4,2,4,17/29);

$$A5 := \begin{bmatrix} 3 & 4 & -2 & 5 \\ 0 & -\frac{29}{3} & -\frac{20}{3} & \frac{17}{3} \\ 0 & 0 & \frac{24}{29} & -\frac{148}{29} \\ 0 & 0 & -\frac{36}{29} & \frac{164}{29} \end{bmatrix}$$

> L[4,2]:= -17/29:

> A6:= addrow(A5, 3, 4, 36/24);

$$A6 := \begin{bmatrix} 3 & 4 & -2 & 5 \\ 0 & -\frac{29}{3} & -\frac{20}{3} & \frac{17}{3} \\ 0 & 0 & \frac{24}{29} & -\frac{148}{29} \\ 0 & 0 & 0 & -2 \end{bmatrix}$$

> L[4,3]:= -36/24:

> evalm(L);

$$\begin{bmatrix} 1 & 0 & 0 & 0 \\ \frac{2}{3} & 1 & 0 & 0 \\ \frac{1}{3} & -\frac{8}{29} & 1 & 0 \\ -\frac{2}{3} & -\frac{17}{29} & -\frac{3}{2} & 1 \end{bmatrix}$$

> evalm(L&*A6);

$$\begin{bmatrix} 3 & 4 & -2 & 5 \\ 2 & -7 & -8 & 9 \\ 1 & 4 & 2 & -5 \\ -2 & 3 & 4 & -1 \end{bmatrix}$$

We now have factored the matrix A into the product of lower and upper triangular matrices:

$$
\begin{bmatrix}
3 & 4 & -2 & 5 \\
2 & -7 & -8 & 9 \\
1 & 4 & 2 & -5 \\
-2 & 3 & 4 & -1
\end{bmatrix}
=
\begin{bmatrix}
1 & 0 & 0 & 0 \\
\frac{2}{3} & 1 & 0 & 0 \\
\frac{1}{3} & -\frac{8}{29} & 1 & 0 \\
-\frac{2}{3} & -\frac{17}{29} & -\frac{3}{2} & 1
\end{bmatrix}
\begin{bmatrix}
3 & 4 & -2 & 5 \\
0 & -\frac{29}{3} & -\frac{20}{3} & \frac{17}{3} \\
0 & 0 & \frac{24}{29} & -\frac{148}{29} \\
0 & 0 & 0 & -2
\end{bmatrix}
$$

Now, we must ask the question: Why is this factorization of any use? As always, our focus is on solving systems of equations of the form $AX = B$. What we now have is a system of the form $LUX = B$. There must be an advantage to this form even though it appears to be potentially more complicated due to extra matrix multiplications. To solve for X in $LUX = B$ we perform a two part process. The first step in the process is to solve the system $LY = B$ for the unknown column matrix Y. The second step is to solve the system $UX = Y$. The second step arises from the fact that we simply performed the substitution $Y = UX$ in the original equation to end up with $LY = B$.

We apply this technique to our example. First, we will construct a linear system of equations involving our matrix A by picking a particular value of B:

$$
\begin{bmatrix}
3 & 4 & -2 & 5 \\
2 & -7 & -8 & 9 \\
1 & 4 & 2 & -5 \\
-2 & 3 & 4 & -1
\end{bmatrix}
\begin{bmatrix}
x_1 \\
x_2 \\
x_3 \\
x_4
\end{bmatrix}
=
\begin{bmatrix}
-1 \\
2 \\
-3 \\
-1
\end{bmatrix}
$$

As instructed by the first step in this process, we start by solving the system:

$$
\begin{bmatrix}
1 & 0 & 0 & 0 \\
\frac{2}{3} & 1 & 0 & 0 \\
\frac{1}{3} & -\frac{8}{29} & 1 & 0 \\
-\frac{2}{3} & -\frac{17}{29} & -\frac{3}{2} & 1
\end{bmatrix}
\begin{bmatrix}
y_1 \\
y_2 \\
y_3 \\
y_4
\end{bmatrix}
=
\begin{bmatrix}
-1 \\
2 \\
-3 \\
-1
\end{bmatrix}
$$

The method of solving the above system is known as *forward substitution*. The first row of the above system of equations is simple y_1 is $y_1 = -1$. Now that we know the value of y_1, notice that the second row is $\frac{2}{3}y_1 + y_2 = 2$. We *forward* the value of y_1 into the second row to obtain y_2, yielding $y_2 = \frac{8}{3}$. In a similar fashion, now that y_1 and y_2 are known, y_3 can be solved for in row 3. The last variable y_4 can be solved for from row 4 to give a final solution set of

$$
\left\{ y_1 = -1,\ y_2 = \frac{8}{3},\ y_3 = -\frac{56}{29},\ y_4 = -3 \right\}
$$

Onto the second step in the process, we have the system

$$\begin{bmatrix} 3 & 4 & -2 & 5 \\ 0 & -\frac{29}{3} & -\frac{20}{3} & \frac{17}{3} \\ 0 & 0 & \frac{24}{29} & -\frac{148}{29} \\ 0 & 0 & 0 & -2 \end{bmatrix} \begin{bmatrix} x_1 \\ x_2 \\ x_3 \\ x_4 \end{bmatrix} = \begin{bmatrix} -1 \\ \frac{8}{3} \\ -\frac{56}{29} \\ -3 \end{bmatrix}$$

which can be solved in a similar fashion, except by *back substitution*. We start with the last equation, which is $-2x_4 = -3$, solve for x_4, *back* up one row, determine x_3 and so on. Our final solution is given by

$$\left\{ x_1 = \frac{22}{3}, \ x_2 = -\frac{25}{6}, \ x_3 = \frac{83}{12}, \ x_4 = \frac{3}{2} \right\}$$

We will perform all of these steps in *Maple* next.

```
> X:= matrix(4,1,[x1,x2,x3,x4]):
> Y:= matrix(4,1,[y1,y2,y3,y4]):
> B:= matrix(4,1,[-1,2,-3,-1]):
> lhsL:= evalm(L&*Y);
```

$$lhsL := \begin{bmatrix} y1 \\ \frac{2}{3}y1 + y2 \\ \frac{1}{3}y1 - \frac{8}{29}y2 + y3 \\ -\frac{2}{3}y1 - \frac{17}{29}y2 - \frac{3}{2}y3 + y4 \end{bmatrix}$$

```
> ysolns:= solve({seq(lhsL[k,1]=B[k,1],k=1..4)}, {y1,y2,y3,y4});
```

$$ysolns := \left\{ y_1 = -1, \ y_2 = \frac{8}{3}, \ y_3 = -\frac{56}{29}, \ y_4 = -3 \right\}$$

```
> Ys:= matrix(4,1, [seq(rhs(ysolns[j]), j=1..4)]);
```

$$Ys := \begin{bmatrix} -1 \\ \frac{8}{3} \\ -\frac{56}{29} \\ -3 \end{bmatrix}$$

```
> lhsU:= evalm(A6&*X);
```

$$lhsU := \begin{bmatrix} 3x1 + 4x2 - 2x3 + 5x4 \\ -\frac{29}{3}x2 - \frac{20}{3}x3 + \frac{17}{3}x4 \\ \frac{24}{29}x3 - \frac{148}{29}x4 \\ -2x4 \end{bmatrix}$$

> xsolns:= solve({seq(lhsU[k,1]=Ys[k,1], k=1..4)}, {x1,x2,x3,x4});

$$xsolns := \left\{ x_1 = \frac{22}{3}, \ x_2 = -\frac{25}{6}, \ x_3 = \frac{83}{12}, \ x_4 = \frac{3}{2} \right\}$$

We verify the above solution, with that found by the *rref* command, to the augmented matrix of the original system.

> rref(augment(A,B));

$$\begin{bmatrix} 1 & 0 & 0 & 0 & \frac{22}{3} \\ 0 & 1 & 0 & 0 & -\frac{25}{6} \\ 0 & 0 & 1 & 0 & \frac{83}{12} \\ 0 & 0 & 0 & 1 & \frac{3}{2} \end{bmatrix}$$

■

It should come as no surprise that *Maple* has a built in LU factorization command, *LUdecomp*, which will return an upper triangular matrix. If one wishes to know the corresponding lower triangular matrix as well, an additional parameter is needed. In the following command, the matrix A is factored into a lower triangular matrix LA, and upper triangular matrix UA. The command itself will display UA, but will not automatically display the lower triangular counterpart LA.

> LUdecomp(A,L='LA');

$$\begin{bmatrix} 3 & 4 & -2 & 5 \\ 0 & -\frac{29}{3} & -\frac{20}{3} & \frac{17}{3} \\ 0 & 0 & \frac{24}{29} & -\frac{148}{29} \\ 0 & 0 & 0 & -2 \end{bmatrix}$$

> evalm(LA);

$$\begin{bmatrix} 1 & 0 & 0 & 0 \\ \frac{2}{3} & 1 & 0 & 0 \\ \frac{1}{3} & -\frac{8}{29} & 1 & 0 \\ -\frac{2}{3} & -\frac{17}{29} & -\frac{3}{2} & 1 \end{bmatrix}$$

LU factorization is commonly used in numerical applications due to its efficiency and ease of implementation. For instance, forward or backward substitution can be performed in a very systematic and programmatically simple way. Furthermore, all information regarding the L and U matrices can be stored in one matrix, which can be a very important factor when dealing with large matrices and when memory becomes an issue. Consider our previous example, under the assumption that we remember that the diagonal of L consists of all ones, both matrices L and U can be stored in the single matrix:

$$\begin{bmatrix} 3 & 4 & -2 & 5 \\ \frac{2}{3} & -\frac{29}{3} & -\frac{20}{3} & \frac{17}{3} \\ \frac{1}{3} & -\frac{8}{29} & \frac{24}{29} & -\frac{148}{29} \\ -\frac{2}{3} & -\frac{17}{29} & -\frac{3}{2} & -2 \end{bmatrix}$$

The one problem that we have not dealt with is when pivoting is required to transform our matrix to a triangular one. This would require a type I elementary matrix. There are methods of dealing with this problem, which usually involve a pivoting matrix P and the resulting equation is given by $PA = LU$. This idea is to multiply A on the left by an elementary matrix of type I, or a combination of elementary matrices of type I, so that the resulting matrix can be decomposed into LU form.

Homework Problems

1. Solve the following systems by forward or backward substitution:

(a) $\begin{bmatrix} 1 & 0 & 0 \\ -2 & 1 & 0 \\ 7 & 8 & 1 \end{bmatrix} \begin{bmatrix} y_1 \\ y_2 \\ y_3 \end{bmatrix} = \begin{bmatrix} -1 \\ 2 \\ -3 \end{bmatrix}$

(b) $\begin{bmatrix} -4 & -6 & 3 \\ 0 & 8 & 2 \\ 0 & 0 & 1 \end{bmatrix} \begin{bmatrix} x_1 \\ x_2 \\ x_3 \end{bmatrix} = \begin{bmatrix} 7 \\ -9 \\ 0 \end{bmatrix}$

(c) $\begin{bmatrix} -1 & 0 & -3 & 5 \\ 0 & 2 & 7 & -9 \\ 0 & 0 & -2 & 10 \\ 0 & 0 & 0 & 5 \end{bmatrix} \begin{bmatrix} x_1 \\ x_2 \\ x_3 \\ x_4 \end{bmatrix} = \begin{bmatrix} -8 \\ 3 \\ 7 \\ -1 \end{bmatrix}$

(d) $\begin{bmatrix} 1 & 0 & 0 & 0 \\ -2 & 1 & 0 & 0 \\ 3 & 2 & 1 & 0 \\ -4 & -3 & 5 & 1 \end{bmatrix} \begin{bmatrix} y_1 \\ y_2 \\ y_3 \\ y_4 \end{bmatrix} = \begin{bmatrix} 1 \\ 0 \\ -4 \\ 3 \end{bmatrix}$

2. Compute the LU factorization of the following matrices.

(a) $\begin{bmatrix} 1 & 3 & 1 & -7 \\ -2 & 1 & -2 & 6 \\ 0 & 8 & 1 & -1 \\ -2 & 1 & 0 & -1 \end{bmatrix}$
(b) $\begin{bmatrix} 8 & -4 & -8 & -9 \\ 4 & 1 & 2 & 8 \\ -1 & 0 & 1 & -1 \\ 0 & -2 & 0 & -9 \end{bmatrix}$

(c) $\begin{bmatrix} -1 & 0 & -3 & 5 \\ -2 & 2 & 7 & -9 \\ 3 & -5 & -2 & 10 \\ 6 & 8 & -1 & 5 \end{bmatrix}$
(d) $\begin{bmatrix} 1 & 0 & 0 & 0 \\ -2 & 1 & 0 & 0 \\ 3 & 2 & 1 & 0 \\ -4 & -3 & 5 & 1 \end{bmatrix}$

(e) $\begin{bmatrix} 1 & -3 & 5 & 8 \\ -2 & 1 & -7 & -2 \\ 3 & 0 & 1 & -7 \\ 0 & -3 & 5 & 1 \end{bmatrix}$
(f) $\begin{bmatrix} 1 & -6 & 8 & -1 \\ 0 & 1 & -5 & -2 \\ 3 & 0 & 1 & -3 \\ -4 & -3 & 5 & 1 \end{bmatrix}$

3. Compute the determinant of each of the matrices from problem 2.

4. Solve the following systems of equations by *LU* factorization, performing forward and backward substitution.

(a) $\begin{bmatrix} 1 & 3 & 1 & -7 \\ -2 & 1 & -2 & 6 \\ 0 & 8 & 1 & -1 \\ -2 & 1 & 0 & -1 \end{bmatrix} \begin{bmatrix} x_1 \\ x_2 \\ x_3 \\ x_4 \end{bmatrix} = \begin{bmatrix} -1 \\ 3 \\ -7 \\ 5 \end{bmatrix}$

(b) $\begin{bmatrix} 8 & -4 & -8 & -9 \\ 4 & 1 & 2 & 8 \\ -1 & 0 & 1 & -1 \\ 0 & -2 & 0 & -9 \end{bmatrix} \begin{bmatrix} x_1 \\ x_2 \\ x_3 \\ x_4 \end{bmatrix} = \begin{bmatrix} 5 \\ 0 \\ 1 \\ -1 \end{bmatrix}$

(c) $\begin{bmatrix} -1 & 0 & -3 & 5 \\ -2 & 2 & 7 & -9 \\ 3 & -5 & -2 & 10 \\ 6 & 8 & -1 & 5 \end{bmatrix} \begin{bmatrix} x_1 \\ x_2 \\ x_3 \\ x_4 \end{bmatrix} = \begin{bmatrix} 2 \\ 3 \\ 4 \\ -5 \end{bmatrix}$

Maple Problems

1. Verify your answer to *Homework* problem 2 by using the *LUdecomp* command.

2. Given $L, U \in \mathbb{R}^{n \times n}$, and $X, Y, B \in R^{n \times 1}$, construct a *Maple* procedure to perform forward substitution for the system $LY = B$ and backward substitution for the system $UX = Y$.

3. Using your algorithm from *Maple* problem 2, solve each of the systems from *Homework* problem 1.

4. Construct an algorithm that will perform LU decomposition on a square matrix A, with the results of both L and U stored in a single matrix.

5. Using your algorithms from *Maple* problems 2 and 4, solve each of the systems from *Homework* problem 4.

5.5 Inverses from *rref*

The purpose of this very short section is to find a very fast way of computing inverses to square matrices. The adjoint formula is very useful theoretically, but too cumbersome to be practical as a general computational tool. If we start with a square matrix A of dimension $n \times n$, then by the adjoint formula, assuming $\det(A) \neq 0$, A has an inverse A^{-1}, with $AA^{-1} = I_n$. Defining A_j^{-1} as the jth column of A^{-1}, then the result of the matrix multiplication AA_j^{-1} is simply the jth column of I_n:

$$
\begin{bmatrix} A_{1,1} & A_{1,2} & \cdots & A_{1,n} \\ A_{2,1} & A_{2,2} & \cdots & A_{2,n} \\ \vdots & & \ddots & \vdots \\ A_{n,1} & A_{n,2} & \cdots & A_{n,n} \end{bmatrix} \begin{bmatrix} A_{1,j}^{-1} \\ \vdots \\ A_{j,j}^{-1} \\ \vdots \\ A_{n,j}^{-1} \end{bmatrix} = \begin{bmatrix} 0 \\ \vdots \\ 1 \\ \vdots \\ 0 \end{bmatrix} \tag{5.16}
$$

As a result of the above matrix equation, we end up with the following n equations in the n unknowns, $A_{i,j}^{-1}$, for $1 \leq i \leq n$:

$$A_{1,1}A_{1,j}^{-1} + A_{1,2}A_{2,j}^{-1} + \cdots + A_{1,n}A_{n,j}^{-1} = 0$$
$$A_{2,1}A_{1,j}^{-1} + A_{2,2}A_{2,j}^{-1} + \cdots + A_{2,n}A_{n,j}^{-1} = 0$$

$$\vdots$$

$$A_{j,1}A_{1,j}^{-1} + A_{j,2}A_{2,j}^{-1} + \cdots + A_{j,n}A_{n,j}^{-1} = 1$$

$$\vdots$$

$$A_{n,1}A_{1,j}^{-1} + A_{n,2}A_{2,j}^{-1} + \cdots + A_{n,n}A_{n,j}^{-1} = 0$$

We can now use *rref* to solve this system of equations for the unknown entries in the column matrix A_j^{-1} if we apply *rref* to the augmented matrix $(A|j$th column of $I_n)$. This process of finding each of the columns of A^{-1} using *rref* can be combined into doing *rref* once on the single augmented matrix $(A|I_n)$. We get that

$$\text{rref}(A|I_n) = (I_n|A^{-1}) \tag{5.17}$$

We can start with the simple case of $A \in \mathbb{R}^{2 \times 2}$. Given $A = \begin{bmatrix} a & b \\ c & d \end{bmatrix}$, we know that

$$A^{-1} = \frac{1}{ad - bc} \begin{bmatrix} d & -b \\ -c & a \end{bmatrix} \tag{5.18}$$

So we will attempt to compute the inverse of A by starting with the augmented matrix:

$$\begin{bmatrix} a & b & 1 & 0 \\ c & d & 0 & 1 \end{bmatrix}$$

and performing elementary row operations on $(A|I_2)$ to arrive at the RHS of equation (5.17).

First, under the assumption that $a \neq 0$, we can divide row 1 by a to get

$$\begin{bmatrix} 1 & \dfrac{b}{a} & \dfrac{1}{a} & 0 \\ c & d & 0 & 1 \end{bmatrix}$$

Next, we take $-c \cdot$ row 1 and add it to row 2 giving:

$$\begin{bmatrix} 1 & \dfrac{b}{a} & \dfrac{1}{a} & 0 \\ 0 & \dfrac{ad - bc}{a} & -\dfrac{c}{a} & 1 \end{bmatrix}$$

We now need to make the entry in row 2 column 2 a value of 1. To do this, we multiply row 2 by $\dfrac{a}{ad - bc}$:

$$\begin{bmatrix} 1 & \dfrac{b}{a} & \dfrac{1}{a} & 0 \\ 0 & 1 & -\dfrac{c}{ad - bc} & \dfrac{a}{ad - bc} \end{bmatrix}$$

Next, we take $-\dfrac{b}{a}$ times row 2 and add it to row 1:

$$\begin{bmatrix} 1 & 0 & \dfrac{1}{a} + \dfrac{bc}{a(ad - bc)} & -\dfrac{b}{ad - bc} \\ 0 & 1 & -\dfrac{c}{ad - bc} & \dfrac{a}{ad - bc} \end{bmatrix}$$

with a little algebraic manipulation, we see that $\dfrac{1}{a} + \dfrac{bc}{a(ad - bc)} = \dfrac{d}{ad - bc}$.
Therefore, we have

$$
\begin{bmatrix}
1 & 0 & \dfrac{d}{ad - bc} & -\dfrac{b}{ad - bc} \\[2mm]
0 & 1 & -\dfrac{c}{ad - bc} & \dfrac{a}{ad - bc}
\end{bmatrix}
\tag{5.19}
$$

Notice that after removing the identity from the above matrix (i.e., the first two columns), what remains is A^{-1}, giving us the identity in equation (5.17).

Once we know how to perform this process by hand, we revert to *Maple* for more complicated examples. We will be making use of the *delcols* command from the *linalg* package. This command will allow us to remove the identity from $\left(I_n | A^{-1}\right)$ to get A^{-1}.

Example 5.5.1. We will use the 7×7 matrix A from Example 5.3.4 of Section 5.3 to illustrate the method of inverses from *rref* on a matrix with square dimension larger than 2×2. The *diag* command in the *linalg* package gives us an easy way to create the 7×7 identity matrix I_7.

> with(linalg):
> A:= matrix([[-9,3,0,2,-4,5,1], [6,8,-7,-1,2,10,3], [2,5,-1,-15,6,2,-4], [11,7,5,4,-8,-1,6], [-7,1,1,2,5,-3,8], [1,-12,9,0,4,-2,13], [0,-9,7,2,6,3,-4]]);

$$
A := \begin{bmatrix}
-9 & 3 & 0 & 2 & -4 & 5 & 1 \\
6 & 8 & -7 & -1 & 2 & 10 & 3 \\
2 & 5 & -1 & -15 & 6 & 2 & -4 \\
11 & 7 & 5 & 4 & -8 & -1 & 6 \\
-7 & 1 & 1 & 2 & 5 & -3 & 8 \\
1 & -12 & 9 & 0 & 4 & -2 & 13 \\
0 & -9 & 7 & 2 & 6 & 3 & -4
\end{bmatrix}
$$

> identity7:= diag(1,1,1,1,1,1,1);

$$
identity7 := \begin{bmatrix}
1 & 0 & 0 & 0 & 0 & 0 & 0 \\
0 & 1 & 0 & 0 & 0 & 0 & 0 \\
0 & 0 & 1 & 0 & 0 & 0 & 0 \\
0 & 0 & 0 & 1 & 0 & 0 & 0 \\
0 & 0 & 0 & 0 & 1 & 0 & 0 \\
0 & 0 & 0 & 0 & 0 & 1 & 0 \\
0 & 0 & 0 & 0 & 0 & 0 & 1
\end{bmatrix}
$$

> AI7:= rref(augment(A, identity7)):

> delcols(AI7, 1..7);

$$\begin{bmatrix}
-\frac{46491}{854627} & \frac{19218}{854627} & -\frac{5100}{854627} & \frac{16811}{854627} & -\frac{20348}{854627} & \frac{3743}{854627} & \frac{4576}{854627} \\[4pt]
\frac{167953}{20511048} & -\frac{3487}{5127762} & \frac{139939}{5127762} & \frac{263815}{5127762} & \frac{719173}{10255524} & -\frac{1063685}{20511048} & \frac{474377}{20511048} \\[4pt]
\frac{287649}{6837016} & -\frac{62953}{1709254} & \frac{67669}{1709254} & \frac{114941}{1709254} & \frac{102269}{3418508} & -\frac{67277}{6837016} & \frac{492449}{6837016} \\[4pt]
-\frac{78861}{3418508} & \frac{42124}{2563881} & -\frac{143255}{2563881} & \frac{11334}{854627} & \frac{220933}{5127762} & -\frac{132403}{3418508} & \frac{145687}{3418508} \\[4pt]
-\frac{277067}{5127762} & \frac{67468}{2563881} & \frac{23186}{2563881} & -\frac{9733}{2563881} & \frac{215842}{2563881} & -\frac{160049}{5127762} & \frac{299573}{5127762} \\[4pt]
\frac{315005}{5127762} & \frac{43409}{854627} & \frac{8399}{2563881} & -\frac{9755}{2563881} & -\frac{96527}{2563881} & \frac{101957}{5127762} & \frac{173281}{5127762} \\[4pt]
\frac{59431}{6837016} & \frac{117211}{5127762} & -\frac{20699}{5127762} & \frac{9}{1709254} & \frac{145177}{10255524} & \frac{329485}{6837016} & -\frac{285137}{6837016}
\end{bmatrix}$$

> inverse(A);

$$\begin{bmatrix}
-\frac{46491}{854627} & \frac{19218}{854627} & -\frac{5100}{854627} & \frac{16811}{854627} & -\frac{20348}{854627} & \frac{3743}{854627} & \frac{4576}{854627} \\[4pt]
\frac{167953}{20511048} & -\frac{3487}{5127762} & \frac{139939}{5127762} & \frac{263815}{5127762} & \frac{719173}{10255524} & -\frac{1063685}{20511048} & \frac{474377}{20511048} \\[4pt]
\frac{287649}{6837016} & -\frac{62953}{1709254} & \frac{67669}{1709254} & \frac{114941}{1709254} & \frac{102269}{3418508} & -\frac{67277}{6837016} & \frac{492449}{6837016} \\[4pt]
-\frac{78861}{3418508} & \frac{42124}{2563881} & -\frac{143255}{2563881} & \frac{11334}{854627} & \frac{220933}{5127762} & -\frac{132403}{3418508} & \frac{145687}{3418508} \\[4pt]
-\frac{277067}{5127762} & \frac{67468}{2563881} & \frac{23186}{2563881} & -\frac{9733}{2563881} & \frac{215842}{2563881} & -\frac{160049}{5127762} & \frac{299573}{5127762} \\[4pt]
\frac{315005}{5127762} & \frac{43409}{854627} & \frac{8399}{2563881} & -\frac{9755}{2563881} & -\frac{96527}{2563881} & \frac{101957}{5127762} & \frac{173281}{5127762} \\[4pt]
\frac{59431}{6837016} & \frac{117211}{5127762} & -\frac{20699}{5127762} & \frac{9}{1709254} & \frac{145177}{10255524} & \frac{329485}{6837016} & -\frac{285137}{6837016}
\end{bmatrix}$$

The inverse matrix to A is the same as the matrix we get from $\mathrm{rref}(A|I_n)$ above after deleting its first seven columns.

The following theorem is a continuation of Theorem 5.2.1 on solutions to square linear systems, and it follows nicely from the material we have developed in this section.

Theorem 5.5.1. *Given the system $AX = B$, where $A \in \mathbb{R}^{n \times n}$, $\det(A) \neq 0$ if and only if there exists a unique solution to the system for all B. In this case the unique solution to the system is $X = A^{-1}B$. As a consequence, the system has either no solution, or an infinite number of solutions, exactly when $\det(A) = 0$.*

Proof. If you look at the statement of Theorem 5.2.1, it tells us half of what we want to prove, namely, if $\det(A) \neq 0$, then the system has a unique solution $X = A^{-1}B$. This means that half of our work is already done, since we now only need show that if the system has a unique solution for all B, then $\det(A) \neq 0$. Now, let B be any column out of the identity matrix I_n. Then we can solve all of the equations $AX = B$ for their unique solutions using $\mathrm{rref}(A|I_n)$. The only possible way we can have unique solutions in this situation is when $\mathrm{rref}(A|I_n) = (I_n|C)$, for some $n \times n$ matrix C. This last *rref* computation also tells us that $\mathrm{rref}(A) = I_n$. Then $\det(\mathrm{rref}(A)) = 1$, and so $\det(A) \neq 0$, since

$\det(\text{rref}(A))$ is a non-zero multiple of $\det(A)$ from the proof of Theorem 5.3.2. That is, $\det(A) \neq 0$ exactly when $\det(\text{rref}(A)) \neq 0$. □

Homework Problems

1. Use the method of row reducing $(A|I_n)$ to compute the inverse to each of the following matrices:

(a) $\begin{bmatrix} 1 & -1 & 2 \\ -1 & 3 & -1 \\ 1 & 2 & -2 \end{bmatrix}$
(b) $\begin{bmatrix} 3 & -4 & 1 \\ 3 & 3 & -1 \\ 1 & 2 & -2 \end{bmatrix}$
(c) $\begin{bmatrix} 2 & 3 & 1 \\ -1 & -5 & 2 \\ 1 & -7 & 2 \end{bmatrix}$

(d) $\begin{bmatrix} -1 & 2 & 1 \\ 1 & -2 & 1 \\ 0 & 1 & 0 \end{bmatrix}$
(e) $\begin{bmatrix} 1 & 3 & -5 \\ 1 & 4 & 6 \\ 2 & 3 & 7 \end{bmatrix}$
(f) $\begin{bmatrix} -1 & 2 & 0 \\ 5 & 0 & 2 \\ -4 & 2 & 0 \end{bmatrix}$

(g) $\begin{bmatrix} -1 & 2 & 1 & 0 \\ -2 & 2 & 1 & 1 \\ -3 & -1 & 2 & 1 \\ 2 & -5 & 6 & 1 \end{bmatrix}$
(h) $\begin{bmatrix} 1 & 3 & -5 & 2 \\ 1 & 4 & 6 & 0 \\ 2 & 3 & 7 & 1 \\ -3 & 6 & -4 & 3 \end{bmatrix}$
(i) $\begin{bmatrix} -1 & 2 & 0 & 1 \\ 3 & 0 & 2 & 2 \\ -4 & 2 & 1 & -3 \\ 0 & 2 & 0 & 4 \end{bmatrix}$

2. In this section, it was shown that row reducing the matrix $(A|I_2)$ resulted in the correct value of A^{-1}. The first step in this process required that $a \neq 0$. Repeat this procedure, but this time assume that $a = 0$. You may assume that $b \neq 0$ and $c \neq 0$. Remember that you cannot swap rows.

3. If A is square but has no inverse, then what does $\text{rref}(A|I_n)$ produce and how does it tell us that A has no inverse? Give some examples.

4. Explain why if E is an elementary matrix, E^{-1} is also.

5. Explain why for any square matrix A that is invertible, A^{-1} is a product of elementary matrices.

6. Explain why for any square matrix A that is invertible, A is a product of elementary matrices.

7. Express the matrix in problem 1 part (c), and its inverse, as a product of elementary matrices.

8. Is Theorem 5.5.1 still true if we replace "for all B" with "for at least one B"?

Maple Problems

1. Verify your answers to *Homework* problem 1 by using the *rref* command on the matrices $(A|I_n)$.

2. Compute AA^{-1} for each of the inverses found from *Maple* problem 1, verifying that row reducing $(A|I_n)$ yields $(I_n|A^{-1})$.

3. Let A be a 3×3 matrix and B be a 4×4 matrix, both from *Homework* problem 1. Determine if the following identity holds:

$$rref \left(\begin{array}{cc|c} A & 0 & \\ 0 & B & I_7 \end{array} \right) = rref \left(\begin{array}{c|cc} & A^{-1} & 0 \\ I_7 & 0 & B^{-1} \end{array} \right)$$

Explain why this should or should not work.

5.6   Cramer's Rule

We now discuss a final method of solving linear systems of equations. *Cramer's rule* will allow us to solve a square system of linear equations $AX = B$ when the matrix of coefficients A is invertible. Remember that to be invertible, we require that $\det(A) \neq 0$. Cramer's rule will allow us to compute the solutions to the system using only determinants. We will explore the 2×2 case to see how this is possible. Let $A = \begin{bmatrix} a & b \\ c & d \end{bmatrix}$ and $B = \begin{bmatrix} \alpha \\ \beta \end{bmatrix}$, then the solution is given by $X = A^{-1}B$. Let us look carefully at this solution X using *Maple*, and see if it is possible to write this solution in a form that only involves determinants.

```
> with(linalg):
> A:= matrix([[a, b], [c, d]]);
```

$$A := \begin{bmatrix} a & b \\ c & d \end{bmatrix}$$

```
> B:= matrix([[alpha], [beta]]);
```

$$B := \begin{bmatrix} \alpha \\ \beta \end{bmatrix}$$

```
> X:= evalm(inverse(A)&*B);
```

$$X := \begin{bmatrix} \dfrac{d\alpha}{ad - bc} - \dfrac{b\beta}{ad - bc} \\ \\ -\dfrac{c\alpha}{ad - bc} + \dfrac{a\beta}{ad - bc} \end{bmatrix}$$

> simplify(evalm(det(A)*X));

$$\begin{bmatrix} d\alpha - b\beta \\ -c\alpha + a\beta \end{bmatrix}$$

Notice that if we multiply the solution X by $\det(A)$, then the two entries of this new column each look like determinants of 2×2 matrices. In fact, notice that the first entry in our solution column matrix is

$$d\alpha - b\beta = \det\left(\begin{bmatrix} \alpha & b \\ \beta & d \end{bmatrix}\right)$$

while the second is

$$-c\alpha + a\beta = \det\left(\begin{bmatrix} a & \alpha \\ c & \beta \end{bmatrix}\right)$$

Thus, our solution X can be written using determinants as $X = \begin{bmatrix} x \\ y \end{bmatrix}$ where,

$$x = \frac{\det\left(\begin{bmatrix} \alpha & b \\ \beta & d \end{bmatrix}\right)}{\det(A)}, \quad y = \frac{\det\left(\begin{bmatrix} a & \alpha \\ c & \beta \end{bmatrix}\right)}{\det(A)} \tag{5.20}$$

This method of using determinants to solve square linear systems is called Cramer's rule and it works on any size square system. Let us now check this with an example.

Example 5.6.1. We wish to solve the linear system

$$5x + 13y = -8$$
$$4x - 7y = 2$$

whose matrix of coefficients is $A = \begin{bmatrix} 5 & 13 \\ 4 & -7 \end{bmatrix}$, with column of RHS values $B = \begin{bmatrix} -8 \\ 2 \end{bmatrix}$.

> A:= matrix([[5, 13], [4, -7]]): B:= matrix([[-8], [2]]):
> xnum:= augment(B, delcols(A,1..1));

$$xnum := \begin{bmatrix} -8 & 13 \\ 2 & -7 \end{bmatrix}$$

> ynum:= augment(delcols(A,2..2), B);

$$ynum := \begin{bmatrix} 5 & -8 \\ 4 & 2 \end{bmatrix}$$

> X:= matrix([[det(xnum)/det(A)], [det(ynum)/det(A)]]);

$$X := \begin{bmatrix} -\frac{10}{29} \\ -\frac{14}{29} \end{bmatrix}$$

> evalm(inverse(A)&*B);

$$\begin{bmatrix} -\frac{10}{29} \\ -\frac{14}{29} \end{bmatrix}$$

This example verifies that Cramer's rule does indeed work in the 2×2 case. Let us now see if it works in the 3×3 case with a general computation, and then a numerical example.

> A:= matrix([[a, b, c], [d, e, f], [g, h, i]]);

$$A := \begin{bmatrix} a & b & c \\ d & e & f \\ g & h & i \end{bmatrix}$$

> B:= matrix([[alpha], [beta], [delta]]);

$$B := \begin{bmatrix} \alpha \\ \beta \\ \delta \end{bmatrix}$$

> X:= evalm(inverse(A)&*B);

$$X := \begin{bmatrix} \frac{(ei-fh)\alpha}{aei-afh-dbi+dch+gbf-gce} - \frac{(bi-ch)\beta}{aei-afh-dbi+dch+gbf-gce} + \frac{(bf-ce)\delta}{aei-afh-dbi+dch+gbf-gce} \\ -\frac{(di-fg)\alpha}{aei-afh-dbi+dch+gbf-gce} + \frac{(ai-cg)\beta}{aei-afh-dbi+dch+gbf-gce} - \frac{(af-cd)\delta}{aei-afh-dbi+dch+gbf-gce} \\ \frac{(dh-eg)\alpha}{aei-afh-dbi+dch+gbf-gce} - \frac{(ah-bg)\beta}{aei-afh-dbi+dch+gbf-gce} + \frac{(ae-bd)\delta}{aei-afh-dbi+dch+gbf-gce} \end{bmatrix}$$

> simplify(evalm(det(A)*X));

$$\begin{bmatrix} \alpha ei - \alpha fh - \beta bi + \beta ch + \delta bf - \delta ce \\ -\alpha di + \alpha fg + \beta ai - \beta cg - \delta af + \delta cd \\ \alpha dh - \alpha eg - \beta ah + \beta bg + \delta ae - \delta bd \end{bmatrix}$$

> det(matrix([[alpha, b, c], [beta, e, f], [delta, h, i]]));

$$\alpha ei - \alpha fh - \beta bi + \beta ch + \delta bf - \delta ce$$

> det(matrix([[a, alpha, c], [d, beta, f], [g, delta, i]]));

$$-\alpha di + \alpha fg + \beta ai - \beta cg - \delta af + \delta cd$$

> det(matrix([[a, b, alpha], [d, e, beta], [g, h, delta]]));

$$\alpha dh - \alpha eg - \beta ah + \beta bg + \delta ae - \delta bd$$

Notice that if we multiply the solution X by $\det(A) = aei - afh - dbi + dch + gbf - gce$, then the three entries of this new column matrix each look like determinants of 3×3 matrices:

$$\alpha ei - \alpha fh - \beta bi + \beta ch + \delta bf - \delta ce = \det\left(\begin{bmatrix} \alpha & b & c \\ \beta & e & f \\ \delta & h & i \end{bmatrix}\right)$$

$$-\alpha di + \alpha fg + \beta ai - \beta cg - \delta af + \delta cd = \det\left(\begin{bmatrix} a & \alpha & c \\ d & \beta & f \\ g & \delta & i \end{bmatrix}\right)$$

$$\alpha dh - \alpha eg - \beta ah + \beta bg + \delta ae - \delta bd = \det\left(\begin{bmatrix} a & b & \alpha \\ d & e & \beta \\ g & h & \delta \end{bmatrix}\right)$$

Thus, our solution $X = [x, y, z]^T$ can be written using determinants as

$$x = \frac{\det\left(\begin{bmatrix} \alpha & b & c \\ \beta & e & f \\ \delta & h & i \end{bmatrix}\right)}{\det(A)}, \quad y = \frac{\det\left(\begin{bmatrix} a & \alpha & c \\ d & \beta & f \\ g & \delta & i \end{bmatrix}\right)}{\det(A)}$$

$$z = \frac{\det\left(\begin{bmatrix} a & b & \alpha \\ d & e & \beta \\ g & h & \delta \end{bmatrix}\right)}{\det(A)} \tag{5.21}$$

This is again Cramer's rule at work on a 3×3 square linear system. Now, let us check this 3×3 case with an example.

Example 5.6.2. We want to solve the linear system

$$5x + 13y - 9z = -8$$
$$4x - 7y + 6z = 2$$
$$-x + 11y - 4z = 0$$

Its matrix of coefficients A, and column of RHS values B are given by

$$A = \begin{bmatrix} 5 & 13 & -9 \\ 4 & -7 & 6 \\ -1 & 11 & -4 \end{bmatrix}, B = \begin{bmatrix} -8 \\ 2 \\ 0 \end{bmatrix}$$

> A:= matrix([[5, 13, -9], [4, -7, 6], [-1, 11, -4]]): B:= matrix([[-8], [2], [0]]):
> xnum:= delcols(swapcol(augment(B,A), 1,2), 1..1);

$$xnum := \begin{bmatrix} -8 & 13 & -9 \\ 2 & -7 & 6 \\ 0 & 11 & -4 \end{bmatrix}$$

> ynum:= delcols(swapcol(augment(B,A), 1,3), 1..1);

$$ynum := \begin{bmatrix} 5 & -8 & -9 \\ 4 & 2 & 6 \\ -1 & 0 & -4 \end{bmatrix}$$

> znum:= delcols(swapcol(augment(B,A), 1,4), 1..1);

$$znum := \begin{bmatrix} 5 & 13 & -8 \\ 4 & 7 & -2 \\ -1 & 11 & 0 \end{bmatrix}$$

> X:= matrix([[det(xnum)/det(A)],[det(ynum)/det(A)],[det(znum)/det(A)]]);

$$X := \begin{bmatrix} -\frac{70}{131} \\ \frac{46}{131} \\ \frac{144}{131} \end{bmatrix}$$

> evalm(inverse(A)&*B);

$$\begin{bmatrix} -\frac{70}{131} \\ \frac{46}{131} \\ \frac{144}{131} \end{bmatrix}$$

■

In both the 2×2 and 3×3 cases, it was shown that the kth variable's solution could be expressed as a ratio of determinants. The numerator is the determinant of the matrix obtained by replacing column k of A with B. The denominator is simply the determinant of A.

To generalize this process, given the system $AX = B$ with $A \in \mathbb{R}^{n \times n}$, $B \in \mathbb{R}^{n \times 1}$ and $X = [x_1, x_2, \ldots, x_n]^T$, the solution is given by

$$x_1 = \frac{\det(A_{1,B})}{\det(A)}, \quad x_2 = \frac{\det(A_{2,B})}{\det(A)}, \quad \ldots, x_n = \frac{\det(A_{n,B})}{\det(A)} \tag{5.22}$$

where $A_{k,B}$ is the matrix obtained by replacing column k of A with B. To see exactly how this works, we reexamine the adjoint formula. Under the assumption that A is invertible, then the solution to $AX = B$, with the help of the adjoint formula (5.7), is given by

$$X = \frac{1}{\det(A)} C^T B$$

If we explicitly write out the formula for the ith entry, X_i, in the column matrix X, we have that

$$
\begin{aligned}
X_i &= \frac{1}{\det(A)} \left[C_{i,1}^T B_1 + C_{i,2}^T B_2 + \cdots + C_{i,n}^T B_n \right] \\
&= \frac{1}{\det(A)} \left[C_{1,i} B_1 + C_{2,i} B_2 + \cdots + C_{n,i} B_n \right]
\end{aligned}
\tag{5.23}
$$

Notice that the second line in the above expression is also equivalent to computing the determinant of the matrix A with column i replaced with the column matrix B. This is how we end up with equation (5.22), Cramer's rule.

Example 5.6.3. We do one last example with a 4×4 system:

$$
\begin{bmatrix} -7 & -1 & 2 & 3 \\ -2 & 0 & -5 & 4 \\ -2 & 3 & 0 & 1 \\ 0 & 4 & -3 & 1 \end{bmatrix}
\begin{bmatrix} x_1 \\ x_2 \\ x_3 \\ x_4 \end{bmatrix}
=
\begin{bmatrix} -2 \\ 3 \\ 1 \\ -2 \end{bmatrix}
$$

Using the formula given in (5.22), we have that

$$
\begin{bmatrix} x_1 \\ x_2 \\ x_3 \\ x_4 \end{bmatrix}
= \frac{1}{\det(A)}
\begin{bmatrix} \det(A_{1,B}) \\ \det(A_{2,B}) \\ \det(A_{3,B}) \\ \det(A_{4,B}) \end{bmatrix}
$$

where $\det(A) = 61$ and:

$$
\det(A_{1,B}) = \det \left(\begin{bmatrix} -2 & -1 & 2 & 3 \\ 3 & 0 & -5 & 4 \\ 1 & 3 & 0 & 1 \\ -2 & 4 & -3 & 1 \end{bmatrix} \right) = 399
$$

$$
\det(A_{2,B}) = \det \left(\begin{bmatrix} -7 & -2 & 2 & 3 \\ -2 & 3 & -5 & 4 \\ -2 & 1 & 0 & 1 \\ 0 & -2 & -3 & 1 \end{bmatrix} \right) = 60
$$

$$
\det(A_{3,B}) = \det \left(\begin{bmatrix} -7 & -1 & -2 & 3 \\ -2 & 0 & 3 & 4 \\ -2 & 3 & 1 & 1 \\ 0 & 4 & -2 & 1 \end{bmatrix} \right) = 347
$$

$$\det\left(A_{4,B}\right) = \det\left(\begin{bmatrix} -7 & -1 & 2 & -2 \\ -2 & 0 & -5 & 3 \\ -2 & 3 & 0 & 1 \\ 0 & 4 & -3 & -2 \end{bmatrix}\right) = 679$$

After computing the determinants of the four $A_{i,B}$ matrices, we arrive at our solution:

$$X = \frac{1}{61}\begin{bmatrix} 399 \\ 60 \\ 347 \\ 679 \end{bmatrix}$$

Once again, we will have *Maple* compute the $A_{i,B}$'s, but before we do, we should point out that each of the $A_{i,B}$'s can be computed using several different sequences of commands. In order to make the process simple to program, we use the following steps. If $A \in \mathbb{R}^{n\times n}$, and $B \in \mathbb{R}^{n\times 1}$, then we augment B as the first column of A, resulting in the matrix $(B|A) \in \mathbb{R}^{n\times(n+1)}$. Since we wish to replace the ith column of A with B, which is now column $i+1$ of $(B|A)$, we swap the column $i+1$ and the first column of $(B|A)$. Finally, we delete this first column, which corresponds to the ith column of the original matrix A. This will give us $A_{i,B}$. We use this process in the commands below.

> A:= matrix(4,4,[-7,-1,2,3,-2,0,-5,4,-2,3,0,1,0,4,-3,1]):
> B:= matrix(4,1,[-2,3,1,-2]):
> A1B:= delcols(swapcol(augment(B,A), 1,2), 1..1);

$$A1B := \begin{bmatrix} -2 & -1 & 2 & 3 \\ 3 & 0 & -5 & 4 \\ 1 & 3 & 0 & 1 \\ -2 & 4 & -3 & 1 \end{bmatrix}$$

> A2B:= delcols(swapcol(augment(B,A), 1,3), 1..1);

$$A2B := \begin{bmatrix} -7 & -2 & 2 & 3 \\ -2 & 3 & -5 & 4 \\ -2 & 1 & 0 & 1 \\ 0 & -2 & -3 & 1 \end{bmatrix}$$

> A3B:= delcols(swapcol(augment(B,A), 1,4), 1..1);

$$A3B := \begin{bmatrix} -7 & -1 & -2 & 3 \\ -2 & 0 & 3 & 4 \\ -2 & 3 & 1 & 1 \\ 0 & 4 & -2 & 1 \end{bmatrix}$$

> A4B:= delcols(swapcol(augment(B,A), 1,5), 1..1);

$$A4B := \begin{bmatrix} -7 & -1 & 2 & -2 \\ -2 & 0 & -5 & 3 \\ -2 & 3 & 0 & 1 \\ 0 & 4 & -3 & -2 \end{bmatrix}$$

> x1:= det(A1B)/det(A);

$$x1 := \frac{399}{61}$$

> x2:= det(A2B)/det(A);

$$x2 := \frac{60}{61}$$

> x3:= det(A3B)/det(A);

$$x3 := \frac{347}{61}$$

> x4:= det(A4B)/det(A);

$$x4 := \frac{679}{61}$$

> evalm(inverse(A)&*B);

$$\begin{bmatrix} \frac{399}{61} \\ \frac{60}{61} \\ \frac{347}{61} \\ \frac{679}{61} \end{bmatrix}$$

■

Homework Problems

1. Solve the following systems of equations by using Cramer's rule:

$$\text{(a)} \quad \begin{bmatrix} 3 & 1 & -7 \\ 1 & -2 & 6 \\ 8 & 1 & -1 \end{bmatrix} \begin{bmatrix} x_1 \\ x_2 \\ x_3 \end{bmatrix} = \begin{bmatrix} 3 \\ -7 \\ 5 \end{bmatrix}$$

$$\text{(b)} \quad \begin{bmatrix} 8 & -8 & -9 \\ 4 & 2 & 8 \\ -1 & 1 & -1 \end{bmatrix} \begin{bmatrix} x_1 \\ x_2 \\ x_3 \end{bmatrix} = \begin{bmatrix} 5 \\ 1 \\ -1 \end{bmatrix}$$

$$
(c) \quad
\begin{bmatrix}
1 & -1 & 5 & 7 \\
3 & 2 & 4 & 7 \\
-1 & 2 & 3 & -2 \\
-2 & 0 & 6 & 4
\end{bmatrix}
\begin{bmatrix}
x_1 \\ x_2 \\ x_3 \\ x_4
\end{bmatrix}
=
\begin{bmatrix}
2 \\ 3 \\ 4 \\ -5
\end{bmatrix}
$$

$$
(d) \quad
\begin{bmatrix}
1 & 2 & 3 & -4 \\
1 & 2 & -3 & 4 \\
-1 & 3 & 7 & 8 \\
1 & -2 & -2 & 5
\end{bmatrix}
\begin{bmatrix}
x_1 \\ x_2 \\ x_3 \\ x_4
\end{bmatrix}
=
\begin{bmatrix}
0 \\ 2 \\ 2 \\ 1
\end{bmatrix}
$$

2. Given the following matrices,

$$
A =
\begin{bmatrix}
2 & -1 & 2 \\
3 & 1 & -1 \\
1 & 5 & 4
\end{bmatrix},
\quad
B =
\begin{bmatrix}
-1 \\ 3 \\ 7
\end{bmatrix}
$$

define $A_{i,B}$ to be as specified in this section, and $(A_i|B)$ to be the matrix found after removing column i from A and then augmenting the resulting matrix with B. Compute the determinants of the following matrices.

(a) A (b) $A_{1,B}$ (c) $A_{2,B}$ (d) $A_{3,B}$

(e) $(A_1|B)$ (f) $(A_2|B)$ (g) $(A_3|B)$

3. Determine the relationship between $\det(A_{i,B})$ and $\det((A_i|B))$, and use it to verify your results from problem 2.

4. Let a be a scalar in the matrix equation

$$
\begin{bmatrix}
a & 1 \\
2 & 3
\end{bmatrix}
\begin{bmatrix}
x \\ y
\end{bmatrix}
=
\begin{bmatrix}
4 \\ 5
\end{bmatrix}
$$

For what values of a does this system have one solution? For what values of a does this system have no solution? For what values of a does this system have infinitely many solutions?

5. Let a be a scalar, in the matrix equation

$$
\begin{bmatrix}
4 & 1 & 2 \\
a & 3 & a \\
0 & -1 & 5
\end{bmatrix}
\begin{bmatrix}
x \\ y \\ z
\end{bmatrix}
=
\begin{bmatrix}
4 \\ 5 \\ 6
\end{bmatrix}
$$

For what values of a does this system have one solution? For what values of a does this system have no solution? For what values of a does this system have infinitely many solutions?

6. For the matrices $A = \begin{bmatrix} 3-i & 1+2i \\ 5+i & 1-i \end{bmatrix}$ and $B = \begin{bmatrix} 3+5i \\ 2-6i \end{bmatrix}$, use Cramer's Rule to solve the system $AX = B$.

Maple Problems

1. Solve the following systems of equations by using Cramer's rule:

(a) $\begin{bmatrix} 3 & 1 & -7 & 4 & 3 \\ 1 & -2 & 6 & 2 & -2 \\ 8 & 1 & -1 & 1 & 0 \\ -2 & 3 & -1 & 1 & 2 \\ 4 & 7 & 3 & 2 & -1 \end{bmatrix} \begin{bmatrix} x_1 \\ x_2 \\ x_3 \\ x_4 \\ x_5 \end{bmatrix} = \begin{bmatrix} 3 \\ 2 \\ 9 \\ -1 \\ 2 \end{bmatrix}$

(b) $\begin{bmatrix} 8 & -8 & -9 & 5 & 1 \\ 4 & 2 & 8 & 0 & 2 \\ -1 & 1 & -1 & -1 & 7 \\ 0 & 1 & 4 & 7 & 7 \\ 3 & -1 & -1 & 8 & 0 \end{bmatrix} \begin{bmatrix} x_1 \\ x_2 \\ x_3 \\ x_4 \\ x_5 \end{bmatrix} = \begin{bmatrix} 9 \\ 1 \\ -1 \\ 0 \\ -1 \end{bmatrix}$

(c) $\begin{bmatrix} 1 & -1 & 5 & 7 & -3 & 2 \\ 3 & 2 & 4 & 7 & 3 & -4 \\ -1 & 2 & 3 & -2 & 1 & 3 \\ -2 & 0 & 6 & 4 & 9 & 9 \\ 6 & -2 & 9 & 2 & 0 & 3 \\ -2 & 0 & 3 & -5 & 9 & -8 \end{bmatrix} \begin{bmatrix} x_1 \\ x_2 \\ x_3 \\ x_4 \\ x_5 \\ x_6 \end{bmatrix} = \begin{bmatrix} 3 \\ -3 \\ 4 \\ 0 \\ 4 \\ 1 \end{bmatrix}$

(d) $\begin{bmatrix} 1 & 2 & 3 & -4 & -3 & 0 \\ 5 & -3 & 0 & 2 & 2 & -3 \\ 1 & 2 & -3 & 4 & 1 & 2 \\ -1 & 3 & 7 & 8 & 4 & -1 \\ 1 & -2 & -2 & 5 & -5 & 5 \\ 0 & -2 & -2 & 9 & -5 & 7 \end{bmatrix} \begin{bmatrix} x_1 \\ x_2 \\ x_3 \\ x_4 \\ x_5 \\ x_6 \end{bmatrix} = \begin{bmatrix} -2 \\ 3 \\ 5 \\ 1 \\ -2 \\ 1 \end{bmatrix}$

2. Using your answer to *Homework* problem 3, solve the systems of problem 1 without the use of the *swapcol* command.

3. Use Cramer's rule to solve the system $AX = B$, for A and B defined as follows:

$$A = \begin{bmatrix} 3-i & 1+2i & i & 1+i \\ 5+i & 1-i & -i & 7-i \\ 3+i & 7+2i & 10 & 4+i \\ 6-i & 2-i & 8-i & 7+i \end{bmatrix}, \quad B = \begin{bmatrix} 3+5i \\ 2-6i \\ -2i \\ -4+i \end{bmatrix}$$

Chapter 6

Basic Linear Algebra Topics

6.1 Vectors

The mathematical idea of a vector comes from physics and the ideas of force, velocity, and acceleration. A force (or velocity and acceleration) in physics has two characteristics, the first is strength (or magnitude), which must be non-negative, and the second is direction. If the acceleration on an object is due only to the force of gravity, then the direction of gravity's acceleration is straight down toward the center of the Earth and its magnitude is almost constant near the surface of the Earth, at 32.2 ft/s^2. A vector is any quantity for which the two defining traits of magnitude (norm) and direction can be realized mathematically.

We will denote a vector as a variable with an arrow over it (e.g., $\vec{v}$). When specifically defining a vector, we will use angled brackets. As an example, $\vec{v} = \langle 1, -1, 2, 1 \rangle$ is a vector in $\mathbb{R}^4$. You should think of $\vec{v}$ as the arrow starting at the origin $(0,0,0,0)$ as its base point in $\mathbb{R}^4$ and stopping at the point $(1, -1, 2, 1)$ in $\mathbb{R}^4$. The magnitude of this vector $\vec{v}$, denoted $|\vec{v}|$, is the length of the arrow which is

$$|\vec{v}| = \sqrt{1^2 + (-1)^2 + 2^2 + 1^2} = \sqrt{7}$$

The direction of this vector $\vec{v}$ is given by the arrow that represents it. We have used here that the distance $d(P, Q)$ between two points $P, Q \in \mathbb{R}^n$ is given by

$$d(P, Q) = \sqrt{(p_1 - q_1)^2 + (p_2 - q_2)^2 + \cdots + (p_n - q_n)^2} \qquad (6.1)$$

if $P = (p_1, p_2, \ldots, p_n)$ and $Q = (q_1, q_2, \ldots, q_n)$, which generalizes the distance formula in $\mathbb{R}^2$.

This example also shows that we can have vectors in dimensions greater than three. One of the first things to note is that there is only one vector of magnitude 0, no matter what dimension our vector may be in. This vector is called the zero vector and denoted $\overrightarrow{0}$, given as $\langle 0, 0 \rangle$ in $\mathbb{R}^2$ and $\langle 0, 0, 0, 0 \rangle$ in $\mathbb{R}^4$, for example. The zero vector $\langle 0, 0, \ldots, 0 \rangle \in \mathbb{R}^n$ is the only vector for which direction does not play a role.

Now that we have a definition of vectors, we can think of real Euclidean n-space as a vector space. Instead of n-tuples of points, we will have vectors of dimension n.

Definition 6.1.1. The *vector space* $\mathbb{R}^n$ is the set of vectors $\overrightarrow{v}$ represented by arrows starting at the base point of the origin $(0, 0, \ldots, 0)$ in the point set $\mathbb{R}^n$ and stopping at the point $P = (p_1, p_2, \ldots, p_n)$ in the point set $\mathbb{R}^n$, and $\overrightarrow{v}$ is written as $\langle p_1, p_2, \ldots, p_n \rangle$ with n components $p_1, p_2, \ldots, p_n$.

$$\mathbb{R}^n = \left\{ \langle p_1, p_2, \ldots, p_n \rangle \,\middle|\, (p_1, p_2, \ldots, p_n) \in \mathbb{R}^n \right\}$$

The notation and context will tell us whether $\mathbb{R}^n$ is the vector space or the point set. The vector space $\mathbb{R}^n$ mathematically turns out to be the same as the set of column (or row) matrices in $\mathbb{R}^{n \times 1}$ (or $\mathbb{R}^{1 \times n}$) as we shall see shortly since in the vector space $\mathbb{R}^n$ we can add two vectors and also multiply them by real scalars to get new vectors in $\mathbb{R}^n$. For a more precise description of properties of a vector space, we refer you to the table in Section 8.1.

Definition 6.1.2. For $\overrightarrow{v} = \langle p_1, p_2, \ldots, p_n \rangle \in \mathbb{R}^n$, the *magnitude*, or *length*, of $\overrightarrow{v}$ is defined to be

$$|\overrightarrow{v}| = \sqrt{p_1^2 + p_2^2 + \cdots + p_n^2}$$

A vector $\overrightarrow{v}$ is called a *unit vector* if $|\overrightarrow{v}| = 1$. $\mathbb{R}^n$ is called a vector space because we will be able to do arithmetic in it. The vector space $\mathbb{R}^n$ is said to have *dimension* n since it takes n independent components to form each element of $\mathbb{R}^n$.

We now return to our discussion of vectors. The easiest way to geometrically represent a vector $\overrightarrow{v}$ in $\mathbb{R}^2$ or $\mathbb{R}^3$, with the two properties of magnitude and direction, is that of an arrow with the length of the arrow the magnitude of $\overrightarrow{v}$, and the direction of the arrow the direction of $\overrightarrow{v}$. Two vectors are said to have the same direction if their arrows are parallel, and when their arrows are drawn from a common starting or base point, one of the arrows follows exactly over the other. Two vectors are considered equal if they have the same identical arrows representing them when started at the same base point. Two vectors are said to have opposite directions to each other if when each is represented by arrows with the same base point, then they point in opposite directions along the same line. If $\overrightarrow{v}$ is a vector, then $-\overrightarrow{v}$ is the vector of the same magnitude as $\overrightarrow{v}$, but pointing in the opposite direction.

If c is a real number and $\vec{v}$ is a vector, then we define $c\vec{v}$ to be the vector of magnitude $|c\vec{v}| = |c|\,|\vec{v}|$, where $\vec{v}$ and $c\vec{v}$ have the same direction if $c > 0$ and have opposite directions if $c < 0$. When $c = 0$, then $c\vec{v}$ is the zero vector for any vector $\vec{v}$. Thus, $5\vec{v}$ is a vector five times the magnitude of $\vec{v}$ and in the same direction as $\vec{v}$, while $-5\vec{v}$ is a vector five times the magnitude of $\vec{v}$ and in the opposite direction to $\vec{v}$. Two vectors $\vec{v}$ and $\vec{w}$ are said to be parallel if they are multiples of each other, that is, there is a real number c so that $\vec{w} = c\vec{v}$ (or $\vec{v} = c\vec{w}$). This also implies that two vectors are parallel exactly when their respective arrows are parallel, or equivalently, they are parallel exactly when they determine a pair of parallel lines in $\mathbb{R}^n$.

Now, let us illustrate these concepts. Remember that an arrow represents the same vector no matter its starting or base point, as long as it has the same magnitude (length) and direction. The *Maple* command *arrow* in the *plottools* package allows us to plot arrows to represent vectors, while the command *vector* from the *linalg* package creates a vector in *Maple* as a one-dimensional array or column matrix.

Example 6.1.1. Let $\vec{v} = \langle 13, 0 \rangle \in \mathbb{R}^2$ and $\vec{w} = \langle 9, 15 \rangle \in \mathbb{R}^2$. Then $2\vec{v} = \langle 26, 0 \rangle$ and $-3\vec{w} = \langle -27, -45 \rangle$. The graphs of the vectors $\vec{w}$, $\vec{w}$ and $2\vec{v}$ with base point $P(-7, -19)$, and $\vec{v}$ and $-3\vec{w}$ with base point $Q(20, 14)$ are depicted in Figure 6.1.

```
> with(plots): with(plottools): with(linalg):
> P:= [-7,-19]:
> Q:= [20,14]:
> W:= vector([9,15]):
> V:= vector([13,0]):
> TwiceV:= vector([26,0]):
> MinusThreeW:= vector([-27,-45]):
> arrow_W:= arrow([0,0], W, .4, 2, .2, color = blue):
> arrow_WfromP:= arrow(P, W, .4, 2, .2, color = blue):
> arrow_V:= arrow(Q, V, .4, 2, .2, color = red):
> arrow_TwiceV:= arrow(P, TwiceV, .4, 2, .2, color = red):
> arrow_MinusThreeW:= arrow(Q, MinusThreeW, .4,2,.2, color=blue):
> LabelW:= textplot([4,7,"W"], align={ABOVE,LEFT}, color=black):
> LabelV:= textplot([25,12,"V"], align=BELOW, color=black):
> LabelTwiceV:= textplot([6,-20,"2*V"], align=BELOW, color=black):
> LabelMinusThreeW:= textplot([14,-2,"-3*W"], align = BELOW, color = black):
```

> LabelWfromP:= textplot([-2,-10,"W"], align = {ABOVE,LEFT}, color = black):

> display({arrow_W, arrow_V, arrow_WfromP, arrow_TwiceV, arrow_Minus-ThreeW, LabelW, LabelV, LabelTwiceV, LabelMinusThreeW, LabelWfromP}, scaling=constrained, axes=none);

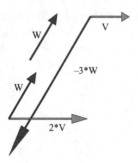

Figure 6.1: Graphical depiction of several vectors at three base points.

Now that we have the geometric interpretation of vectors well in hand, let us move on to vector addition and subtraction. The key to these two operations is the parallelogram law that states geometrically how to add vectors together, which is based on how, physically, two forces are combined to create a single force. In order to add two vectors, $\vec{v}$ and $\vec{w}$, we represent them as arrows starting at the same base point and then the arrow representing their sum $\vec{v} + \vec{w}$ starts at this common base point and stops at the point that is at the opposite end of a diagonal for the parallelogram with $\vec{v}$ and $\vec{w}$ as two adjacent sides. The graph depicted in Figure 6.2 will clarify how this works.

Example 6.1.2. Let $\vec{v} = \langle -10, 17 \rangle \in \mathbb{R}^2$ and $\vec{w} = \langle 35, 8 \rangle \in \mathbb{R}^2$. Then, $\vec{v} + \vec{w} = \langle 25, 25 \rangle$. We will plot these vectors with *Maple*.

> V:= vector([-10,17]):

> W:= vector([35,8]):

> arrow_V:= arrow([0,0], V, .3, 2, .2, color = black):

> LabelV:= textplot([-5,11,"V"], align = ABOVE, color = black):

> arrow_W:= arrow([0,0], W, .3, 2, .2, color = blue):

> LabelW:= textplot([16,3,"W"], align = BELOW, color = black):

> arrow_VplusW:= arrow([0,0], evalm(V+W), .3, 2, .2, color = red):

> LabelVplusW:= textplot([15,12,"V+W"],align=BELOW,color=black):

> Line1:= line([-10,17], [25,25], color = black):

> Line2:= line([35,8], [25,25], color = black):

> display({arrow_W, arrow_V, LabelW, LabelV, arrow_VplusW, LabelVplusW, Line1, Line2}, scaling = constrained, axes = none);

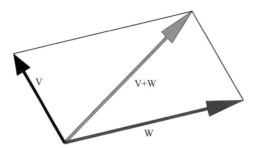

Figure 6.2: The parallelogram formed by the two vectors $\vec{v}$ and $\vec{w}$, and the resulting sum of the two vectors.

Note that for any vector $\vec{v}$, we should define $\vec{v} + (-\vec{v}) = \vec{0}$ since, physically, if you combine two opposite but equal forces, they cancel each other out to a force of $\vec{0}$. You may begin to notice that many of the rules of applying operations to real numbers have corresponding counterparts when dealing with vectors. Try to keep this in mind as you read through this chapter.

Now in order to subtract vectors, we write $\vec{v} - \vec{w} = \vec{v} + (-\vec{w})$, and then perform addition that we already know how to do.

Example 6.1.3. Let $\vec{v} = \langle -10, 17 \rangle \in \mathbb{R}^2$ and $\vec{w} = \langle 35, 8 \rangle \in \mathbb{R}^2$ as in Example 6.1.2. Then $\vec{v} - \vec{w} = \langle -45, 5 \rangle$. We will plot both $\vec{v}$ and $\vec{w}$, along with $\vec{v} - \vec{w}$ and the diagonal of the parallelogram formed by $\vec{v}$ and $\vec{w}$. We will use the same first five commands as those of Example 6.1.2, which we will not reproduce here.

> LabelW:= textplot([16,2, "W"], align = BELOW, color = black):

> arrow_MinusW:= arrow([0,0], evalm(-W), .3, 2, .2, color = blue):

> LabelMinusW:= textplot([-16,-5, "-W"], align=BELOW, color=black):

> arrow_VminusW:= arrow([0,0], evalm(V-W), .3,2,.2, color = red):

> LabelVminusW:= textplot([-21,3, "V-W"],align=BELOW,color=black):

> Line1:= line([-10,17], [-45,9], color = black):

> arrow_VminusWshifted:= arrow([35,8], evalm(V-W), .3, 2, .2, color = red):

> LabelVminusWshifted:= textplot([14,11, "V-W"], align = BELOW, color = black):

> Line2:= line([-35,-8], [-45,9], color = black):

> display({arrow_W, arrow_V, LabelW, LabelV, arrow_VminusW, LabelVmi-nusW, arrow_MinusW, LabelMinusW, arrow_VminusWshifted, LabelVmi-nusWshifted, Line1, Line2}, axes = none);

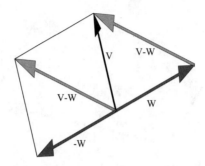

Figure 6.3: The difference of two vectors $\vec{v}$ and $\vec{w}$.

In Figure 6.3 above, notice that the vector $\vec{v} - \vec{w}$ completes the third side of the triangle which has the two vectors $\vec{v}$ and $\vec{w}$ as adjacent sides. Depicted here are two vectors corresponding to $\vec{v} - \vec{w}$, having the same magnitude and direction, and are therefore equivalent.

We now move onto the discussion of linear combinations of vectors. Recall, we defined a linear equation in definition 2.1.2. In a similar fashion we define a linear combination of vectors.

Definition 6.1.3. A *linear combination* $\vec{v}$ of vectors $\vec{v_1}, \vec{v_2}, \ldots, \vec{v_k} \in \mathbb{R}^n$ is another vector in $\mathbb{R}^n$ of the form

$$\vec{v} = a_1 \vec{v_1} + a_2 \vec{v_2} + \cdots + a_k \vec{v_k}$$

for real scalars $a_1, a_2, \ldots, a_k$. Linear combinations are similarly defined for vectors from $\mathbb{C}^n$.

The simplest case to consider is a linear combination, $a\vec{v} + b\vec{w}$, of two vectors $\vec{v}$ and $\vec{w}$, for real constants a and b. Note that when discussing vectors, it is traditional to call real constants *scalars* and to call multiplication of a vector by a real constant *scalar multiplication*.

Any two nonzero vectors $\vec{v}$ and $\vec{w}$ that are not parallel determine the unique plane P that passes through both $\vec{v}$ and $\vec{w}$ as arrows starting at the same base point of the origin, and this plane P consists of all the arrows starting at the origin formed by the vectors of their linear combinations $a\vec{v} + b\vec{w}$, that is, $P = \{a\vec{v} + b\vec{w} \mid a, b \in \mathbb{R}\}$. This plane P as a point set is the unique plane through the three points {origin, endpoint of $\vec{v}$'s arrow starting at the origin, endpoint of $\vec{w}$'s arrow starting at the origin}.

Therefore, any linear combination $a\vec{v} + b\vec{w}$ of the two vectors $\vec{v}$ and $\vec{w}$ will also lie in this plane. Moreover, if $\vec{u}$ is any vector in this plane, then there are unique scalars a and b so that $\vec{u} = a\vec{v} + b\vec{w}$. This fact has different meanings depending on what dimension the vectors are elements of. For instance, if the two vectors are elements of $\mathbb{R}^2$, then the plane formed by the two vectors contains all of $\mathbb{R}^2$. However, if the vectors are elements of $\mathbb{R}^3$, then they form a plane in $\mathbb{R}^3$, which is simply a subset of $\mathbb{R}^3$. We will now have *Maple* illustrate this concept for us.

Example 6.1.4. We will once again consider the vectors $\vec{v} = \langle -10, 17 \rangle \in \mathbb{R}^2$ and $\vec{w} = \langle 35, 8 \rangle \in \mathbb{R}^2$. The linear combination we will consider here is $2.5\vec{v} + 3.7\vec{w} = \langle 104.5, 72.1 \rangle$, which is depicted in Figure 6.4.

> evalm(2.5*V+3.7*W);

$$\begin{bmatrix} 104.5 & 72.1 \end{bmatrix}$$

> arrow_aV:= arrow([0,0], evalm(2.5*V), .4, 4, .2, color = black):

> LabelaV:= textplot([-10,30, "a*V"], align = BELOW, color = black):

> arrow_bW:= arrow([0,0], evalm(3.7*W), .4, 4, .2, color = blue):

> LabelbW:= textplot([120,25, "b*W"], align = BELOW, color = black):

> arrow_aVplusbW:= arrow([0, 0], evalm(2.5*V + 3.7*W), .4, 4, .2, color = red):

> LabelaVplusbW:= textplot([95,55, "a*V+b*W"], align = BELOW, color = black):

> Line1:= line([2.5*(-10),2.5*17], [104.5,72.1], color = black):

> Line2:= line([3.7*35,3.7*8], [104.5,72.1], color = black):

> display({arrow_W, arrow_V, LabelW, LabelV, arrow_aV, arrow_bW, LabelaV, LabelbW, arrow_aVplusbW, LabelaVplusbW, Line1, Line2}, axes = boxed);

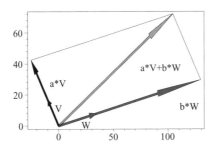

Figure 6.4: The graphical depiction of a linear combination of vectors.

For the next set of operations, we will restrict ourselves to working in the xy-plane (i.e. $\mathbb{R}^2$), however, note that all the material presented here can be generalized to higher dimensions. First, we will let all of our vectors be represented by arrows starting at the origin. Next, we define $\vec{i}$ to be the unit vector from the origin to the point $(1,0)$. We can write $\vec{i}$ as its endpoint, or $\vec{i} = \langle 1, 0 \rangle$. Similarly, let $\vec{j}$ be the unit vector from the origin to the point $(0,1)$, and thus $\vec{j} = \langle 0, 1 \rangle$. Then for any scalars a and b, $a\vec{i}$ is the vector from the origin to the point $(a,0)$, while $b\vec{j}$ is the vector from the origin to the point $(0,b)$. As well, the vector $a\vec{i} + b\vec{j}$ is the vector from the origin to the point (a,b). It now makes sense to represent any vector, $\vec{v} = a\vec{i} + b\vec{j}$ in the xy-plane, by its endpoint (a,b) when all vectors start at the origin. That is, $\vec{v} = \langle a, b \rangle$. The first *component* of $\vec{v}$ is a, and the second component of $\vec{v}$ is b. A vector $\vec{v} = \langle a, b \rangle$, with two components, is said to be a two-dimensional vector and an element of $\mathbb{R}^2$.

Definition 6.1.4. Given a scalar α, and vector $\vec{v} = \langle p_1, p_2, \ldots, p_n \rangle \in \mathbb{R}^n$, the *scalar multiplication* $\alpha\vec{v}$ is defined to be

$$\alpha\vec{v} = \langle \alpha\, p_1, \, \alpha\, p_2, \ldots, \alpha\, p_n \rangle$$

Definition 6.1.5. Given two vectors $\vec{v} = \langle p_1, p_2, \ldots, p_n \rangle$ and $\vec{w} = \langle q_1, q_2, \ldots, q_n \rangle$ of $\mathbb{R}^n$, we define the *vector addition* of $\vec{v} + \vec{w}$ to be the addition of components, that is

$$\vec{v} + \vec{w} = \langle p_1 + q_1, p_2 + q_2, \ldots, p_n + q_n \rangle$$

Putting these two definitions together tells us that if $\vec{v} = a\vec{i} + b\vec{j}$ and $\vec{w} = c\vec{i} + d\vec{j}$, for any two vectors $\vec{v}$ and $\vec{w}$ in the xy-plane, then

$$\alpha\vec{v} + \beta\vec{w} = (\alpha a + \beta c)\,\vec{i} + (\alpha b + \beta d)\,\vec{j}$$

or,

$$\alpha\langle a, b \rangle + \beta\langle c, d \rangle = \langle \alpha a + \beta c, \, \alpha b + \beta d \rangle$$

Example 6.1.5. We will now draw a picture to illustrate how the vector $\vec{v} = 5\vec{i} + 3\vec{j} = \langle 5, 3 \rangle$ relates to the two unit vectors $\vec{i}$ and $\vec{j}$. Refer to Figure 6.5.

```
> V:= vector([5,3]): Vector_I:= vector([1,0]): Vector_5I:= vector([5,0]):
> J:= vector([0,1]): Vector_3J:= vector([0,3]):
> arrow_V:= arrow([0,0], V, .1, .5, .2, color = red):
> LabelV:= textplot([3,2,"V"], align = ABOVE, font = [Times, 20], color = black):
```

> arrow_I:= arrow([0,0], Vector_I, .1, .5, .2, color = blue):
> LabelI:= textplot([1.1,.25,"I"], font = [Times, 20], color = black):
> arrow_J:= arrow([0,0], J, .1, .5, .2, color = black):
> LabelJ:= textplot([.3,1.1,"J"], font = [Times, 20], color = black):
> arrow_5I:= arrow([0,0], Vector_5I, .1, .5, .2, color = blue):
> Label5I:= textplot([3.5,.25,"5*I"], font=[Times, 20], color=black):
> arrow_3J:= arrow([0,0], Vector_3J, .1, .5, .2, color = black):
> Label3J:= textplot([.35,2,"3*J"], font=[Times, 20], color=black):
> display({arrow_V, arrow_I, LabelV, LabelI, arrow_J, LabelJ, arrow_5I, Label5I, arrow_3J, Label3J}, axes = boxed);

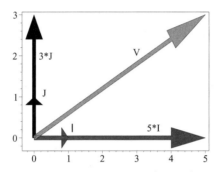

Figure 6.5: The vector $\vec{v} = 5\,\vec{i} + 3\,\vec{j} = \langle 5, 3 \rangle$.

We next move on to three-dimensional vectors. In $\mathbb{R}^3$, with the xyz-coordinate system, we have three unit vectors $\vec{i} = \langle 1, 0, 0 \rangle$, $\vec{j} = \langle 0, 1, 0 \rangle$, and $\vec{k} = \langle 0, 0, 1 \rangle$ moving along the positive coordinate axes. Every three-dimensional vector $\vec{v}$ of $\mathbb{R}^3$ can be written as a linear combination of $\vec{i}$, $\vec{j}$, and $\vec{k}$, that is, $\vec{v} = a\,\vec{i} + b\,\vec{j} + c\,\vec{k} = \langle a, b, c \rangle$.

Example 6.1.6. Let us plot the vector $\vec{v} = 4\,\vec{i} - 7\,\vec{j} - 3\,\vec{k} = \langle 4, -7, -3 \rangle$, which is depicted in Figure 6.6.

> V:= vector([4,-7,-3]):
> arrow_V:= arrow([0,0,0], V, .1, .5, .2, cylindrical_arrow, color = red):
> LabelV:= textplot3d([2,-4,-1.5,"V"], align = ABOVE, color = black):
> Vector_I:= vector([1,0,0]): J:= vector([0,1,0]): K:= vector([0,0,1]):
> arrow_I:= arrow([0,0,0], Vector_I, .1, .5, .2, color = blue):
> arrow_J:= arrow([0,0,0], J, .1, .5, .2, color = blue):

> arrow_K:= arrow([0,0,0], K, .1, .5, .2, color = blue):
> LabelI:= textplot3d([1.2,0,0,"I"], align = BELOW, color = black):
> LabelJ:= textplot3d([0,1.2,0,"J"], align = BELOW, color = black):
> LabelK:= textplot3d([0,0,1.2,"K"], align = BELOW, color = black):
> Vector_4I:= vector([4,0,0]):
> Vector_7J:= vector([0,-7,0]):
> Vector_3K:= vector([0,0,-3]):
> arrow4I:= arrow([0,0,0], Vector_4I, .1, .5, .2, color = blue):
> arrow_7J:= arrow([0,0,0], Vector_7J, .1, .5, .2, color = blue):
> arrow_3K:= arrow([0,0,0], Vector_3K, .1, .5, .2, color = blue):
> Label4I:= textplot3d([3,0,0,"4*I"], align = ABOVE, color = black):
> Label_7J:= textplot3d([0,-5,0,"-7*J"], align = ABOVE, color = black):
> Label_3K:= textplot3d([0,0,-2,"-3*K"], align=ABOVE, color=black):
> display({arrow_V, arrow_I, LabelV, LabelI, arrow_J, LabelJ, arrow_K, LabelK, arrow4I, Label4I, arrow_7J, Label_7J, arrow_3K, Label_3K}, labels = [x,y,z], axes = frame);

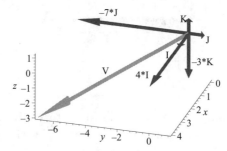

Figure 6.6: Graphical depiction of a three-dimensional vector $\vec{v}$ expressed as a linear combination of $\vec{i}$, $\vec{j}$, and $\vec{k}$.

Unfortunately, we cannot graphically depict vectors that are of dimension four or greater, but we can express them mathematically. $\mathbb{R}^n$ is the set of n-dimensional vectors $\vec{v}$ that have n components (i.e., $\vec{v} = \langle a_1, a_2, \ldots, a_n \rangle$, for n scalar components $a_1, a_2, \ldots, a_n$). Vectors work in exactly the same manner no matter their dimension. That is, they are treated like (row or column) matrices for the purposes of computation.

We continue this section with some terminology. Given a point P in $\mathbb{R}^n$, the vector $\vec{v_P}$, which is represented by an arrow starting at the origin and stopping at a point P, is called the *position vector* for the point P with $\vec{v_P} = \langle P \rangle$. If we have an arrow representing a vector $\vec{v}$ that starts at the point P and stops at the point Q, then the vector $\vec{v}$ is written as $\vec{v} = \overrightarrow{PQ}$, and it is computed by $\vec{v} = \overrightarrow{v_Q} - \overrightarrow{v_P}$. This is an example of a *displacement vector*. We can write any position vector as a displacement vector by letting P be the origin, that is, the position vector $\overrightarrow{v_Q} = \overrightarrow{0Q}$. We now do an example of this in order to see how this works.

Example 6.1.7. Let us compute and plot the displacement vector $\vec{v} = \overrightarrow{PQ}$, which is represented by the arrow starting at the point $P(-8, 11)$ and stopping at the point $Q(7, 4)$. This vectors in this example are illustrated in Figure 6.7.

> P:= [-8,11]: Q:= [7,4]:
> V:= evalm(Q-P);

$$V := \begin{bmatrix} 15 & -7 \end{bmatrix}$$

> arrow_V:= arrow([0,0], V, .2, 1, .2, color = red):
> LabelV:= textplot([op(convert(V,list)),"V"], align = {ABOVE, RIGHT}, font = [Times, 10], color = black):
> arrowP:= arrow([0,0], P, .2, 1, .2, color = blue):
> LabelP:= textplot([op(P),"P"], font = [Times, 10], align = ABOVE, color = black):
> arrowQ:= arrow([0,0], Q, .2, 1, .2, color = blue):
> LabelQ:= textplot([op(Q),"Q"], font = [Times, 10], align = {ABOVE, RIGHT}, color = black):
> arrowPQ:= arrow(P, Q, .2, 1, .2, color = red):
> LabelPQ:= textplot([op(convert(evalm((P+Q)/2),list)), "PQ"], align = {ABOVE, RIGHT}, font = [Times, 10], color = black):
> display({arrow_V, LabelV, arrowP, LabelP, arrowQ, LabelQ, arrowPQ, LabelPQ}, axes = boxed, view = [-10..16, -8..13]);

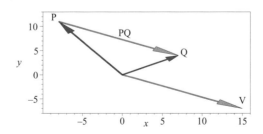

Figure 6.7: Displacement vector $\overrightarrow{PQ}$ and position vector $\vec{v}$ in relation to $\vec{P}$ and $\vec{Q}$. Notice that $\overrightarrow{PQ}$ and $\vec{v}$ have the same direction and magnitude.

Vectors in mathematics are very useful for doing geometry, so we need to be able to define the angle between two vectors. The angle θ between two nonzero vectors $\vec{v}$ and $\vec{w}$ is the angle between 0 and π radians that is the smallest angle formed by placing the arrows representing $\vec{v}$ and $\vec{w}$ at the same starting point. Again, we illustrate this idea with the help of *Maple*, as depicted in Figure 6.8.

> V:= vector([-10,17]):

> W:= vector([35,8]):

> arrow_V:= arrow([0,0], V, .3, 2, .2, color = black):

> LabelV:= textplot([-5,11, "V"], align = ABOVE, color = black):

> arrow_W:= arrow([0,0], W, .3, 2, .2, color = blue):

> LabelW:= textplot([16,3, "W"], align = BELOW, color = black):

> LabelAngle:= textplot([1.5,3, q], align = ABOVE, color = black, font = [SYMBOL, 10]):

> Arc:= arc([0,0], 7, arctan(8/35)..Pi-arctan(17/10), color = black):

> display({arrow_W, arrow_V, LabelW, LabelV, LabelAngle, Arc}, scaling = constrained, axes = none);

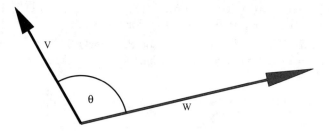

Figure 6.8: The angle θ between two vectors.

We shall be able to compute the angle between two vectors $\vec{v}$ and $\vec{w}$ using the dot product of two vectors that is defined in Section 6.2.

	Basic Vector Addition and Scalar Multiplication Properties
Vector Addition	$\vec{u} + \vec{v} = \vec{v} + \vec{u}$ $(\vec{u} + \vec{v}) + \vec{w} = \vec{u} + (\vec{v} + \vec{w})$ $\vec{u} + (-\vec{v}) = \vec{u} - \vec{v}$ $\vec{u} + (-\vec{u}) = \vec{0}$ $\vec{u} + \vec{0} = \vec{u}$
Scalar Multiplication	$a(\vec{u} + \vec{v}) = a\vec{u} + a\vec{v}$ $(a + b)\vec{u} = a\vec{u} + b\vec{u}$ $(ab)\vec{u} = a(b\vec{u})$ $1\vec{u} = \vec{u}$ $0\vec{u} = \vec{0}$ $|a\vec{u}| = |a||\vec{u}|$

Homework Problems

1. If $\vec{u} = \langle 1, 3 \rangle$, $\vec{v} = \langle 5, -7 \rangle$, and $\vec{w} = \langle -4, 5 \rangle$, perform the following operations:

(a) $\vec{u} - \vec{v}$ (b) $4\vec{u} + 3\vec{w}$ (c) $-6\vec{v} - 2\vec{w}$

(d) $\vec{u} + \vec{v} - \vec{w}$ (e) $2\vec{u} - 3\vec{v} + 6\vec{w}$ (f) $-3\vec{w} + 2\vec{v} - 7\vec{u}$

(g) $\alpha\vec{u} - \beta\vec{v}$ (h) $\alpha\vec{u} - \beta\vec{v} + \gamma\vec{w}$ (i) $\alpha\vec{w} + \beta\vec{v}$

2. If $\vec{u} = \langle -1, 3, 2 \rangle$, $\vec{v} = \langle 5, 0, -7 \rangle$, and $\vec{w} = \langle -4, -2, 2 \rangle$, perform the following operations:

(a) $3\vec{u} + 2\vec{v}$ (b) $6\vec{v} - 2\vec{w}$ (c) $7\vec{v} + 3\vec{w}$

(d) $2\vec{u} - \vec{v} + 3\vec{w}$ (e) $-5\vec{u} + 2\vec{w} - 3\vec{v}$ (f) $-6\vec{w} - \vec{v} + 8\vec{u}$

(g) $\alpha\vec{u} + \beta\vec{w}$ (h) $\alpha\vec{u} - \beta\vec{v} + \gamma\vec{w}$ (i) $\alpha\vec{u} + \beta\vec{v}$

3. If $\vec{u} = \langle -2, 3 \rangle$ and $\vec{v} = \langle 1, 2 \rangle$, find values of α and β for each of the following vectors $\vec{w}$ so that $\vec{w} = \alpha \vec{u} + \beta \vec{v}$.

(a) $\langle 1, 3 \rangle$ (b) $\langle -1, 2 \rangle$ (c) $\langle -5, 6 \rangle$

(d) $\langle -1, 5 \rangle$ (e) $\langle 4, 4 \rangle$ (f) $\langle -8, 12 \rangle$

4. If $\vec{u} = \langle -1, 1, 0 \rangle$, $\vec{v} = \langle 1, 1, 0 \rangle$, and $\vec{w} = \langle 0, 0, 1 \rangle$, find values of α, β, and γ for each of the following vectors $\vec{x}$, so that $\vec{x} = \alpha \vec{u} + \beta \vec{v} + \gamma \vec{w}$:

(a) $\langle 1, 3, -1 \rangle$ (b) $\langle -1, 2, 4 \rangle$ (c) $\langle -5, 6, -4 \rangle$

5. If $\vec{u} = \langle -1, 1, 1 \rangle$ and $\vec{v} = \langle 1, 0, 2 \rangle$, find values of α and β (if possible) for each of the following vectors $\vec{w}$ so that $\vec{w} = \alpha \vec{u} + \beta \vec{v}$:

(a) $\langle -5, 2, 4 \rangle$ (b) $\langle 1, 4, 14 \rangle$ (c) $\langle 0, 0, 1 \rangle$

(d) $\langle 0, 1, 2 \rangle$ (e) $\langle 1, 4, 1 \rangle$ (f) $\langle -1, 2, 4 \rangle$

6. For each of the vectors $\vec{w}$ of problem 5, construct a matrix $A \in \mathbb{R}^{3 \times 3}$ whose columns are the vectors $\vec{u}$, $\vec{v}$, and $\vec{w}$, then compute $\det(A)$.

7. Interpret your results from problem 6.

8. Let $\vec{v} = \langle a, b \rangle \in \mathbb{R}^2$. The slope of $\vec{v}$ is $m_{\vec{v}} = \dfrac{b}{a}$ if $a \neq 0$ and otherwise it does not exist.

(a) Explain why $\vec{v} = \langle a, b \rangle$ and $\vec{w} = \langle -b, a \rangle$ are perpendicular (or orthogonal) vectors.

(b) Let $\vec{w} \in \mathbb{R}^2$. Explain why $\vec{v}$ and $\vec{w}$ are parallel exactly when $m_{\vec{v}} = m_{\vec{w}}$, or both slopes do not exist.

9. Find and plot the vector $\vec{w}$ with the original vector $\vec{v}$:

(a) A unit vector $\vec{w}$ in the opposite direction to $\vec{v} = \langle -2, 5 \rangle$.

(b) A vector $\vec{w}$ of length seven in the same direction as $\vec{v} = \langle 4, -7 \rangle$.

(c) A vector $\vec{w}$ of length ten in the opposite direction to $\vec{v} = \langle 1, 6 \rangle$.

(d) A vector $\vec{w}$ of length three parallel to $\vec{v} = \langle -4, 8 \rangle$ starting at $P(5, 8)$.

(e) A vector $\overrightarrow{w}$ of length thirteen perpendicular to $\overrightarrow{v} = \langle -7, -3 \rangle$.

(f) A vector $\overrightarrow{w}$ of length two perpendicular to $\overrightarrow{v} = \langle 4, -8 \rangle$ starting at $P(3, 10)$.

(g) A vector $\overrightarrow{w}$ of length eight parallel to the line $4x + 7y = 10$.

10. Using trigonometry, find the angle between the two vectors $\overrightarrow{v} = \langle -2, 5 \rangle$ and $\overrightarrow{w} = \langle 7, 3 \rangle$.

11. Let $\overrightarrow{v} = \langle -2, 5 \rangle$ and $\overrightarrow{w} = \langle 7, 3 \rangle$ be two adjacent sides of a triangle. Find the length of the third side of this triangle using the length of a vector.

12. Find the distance between the two points $P(1, -4, 7, 0, 2)$, $Q(-9, 3, 5, -6, 8)$.

Maple Problems

1. Graph each set of three vectors from *Homework* problem 3.

2. Graph each set of three vectors from *Homework* problem 5. Is it easy to see from the visual depiction of the sets of vectors why your answers to *Homework* problem 5 are what they are?

3. Find and plot the vector $\overrightarrow{w}$ with the original vector $\overrightarrow{v}$:

(a) A unit vector $\overrightarrow{w}$ in the opposite direction to $\overrightarrow{v} = \langle -2, 5, 9 \rangle$.

(b) A vector $\overrightarrow{w}$ of length seven in the same direction as $\overrightarrow{v} = \langle 4, -7, -3 \rangle$.

(c) A vector $\overrightarrow{w}$ of length ten in the opposite direction to $\overrightarrow{v} = \langle 1, 6, 4 \rangle$.

(d) A vector $\overrightarrow{w}$ of length three parallel to $\overrightarrow{v} = \langle -4, 8, -5 \rangle$ starting at the point $P(5, 8, 2)$.

4. Using trigonometry, find the angle between the two vectors $\overrightarrow{v} = \langle -2, 5, 7 \rangle$ and $\overrightarrow{w} = \langle 6, -3, 9 \rangle$. Plot both of these vectors.

5. Find the lengths of the following vectors:

$\qquad$ (a) $\langle -2, 3, -5, 7, 0, 4 \rangle$, $\qquad$ (b) $\langle -9, -5, 2, -7, 1, -3, 4, 0, 6 \rangle$

6.2   Dot Product

We spent the entire last section developing the basic arithmetic concepts of addition, subtraction, and scalar multiplication for vectors, which also gave rise to linear combinations of vectors. The next concept from arithmetic that we wish to develop for vectors is "multiplication" of two vectors. The questions remain, how and what is the end result? One form of vector multiplication is known as the *dot product*, which takes two vectors and returns a real scalar.

Definition 6.2.1. The *dot product* of two n-dimensional column vectors $\vec{v}$, $\vec{w} \in \mathbb{R}^n$ is a real number, denoted $\vec{v} \cdot \vec{w}$, and defined in terms of matrix multiplication by $\vec{v} \cdot \vec{w} = \vec{v}^T \vec{w}$. More specifically, if $\vec{v} = \langle v_1, v_2, \ldots, v_n \rangle$ and $\vec{w} = \langle w_1, w_2, \ldots, w_n \rangle$, then

$$\vec{v} \cdot \vec{w} = \begin{bmatrix} v_1 & v_2 & \cdots & v_n \end{bmatrix} \begin{bmatrix} w_1 \\ w_2 \\ \vdots \\ w_n \end{bmatrix} = \sum_{j=1}^{n} v_j w_j \qquad (6.2)$$

As we will find out shortly, the dot product of two vectors is a very important concept since it incorporates, in a single construction, the two ideas of the length of a vector and the angle between two vectors. Since the dot product $\vec{v} \cdot \vec{v}$ of a vector $\vec{v}$ with itself is the square $|\vec{v}|^2$ of $\vec{v}$'s length, the dot product is actually a generalization of the basic geometric concept of length. Amazingly, the dot product of two vectors also allows us to determine the angle between them, and so the dot product encompasses all of basic geometry, since it is the embodiment of both length and angle.

It should be fairly clear from this definition of dot product that it has the following properties. First, it is symmetric in that $\vec{v} \cdot \vec{w} = \vec{w} \cdot \vec{v}$ and, second, we have the length property $\vec{v} \cdot \vec{v} = |\vec{v}|^2$, which leads us to the following definition:

Definition 6.2.2. The *length (norm, magnitude)* of a vector $\vec{v} \in \mathbb{R}^n$, denoted $|\vec{v}|$ is given, in terms of the dot product, by the formula

$$|\vec{v}| = \sqrt{\vec{v} \cdot \vec{v}} \qquad (6.3)$$

One more important property of the dot product is that it is distributive, in the sense that for any real numbers a and b and any vectors $\vec{u}$, $\vec{v}$, and $\vec{w}$ in $\mathbb{R}^n$:

$$\vec{v} \cdot (a\vec{u} + b\vec{w}) = a(\vec{v} \cdot \vec{u}) + b(\vec{v} \cdot \vec{w}) \qquad (6.4)$$

The identity given above is simple to show, and even simpler if you consider the dot product as a matrix multiplication, and we know that matrix multiplication is distributive.

Example 6.2.1. *Maple* has a *dotprod* command in the *linalg* package. We will now use it to find the dot product of the two vectors $\vec{v} = \langle 5, -1, 2, 3 \rangle$ and $\vec{w} = \langle -4, 0, -5, 7 \rangle$.

```
> with(linalg):
> V:= vector([5, -1, 2, 3]):
> W:= vector([-4, 0, -5, 7]):
> dotprod(V,W);
```
$$-9$$

Let us now see how the dot product of two vectors is related to the vector lengths of each and the angle θ between them for vectors of dimension two or three. The angle θ between a pair of two- or three-dimensional vectors $\vec{v}$ and $\vec{w}$ is the angle between 0 and π radians gotten by placing them as arrows with the same base point and rotating one of them to the other through the smallest possible positive angle. We will also generalize this concept as to define the angle between any two vectors of the same dimension.

The key to seeing this relationship is writing out the law of cosines for the triangle formed by the three vectors $\vec{v}$, $\vec{w}$, and $\vec{v} - \vec{w}$, using the angle θ between $\vec{v}$ and $\vec{w}$. Applying the *law of cosines* to this triangle we get

$$|\vec{v} - \vec{w}|^2 = |\vec{v}|^2 + |\vec{w}|^2 - 2|\vec{v}||\vec{w}|\cos(\theta) \tag{6.5}$$

For simplicity, we will next assume that the vectors are of dimension two so that $\vec{v} = \langle v_1, v_2 \rangle$ and $\vec{w} = \langle w_1, w_2 \rangle$, in terms of components. Then,

$$|\vec{v} - \vec{w}|^2 - |\vec{v}|^2 - |\vec{w}|^2 = (v_1 - w_1)^2 + (v_2 - w_2)^2 - \left(v_1^2 + v_2^2 + w_1^2 + w_2^2\right)$$
$$= -2\left(v_1 w_1 + v_2 w_2\right)$$

Now putting this into equation (6.5) above gives

$$-2\left(v_1 w_1 + v_2 w_2\right) = -2|\vec{v}||\vec{w}|\cos(\theta)$$

or

$$v_1 w_1 + v_2 w_2 = |\vec{v}||\vec{w}|\cos(\theta)$$

The LHS of the above equation is simply the dot product of $\vec{v}$ and $\vec{w}$, so we get

$$\vec{v} \cdot \vec{w} = |\vec{v}||\vec{w}|\cos(\theta) \tag{6.6}$$

This formula now tells us that taking the dot product of two two- or three-dimensional vectors is a way of incorporating length and angle for these two vectors into a single quantity. We now arrive at the following definition:

Definition 6.2.3. The general angle θ between any two nonzero vectors $\vec{v}$ and $\vec{w}$ of $\mathbb{R}^n$ is given by the formula

$$\theta = \cos^{-1}\left(\frac{\vec{v} \cdot \vec{w}}{|\vec{v}||\vec{w}|}\right) \tag{6.7}$$

Once again, note that equation (6.7) is the definition of the angle θ between any two nonzero vectors in any dimension, instead of just the dimensions two or three where we can picture things geometrically. Recall that $\cos^{-1}(x)$ is the unique angle θ between 0 and π radians for which $\cos(\theta) = x$ for any $x \in [-1, 1]$. This tells us that our general angle definition will only work if $|\vec{v} \cdot \vec{w}| \leq |\vec{v}||\vec{w}|$, for any two vectors $\vec{v}$ and $\vec{w}$ in $\mathbb{R}^n$. This inequality is called the *Cauchy-Schwarz Inequality* and you are asked to prove it in *Homework* problem 5.

Example 6.2.2. As an example, we draw the triangle (Figure 6.9), with adjacent sides the vectors $\vec{v}$ and $\vec{w}$, with its altitude from the tip of the vector $\vec{v} = \langle 12, 19 \rangle$ to the base vector $\vec{w} = \langle 35, 0 \rangle$.

```
> with(plots): with(plottools):
> V:= vector([12,19]):
> arrow_V:= arrow([0,0], V, .3,2,.2, color = black):
> LabelV:= textplot([6,11,"V"], align = ABOVE, color = black):
> W:= vector([35,0]):
> arrow_W:= arrow([0,0], W, .3, 2, .2, color = blue):
> LabelW:= textplot([18,-.5,"W"], align = BELOW, color = black):
> vector_VminusW:= evalm(V-W):
> arrow_VminusW:= arrow([35,0], vector_VminusW, .3,2,.2, color = red):
> LabelVminusW:= textplot([24,10,"V-W"], align = {ABOVE, RIGHT}, color = black):
> LabelAngle:= textplot([5,1.5,q], align = ABOVE, color = black, font = [SYMBOL, 10]):
> Arc:= arc([0,0], 9, 0..arctan(19/12), color = black):
> Altitude:= line([12,0],[12,19],color = black):
> LabelAlt:= textplot([13,10,"H"], align = RIGHT, color = black):
> RightAngle:= polygon([[12,0], [12,2], [14,2], [14,0]], color = blue):
```

> display({arrow_W, arrow_V, arrow_VminusW, LabelW, LabelV,
LabelVminusW, LabelAngle, Arc, Altitude, LabelAlt, RightAngle},
scaling=constrained, axes=none);

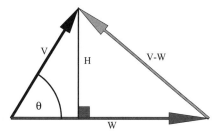

Figure 6.9: Vectors $\overrightarrow{v}$ and $\overrightarrow{w}$ and the angle between them.

The dot product formula can also be used to define the area of the above triangle with sides given by the vectors $\overrightarrow{v}$, $\overrightarrow{w}$ and $\overrightarrow{v} - \overrightarrow{w}$. The area of the triangle is given by $A = \dfrac{1}{2} b h$, where h is the length of our altitude and b is the length of the base. In our situation, we have $A = \frac{1}{2} |\overrightarrow{w}| |\overrightarrow{v}| \sin(\theta)$, since $b = |\overrightarrow{w}|$ and $h = |\overrightarrow{v}| \sin(\theta)$. Squaring this equation gives

$$A^2 = \frac{1}{4} |\overrightarrow{w}|^2 |\overrightarrow{v}|^2 \sin(\theta)^2$$
$$= \frac{1}{4} |\overrightarrow{w}|^2 |\overrightarrow{v}|^2 \left(1 - \cos(\theta)^2\right)$$
$$= \frac{1}{4} \left(|\overrightarrow{w}|^2 |\overrightarrow{v}|^2 - (\overrightarrow{v} \cdot \overrightarrow{w})^2\right)$$

Finally, we arrive at the formula for the area of the triangle in terms of the dot product:

$$A = \frac{1}{2} \sqrt{|\overrightarrow{w}|^2 |\overrightarrow{v}|^2 - (\overrightarrow{v} \cdot \overrightarrow{w})^2} \tag{6.8}$$

Notice that in order to find the area of this triangle, all we need are its two adjacent side vectors $\overrightarrow{v}$ and $\overrightarrow{w}$.

Example 6.2.3. Let us now do an example of using our dot product formula to find the three interior angles and the area of the triangle with vertices at $P(-8, 3, 17)$, $Q(10, -6, -9)$, and $R(30, 13, 25)$, which is depicted in Figure 6.10. We will compute all three angles directly from our formula and then verify that they add up to π. As we proceed, pay special attention to how we compute the magnitude of a vector in *Maple* using the *norm* command. The concept of taking the norm of a vector is much more general than what we have discussed here, hence there are many different ways to compute the norm of a vector. To ensure that the magnitude of the vector is computed, you must use the form "norm(vector, frobenius)".

> P:= [-8,3,17]: Q:= [10,-6,-9]: R:= [30,13,25]:
> trianglePQR:= polygon([P,Q,R], color = gold):
> arrow_PQ:= arrow(P, Q, .7, 3, .2, cylindrical_arrow, color = red):
> arrow_PR:= arrow(P, R, .7, 3, .2, cylindrical_arrow, color = black):
> arrow_RQ:= arrow(R, Q, .7, 3, .2, cylindrical_arrow, color = blue):
> Label_PQ:= textplot3d([1,-2,4,"PQ"], color = black, align = LEFT):
> Label_PR:= textplot3d([11,8,21,"PR"], color = black, align = LEFT):
> Label_RQ:= textplot3d([20,4,8,"RQ"], color = black, align = LEFT):
> display({trianglePQR, arrow_PQ, Label_PQ, arrow_PR, Label_PR, arrow_RQ, Label_RQ}, axes = boxed, scaling = constrained);

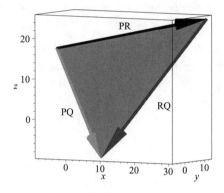

Figure 6.10: Triangle formed by the points P, Q, and R.

> PQ:= evalm(vector(Q) - vector(P)):
> PR:= evalm(vector(R) - vector(P)):
> RQ:= evalm(vector(Q) - vector(R)):
> angle_P:= evalf(arccos(dotprod(PQ, PR)/(norm(PQ, frobenius)*norm(PR, frobenius))));

$$angle\_P := 1.273670405$$

> angle_Q:= evalf(arccos(dotprod(PQ, RQ)/(norm(PQ, frobenius)*norm(RQ, frobenius))));

$$angle\_Q := 1.066955166$$

> angle_R:= evalf(arccos(dotprod(RQ, -PR)/(norm(RQ, frobenius)*norm(PR, frobenius))));

$$angle\_R := 0.8009670824$$

> angle_P + angle_Q + angle_R;

$$3.141592653$$

> Area:= evalf(1/2*sqrt(norm(PQ, frobenius)^2*norm(PR, frobenius)^2 - dot-prod(PQ, PR)^2));

$$Area := 630.3276926$$

As expected, the sum of these three interior angles of the triangle is π, and is easily computable from the dot product formula once we have three vectors that define the sides of the triangle. We have also used equation (6.8) to compute the area of the triangle.

Maple does have a built in command to find the angle between two real or complex valued vectors. The command *VectorAngle* is located in the *Linear-Algebra* package. We reproduce the angle calculations from above using the *VectorAngle* command.

> with(LinearAlgebra):
> VectorAngle(Vector(PQ), Vector(PR));

$$\arccos\left(\frac{193}{434562}\sqrt{1081}\sqrt{402}\right)$$

> evalf(VectorAngle(Vector(PQ), Vector(PR)));

$$1.273670405$$

> evalf(VectorAngle(Vector(PQ), Vector(RQ)));

$$1.066955166$$

> evalf(VectorAngle(Vector(RQ), -1*Vector(PR)));

$$0.8009670824$$

Following in this train of thought, notice that if we go back to our dot product expression (6.6):

$$\vec{v} \cdot \vec{w} = |\vec{v}||\vec{w}|\cos(\theta)$$

we can also easily determine when two vectors are perpendicular (or orthogonal) to each other. Remember that if two vectors are orthogonal, then the angle θ between them must be $\frac{\pi}{2}$. Putting these two pieces of information together tells us that the two vectors $\vec{v}$ and $\vec{w}$ are orthogonal exactly when $\vec{v} \cdot \vec{w} = 0$. This is an extremely convenient way to test for the orthogonality of two vectors as well as a useful tool to create a vector orthogonal to a given one.

Example 6.2.4. Starting with $\vec{v} = \langle -4, 5, 11, -3 \rangle$, we will find two vectors, $\vec{u}$ and $\vec{w}$, orthogonal to $\vec{v}$ by switching components of $\vec{v}$ and altering signs.

> V:= vector([-4,5,11,-3]): U:= vector([5,4,3,11]):
> dotprod(V,U);

$$0$$

> W:= vector([11,3,4,5]):
> dotprod(V,W);

$$0$$

> dotprod(U,W);

$$134$$

> Angle_UandW:= evalf(arccos(dotprod(U, W)/(norm(U, frobenius)*norm(W, frobenius))));

$$Angle\_UandW := 0.6703154594$$

So the pairs $\vec{v}$ and $\vec{w}$ as well as $\vec{v}$ and $\vec{u}$ are orthogonal, while the pair $\vec{u}$ and $\vec{w}$ is not.

So far, the dot product has been explicitly defined for vectors with real components. We now generalize this to vectors in $\mathbb{C}^n$ with complex components, but first we formally define the vector space $\mathbb{C}^n$.

Definition 6.2.4. The *vector space* $\mathbb{C}^n$ is the set of vectors $\vec{v}$ written as $\langle q_1, q_2, \ldots, q_n \rangle$ with n components $q_1, q_2, \ldots, q_n$.

$$\mathbb{C}^n = \left\{ \langle q_1, q_2, \ldots, q_n \rangle \,\middle|\, (q_1, q_2, \ldots, q_n) \in \mathbb{C}^n \right\}$$

In the above definition of the complex vector space $\mathbb{C}^n$, note that the scalars are complex numbers. For the real vector space $\mathbb{R}^n$, the scalars are real. However, the same vector space properties of addition and scalar multiplication hold in $\mathbb{C}^n$ as they do in $\mathbb{R}^n$, just for a different set of scalars.

The complex dot product on $\mathbb{C}^n$ is designed to be the same as the real dot product on $\mathbb{R}^n$ (since $\mathbb{R}^n \subset \mathbb{C}^n$) so that when we use it on a pair of real vectors it agrees with the real dot product. Moreover, the complex dot product of a vector with itself should still give a non-negative real number so that $\vec{v} \cdot \vec{v} = |\vec{v}|^2$ for $v \in \mathbb{C}^n$ can define the length $|\vec{v}|$, where it must also be true that $|\vec{v}| = 0$ if and only if $\vec{v} = \vec{0}$. The complex dot product on $\mathbb{C}^n$ is the simplest extension of the real dot product on $\mathbb{R}^n$ that behaves in this manner.

Definition 6.2.5. The *dot product* of two complex valued n-dimensional column vectors $\vec{u}$ and $\vec{v}$ is a complex number, denoted $\vec{u} \cdot \vec{v}$, and defined in terms of matrix multiplication by $\vec{u} \cdot \vec{v} = \vec{u}^T \overline{\vec{v}}$, where $\overline{\vec{v}}$ denotes the conjugate vector of $\vec{v}$. More specifically, if

$$\vec{u} = \langle a_1 + ib_1, a_2 + ib_2, \ldots, a_n + ib_n \rangle, \quad \vec{v} = \langle c_1 + id_1, c_2 + id_2, \ldots, c_n + id_n \rangle$$

with a_k's, b_k's, c_k's, and d_k's all real, then

$$\vec{u} \cdot \vec{v} = \begin{bmatrix} a_1 + ib_1 & a_2 + ib_2 & \cdots & a_n + ib_n \end{bmatrix} \begin{bmatrix} c_1 - id_1 \\ c_2 - id_2 \\ \vdots \\ c_n - id_n \end{bmatrix} \tag{6.9}$$

$$= \sum_{k=1}^{n} (a_k + ib_k)(c_k - id_k)$$

Notice that the above definition of the dot product does indeed reduce to the already given formulation if $\vec{u}$ and $\vec{v}$ are both real. However, one surprising result is that $\vec{u} \cdot \vec{v} = \overline{\vec{v} \cdot \vec{u}}$. Furthermore, when computing the norm of a complex vector, notice that it will always end up being real valued:

$$|\vec{u}|^2 = \vec{u} \cdot \vec{u}$$

$$= \sum_{k=1}^{n} (a_k + ib_k)(a_k - ib_k)$$

$$= \sum_{k=1}^{n} (a_k^2 + b_k^2)$$

Example 6.2.5. We let *Maple* do one example for us, choosing the following two vectors in $\mathbb{C}^3$, $\vec{u} = \langle 2 - i, 1 - 3i, i \rangle$ and $\vec{v} = \langle 2 - 5i, 6 + 7i, -3 \rangle$. We will compute $\vec{u} \cdot \vec{v}$, $\vec{v} \cdot \vec{u}$, and $|\vec{u}|$, making sure that the properties of the complex dot product just described do indeed hold.

> u:= vector([2-I,1-3*I,I]);

$$\alpha := \begin{bmatrix} 2 - I & 1 - 3I & I \end{bmatrix}$$

> v:= vector([2-5*I,6+7*I,-3]);

$$\beta := \begin{bmatrix} 2 - 5I & 6 + 7I & -3 \end{bmatrix}$$

> dotprod(u,v);

$$-6 - 20I$$

> sum(u[k]*conjugate(v[k]), k=1..3);

$$-6 - 20I$$

> dotprod(v, u);

$$-6 + 20I$$

> norm(u,frobenius);

$$4$$

> sqrt(sum(u[k]*conjugate(u[k]), k=1..3));

$$4$$

We conclude this section with a table of dot product properties. Note the subtle differences between the real and the complex vector cases. It was shown previously that $\vec{u} \cdot \vec{v} = \overline{\vec{v} \cdot \vec{u}}$ for complex valued vectors $\vec{u}$ and $\vec{v}$. However, a more subtle difference can be seen in the very last property. For the real valued case, a and b are real scalars and they can be moved around in the dot products in any way we wish. In the complex case, where α and β are complex scalars, notice that to pull the scalar β, which is multiplied by the second vector in the dot product, out of the dot product, requires us to use the complex conjugate of β.

Dot Product Properties	
Real dot product $\vec{u}, \vec{v}, \vec{w} \in \mathbb{R}^n$ $a, b \in \mathbb{R}$	$\vec{u} \cdot \vec{v} = \vec{v} \cdot \vec{u}$ $(\vec{u} + \vec{v}) \cdot \vec{w} = \vec{u} \cdot \vec{w} + \vec{v} \cdot \vec{w}$ $\vec{u} \cdot \vec{u} = |\vec{u}|^2$ $\vec{u} \cdot \vec{0} = 0$ $(a\vec{u}) \cdot \vec{v} = a(\vec{u} \cdot \vec{v})$ $\vec{u} \cdot (b\vec{v}) = b(\vec{u} \cdot \vec{v})$
Complex dot product $\vec{u}, \vec{v}, \vec{w} \in \mathbb{C}^n$ $\alpha, \beta \in \mathbb{C}$	$\vec{u} \cdot \vec{v} = \overline{\vec{v} \cdot \vec{u}}$ $\vec{u} \cdot \vec{u} = |\vec{u}|^2$ $(\vec{u} + \vec{v}) \cdot \vec{w} = \vec{u} \cdot \vec{w} + \vec{v} \cdot \vec{w}$ $\vec{u} \cdot \vec{0} = 0$ $(\alpha\vec{u}) \cdot \vec{v} = \alpha(\vec{u} \cdot \vec{v})$ $\vec{u} \cdot (\beta\vec{v}) = \overline{\beta}(\vec{u} \cdot \vec{v})$

Homework Problems

1. Compute the dot products of the following pairs of vectors.

(a) $\langle -5, 2\rangle, \ \langle 3, -2\rangle$ (b) $\langle 1, 4\rangle, \ \langle -6, 3\rangle$

(c) $\langle -2 - i, 1\rangle, \ \langle 3, -2i\rangle$ (d) $\langle 3, 2, -2\rangle, \ \langle 4, 3, -2\rangle$

(e) $\langle 1, 0, 4\rangle, \ \langle 2, -6, 0\rangle$ (f) $\langle 4, 2, 5\rangle, \ \langle 1, 3, -2\rangle$

2. Compute the norms of the following vectors.

(a) $\langle -5, 2\rangle$ (b) $\langle -3, 3\rangle$ (c) $\langle 5, 3 - 2i\rangle$

(d) $\langle 1, 0, -2\rangle$ (e) $\left\langle \frac{1}{2}, 0, 4\right\rangle$ (f) $\left\langle 3 - 6i, -\frac{1}{\sqrt{3}}, 2 + i\right\rangle$

(g) $\langle 1, 2, -1, 1\rangle$ (h) $\langle -1, 2, 5, 3, 2\rangle$ (i) $\langle 1 - i, 8 + 2i, 3 + 2i, 1 - i\rangle$

3. Determine the angle between the following pairs of vectors.

(a) $\langle -3, 6\rangle, \ \langle 4, 2\rangle$ (b) $\langle 1, -4\rangle, \ \langle -2, 2\rangle$ (c) $\langle -2, 1\rangle, \ \langle 2, -1\rangle$

(d) $\langle 1, 0, -2\rangle, \ \langle 0, 3, 0\rangle$ (e) $\langle 3, -2, 4\rangle, \ \langle 2, -6, 2\rangle$

4. For each of the following vectors $\vec{v}$, find a second vector $\vec{w}$ such that $\vec{v} \perp \vec{w}$.

(a) $\langle 2, 3\rangle$ (b) $\langle 5, -1\rangle$ (c) $\langle 1, 0, 5\rangle$

(d) $\langle 2, 1, 3\rangle$ (e) $\langle 7, 0, -2, 2\rangle$ (f) $\langle -2, 3, 1, 5\rangle$

5. Prove the *Cauchy-Schwarz inequality*, which states that if $\vec{u}, \vec{v} \in \mathbb{R}^n$, then

$$|\vec{u} \cdot \vec{v}| \leq |\vec{u}||\vec{v}|$$

6. Use the *Cauchy-Schwarz inequality* to prove the *Triangle inequality*, which states that for $\vec{u}, \vec{v} \in \mathbb{R}^n$:

$$|\vec{u} + \vec{v}| \leq |\vec{u}| + |\vec{v}|$$

Hint: Start with the fact that $|\vec{u} + \vec{v}|^2 = (\vec{u} + \vec{v}) \cdot (\vec{u} + \vec{v})$, and expand the righthand side using the distributivity of the dot product.

7. Find the area and the interior angles of the triangle with the two adjacent sides $\vec{v} = \langle -5, -9 \rangle$ and $\vec{w} = \langle 7, 3 \rangle$.

8. Find the area and the interior angles of the triangle with the two adjacent sides $\vec{v} = \langle -5, -9, 12 \rangle$ and $\vec{w} = \langle 7, 3, -6 \rangle$.

9. Consider the following two vectors in $\mathbb{C}^4$

$$\vec{v} = \langle -5 + 4i, -9 - i, 12, 7i \rangle, \ \vec{w} = \langle 7 - 2i, 3, -6 + 5i, 1 + i \rangle$$

For parts (c) and (d), you may wish to refer back to problems 5 and 6, and make use of the Triangle and Cauchy-Schwarz Inequalities.

(a) Compute $\vec{v} \cdot \vec{w}$ and $\vec{w} \cdot \vec{v}$.

(b) Find the norms of the two vectors $\vec{v}$ and $\vec{w}$.

(c) Is it true that $|\vec{v} + \vec{w}| \le |\vec{v}| + |\vec{w}|$?

(d) Is it true that $|\vec{u} \cdot \vec{v}| \le |\vec{u}||\vec{v}|$?

10. Find a vector $\vec{w}$ of norm seven perpendicular to $\vec{v} = \langle -5, -9 \rangle$.

11. Find two different vectors $\vec{w}$ of norm four perpendicular to $\vec{v} = \langle 7, 3, -6 \rangle$.

12. Find two nonzero vectors $\vec{w_1}$ and $\vec{w_2}$ perpendicular to $\vec{v} = \langle 7, 3, -6 \rangle$, where $\vec{w_1}$ and $\vec{w_2}$ are also perpendicular.

13. Let $\vec{v} = \langle a, b, c \rangle \in \mathbb{R}^3$ be fixed. What equation must $\vec{w} = \langle x, y, z \rangle \in \mathbb{R}^3$ satisfy for $\vec{w}$ and $\vec{v}$ to be perpendicular, and what does this equation represent in $\mathbb{R}^3$?

14. Find the equation of the plane in $\mathbb{R}^3$ that is perpendicular to the vector $\vec{v} = \langle 7, 3, -6 \rangle$ and goes through the origin $\langle 0, 0, 0 \rangle$.

15. Find the equation of the plane in $\mathbb{R}^3$ that is perpendicular to the vector $\vec{v} = \langle 7, 3, -6 \rangle$ and goes through the point $P(-2, 5, 8)$.

16. Use the dot product to show that the diagonals of a square are perpendicular.

17. Find a formula for the angle between the two diagonals of a parallelogram in terms of the lengths of any two of its adjacent sides.

18. Let $f(x)$ and $g(x)$ be any two real continuous functions for $x \in [a, b]$. Define the dot product of $f(x)$ and $g(x)$ by

$$f(x) \cdot g(x) = \int_a^b f(x)\, g(x) dx$$

(a) Find $f(x) \cdot g(x)$, $f(x) \cdot f(x)$, and $g(x) \cdot g(x)$ for $f(x) = e^x$ and $g(x) = \cos(x)$ on the interval $[0, \pi]$.

(b) Show that this dot product of two functions satisfies the usual properties of a real dot product.

(c) Find all possible dot products of the functions

$$\{1, \cos(x), \cos(2x), \sin(x), \sin(2x)\}$$

with each other and themselves on the interval $[0, 2\pi]$. Do you see a pattern here, what is it?

(d) Find all possible dot products of the functions

$$\{1, x, x^2, x^3\}$$

with each other and themselves on the interval $[-1, 1]$.

19. Let A be a real $n \times n$ matrix with $\vec{u}, \vec{v} \in \mathbb{R}^n$ written as column vectors in $\mathbb{R}^n$. Show that the real dot product satisfies $(A\vec{u}) \cdot \vec{v} = \vec{u} \cdot (A^T \vec{v})$. Give an example when $n = 2$ to illustrate this formula.

20. Let A be a complex $n \times n$ matrix with $\vec{u}, \vec{v} \in \mathbb{C}^n$ written as column vectors in $\mathbb{C}^n$. Show that the complex dot product satisfies $(A\vec{u}) \cdot \vec{v} = \vec{u} \cdot (\overline{A}^T \vec{v})$. Give an example when $n = 2$ to illustrate this formula.

Maple Problems

1. Compute the angle between each of the following pairs of vectors $\vec{v}$ and $\vec{w}$.

   (a) $\vec{v} = \langle -5, 2 \rangle$, $\vec{w} = \langle 3, -2 \rangle$ (b) $\vec{v} = \langle 1, 4 \rangle$, $\vec{w} = \langle -6, 3 \rangle$

   (c) $\vec{v} = \langle -2, 3 \rangle$, $\vec{w} = \langle 5, 0 \rangle$ (d) $\vec{v} = \langle 6, -2 \rangle$, $\vec{w} = \langle -7, -9 \rangle$

2. Starting with each of the vectors $\vec{v}$ from problem 1, use rotation matrices and scalar multiplication to transform them to the vectors $\vec{w}$. Treating the vectors as column matrices, you should be able to express each vector $\vec{w}$ as:

$$\vec{w} = r A_\theta \, \vec{v}$$

3. Graph each pair of vectors from problem 1.

4. Compute the angle between each of the following pairs of vectors $\vec{v}$ and $\vec{w}$.

    (a) $\vec{v} = \langle 1, -5, 0 \rangle$, $\vec{w} = \langle 0, 5, 2 \rangle$ (b) $\vec{v} = \langle 1, 0, 4 \rangle$, $\vec{w} = \langle -6, 3, 4 \rangle$

5. The angles between each pair of vectors in problem 4 do not necessarily correspond to an angle in the xy-, xz-, or yz-plane. Can you come up with a process similar to that of the two-dimensional case in which the initial vector $\vec{v}$ was rotated and scaled appropriately to become the vector $\vec{w}$? You may wish to refer to *Homework* problem 10 from Section 4.4, where the 3×3 matrices for rotations about the x-, y-, and z-axes were given.

6. Let $P(-1, 3)$, $Q(2, -5)$, $R(7, 11)$, and $T(1, 15)$ be the counterclockwise vertices of a quadrilateral.

    (a) Find all four interior angles of this quadrilateral and the lengths of all four sides.

    (b) What is the sum of these four interior angles?

    (c) Find the area of this quadrilateral.

7. Find the area and the three interior angles for the triangle with vertices at $P(-2, 7, 3)$, $Q(8, -10, 4)$, and $R(6, 1, 9)$.

8. Plot the results of *Homework* problems 14 and 15, along with their corresponding perpendicular vectors.

9. Verify your answers to *Homework* problems 18 (a), (c), and (d).

10. Give examples when $n = 4$ to illustrate the formulas of *Homework* problems 19 and 20.

6.3 Cross Product

As discussed in Section 6.2, the dot product of two vectors resulted in a scalar. The next vector operation we will consider is the *cross product*, which can only be applied to vectors in $\mathbb{R}^3$. The cross product of two vectors in $\mathbb{R}^3$, denoted $\overrightarrow{u} \times \overrightarrow{v}$, is designed to produce another vector in $\mathbb{R}^3$ that is orthogonal to both $\overrightarrow{u}$ and $\overrightarrow{v}$.

Definition 6.3.1. The *cross product* of two vectors $\overrightarrow{u} = \langle u_1, u_2, u_3 \rangle$, $\overrightarrow{v} = \langle v_1, v_2, v_3 \rangle$ in $\mathbb{R}^3$ is denoted $\overrightarrow{u} \times \overrightarrow{v}$, and is computed by the following formula:

$$\overrightarrow{u} \times \overrightarrow{v} = \det \left(\begin{bmatrix} \overrightarrow{i} & \overrightarrow{j} & \overrightarrow{k} \\ u_1 & u_2 & u_3 \\ v_1 & v_2 & v_3 \end{bmatrix} \right)$$

$$= (u_2 v_3 - v_2 u_3) \overrightarrow{i} - (u_1 v_3 - v_1 u_3) \overrightarrow{j} + (u_1 v_2 - v_1 u_2) \overrightarrow{k}$$

$$= \langle u_2 v_3 - v_2 u_3, -u_1 v_3 + v_1 u_3, u_1 v_2 - v_1 u_2 \rangle$$

In the above formulation, notice that the determinant was expanded along the first row of the matrix, while $\overrightarrow{u}$ and $\overrightarrow{v}$ make up the second and third rows of the matrix, respectively.

The key to defining the cross product is realizing the following two facts. First, the determinant of a square matrix is 0 if two rows are the same. Second, if we write $\overrightarrow{s} = \langle s_1, s_2, s_3 \rangle$ and $\overrightarrow{t} = t_1 \overrightarrow{i} + t_2 \overrightarrow{j} + t_3 \overrightarrow{k}$, where $\overrightarrow{i} = \langle 1, 0, 0 \rangle$, $\overrightarrow{j} = \langle 0, 1, 0 \rangle$ and $\overrightarrow{k} = \langle 0, 0, 1 \rangle$, then $\overrightarrow{s} \cdot \overrightarrow{t} = t_1 s_1 + t_2 s_2 + t_3 s_3$ where the three unit vectors in writing out $\overrightarrow{t}$ have been replaced by the three components of $\overrightarrow{s}$.

Theorem 6.3.1. *The vector $\overrightarrow{u} \times \overrightarrow{v}$ is orthogonal to both of the vectors $\overrightarrow{u}$ and $\overrightarrow{v}$.*

Proof. Note that for any vector $\overrightarrow{s} = \langle s_1, s_2, s_3 \rangle$, we have

$$\overrightarrow{s} \cdot (\overrightarrow{u} \times \overrightarrow{v}) = \det \left(\begin{bmatrix} s_1 & s_2 & s_3 \\ u_1 & u_2 & u_3 \\ v_1 & v_2 & v_3 \end{bmatrix} \right)$$

By setting $\overrightarrow{s} = \overrightarrow{u}$ or $\overrightarrow{s} = \overrightarrow{v}$ in the above formula, note that the resulting matrix will have a repeated row, yielding a determinant of 0. Therefore, both $\overrightarrow{u} \cdot (\overrightarrow{u} \times \overrightarrow{v}) = 0$ and $\overrightarrow{v} \cdot (\overrightarrow{u} \times \overrightarrow{v}) = 0$ and we can conclude that $\overrightarrow{u} \times \overrightarrow{v}$ is orthogonal to both $\overrightarrow{u}$ and $\overrightarrow{v}$. $\square$

One of the simplest properties to see of the cross product is that

$$\overrightarrow{u} \times \overrightarrow{v} = -(\overrightarrow{v} \times \overrightarrow{u}) \tag{6.10}$$

This is due to the fact that swapping $\overrightarrow{u}$ and $\overrightarrow{v}$ in the matrix used to compute the cross product can be achieved through one multiplication by a type I elementary matrix, which changes the sign of the determinant.

Example 6.3.1. Again, we need an example to illustrate these ideas. Let us find the cross product of $\vec{u} = \langle -3, 7, 11 \rangle$ and $\vec{v} = \langle 6, -1, 9 \rangle$ and also test that $\vec{u} \times \vec{v}$ is orthogonal to both $\vec{u}$ and $\vec{v}$ by using the dot product. We shall also plot all three vectors.

> with(linalg): with(plots): with(plottools):
> U:= [-3,7,11]:
> V:= [6,-1,9]:
> det(array([[I, J, K], U, V]));

$$74I + 93J - 39K$$

> UxV:= crossprod(vector(U), vector(V));

$$UxV := \begin{bmatrix} 74 & 93 & -39 \end{bmatrix}$$

> VxU:= crossprod(vector(V), vector(U));

$$VxU := \begin{bmatrix} -74 & -93 & 39 \end{bmatrix}$$

> dotprod(vector(U), UxV);

$$0$$

> dotprod(vector(V), UxV);

$$0$$

> W:= [18,-10,-15]:
> det(array([W, U, V]));

$$987$$

> dotprod(vector(W), UxV);

$$987$$

Now we plot the vectors $\vec{u}$, $\vec{v}$, and $\frac{1}{4}(\vec{u} \times \vec{v})$ in order to scale back the length of $\vec{u} \times \vec{v}$ so it is close to the lengths of $\vec{u}$ and $\vec{v}$.

> Origin:= [0,0,0]:
> arrow_U:= arrow(Origin, U, .3, 2, .2, cylindrical_arrow, color = black):
> Label_U:= textplot3d([-1,2,5,"U"], align = ABOVE, color = black):
> arrow_V:= arrow(Origin, V, .3, 2, .2, cylindrical_arrow, color = blue):
> Label_V:= textplot3d([3,-.5,5,"V"], align = BELOW, color = black):
> arrow_UxV:= arrow(Origin, evalm(.25*UxV), .3, 2, .2, cylindrical_arrow, color = red):

> Label_UxV:= textplot3d([10,12,-5,"UxV"], align=ABOVE, color=black):

> RightAngle_U:= polygon([Origin, [-3/4, 7/4, 11/4], [-3/4+74/40, 7/4+93/40, 11/4-39/40], [74/40, 93/40, -39/40]], color = black):

> RightAngle_V:= polygon([Origin, [6/4, -1/4, 9/4], [6/4+74/40, -1/4+93/40, 9/4-39/40], [74/40, 93/40, -39/40]], color = blue):

> display({arrow_U, arrow_V, arrow_UxV, Label_U, Label_V, Label_UxV, RightAngle_U, RightAngle_V}, axes = boxed, scaling = constrained);

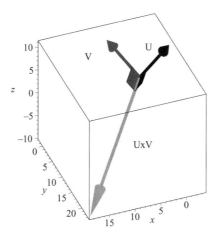

Figure 6.11: The cross product yields a vector orthogonal to both $\overrightarrow{U}$ and $\overrightarrow{V}$.

From Figure 6.11, we should recognize that the cross product satisfies the *right-hand rule*, which states that if you place your right hand so that your fingers bend from $\overrightarrow{u}$ to $\overrightarrow{v}$, then your thumb will point in the direction of $\overrightarrow{u} \times \overrightarrow{v}$.

As stated previously, vectors have two properties: direction and magnitude. We have the direction of the new vector, which is perpendicular to both $\overrightarrow{u}$ and $\overrightarrow{v}$. We next determine a formula relating the magnitude of $\overrightarrow{u} \times \overrightarrow{v}$ to the magnitudes of $\overrightarrow{u}$ and $\overrightarrow{v}$. If $\overrightarrow{u} = \langle u_1, u_2, u_3 \rangle$ and $\overrightarrow{v} = \langle v_1, v_2, v_3 \rangle$, then

$$\overrightarrow{u} \times \overrightarrow{v} = \langle u_2 v_3 - v_2 u_3, -(u_1 v_3 - v_1 u_3), u_1 v_2 - v_1 u_2 \rangle$$

and

$$|\overrightarrow{u} \times \overrightarrow{v}|^2 = (u_2 v_3 - v_2 u_3)^2 + (u_1 v_3 - v_1 u_3)^2 + (u_1 v_2 - v_1 u_2)^2$$

We will now use *Maple* to perform a few algebraic manipulations to help prove the identity:

$$|\overrightarrow{u} \times \overrightarrow{v}|^2 = |\overrightarrow{u}|^2 |\overrightarrow{v}|^2 - (\overrightarrow{u} \cdot \overrightarrow{v})^2 \qquad (6.11)$$

\> simplify((u[2]*v[3] - v[2]*u[3])^2 + (u[1]*v[3] - v[1]*u[3])^2 + (u[1]*v[2] - v[1]*u[2])^2);

$$u_2^2 v_3^2 - 2u_2 v_3 v_2 u_3 + v_2^2 u_3^2 + u_1^2 v_3^2 - 2u_1 v_3 v_1 u_3 + v_1^2 u_3^2 + u_1^2 v_2^2 - 2u_1 v_2 v_1 u_2 + v_1^2 u_2^2$$

\> U:= vector([u[1], u[2], u[3]]): V:= vector([v[1], v[2], v[3]]):
\> NormSqr_U:= evalm(transpose(U)&*U);

$$NormSqr\_U := u_1^2 + u_2^2 + u_3^2$$

\> NormSqr_V:= evalm(transpose(V)&*V);

$$NormSqr\_V := v_1^2 + v_2^2 + v_3^2$$

\> RealDotProd_UV:= evalm(transpose(U)&*V);

$$RealDotProd\_UV := u_1 v_1 + u_2 v_2 + u_3 v_3$$

\> simplify(NormSqr_U*NormSqr_V - RealDotProd_UV^2);

$$u_2^2 v_3^2 - 2u_2 v_3 v_2 u_3 + v_2^2 u_3^2 + u_1^2 v_3^2 - 2u_1 v_3 v_1 u_3 + v_1^2 u_3^2 + u_1^2 v_2^2 - 2u_1 v_2 v_1 u_2 + v_1^2 u_2^2$$

Now we know from this last *Maple* calculation that equation (6.11) holds true. With only a simple substitution of equation (6.6), $\vec{u} \cdot \vec{v} = |\vec{u}||\vec{v}| \cos(\theta)$, into equation (6.11), we arrive at

$$|\vec{u} \times \vec{v}|^2 = |\vec{u}|^2 |\vec{v}|^2 \left(1 - \cos^2(\theta)\right)$$
$$= |\vec{u}|^2 |\vec{v}|^2 \sin^2(\theta)$$

Now we have found an expression for the norm of $\vec{u} \times \vec{v}$ involving the magnitudes of $\vec{u}$, $\vec{v}$ and the angle θ between them:

$$|\vec{u} \times \vec{v}| = |\vec{u}||\vec{v}| \sin(\theta) \tag{6.12}$$

To interpret this, notice that the maximum magnitude of $\vec{u} \times \vec{v}$ occurs when the vectors are orthogonal, i.e. $\theta = \frac{\pi}{2}$, and thus $\sin(\theta) = 1$, yielding $|\vec{u} \times \vec{v}| = |\vec{u}||\vec{v}|$. As the vectors $\vec{u}$ and $\vec{v}$ become closer to lying on the same line ($\theta = 0$ or $\theta = \pi$), the magnitude of their cross product tends to 0.

     Next we turn to an application of the cross product. Similar to the dot product, the cross product can give us a very nice way of computing the area of a triangle, albeit in $\mathbb{R}^2$ or $\mathbb{R}^3$ only. Since the cross product only applies to vectors in $\mathbb{R}^3$, if the triangle is in $\mathbb{R}^2$, adding a third component with value 0 to each vector in order to move it to the xy-plane in $\mathbb{R}^3$ will allow the following computations to hold. From Section 6.2, we know that the area A of a triangle in $\mathbb{R}^3$ with two adjacent side vectors $\vec{v}$ and $\vec{w}$ is given by (6.8):

$$A = \frac{1}{2}\sqrt{|\vec{w}|^2 |\vec{v}|^2 - (\vec{v} \cdot \vec{w})^2}$$

We just proved (6.11), whose RHS is the expression under the radical of (6.8). Therefore, we now have an expression for the area of a triangle involving the cross product:

$$A = \frac{1}{2}|\vec{v} \times \vec{w}| \qquad (6.13)$$

This area formula is quite simple and straightforward to compute, however as discussed, it only applies in dimensions two and three, as opposed to the previous formula with dot products that applies in any dimension.

Example 6.3.2. Now let us use all of our information, including equations (6.8) and (6.13), to test each other's correctness for computing the area of the triangle in $\mathbb{R}^3$ with two adjacent side vectors $\vec{v} = \langle 9, -5, -13 \rangle$ and $\vec{w} = \langle -4, 7, 2 \rangle$.

> V:= vector([9,-5,-13]): W:= vector([-4,7,2]):
> VxW:= crossprod(V,W);

$$VxW := \begin{bmatrix} 81 & 34 & 43 \end{bmatrix}$$

> Area_1:= 1/2*norm(VxW,frobenius);

$$Area\_1 := \frac{1}{2}\sqrt{9566}$$

> Area_2:= 1/2*sqrt(norm(V, frobenius)^2*norm(W, frobenius)^2 -dotprod(V, W)^2);

$$Area\_2 := \frac{1}{2}\sqrt{9566}$$

> theta:= evalf(arccos(dotprod(V, W)/(norm(V, frobenius)*norm(W, frobenius))));

$$\theta := 2.352057421$$

> evalf(norm(VxW,frobenius));

$$97.80593029$$

> Length_VxW:= evalf(norm(V,frobenius)*norm(W,frobenius)*sin(theta));

$$Length\_VxW := 97.80593024$$

The above calculations verify the correctness of all of our formulas involving the dot and cross products. Finally, we include a small table of cross product properties, including two relationships involving both the dot product and cross product.

Cross Product Properties

Basic Properties	$\vec{u} \times \vec{v} = -(\vec{v} \times \vec{u})$
	$\vec{u} \times (\vec{v} + \vec{w}) = (\vec{u} \times \vec{v}) + (\vec{u} \times \vec{w})$
	$(a\vec{u}) \times \vec{v} = \vec{u} \times (a\vec{v}) = a(\vec{u} \times \vec{v})$
Cross and Dot Products	$(\vec{u} \times \vec{v}) \cdot \vec{w} = \vec{u} \cdot (\vec{v} \times \vec{w})$
	$\vec{u} \times (\vec{v} \times \vec{w}) = (\vec{u} \cdot \vec{w})\vec{v} - (\vec{u} \cdot \vec{v})\vec{w}$

Homework Problems

1. Given $\vec{u} = \langle 1, 2, -1 \rangle$, $\vec{v} = \langle -3, -1, 4 \rangle$, and $\vec{w} = \langle 5, -1, 0 \rangle$, compute the following. As previously defined, $\vec{i} = \langle 1, 0, 0 \rangle$, $\vec{j} = \langle 0, 1, 0 \rangle$, and $\vec{k} = \langle 0, 0, 1 \rangle$ are the three standard unit vectors of $\mathbb{R}^3$.

   (a) $\vec{v} \times \vec{w}$ (b) $\vec{w} \times \vec{u}$ (c) $\vec{u} \times \vec{v}$

   (d) $\vec{w} \times \vec{v}$ (e) $(\vec{u} \times \vec{v}) \times \vec{w}$ (f) $\vec{u} \times (\vec{v} \times \vec{w})$

   (g) $\vec{i} \times \vec{v}$ (h) $\vec{u} \times \vec{k}$ (i) $\vec{j} \times \vec{w}$

2. Compute the areas of each of the triangles defined by the following sets of points.

   (a) $\{(0, 1), (2, 3), (6, 2)\}$ (b) $\{(0, 1, 2), (2, 3, 1), (1, 6, 2)\}$

3. Prove the following properties of the cross product.

   (a) $\vec{i} \times \vec{j} = \vec{k}$ (b) $\vec{j} \times \vec{k} = \vec{i}$ (c) $\vec{k} \times \vec{i} = \vec{j}$

   (d) $\vec{j} \times \vec{i} = -\vec{k}$ (e) $\vec{k} \times \vec{j} = -\vec{i}$ (f) $\vec{i} \times \vec{k} = -\vec{j}$

4. If $\vec{u}, \vec{v}, \vec{w} \in \mathbb{R}^3$, and $\alpha \in \mathbb{R}$, verify the following identities: (Property (c) is an example of a *Jacobi identity*).

   (a) $\vec{u} \times (\vec{v} + \vec{w}) = (\vec{u} \times \vec{v}) + (\vec{u} \times \vec{w})$

(b) $(\alpha \overrightarrow{u}) \times \overrightarrow{v} = \overrightarrow{u} \times (\alpha \overrightarrow{v})$

(c) $\overrightarrow{u} \times (\overrightarrow{v} \times \overrightarrow{w}) + \overrightarrow{v} \times (\overrightarrow{w} \times \overrightarrow{u}) + \overrightarrow{w} \times (\overrightarrow{u} \times \overrightarrow{v}) = 0$

5. If $\overrightarrow{v(t)} = \langle \cos(t), \sin(t), 0 \rangle$, show that the angle between $\overrightarrow{v(t)}$ and $\overrightarrow{i}$ is $\theta = t$, and that $\left| \overrightarrow{v(t)} \times \overrightarrow{i} \right|^2 = \sin^2(t)$.

6. Verify the two cross product and dot product properties in the table at the end of the section.

7. (a) Let $ax + by + cz = d$ be the equation of a plane. Show that for any two points P and Q in this plane, the displacement vector $\overrightarrow{PQ}$ is perpendicular to $\overrightarrow{v} = \langle a, b, c \rangle$, that is, $\overrightarrow{v} = \langle a, b, c \rangle$ is perpendicular to the plane $ax + by + cz = d$.

(b) Let $ax + by + cz = d$ and $ex + fy + gz = h$ be two intersecting planes. Find the equation of the plane through the origin perpendicular to these two given planes.

(c) Let $5x + 2y + 7z = -9$ and $3x - 4y + 11z = 1$ be two intersecting planes. Find the equation of the plane through the origin perpendicular to these two given planes. Find the equation of the plane through $P(-1, 0, 4)$ perpendicular to these two given planes.

8. (a) Let $x = at + \alpha$, $y = bt + \beta$, $z = ct + \gamma$ be the parametric equation of a line in space for $t \in \mathbb{R}$. Show that for any two points P and Q on this line that the displacement vector $\overrightarrow{PQ}$ is parallel to $\overrightarrow{v} = \langle a, b, c \rangle$, that is, $\overrightarrow{v} = \langle a, b, c \rangle$ is parallel to the line $x = at + \alpha$, $y = bt + \beta$, $z = ct + \gamma$. Note that when $t = 0$, we have that (α, β, γ) is a point on this line. As well, this parametric equation for a line in space is the spacial version of the point-slope formula for a line in the xy-plane.

(b) Let $px + qy + rz = s$ and $ex + fy + gz = h$ be two intersecting planes. Find a vector parallel to their line of intersection.

(c) Let $5x + 2y + 7z = -9$ and $3x - 4y + 11z = 1$ be two intersecting planes. Find the parametric equation for their line of intersection.

9. Using problem 7, find the equation of the plane through the three points $P(1, 5, 9)$, $Q(-3, 4, -8)$, and $R(7, -2, 6)$.

10. (See problem 8.) (a) Let $x = at + \alpha$, $y = bt + \beta$, $z = ct + \gamma$ and $x = dt + \delta$, $y = et + \theta$, $z = ft + \lambda$ be two intersecting lines in space. Find a vector perpendicular to the plane through these two points.

(b) Find the equation of the plane through the two intersecting lines $x = 5t + 2, y = -3t + 1, z = 4t - 5$ and $x = -7t + 2, y = 2t + 1, z = 9t - 5$.

11. Explain why the associative property of the cross product is false, that is, explain why

$$\vec{u} \times (\vec{v} \times \vec{w}) \neq (\vec{u} \times \vec{v}) \times \vec{w}$$

Give a general example to illustrate that this is correct.

12. Can the cross product be defined for complex vectors in $\mathbb{C}^3$? If no, explain why not. If yes, then do all of the properties of the cross product still hold if we switch to complex dot product?

Maple Problems

1. Run the following set of *Maple* commands and explain the change in direction and magnitude of the cross product vector depicted in red.

```
> restart: with(plots): with(linalg): with(LinearAlgebra):
> assume(t, real):
> Origin:=[0,0,0]:
> XVec:=[1,0,0]:
> arrowXVec:= plottools[arrow](Origin, XVec, .1, .2, .3, cylindrical_arrow,
color=black):
> RotVec:= t-> [cos(t),sin(t),0]:
> CrossVec:= t-> crossprod(XVec,RotVec(t)):
> for k from 1 to 20 do
     arrowRotVec[k]:= plottools[arrow](Origin, RotVec(k*Pi/10),.1,.2,.3, cylin-
drical_arrow, color = blue):
     arrowCrossVec[k]:= plottools[arrow](Origin, CrossVec(k*Pi/10), .1, .2, .3,
cylindrical_arrow, color=red):
     CrossPlot[k]:= display({arrowRotVec[k],arrowCrossVec[k],arrowXVec}, ori-
entation = [60,55], view = [-1.5..1.5, -1.5..1.5, -1.5..1.5], axes = boxed, labels
= [x,y,z], title = typeset(theta = k*Pi/10, ".")):
     end do:
> display([seq(CrossPlot[k], k=1..20)], insequence=true);
```

2. Given the vector $\vec{w} = \langle 0, 1, 1 \rangle$ and the vector function $\vec{v}(t) = \langle \cos(t), 1 + 2\sin(t), -\sin(t) \rangle$, perform the following:

(a) Plot the angle θ between $\vec{w}$ and $\vec{v}(t)$ as a function of t for $0 \le t \le 2\pi$.

(b) Find any values t such that $\vec{w} \perp \vec{v}(t)$.

(c) Plot $|\vec{w} \times \vec{v}(t)|$.

(d) Find the maximum value of $|\vec{w} \times \vec{v}(t)|$ for $0 \le t \le 2\pi$.

(e) Plot both $|\vec{w} \times \vec{v}(t)|$ and $\theta(t)$ on the same graph and compare results.

(f) Graph the vectors $\vec{w}$, $\vec{v}(t)$ and $\vec{w} \times \vec{v}(t)$ for $0 \le t \le 2\pi$ using the code from problem 1.

3. Given the following vector functions:

$$\vec{w}(t) = \langle \sin(t), \cos(t), 0 \rangle, \qquad \vec{v}(t) = \langle \cos(t), 2\sin(t), -\sin(t) \rangle$$

perform the following:

(a) Plot the angle θ between $\vec{w}(t)$ and $\vec{v}(t)$ as a function of t for $0 \le t \le 2\pi$.

(b) Find any values t such that $\vec{w}(t) \perp \vec{v(t)}$.

(c) Plot $|\vec{w}(t) \times \vec{v}(t)|$.

(d) Find the maximum value of $|\vec{w}(t) \times \vec{v}(t)|$ for $0 \le t \le 2\pi$.

(e) Plot both $|\vec{w}(t) \times \vec{v}(t)|$ and $\theta(t)$ on the same graph and compare results.

(f) Graph the vectors $\vec{w}(t)$, $\vec{v}(t)$ and $\vec{w}(t) \times \vec{v}(t)$ for $0 \le t \le 2\pi$. Modifying the code from problem 1 may be the easiest method to solving this problem.

4. Do *Homework* problem 7 (c).

5. Do *Homework* problem 8 (c).

6. Do *Homework* problem 9.

7. Do *Homework* problem 10 (b).

6.4   Vector Projection

In this section, we are interested in the idea of *projecting* one vector $\vec{w}$ perpendicularly onto another vector $\vec{v}$, resulting in a third vector in the direction of $\vec{v}$. At first glance, this process may appear to be a mental exercise only, however, there are many situations in which it can be applied. First, we start by letting $\vec{v}$ and $\vec{w}$ be two nonzero vectors of the same dimension that are not scalar multiples of each other, in which case they are said to be *independent vectors*. These vectors thus determine different lines, and so together they determine a unique plane that contains both of them. We define the vector projection of the vector $\vec{w}$ onto the vector $\vec{v}$, denoted $\text{proj}_{\vec{v}}(\vec{w})$, to be the vector that, as an arrow, starts at the common starting point of $\vec{v}$ and $\vec{w}$'s arrows, while stopping at the point on $\vec{v}$'s arrow that is perpendicularly below the endpoint of $\vec{w}$'s arrow.

Example 6.4.1. Before we get into the mathematics behind projection, we will project the vector $\vec{w} = \langle 9, 8 \rangle$ onto the vector $\vec{v} = \langle 17, 0 \rangle$.

> with(plots): with(plottools): with(linalg):

> W:= vector([9,8]):

> V:= vector([17,0]):

> proj_WtoV:= vector([9,0]):

> arrow_W:= arrow([0,0], W, .2, 1, .2, color = black):

> arrow_V:= arrow([0,0], V, .2, 1, .2, color = blue):

> arrow_proj:= arrow([0,0], proj_WtoV, .2, 1, .2, color = red):

> LabelW:= textplot([9,8,"W"], align={ABOVE,RIGHT}, color=black):

> LabelV:= textplot([17,.5,"V"], align = ABOVE, color = black):

> LabelProj:= textplot([4,-1.5,"ProjWtoV"],align=ABOVE, color=black):

> VerticalLeg:= line([9,0], [9,8], color = black):

> RightAngle:= polygon([[9,0],[10,0],[10,1],[9,1]], color = blue):

> LabelAngle:= textplot([3.75,1, q], align=ABOVE, color=black, font = [SYMBOL, 10]):

> Arc:= arc([0,0], 5.5, 0..arctan(8/9), color = black):

> display({arrow_W, arrow_V, arrow_proj, LabelW, LabelV, LabelProj, Ver-
ticalLeg, RightAngle, LabelAngle, Arc}, scaling = constrained, axes = none);

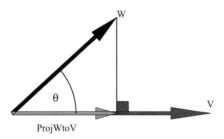

Figure 6.12: The projection of $\overrightarrow{w}$ onto $\overrightarrow{v}$, denoted $\mathrm{proj}_{\overrightarrow{v}}(\overrightarrow{w})$.

The example above is quite simple. Notice that $\overrightarrow{v} = \langle 17, 0 \rangle$ lies along the x-axis; so to project the vector $\overrightarrow{w} = \langle 9, 8 \rangle$ onto $\overrightarrow{v}$, we simply take the x-component of $\overrightarrow{w}$, since the direction perpendicular to $\overrightarrow{v}$ is the y-direction. Therefore, $\mathrm{proj}_{\overrightarrow{v}}(\overrightarrow{w}) = \langle 9, 0 \rangle$. From Figure 6.12 above, it should also be clear that the vector $\overrightarrow{w} - \mathrm{proj}_{\overrightarrow{v}}(\overrightarrow{w})$ is orthogonal to $\overrightarrow{v}$, that is,

$$\overrightarrow{v} \cdot (\overrightarrow{w} - \mathrm{proj}_{\overrightarrow{v}}(\overrightarrow{w})) = 0$$

Now, we want to get a formula for the vector projection $\mathrm{proj}_{\overrightarrow{v}}(\overrightarrow{w})$ of $\overrightarrow{w}$ onto $\overrightarrow{v}$ in terms of arbitrary vectors $\overrightarrow{w}$ and $\overrightarrow{v}$ of the same dimension. Upon further examination of Figure 6.12, above, we see that $\mathrm{proj}_{\overrightarrow{v}}(\overrightarrow{w})$ is a positive multiple of $\overrightarrow{v}$ since its arrow moves along $\overrightarrow{v}$'s arrow in the same direction as $\overrightarrow{v}$. As a result of this observation, the following equality must hold:

$$\frac{\mathrm{proj}_{\overrightarrow{v}}(\overrightarrow{w})}{|\mathrm{proj}_{\overrightarrow{v}}(\overrightarrow{w})|} = \frac{\overrightarrow{v}}{|\overrightarrow{v}|} \tag{6.14}$$

This is true since both vectors in the above expression are unit vectors and point in the same direction. Solving for $\mathrm{proj}_{\overrightarrow{v}}(\overrightarrow{w})$ in the equation (6.14) gives

$$\mathrm{proj}_{\overrightarrow{v}}(\overrightarrow{w}) = \frac{|\mathrm{proj}_{\overrightarrow{v}}(\overrightarrow{w})|}{|\overrightarrow{v}|} \overrightarrow{v} \tag{6.15}$$

Also, from the right triangle with sides $\overrightarrow{v}$ and $\overrightarrow{w}$ and angle θ between them:

$$|\mathrm{proj}_{\overrightarrow{v}}(\overrightarrow{w})| = |\overrightarrow{w}| \cos(\theta) \tag{6.16}$$

Remember, magnitude and a direction are the two properties that define a vector, and we now have both. By putting these together, we see that

$$\mathrm{proj}_{\overrightarrow{v}}(\overrightarrow{w}) = \frac{|\overrightarrow{w}|}{|\overrightarrow{v}|} \cos(\theta) \overrightarrow{v}$$

This can be simplified even further if we multiply the RHS by $\frac{|\vec{v}|}{|\vec{v}|}$:

$$\text{proj}_{\vec{v}}(\vec{w}) = \frac{|\vec{w}||\vec{v}|}{|\vec{v}|^2}\cos(\theta)\vec{v}$$

Remember that the dot product formula states that $\vec{w} \cdot \vec{v} = |\vec{w}||\vec{v}|\cos(\theta)$, therefore we arrive at the final formulation of the projection of $\vec{w}$ onto $\vec{v}$:

$$\text{proj}_{\vec{v}}(\vec{w}) = \frac{\vec{w} \cdot \vec{v}}{|\vec{v}|^2}\vec{v} \tag{6.17}$$

Note that we could also write the above expression as

$$\text{proj}_{\vec{v}}(\vec{w}) = \left(\frac{\vec{w} \cdot \vec{v}}{|\vec{v}|}\right)\frac{\vec{v}}{|\vec{v}|}$$

where the vector $\frac{\vec{v}}{|\vec{v}|}$ is the unit vector pointing in the direction of $\vec{v}$. By computing the unit vector in the direction of $\vec{v}$ first, the projection formula becomes much simpler. If we set $\vec{u} = \frac{\vec{v}}{|\vec{v}|}$ in the above formula, we have

$$\text{proj}_{\vec{v}}(\vec{w}) = (\vec{w} \cdot \vec{u})\vec{u} \tag{6.18}$$

Furthermore, notice that the magnitude of the projection of $\vec{w}$ onto $\vec{v}$ is simply

$$|\text{proj}_{\vec{v}}(\vec{w})| = \left|\frac{\vec{w} \cdot \vec{v}}{|\vec{v}|^2}\vec{v}\right| = \left|\frac{\vec{w} \cdot \vec{v}}{|\vec{v}|}\right| = \frac{|\vec{w} \cdot \vec{v}|}{|\vec{v}|}$$

The expression that lies inside the magnitude signs above is referred to as the *component* of $\vec{w}$ onto $\vec{v}$, denoted $\text{comp}_{\vec{v}}(\vec{w})$:

$$\text{comp}_{\vec{v}}(\vec{w}) = \frac{\vec{w} \cdot \vec{v}}{|\vec{v}|} \tag{6.19}$$

The component can be thought of as the signed magnitude of the projection vector. Note that if the vector $\vec{v}$ should point in the opposite direction to the general direction of $\vec{w}$, then the formula for computing the projection vector $\text{proj}_{\vec{v}}(\vec{w})$ of $\vec{w}$ onto $\vec{v}$ still holds. The projection is performed, not with respect to $\vec{v}$, but instead to the line that $\vec{v}$ determines. This situation occurs when $\theta > \frac{\pi}{2}$, as Figure 6.13 illustrates.

Example 6.4.2. Once again we will use the vector $\vec{w} = \langle 9, 8 \rangle$, this time projecting it onto the vector $\vec{v} = \langle -17, 0 \rangle$.

> W:= vector([9,8]): V:= vector([-17,0]):
> arrow_V:= arrow([0,0], V, .2, 1, .2, color=blue):
> LabelV:= textplot([-17,.5,"V"], align=ABOVE, color=black):
> LabelAngle:= textplot([-1.5,1.5, q], align=ABOVE, color=black, font = [SYMBOL, 10]):
> Arc:= arc([0,0], 5.5, arctan(8/9)..Pi, color = black):
> display({arrow_W, arrow_V, arrow_proj, LabelW, LabelV, LabelProj, VerticalLeg, RightAngle, LabelAngle, Arc}, scaling = constrained, axes = none);

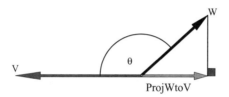

Figure 6.13: Projection of $\vec{w}$ onto $\vec{v}$ with $\theta > \dfrac{\pi}{2}$.

Example 6.4.3. Now let us do a vector projection computation in $\mathbb{R}^3$, with $\vec{v} = \langle -5, 9, -12 \rangle$ and $\vec{w} = \langle 7, 4, 8 \rangle$. We will compute the vector projection $\text{proj}_{\vec{v}}(\vec{w})$ of $\vec{w}$ onto $\vec{v}$ and plot all three vectors. This is illustrated in Figure 6.14.

> V:= vector([-5,9,-12]): W:= vector([7,4,8]):
> Proj_WontoV:= evalm(dotprod(W,V)/norm(V,frobenius)^2*V);

$$Proj\_WontoV := \left[\begin{array}{ccc} \dfrac{19}{10} & -\dfrac{171}{50} & \dfrac{114}{25} \end{array} \right]$$

> arrow_W:= arrow([0,0,0], W, .2, 1, .2, cylindrical_arrow, color = black):
> arrow_V:= arrow([0,0,0], V, .2, 1, .2, cylindrical_arrow, color = blue):
> arrow_proj:= arrow([0,0,0], Proj_WontoV, .2, 1, .2, cylindrical_arrow, color = red):
> LabelW:= textplot3d([4, 2, 4, "W"], align = {ABOVE, RIGHT}, color = black):

> LabelV:= textplot3d([-2,5,-6,"V"], align = ABOVE, color = black):

> LabelProj:= textplot3d([2, -5, 5, "ProjWtoV"], align = BELOW, color = black):

> display({arrow_W, arrow_V, arrow_proj, LabelW, LabelV, LabelProj}, axes = boxed);

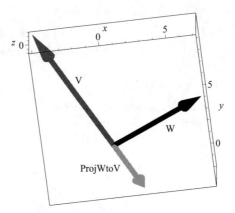

Figure 6.14: The projection process applied to a pair of three-dimensional vectors.

Vector projections have many useful applications. One of the simplest is deriving a formula for the shortest (perpendicular) distance between a point $P(x_0, y_0)$ and a line L given by the equation $ax + by = c$ in the xy-plane. We will denote this distance as $D(P, L)$. To start, we need to find a vector $\vec{n}$ perpendicular to the line L. Note that this also implies that $\vec{n}$ would be orthogonal to any vector parallel to the line L. This orthogonal vector is also sometimes referred to as a *normal vector* to the line L. The vector $\vec{n} = \langle a, b \rangle$ is such a vector since if we take any two points $Q(x, y)$ and $R(s, t)$ on this line, then the vector $\overrightarrow{RQ} = \langle x - s, y - t \rangle$ is parallel to the line L:

$$\vec{n} \cdot \overrightarrow{RQ} = a(x - s) + b(y - t)$$
$$= ax + by - (as + bt)$$
$$= c - c = 0$$

Second, we need to find a vector that moves from the point $P(x_0, y_0)$ to the line L. We take the vector $\vec{w} = \overrightarrow{PQ} = \langle x - x_0, y - y_0 \rangle$ where $Q(x, y)$ is an arbitrary point on L. Now thinking of the vector $\vec{n}$, which is orthogonal to the line L, also starting at the point $P(x_0, y_0)$, we see that the vector projection $\text{proj}_{\vec{n}}(\vec{w})$ of $\vec{w}$ onto $\vec{n}$ is the vector starting at the point $P(x_0, y_0)$ and ending at the point on the line L closest to $P(x_0, y_0)$. The length of the vector projection

$|\text{proj}_{\vec{n}}(\vec{w})|$ will be the shortest distance from the point $P(x_0, y_0)$ to our line L. This situation is depicted in Figure 6.15.

When we now put all of the above information together, we have that the shortest distance from the point $P(x_0, y_0)$ to our line L with equation $ax + by = c$ is given by $|\text{proj}_{\vec{n}}(\vec{w})|$, for the vectors $\vec{n} = \langle a, b \rangle$ and $\vec{w} = \langle x - x_0, y - y_0 \rangle$ with $Q(x, y)$ being any point on the line L. Putting this into a formula, and with some algebraic manipulations, we get

$$|\text{proj}_{\vec{n}}(\vec{w})| = \left| \frac{\vec{w} \cdot \vec{n}}{|\vec{n}|^2} \vec{n} \right| = \frac{|\vec{w} \cdot \vec{n}|}{|\vec{n}|} = \frac{|a(x - x_0) + b(y - y_0)|}{\sqrt{a^2 + b^2}}$$

$$= \frac{|ax + by - ax_0 - by_0|}{\sqrt{a^2 + b^2}} = \frac{|c - ax_0 - by_0|}{\sqrt{a^2 + b^2}}$$

$$= \frac{|ax_0 + by_0 - c|}{\sqrt{a^2 + b^2}}$$

So the shortest distance $D(P, L)$ from the point $P(x_0, y_0)$ to the line $L : ax + by = c$ is given by

$$D(P, L) = \frac{|ax_0 + by_0 - c|}{\sqrt{a^2 + b^2}} \tag{6.20}$$

Example 6.4.4. Now, let us draw the picture to illustrate equation (6.20) derivation and then do a shortest distance calculation from the point $P(-7, 13)$ to the line L given by $20x - 32y = 165$, refer to Figure 6.15 for this example. Note the formula given in the definition of the variable *LPerpthruP* in the code below. This variable defines the equation of the line perpendicular to L. To understand the formula given, remember that any line perpendicular to L must be of the form $32x + 20y = C$, for arbitrary $C \in \mathbb{R}$ since slopes of perpendicular lines in $\mathbb{R}^2$ are negative reciprocals of one another. In our case, the point P must satisfy the equation $32x + 20y = C$, hence $C = 32(-7) + 20(13)$. This is how *LPerpthruP* is defined.

```
> L:= 20*x-32*y = 165:
> LPerpthruP:= 32*x+20*y = 32*(-7)+20*13:
> solve({L,LPerpthruP},{x,y});
```

$$\left\{ x = \frac{1113}{356}, \ y = -\frac{285}{89} \right\}$$

```
> v:= vector([20,-32]):
> P:= [-7,13]: Q:= [14, 5/8*14-165/32]:
> proj_PQtov:= evalm(dotprod(vector([21, 5/8*14 - 165/32 - 13]), v)/norm(v,
frobenius)^2*v);
```

$$proj\_PQtov := \left[\begin{array}{cc} \dfrac{3605}{356} & -\dfrac{1442}{89} \end{array} \right]$$

> f:= solve(L,y);

$$f := \frac{5}{8}x - \frac{165}{32}$$

> Dist_PtoL:= evalf(abs(dotprod(v, vector(P)) - dotprod(v, vector(Q)))/norm (v,frobenius));

$$Dist\_PtoL := 19.10646179$$

> plotP:= pointplot({P, Q}, symbol = circle, symbolsize=30, color=black):

> plotL:= plot(f, x=-12..20):

> arrow_v:= arrow(P, v, .2, 2, .2, color = black):

> arrow_PQ:= arrow(P, Q, .2, 2, .2, color = blue):

> arrow_ProjPQtov:= arrow(P, proj_PQtov, .4, 2, .2, color = red):

> LabelV:= textplot([8,-10,"V"],align={ABOVE,RIGHT},color=black):

> LabelPQ:= textplot([5,9,"PQ"], align = ABOVE, color = black):

> LabelProj:= textplot([-8,5,"ProjWtoV"], align=ABOVE, color=black):

> LabelL:= textplot([-10, -7, "Line L"], align = {ABOVE, RIGHT}, color = black):

> LabelP:= textplot([-9,13,"P"],align={ABOVE,RIGHT}, color=black):

> LabelQ:= textplot([16,3,"Q"], align = ABOVE, color = black):

> RightAngle:= polygon([[1113/356, -285/89], [1113/356 + 2, -285/89 - 3.2], [1113/356 + 5.2, -285/89 - 1.2], [1113/356 + 3.2, -285/89 + 2]],color=black):

> display({plotP, plotL, arrow_v, arrow_PQ, arrow_ProjPQtov, LabelV, LabelPQ, LabelProj, LabelL, LabelP, LabelQ, RightAngle}, axes = none, scaling = constrained);

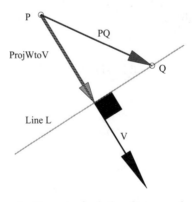

Figure 6.15: Using projections to find the distance from a point to a line.

Now we will take things one step further. What if we desire to know the distance from a point to a plane? Referring back to Section 2.1, we determined that the equation for a plane was given by $ax + by + cz = d$. From our discussion of the distance from a point to a line, we noticed that the vector representing the orthogonal direction to the line was given by the coefficients in front of the x and y variables. A similar argument holds true for the three-dimensional case that gives the vector $\overrightarrow{n} = \langle a, b, c \rangle$ is orthogonal to the plane $ax + by + cz = d$, which we will call R.

Given a point on the plane, call it $P_1(x_1, y_1, z_1)$, it must satisfy $ax_1 + by_1 + cz_1 = d$. Plugging this in for d and moving everything to the LHS of the equation of the plane R gives $a(x - x_1) + b(y - y_1) + c(z - z_1) = 0$. Now we relate this to the dot product that we just recently introduced; if we define the two vectors $\overrightarrow{n} = \langle a, b, c \rangle$ and $\overrightarrow{w} = \langle x - x_1, y - y_1, z - z_1 \rangle$, notice that

$$a(x - x_1) + b(y - y_1) + c(z - z_1) = \overrightarrow{n} \cdot \overrightarrow{w} = 0 \tag{6.21}$$

Geometrically speaking, this states that given a point $P_1(x_1, y_1, z_1)$ and direction $\overrightarrow{n}$, any other point $P(x, y, z)$ that lies on the plane must satisfy the equation

$$\overrightarrow{n} \cdot \left(\overrightarrow{P_1} - \overrightarrow{P_0} \right) = 0$$

Applying this process to planes of arbitrary dimension yields the following definition.

Definition 6.4.1. Given a plane in $n-$dimensional space defined by the expression $a_1 x_1 + a_2 x_2 + \cdots + a_n x_n = d$, the *normal vector* $\overrightarrow{n}$ is independent of the value of d and is given by $\overrightarrow{n} = \langle a_1, a_2, \ldots, a_n \rangle$.

How does the definition of the normal vector help us in our case? Well similar to the solution to the distance of a point from a line, the shortest distance from a point to a plane is found by drawing a line segment perpendicular to the line passing through the point. The direction of this perpendicular line in three dimensions is $\overrightarrow{n} = \langle a, b, c \rangle$. Now once we have this, we notice that the previous two-dimensional approach also describes our situation if we were looking at the plane directly from the side. The entirety of the previous argument still holds for the three-dimensional case. So the shortest distance, $D(P, R)$, from the point $P(x_0, y_0, z_0)$ to the plane $R : ax + by + cz = d$, is given by

$$D(P, R) = \frac{|ax_0 + by_0 + cz_0 - d|}{\sqrt{a^2 + b^2 + c^2}} \tag{6.22}$$

Imagine now how this could be modified for higher dimensional problems.

Example 6.4.5. Now, we will do an example to illustrate this formula for a point and a plane. The plane is given by $2x - 3y + 6z = 1$, and we wish to find the distance from the plane to the point $P(1, 1, 7)$. Furthermore, we will choose the point $Q\left(1, 1, \frac{1}{3}\right)$ on the plane for which to perform our computations.

Before we begin the computations, we should discuss how to set up the equation of a line L in $\mathbb{R}^3$, given a point P and a direction $\vec{n}$. For this example, the point is $P(1, 1, 7)$, and the direction will be $\vec{n} = \langle 2, -3, 6 \rangle$, the normal vector to the plane $2x - 3y + 6z = 1$. If we choose any other point $\langle x, y, z \rangle$ on the L, then the following equation must hold:

$$\langle x, y, z \rangle - \langle 1, 1, 7 \rangle = t \langle 2, -3, 6 \rangle$$

for some real value t. This is true since a line always points in the same direction, hence the difference between two points on the line, yielding a displacement vector, must also point in the direction of the line, and must therefore be a scalar multiple of this direction. Looking at this in terms of components yields the set of three equations $x - 1 = 2t$, $y - 1 = -3t$, $z - 7 = 6t$, or $x = 2t + 1$, $y = -3t + 1$, and $z = 6t + 7$, which is the parametric equation for the line perpendicular to the plane $2x - 3y + 6z = 1$, which passes through the point $P(1, 1, 7)$. Solving each of these for t gives

$$\left\{ t = \frac{x - 1}{2}, \ t = \frac{y - 1}{-3}, \ t = \frac{z - 7}{6} \right\} \tag{6.23}$$

The rectangular equations of this line are

$$\frac{x - 1}{2} = \frac{y - 1}{-3}, \ \text{and} \ \frac{y - 1}{-3} = \frac{z - 7}{6}$$

If we wish to know where the line L intersects the plane $2x - 3y + 6z = 1$, we need a system of three equations to solve for the three unknown coordinates x, y, and z. Setting the first two, and last two, equations equal in system (6.23), and adding the condition that the point must lie in the plane $2x - 3y + 6z = 1$, gives the following system of three equations:

$$\left\{ \frac{x - 1}{2} = \frac{y - 1}{-3}, \ \frac{y - 1}{-3} = \frac{z - 7}{6}, \ 2x - 3y + 6z = 1 \right\} \tag{6.24}$$

Solving system (6.24) for x, y and z gives the point

$$\left\{ x = -\frac{13}{19}, z = -\frac{87}{38}, y = -\frac{102}{19} \right\}$$

This point lies both on the line L and the plane $2x - 3y + 6z = 1$, and must be the closest point on the plane to the point P, since L has direction normal to the plane. *Maple* will verify this below with the help of projections.

> Plane:= 2*x-3*y+6*z=1;

$$Plane := 2x - 3y + 6z = 1$$

> LPerpthruPlane:= (x-1)/2=(y-1)/(-3), (y-1)/(-3)=(z-7)/6:
> solve({Plane,LPerpthruPlane},{x,y,z});

$$\left\{ x = -\frac{31}{49},\ y = \frac{169}{49},\ z = \frac{103}{49} \right\}$$

> subs({x = -13/19, z = -87/38, y = -102/19}, Plane);

$$1 = 1$$

The above calculation verifies that the point found in the *solve* command prior actually does lie on the plane. We can now calculate the distance from the point to the plane.

> v:= vector([2,-3, 6]):
> P:= [1, 1, 7]: Q:= [1, 1, 1/3]:
> proj_PQtov:= evalm(dotprod(vector(Q-P),v)/norm(v,frobenius)^2*v);

$$proj\_PQtov = \left[\ -\frac{80}{49} \quad \frac{120}{49} \quad -\frac{240}{49}\ \right]$$

> f:= solve(Plane,z);

$$f := -\frac{1}{3}x + \frac{1}{2}y + \frac{1}{6}$$

> Dist_PtoPlane:= evalf(abs(dotprod(v,vector(P))-1)/norm(v,frobenius));

$$Dist\_PtoPlane := 5.714285714$$

This is the actual value of the distance from the point to the plane. We now perform some *Maple* commands to plot the resulting vectors and the plane to illustrate these calculations, see Figure 6.16.

> plotP:= pointplot3d({P, Q}, symbol = circle, symbolsize = 20, color = black, thickness=4):
> plotPlane:= plot3d(f, x=-10..10,y=-10..10, style=PATCHNOGRID):
> arrow_v:= arrow(P, v, .2, 2, .2, color = black):
> arrow_PQ:= arrow(P, Q, .2, 2, .2, color = blue):
> evalm(P+ proj_PQtov);

$$\left[\ -\frac{31}{49} \quad \frac{169}{49} \quad \frac{103}{49}\ \right]$$

> plotPlaneIntercept:= pointplot3d([evalm(P+ proj_PQtov)],symbol = circle, symbolsize = 20, color = black, thickness=4):

> arrow_ProjPQtov:= arrow(P, proj_PQtov, .4, 2, .2, color = red):

> LabelV:= textplot3d([2, -2, 10,"V"], align = {ABOVE,RIGHT}, color = black):

> LabelPQ:= textplot3d([2,1,3,"PQ"], align = ABOVE, color = black):

> LabelProj:= textplot3d([-2, 109/49, 5,"ProjWtoV"], align = ABOVE, color = black):

> LabelP:= textplot3d([1, 0.5, 7,"P"], align = {BELOW, RIGHT}, color = black):

> LabelQ:= textplot3d([1, -0.5, 1/3,"Q"], align = {ABOVE, LEFT}, color = black):

> display({plotP, plotPlane, arrow_v, arrow_PQ, arrow_ProjPQtov, LabelV, LabelPQ, LabelProj, LabelP, LabelQ, plotPlaneIntercept}, orientation = [-160,70], axes=boxed);

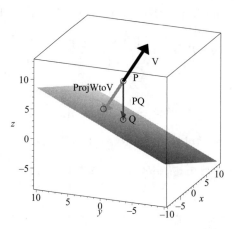

Figure 6.16: Using projections to find the distance from a point to a plane.

Homework Problems

1. For $\vec{v}, \vec{w} \in \mathbb{R}^n$, under what conditions does $\text{proj}_{\vec{v}}(\vec{w}) = \vec{0}$?

2. For $\vec{v}, \vec{w} \in \mathbb{R}^n$, relate the sign of $\text{comp}_{\vec{v}}(\vec{w})$ to the angle θ between $\vec{v}$ and $\vec{w}$.

3. Compute $\text{comp}_{\vec{v}}(\vec{w})$ for the following pairs of vectors:

(a) $\vec{v} = \langle -1, 2 \rangle$, $\vec{w} = \langle 3, 5 \rangle$ (b) $\vec{v} = \langle 4, 6 \rangle$, $\vec{w} = \langle 2, 3 \rangle$

(c) $\vec{v} = \langle 3, 0 \rangle$, $\vec{w} = \langle 5, 1 \rangle$ (d) $\vec{v} = \langle -1, 0, 2 \rangle$, $\vec{w} = \langle 3, 2, -2 \rangle$

(e) $\vec{v} = \langle 1, 1, -1 \rangle$, $\vec{w} = \langle 2, -1, 2 \rangle$ (f) $\vec{v} = \langle 3, 0, 0 \rangle$, $\vec{w} = \langle -1, 1, 1 \rangle$

(g) $\vec{v} = \langle 3, -2, -4 \rangle$, $\vec{w} = \langle -1, 2, 0 \rangle$ (h) $\vec{v} = \langle 1, 1, -1, 1 \rangle$, $\vec{w} = \langle 1, -1, 1, 1 \rangle$

4. Compute $\text{proj}_{\vec{v}}(\vec{w})$ for the following pairs of vectors:

(a) $\vec{v} = \langle -1, 2 \rangle$, $\vec{w} = \langle 3, 5 \rangle$ (b) $\vec{v} = \langle 4, 6 \rangle$, $\vec{w} = \langle 2, 3 \rangle$

(c) $\vec{v} = \langle 3, 0 \rangle$, $\vec{w} = \langle 5, 1 \rangle$ (d) $\vec{v} = \langle -1, 0, 2 \rangle$, $\vec{w} = \langle 3, 2, -2 \rangle$

(e) $\vec{v} = \langle 1, 1, -1 \rangle$, $\vec{w} = \langle 2, -1, 2 \rangle$ (f) $\vec{v} = \langle 3, 0, 0 \rangle$, $\vec{w} = \langle -1, 1, 1 \rangle$

(g) $\vec{v} = \langle 3, -2, -4 \rangle$, $\vec{w} = \langle -1, 2, 0 \rangle$ (h) $\vec{v} = \langle 1, 1, -1, 1 \rangle$, $\vec{w} = \langle 1, -1, 1, 1 \rangle$

5. Compute the normal vectors to each of the following planes:

(a) $3x - 5y = 7$ (b) $2x + 8y = 2$

(c) $x - 5y + 7z = 3$ (d) $2w - 4x + 5y - 7z = 2$

6. Find the distance from the point $P(2, 3)$ to the line $3x - 4y = 6$.

7. Find the distance from the point $P(1, -2, 4)$ to the plane $2x + 5y - 6z = 1$.

8. Find the distance from the point $P(1, -2, 4, -1)$ to the plane $w + 3x - 2y + 2z = -1$.

9. For each of problems 6-8, find the point that lies on the line or plane at the location that minimizes distance between the line or plane and P.

10. Can the ideas of this section be generalized to vector projection of complex vectors in $\mathbb{C}^n$? If yes, then explain how and what still works using the complex dot product. If no, then explain why not specifically stating what fails.

11. Find a formula for the shortest distance between the two parallel planes $ax + by + cz = d$ and $ax + by + cz = e$. Test your formula on an example of two planes.

12. Find a formula for the shortest distance between the line $x = at + \alpha$, $y = bt + \beta$, $z = ct + \gamma$, parallel to the plane $ex + fy + gz = h$. Test your formula on an example of a parallel line and plane. Also, how can we test if a line and plane in $\mathbb{R}^3$ are parallel or not?

13. Find a formula for the angle between two non-parallel planes in space. Also, find a formula for the angle between a line and plane in space which are non-parallel.

Maple Problems

1. Given the vector $\vec{v} = \langle 1, 0 \rangle$ and the vector function $\vec{w}(t) = \langle \cos(t), \sin(t) \rangle$, perform the following:

  (a) Plot $\mathrm{comp}_{\vec{v}}(\vec{w}(t))$ as a function of t for $0 \le t \le 2\pi$.

  (b) Plot $\mathrm{proj}_{\vec{v}}(\vec{w}(t))$ as a function of t for $0 \le t \le 2\pi$ along with both $\vec{v}$ and $\vec{w}(t)$.

  (c) Find the values of t that maximize and minimize $|\mathrm{comp}_{\vec{v}}(\vec{w}(t))|$.

2. Given the vector $\vec{v} = \langle 0, 1, 1 \rangle$ and the vector function $\vec{w}(t) = \langle \cos(t), 1 + 2\sin(t), -\sin(t) \rangle$, perform the following:

  (a) Plot $\mathrm{comp}_{\vec{v}}(\vec{w}(t))$ as a function of t for $0 \le t \le 2\pi$.

  (b) Plot $\mathrm{proj}_{\vec{v}}(\vec{w}(t))$ as a function of t for $0 \le t \le 2\pi$ along with both $\vec{v}$ and $\vec{w}(t)$.

  (c) Find the values of t which maximize and minimize $|\mathrm{comp}_{\vec{v}}(\vec{w}(t))|$.

3. Graph the point, line and the displacement vector with base at P and tip at the point on the line closest to P from *Homework* problem 6.

4. Graph the point, plane and the displacement vector with base at P and tip at the point on the line closest to P from *Homework* problem 7.

Chapter 7

A Few Advanced Linear Algebra Topics

7.1   Rotations in Space

Next, we investigate a topic in computer graphics: That of rotating parametric surfaces and curves in space about any line of rotation not passing through them. For simplicity, we will only look at the case when the line passes through the origin. The procedure that we develop can be generalized to an arbitrary line of rotation, and you are asked to do so in *Homework* problem 5.

In order to rotate about a line in space, we need to understand a few important concepts, the first of which is how to define a unique line in space. Since the line passes through the origin, it is determined uniquely if we know another point P on the line or equivalently any vector $\vec{v} = \overrightarrow{0P}$ parallel to the line. Here $\vec{0}$ refers to the origin. The concept of rotating a single point Q through an angle α about a line L generalizes easily to rotating a parametric curve or surface about the line, so let us begin with just a single point. As the point Q is rotated (or revolved) about the line L (Q is not on L), the point Q will travel along an arc of the circle C orthogonal to L where the center of the circle C is the nearest point of L to Q.

To rotate the single point Q through an angle α about the line L, we will use vector operations and position vectors to find the circle of revolution C in which the point Q rotates about the line L. The circular arc along which Q travels should be perpendicular to the line L, with a radius equal to the shortest (perpendicular) distance from Q to L. The circle's center C corresponds to the point on the line L nearest to Q. The circle itself will be given parametrically in terms of the parameter α that is now the variable angle of rotation. Rotating Q along the arc of the circle through an angle of α will result in a new point N. For ease of computation, we will convert P and Q to vectors, leaving them capitalized to remember that they correspond to points. (Consider the fact that a point is really the terminal point of a vector starting at the origin.)

First, we must find the vector $\vec{C}$, which is the center of the circle of rotation. This vector is found by computing the projection of the vector $\vec{Q}$ (our point) onto the vector $\vec{P}$ (parallel to our line through the origin). Using the vector projection formula from Section 6.4 gives

$$\vec{C} = \frac{\vec{P} \cdot \vec{Q}}{\left|\vec{P}\right|^2} \vec{P} \tag{7.1}$$

The radius R, of the circle of rotation, is just the length of the vector $\vec{Q} - \vec{C}$, and so $R = \left|\vec{Q} - \vec{C}\right|$.

Next, consider the following vector:

$$\vec{U} = \frac{\vec{Q} - \vec{C}}{R} \tag{7.2}$$

which is of unit length, starting at C and traveling in the direction of Q. The vector $\vec{U}$ is in the plane of the circle of rotation.

Another unit vector in the plane of the circle of rotation, perpendicular to both $\vec{U}$ and $\vec{C}$, is given by

$$\vec{V} = \frac{\vec{C} \times \vec{U}}{\left|\vec{C} \times \vec{U}\right|} \tag{7.3}$$

where $\vec{V}$ is simply a scalar multiple of $\vec{C} \times \vec{U}$, and must therefore be perpendicular to both $\vec{C}$ and $\vec{U}$, by definition of the cross product. The vector $\vec{V}$ is made unit length simply by dividing the cross product by its magnitude.

So what has this accomplished? Consider the following idea: In the plane of the circle of rotation, $\vec{U}$ and $\vec{V}$ are acting as the positive x- and y-axes unit vectors where our point Q is always positioned at the coordinates $(1, 0)$ in the $\left(\vec{U}, \vec{V}\right)$ coordinate system. Now the circle of rotation has the vector parametric equation given by

$$\vec{Q}_{\text{rot}}(\alpha) = \vec{C} + R\cos(\alpha)\vec{U} + R\sin(\alpha)\vec{V} \tag{7.4}$$

Example 7.1.1. For our first example, consider the following situation: We wish to rotate the point $Q(3, 5, 9)$, through the angle $\alpha = \dfrac{\pi}{3}$, about the line

L, which passes through the origin and the point $P(-2, 1, 4)$. Upon examining equation (7.4), we see that

$$\overrightarrow{Q}_{\text{rot}}(0) = \overrightarrow{C} + R\overrightarrow{U}$$

which by equation (7.2) is Q. Second, by setting $\alpha = \frac{\pi}{3}$, we arrive at the point N:

$$\overrightarrow{N} = \overrightarrow{Q}_{\text{rot}}\left(\frac{\pi}{3}\right) = \overrightarrow{C} + R\cos\left(\frac{\pi}{3}\right)\overrightarrow{U} + R\sin\left(\frac{\pi}{3}\right)\overrightarrow{V}$$

> with(plots): with(linalg): with(plottools):
> P:= array([-2, 1, 4]): Q:= array([3, 5, 9]):
> C:= evalm(dotprod(P,Q)/dotprod(P,P)*P):
> Rad:= evalm(norm(Q - C, 2)):
> U:= evalm((Q - C)/Rad):
> V:= evalm(crossprod(C, U)/norm(crossprod(C, U), 2)):
> circlerotation:= A -> convert(evalm(C + Rad*cos(A)*U + Rad*sin(A)*V), list):
> spacecircle:= spacecurve(circlerotation(A), A = 0..2*Pi, color = blue):
> N:= evalf(circlerotation(Pi/3)):
> points:= sphere(convert(Q,list),.25,color=black), sphere(convert(N,list), .25, color=black), sphere(convert(C,list), .25, color=black):
> text:= textplot3d([[Q[1], Q[2], Q[3] + 1, "Q"], [N[1], N[2], N[3] + 1, "N"], [C[1], C[2], C[3] - 1, "C"]], align = RIGHT, color = black):
> Line:= spacecurve([t*P[1], t*P[2], t*P[3]], t = 0..1.5*Rad, color = red):
> display3d([points, text, Line, spacecircle], axes = boxed, scaling = constrained);

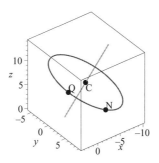

Figure 7.1: Rotating the point Q through the angle α about the line L gives the new point N.

Now that we have the resources required to rotate a point about a line through the origin, as in Figure 7.1, we can determine what else is required to rotate more complicated objects about the line. A point is simply a zero-dimensional object in $\mathbb{R}^3$, a curve will be a one-dimensional object in $\mathbb{R}^3$, and can be parameterized by a single variable. These one-dimensional curves in space are referred to as *space curves*. Space curves are given in vector form by $\langle x(t), y(t), z(t) \rangle$ for $t \in [a, b]$. Upon examining all the work done to revolve a single point Q about a line, it should become apparent that replacing $Q(x, y, z)$ with $\overrightarrow{Q}(x(t), y(t), z(t))$ for $t \in [a, b]$ changes nothing in the formulas. In other words, the formula holds for each fixed $t \in [a, b]$ and we can therefore consider the formulas to hold along the entire space curve for all $t \in [a, b]$. So, when we revolve a space curve about a line L we are really just revolving each of the points of the space curve about L to produce another space curve.

Example 7.1.2. As an example, we will rotate a helix about the line L, which passes through the origin and the point $P(-2, 1, 4)$, which is the same line used in Example 7.1.1. We define our circular helix vectorially by

$$\langle x(t), y(t), z(t) \rangle = \left\langle 6 + \frac{2}{3}\sin(t), 2 + \frac{2}{3}\cos(t), 5 + \frac{t}{4} \right\rangle \qquad (7.5)$$

for $t \in [-2\pi, 3\pi]$. The central axis for this helix is parallel to the z-axis and passes through the point $(6, 2, 5)$, where each point of the helix is at a fixed distance of $\frac{2}{3}$ from this central axis. (See Figure 7.2).

> Line:= spacecurve([t*P[1], t*P[2], t*P[3]], t=-1.5..2.5, color = red, thickness = 2):

> Helix:=[6+2/3*sin(t), 2+2/3*cos(t),5+t/4]:

> HelixOrig:=spacecurve(Helix, t=-2*Pi..3*Pi, color=blue, thickness=4):

> display(HelixOrig, axes=boxed, scaling=constrained);

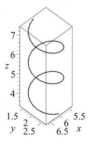

Figure 7.2: Graph of the helix space curve that we will rotate.

We can also think of this helix as the collection of the tips of the position vectors given in equation (7.5). The following code plots, as an animation, the position vector field for this helix as arrows from the origin to the space curve. Figure 7.3 consists of three frames in the animation.

> HelixP:= t-> [6+(2/3)*sin(t), 2+(2/3)*cos(t), 5+(1/4)*t]:

> OriginSphere:= sphere([0,0,0], .15, color = BLACK):

> HelixArrows:= t-> display(arrow([0, 0, 0], HelixP(t), .1, .25, .1, cylindrical_arrow, color = RED)):

> animate(HelixArrows, [t], t=-2*Pi .. 3*Pi, background=display({HelixOrig, OriginSphere}), axes = boxed, scaling = constrained, frames = 50);

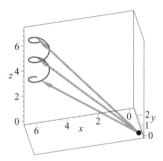

Figure 7.3: Graph of the helix space curve along with three position vectors starting at the origin terminating at the helix.

> SHelix:= array(Helix):

> Qtop:= simplify(subs(t=-2*Pi, Helix));

$$\left[6, \frac{8}{3}, 5 - \frac{1}{2}\pi \right]$$

> Qbot:= simplify(subs(t=3*Pi, Helix));

$$\left[6, \frac{4}{3}, 5 + \frac{3}{4}\pi \right]$$

> Ctop:= evalm(dotprod(P, Qtop)/dotprod(P, P)*P):

> Cbot:= evalm(dotprod(P, Qbot)/dotprod(P, P)*P):

> Radtop:= evalm(norm(Qtop - Ctop, 2)):

> Radbot:= evalm(norm(Qbot - Cbot, 2)):

> Utop:= evalm((Qtop - Ctop)/Radtop):

> Ubot:= evalm((Qbot - Cbot)/Radbot):

> Vtop:= evalm(crossprod(Ctop, Utop)/norm(crossprod(Ctop, Utop), 2)):

> Vbot:= evalm(crossprod(Cbot, Ubot)/norm(crossprod(Cbot, Ubot), 2)):

> circlerotationTop:= A -> convert(evalm(Ctop + Radtop*cos(A)*Utop + Radtop*sin(A)*Vtop),list):

> circlerotationBot:= A -> convert(evalm(Cbot + Radbot*cos(A)*Ubot + Radbot*sin(A)*Vbot),list):

> spacecircleTop:= spacecurve(circlerotationTop(A), A = 0..2*Pi, color = black, thickness=2):

> spacecircleBot:= spacecurve(circlerotationBot(A), A = 0..2*Pi, color = black, thickness=2):

> CH:= evalm(dotprod(P,SHelix)/dotprod(P,P)*P):

> rH:= evalm(norm(SHelix - CH, 2)):

> UH:= evalm(1/rH*(SHelix - CH)):

> VH:= evalm(1/norm(crossprod(CH, UH), 2)*crossprod(CH, UH)):

> Helixrot:= A -> evalf(convert(evalm(CH+rH*cos(A)*UH+rH*sin(A)*VH), list)):

> HelixNews:= seq(spacecurve(Helixrot(k*Pi/3.), t=-2*Pi..3*Pi, color=blue, thickness=4), k=0..5):

> display3d([Line, spacecircleTop, spacecircleBot, display(HelixNews, insequence = true)], axes = boxed, view=[-10..10, -10..10, -6..10], scaling = constrained);

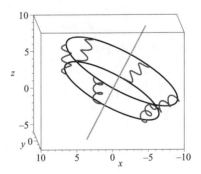

Figure 7.4: The helix was rotated in six steps about the line L, all six frames of the animation are shown above. The two circles are the circles of revolution perpendicular to the line L.

In Figure 7.4, note that we have included two circles of revolution. These circles correspond to the start of the helix, found when $t = -2\pi$, and the end of the helix, at $t = 3\pi$. Clearly, the helix is rotated perpendicularly about the line L, bounded by these two circles of revolution.

The only object left to rotate about a line in $\mathbb{R}^3$ is a two-dimensional surface, which must be parameterized by two variables.

Example 7.1.3. We will take the *torus*(or donut shape) to be our surface $\overrightarrow{T}$ with center at the point $(4, 7, 9)$, inner radius one, and outer radius two, which is plotted in Figure 7.5. Vectorially, the equation defining $\overrightarrow{T}$ is given by

$$\overrightarrow{T}(u, v) = \langle 4 + (2 + \cos(u))\cos(v), 7 + (2 + \cos(u))\sin(v), 9 + \sin(u)\rangle$$

for $u \in [-\pi, \pi]$, $v \in [-\pi, \pi]$. Note that $\overrightarrow{T}$ is parallel to the xy-plane, has three components, and is defined by two independent variables u and v. Therefore, $\overrightarrow{T}$ is a two-dimensional surface in $\mathbb{R}^3$.

```
> T:= [4 + (2 + cos(u))*cos(v), 7 + (2 + cos(u))*sin(v), 9 + sin(u)]:
> S:= array(T):
> torusbegin:= plot3d(T, u = -Pi..Pi, v = -Pi..Pi):
> display3d(torusbegin, axes = boxed, scaling = constrained);
```

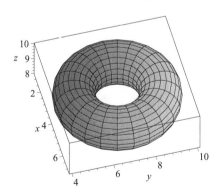

Figure 7.5: The torus as defined vectorially by $\overrightarrow{T}$ given above.

```
> P:= array([-2, 1, 4]):
> C:= evalm(dotprod(P,S)/dotprod(P,P)*P):
> r:= evalm(norm(S - C, 2)):
> U:= evalm(1/r*(S - C)):
> V:= evalm(1/norm(crossprod(C, U), 2)*crossprod(C, U)):
> torusrot:= A -> evalf(convert(evalm(C+r*cos(A)*U+r*sin(A)*V), list)):
```

> TorusNews:= seq(plot3d(torusrot(k*Pi/3.), u=-Pi..Pi, v=-Pi..Pi), k=0..6):

> Line:= spacecurve([t*P[1], t*P[2], t*P[3]], t = -1..4, color = red, thickness=2):

> display3d([Line, display(TorusNews, insequence = true)], axes = boxed, scaling = constrained);

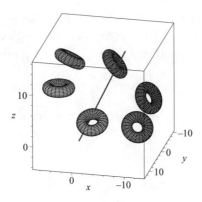

Figure 7.6: The composite image of all six tori in the rotation about the line L, given by the animated *display3d* command above.

The "insequence = true" option in the *display3d* command creates an animation of the figures to be plotted, in the order they appear in the command. In Figure 7.6, we have displayed all six frames in the animation. We can use the following command to display them all, as in Figure 7.7, without the animation:

> display3d([TorusNews, Line],axes = boxed, scaling = constrained);

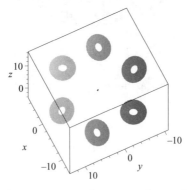

Figure 7.7: Six frames from rotating the torus about the line L. View is from directly above the line.

To make it easier to display as many steps in the rotation as we want, such as in Figure 7.8, we can construct an animation using the *animate3d* command. In the following, notice that the option $A = 0..2\pi$ tells *Maple* to treat the function *torusrot* as a function of the angle A, where A ranges from 0 to 2π. The default number of frames in the *animate3d* command is eight, but can be adjusted as desired to any other positive integer, we use six frames to mimic the results of the previous code.

> movingtori:= animate3d(torusrot(A), u = -Pi..Pi, v = -Pi..Pi, A = 0..2*Pi, style=patchnogrid, frames=6):
> display3d([Line, movingtori], axes = boxed, scaling = constrained);

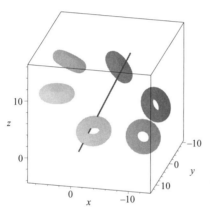

Figure 7.8: Six frames from rotating the torus about the line L, viewed from the side.

Homework Problems

1. Instead of circular rotations about a line L, construct a method for rotating a point Q about a line in an elliptical path. You may assume that the line is at the center of the ellipse and the point Q lies at one of the vertices at an endpoint of the major or minor axis of the ellipse.

2. Generalize the method you have produced in problem 1 replacing circular rotations about a line L to general rotations along any planar closed parametric curve, such as the ellipse of problem 1, where the line L goes through the interior of the planar closed parametric curve. Can this be generalized further so that the rotation curve need not be planar?

3. Express the rotation function given by

$$\vec{Q}_{\mathrm{rot}}(\alpha) = \vec{C} + R\cos(\alpha)\vec{U} + R\sin(\alpha)\vec{V}$$

in terms of matrices and matrix multiplication. The points and vectors should be treated as column matrices.

4. Consider the following situation. If the vector $\vec{P}$ is parallel to the axis of rotation (a line through the origin) and is defined by $\vec{P} = \langle 0, a, b \rangle$ and the position vector $\vec{Q}$ to be rotated is given by $\vec{Q} = \langle c, 0, 0 \rangle$, then the method breaks down since the center $\vec{C}$ is the origin or zero vector and so the unit vector $\vec{V}$ cannot be gotten by $\vec{V} = \dfrac{\vec{C} \times \vec{U}}{\left| \vec{C} \times \vec{U} \right|}$ because $\vec{C} \times \vec{U} = \vec{0}$. How can you get around this problem, which occurs whenever the vectors $\vec{P}$ and $\vec{Q}$ are perpendicular?

5. Generalize the method of this section to rotating about an arbitrary line L, not necessarily going through the origin, but given as passing through two points P and S.

6. Use the method from problem 5 to find the coordinates of the point N that is gotten by rotating the point $Q(1, 9, -5)$ about the line L through the two points $P(7, -2, 4)$ and $S(-10, 3, -5)$ through the angle $\alpha = \frac{3}{4}\pi$.

Maple Problems

1. Rotate the point $Q(1, 1, 1)$ about the line that passes through the origin and the point $P(-1, 2, -2)$.

2. Determine all angles α from problem 1 such that $Q_{\mathrm{rot}}(\alpha)$ lies in the xy-plane, yz-plane or xz-plane.

3. Using your answers to *Homework* problem 1, rotate the point $Q(1, 1, 1)$ about the line that passes through the origin and the point $P(3, 4, 3)$ in an elliptical path. Set the minor axis radius to the value $R = \left| \vec{Q} - \vec{C} \right|$ and major to $3R$.

4. Rotate a sphere of radius 1, with center located at $(1, 2, 1)$, about the line that passes through the origin and the point $P(-6, 7, 3)$.

5. A Mobius surface M is given parametrically by

$$M = \left[6 + \cos(u) + v\sin\left(\frac{u}{2}\right)\cos(u), 5 + \sin(u) + v\sin\left(\frac{u}{2}\right)\sin(u), 4v\cos\left(\frac{u}{2}\right)\right]$$

for $-\pi \le u \le \pi$ and $-\frac{\pi}{2} \le v \le \frac{\pi}{2}$. Rotate M about the line that passes through the origin and the point $P(3, 4, -7)$.

6. Using *Homework* problem 5, animate the rotation of the helix

$$\langle x(t), y(t), z(t)\rangle = \langle 9 + 2\sin(t), \ 2 + 2\cos(t), \ t\rangle, \ t \in [-2\pi, \ 2\pi]$$

about the line L through the two points $P(7, -2, 4)$ and $S(-10, 3, -5)$.

7. Using *Homework* problem 5, animate the rotation of the Mobius surface M, of problem 5, about the line L through the two points $P(7, -2, 4)$ and $S(-10, 3, -5)$.

8. If you did either of *Homework* problems 1 or 2, test to see if you ideas worked by computing and plotting an example using *Maple*.

7.2   "Rolling" a Circle Along a Curve

We now combine calculus, vectors, and parametric curves in an attempt to mathematically roll a circle along a parametric curve. The circles will have a constant radius R in our first example, while for the second example the circles will have a variable radius. Our circles will be tangent to the base curve they are rolling along in both versions, since this will make the rolling look closest to the actual physical process. In order to visualize this process, a vector format for our curves and their tangents is necessary.

If we have a parametric curve in the plane defined by $x = f(t)$ and $y = g(t)$ with t in the interval $[a, b]$ as our base curve, its vector format is $\overrightarrow{X}(t) = \langle f(t), g(t)\rangle$. This is called the *position vector field* of the parametric curve. Its derivative vector field or *tangent vector field* $\overrightarrow{T}(t)$ is

$$\overrightarrow{T}(t) = \frac{d\overrightarrow{X}}{dt} = \left\langle \frac{df}{dt}, \frac{dg}{dt} \right\rangle \tag{7.6}$$

A vector field perpendicular to $\overrightarrow{T}(t)$ is

$$\overrightarrow{P}(t) = \left\langle -\frac{dg}{dt}, \frac{df}{dt} \right\rangle \tag{7.7}$$

since

$$\overrightarrow{T}(t) \cdot \overrightarrow{P}(t) = \left\langle \frac{df}{dt}, \frac{dg}{dt} \right\rangle \cdot \left\langle -\frac{dg}{dt}, \frac{df}{dt} \right\rangle = 0$$

Also, $\vec{P}(t)$ can be made length R, corresponding to the radius of our circles, by first making it a unit vector field and then multiplying by R. Then the center vector field $\vec{C}(t)$ for our circles is given by

$$\vec{C}(t) = \vec{X}(t) + \frac{R}{\left|\vec{P}(t)\right|}\vec{P}(t) \tag{7.8}$$

If we now write $\vec{C}(t) = \langle H(t), K(t) \rangle$, then our rolling circles are given by the parametric curve:

$$\begin{aligned} \langle x(s),\, y(s) \rangle &= \langle R\cos(s) + H(t),\, R\sin(s) + K(t) \rangle \\ &= R\langle \cos(s), \sin(s) \rangle + \vec{C}(t) \end{aligned} \tag{7.9}$$

As s varies in the interval $[0, 2\pi]$ we move around this circle while we get one circle for each fixed value $t \in [a, b]$. In this situation, s is the variable that moves us around each circle while t is the rolling variable that moves us from one circle to the next circle because it controls the location of the circles' centers.

The formula for $\vec{C}(t)$ in (7.8) may appear complicated, however, if we take a minute to examine each piece, we should begin to see how simple it really is. Since $\vec{C}(t)$ is the center of the circle that we wish to roll along the curve, we need to start on the curve, $\vec{X}(t)$ and move R units perpendicular to $\vec{X}(t)$. Since $\vec{P}(t)$ is orthogonal to $\vec{T}(t)$, which in turn is tangent to $\vec{X}(t)$, we have that $\vec{P}(t)$ is orthogonal to $\vec{X}(t)$. To proceed R units in the $\vec{P}(t)$ direction, we must first make $\vec{P}(t)$ unit length by dividing it by its magnitude, then multiplying it by R. This gives the second piece of the sum found in (7.8).

Example 7.2.1. Let us look at an example to see this rolling in action. As the base parametric curve, let us use the parametric version of $y = 3\sin(x)$ given by $x = t$ and $y = 3\sin(t)$ for $t \in [0, 2\pi]$. The plot, given in Figure 7.9, will contain both the base parametric curve and seven circles spaced 1 radian apart, with each circle having a radius of 0.3. A little side note, since we will be taking the dot product of vectors that involve the independent variable t, *Maple* will wish to use the complex definition of the dot product. To ensure that this does not happen, we use the option *orthogonal*, which tells *Maple* that the vectors are real vectors.

> with(plots): with(linalg):

> basex:= t: basey:= 3*sin(t):

> baseplot:= plot([basex, basey, t = -1..7], color = blue):

> basevectorX:= array([basex, basey]):

> tangentvectorT:= map(diff, basevectorX, t);

$$tangentvectorT := \begin{bmatrix} 1 & 3\cos(t) \end{bmatrix}$$

> perpvectorP:= array([- tangentvectorT[2], tangentvectorT[1]]);

$$perpvectorP := \begin{bmatrix} -3\cos(t) & 1 \end{bmatrix}$$

> unitperpvectorU:= perpvectorP/norm(perpvectorP, 2):
> dotprod(tangentvectorT, perpvectorP, orthogonal);

$$0$$

> R:= .3:
> centervectorC:= evalm(basevectorX + R*unitperpvectorU):
> centers:= [evalf(seq([centervectorC[1], centervectorC[2]], t=0..6))];

$$centers :=[-0.2846049894, 0.09486832980], [0.7446797602, 2.681929858],$$
$$[2.234146459, 2.915443435], [3.284316278, 0.5190901362],$$
$$[4.267254760, -2.134117745], [4.805574650, -2.648302294],$$
$$[5.716592493, -0.7398585626]$$

> centerplot:= seq(pointplot(centers[i], color = black, symbol = cross), i=1 ..
nops(centers)):
> circles:= seq(plot([R*cos(s) + centervectorC[1], R*sin(s) + centervectorC[2],
s=0..2*Pi], color = magenta, scaling = constrained), t=0..6):
> display({baseplot, circles, centerplot}, view = [-1..7, -4..4]);

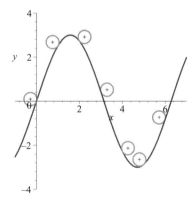

Figure 7.9: Graph displaying circles, centers of circles, and the parametric curve.

Next, we give a few ways in which to display the animations. The first method is to use the "insequence = true" option in the *display* command.

> display({display(baseplot), display([circles], insequence = true), display([centerplot], insequence = true)}, view = [-1..7, -4..4], scaling = constrained);

Another way to create this animation is using the *animate* command directly:

> animation:= animate([R*cos(s) + centervectorC[1], R*sin(s) + centervectorC[2], s=0..2*Pi], t=0..6, frames = 7, color = magenta):

> display({display(baseplot), display([centerplot], insequence = true), animation}, view = [-1..7, -4..4], scaling = constrained);

Next, we modify the above calculations with a nonconstant radius. In particular, we are going to choose the radius to be $\frac{1}{\kappa}$, where κ is the *curvature* of the function. Curvature, loosely defined, is the amount that a function changes direction per unit length. For a parametric space curve $\overrightarrow{r}(t) = \langle x(t), y(t), z(t) \rangle$, the formula for curvature is given by

$$\kappa(t) = \frac{|\overrightarrow{r}'(t) \times \overrightarrow{r}''(t)|}{|\overrightarrow{r}'(t)|^3} \tag{7.10}$$

We omit the derivation, which can be found in most multivariate calculus books. The formula given in equation (7.10) simplifies somewhat in two dimensions. If $\overrightarrow{r}(t) = \langle x(t), y(t) \rangle$, then

$$\kappa(t) = \frac{|x'(t)y''(t) - x''(t)y'(t)|}{((x'(t))^2 + (y'(t))^2)^{\frac{3}{2}}} \tag{7.11}$$

This formula can be computed quite readily from the three-dimensional version by setting $z(t) = 0$. For a circle of radius r, it can be shown that the curvature at any given point is $\frac{1}{r}$. Thus, if we fix a point on our graph and compute its curvature κ and draw a circle with radius $\frac{1}{\kappa}$ touching the function at that point, in essence, we have created a circle of best fit for the function at that specified point. This also implies that the circle should always lie on the concave-in side of the function and should have the same tangent line at the point. A circle that satisfies the above criteria is sometimes referred to as an *osculating circle*.

To see how this works, we will start with a curve in the plane given parametrically by $\overrightarrow{r}(t) = \langle x(t), y(t) \rangle$. We have the formula for curvature $\kappa(t)$ already given. We now need to find the center of the circle that has radius $\frac{1}{\kappa(t)}$. To do this, we need to find the vector orthogonal to the tangent vector that points toward the concave-in side of the curve. So, first we compute the unit tangent vector, defined as

$$\vec{T}(t) = \frac{\vec{r}'(t)}{|\vec{r}'(t)|} = \left\langle \frac{x'(t)}{\sqrt{(x'(t))^2 + (y'(t))^2}}, \frac{y'(t)}{\sqrt{(x'(t))^2 + (y'(t))^2}} \right\rangle \qquad (7.12)$$

With a little bit of work, we can compute the derivative of the unit tangent vector:

$$\frac{d}{dt}\vec{T}(t) = \left\langle \frac{x''(t)(y'(t))^2 - x'(t)y'(t)y''(t)}{(x'(t))^2 + (y'(t))^2}, \frac{y''(t)(x'(t))^2 - x'(t)x''(t)y'(t)}{(x'(t))^2 + (y'(t))^2} \right\rangle \qquad (7.13)$$

It may not be immediately obvious, but the unit tangent vector and its derivative are orthogonal. We could actually compute the dot product between the expressions given in (7.12) and (7.13), or we could make the following observation: since $\vec{T}(t)$ is a unit vector, $\left|\vec{T}(t)\right| = 1$ for all t, the derivative of the magnitude of $\vec{T}(t)$ must be zero:

$$\frac{d}{dt}\left|\vec{T}(t)\right| = 0 \qquad (7.14)$$

However, we know that $\left|\vec{T}(t)\right|^2 = \vec{T}(t) \cdot \vec{T}(t)$, and thus

$$\frac{d}{dt}\left|\vec{T}(t)\right|^2 = 0 \qquad (7.15)$$

Now, we introduce an important identity involving the derivative of the dot product. If $\vec{S}(t)$ and $\vec{T}(t)$ are two vector valued functions, then

$$\frac{d}{dt}\left(\vec{S}(t) \cdot \vec{T}(t)\right) = \vec{T}'(t) \cdot \vec{T}(t) + \vec{S}(t) \cdot \vec{T}'(t) \qquad (7.16)$$

Applying this formula to $\left|\vec{T}(t)\right|^2$ gives us the following:

$$\begin{aligned}
\frac{d}{dt}\left|\vec{T}(t)\right|^2 &= \frac{d}{dt}\left(\vec{S}(t) \cdot \vec{T}(t)\right) \\
&= \vec{T}'(t) \cdot \vec{T}(t) + \vec{T}(t) \cdot \vec{T}'(t) \\
&= 2\vec{T}'(t) \cdot \vec{T}(t)
\end{aligned}$$

Using equation (7.14) and the last line of the above set of equalities gives

$$\vec{T}'(t) \cdot \vec{T}(t) = 0$$

and we can therefore conclude that $\overrightarrow{T}'(t)$ and $\overrightarrow{T}(t)$ are orthogonal vectors. We now define the unit normal vector $\overrightarrow{N}(t)$ as

$$\overrightarrow{N}(t) = \frac{\overrightarrow{T}'(t)}{\left|\overrightarrow{T}'(t)\right|} \tag{7.17}$$

Finally, to locate the center of the circle, we start at the base point $\overrightarrow{r}(t)$ and move a distance of $\dfrac{1}{\kappa(t)}$ in the direction $\overrightarrow{N}(t)$, therefore

$$\overrightarrow{C}(t) = \overrightarrow{r}(t) + \frac{1}{\kappa(t)}\overrightarrow{N}(t). \tag{7.18}$$

Example 7.2.2. Now, we will do an example with a parametric curve in the xy-plane. The curve we will use is in the class of logarithmic spirals. In particular, we choose

$$\overrightarrow{r}(t) = \left\langle 2e^{-0.2t}\cos(t),\ 2e^{-0.2t}\sin(t)\right\rangle$$

We will have *Maple* perform all the computations. Pay close attention to the definitions of *D2X* and *D2Y*, where we use the symbol *D@@2* to express the second derivative operator. We plot the curvature in Figure 7.10, while Figures 7.11 and 7.12 correspond to the process of rolling a circle, with radius inversely proportional to curvature, along the logarithmic spiral.

```
> with(plottools):

> x:= t->2*exp(-0.2*t)*cos(t):

> y:= t->2*exp(-0.2*t)*sin(t):

> DX:= D(x):

> D2X:= (D@@2)(x):

> DY:= D(y):

> D2Y:= (D@@2)(y):

> kappa:= t-> sqrt((DX(t) * D2Y(t) - DY(t) * D2X(t))^2) / ((DX(t)^2 +
DY(t)^2 )^(3/2)):
```

> plot(kappa(t), t=0..5*Pi);

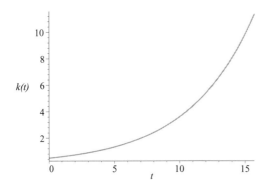

Figure 7.10: Graph of curvature for the parametric function representing the logarithmic spiral.

> PositionV:= t-> <x(t),y(t)>:

> TangentVX:= t->DX(t)/sqrt(DX(t)^2+DY(t)^2):

> TangentVY:= t->DY(t)/sqrt(DX(t)^2+DY(t)^2):

> TangentV:= t-> <TangentVX(t),TangentVY(t)>:

> DTangentVX:= D(TangentVX):

> DTangentVY:= D(TangentVY):

> NormalVX:= t-> DTangentVX(t) / sqrt(DTangentVX(t)^2 + DTangentVY(t)^2):

> NormalVY:= t-> DTangentVY(t) / sqrt(DTangentVX(t)^2 + DTangentVY(t)^2):

> NormalV:= t-> <DTangentVX(t), DTangentVY(t)>:

> EndPointV:= t-> evalm(PositionV(t) + 1/kappa(t)*NormalV(t)):

> plot1:= plot([x(t), y(t), t=0..5*Pi], thickness=2, color=red, numpoints = 2500):

> plot2:= implicitplot((xv - evalf(EndPointV(Pi/2)[1]))^2 + (yv - evalf(EndPointV(Pi/2)[2]))^2 = (1 / evalf(kappa(Pi/2)))^2, xv=-2..2, yv=-2..2, color=blue):

> plot3:= plot([x(Pi/2)*s + (1 - s)*EndPointV(Pi/2)[1], y(Pi/2)*s + (1 - s)*EndPointV(Pi/2)[2], s=0..1], color=black):

> plot4:= pointplot({ evalf([EndPointV(Pi/2)[1], EndPointV(Pi/2)[2]]), evalf([PositionV(Pi/2)[1], PositionV(Pi/2)[2]])}, symbol=CIRCLE, symbolsize=14):

> display({plot1,plot2,plot3,plot4}, scaling=CONSTRAINED);

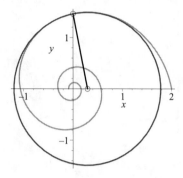

Figure 7.11: Graph of the osculating circle at $t = \frac{1}{2}\pi$ to the logarithmic spiral.

> F:= proc(t)
   plots[display](
      circle([evalf(EndPointV(t))[1], evalf(EndPointV(t))[2]], 1/kappa(t),
      color=BLUE),
      plot([x(t)*s+(1-s)*EndPointV(t)[1], y(t)*s+(1-s)*EndPointV(t)[2],
      s=0..1], color=black),
      pointplot({evalf([EndPointV(t)[1], EndPointV(t)[2]]), evalf([
      PositionV(t)[1], PositionV(t)[2]])}, symbol=CIRCLE, symbolsize=10)):

   end:

> animate(F,[t], t=0..5*Pi, background=plot1, scaling=constrained,
axes=NORMAL, frames=30);

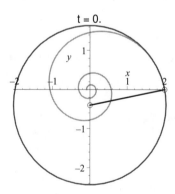

Figure 7.12: Graph of the first frame in the animated osculating circle example.

The above calculations work well for a parametrically defined function in the plane. This method also works for standard functions of one variable of the form $y = f(x)$, since they can be parameterized in the form $\overrightarrow{r}(t) = \langle t, f(t) \rangle$. Now, the question remains: How do we modify this formula for functions in $\mathbb{R}^3$ of the form $\overrightarrow{r}(t) = \langle x(t), y(t), z(t) \rangle$? The first thing that we must do is determine what plane the osculating circle lies in. Fortunately, it lies in the *osculating plane*, which is formed by the unit tangent and unit normal vectors $\overrightarrow{T}(t)$ and $\overrightarrow{N}(t)$, respectively. The work to be done still is to compute the equation of the circle of radius $\dfrac{1}{\kappa}$ with center in the osculating plane. We leave it to the interested reader to modify the code for the osculating circle in the plane to make an animated plot of an osculating circle in $\mathbb{R}^3$.

Homework Problems

1. Verify that a circle of radius r, defined by the position vector field $\overrightarrow{c}(t) = \langle r\cos(t), r\sin(t) \rangle$ has curvature $\kappa = \dfrac{1}{r}$.

2. The the position vector field $\overrightarrow{s}(t) = \langle r\cos(at), r\sin(at) \rangle$, with $a > 0$ still defines a circle of radius of r. One revolution takes $\dfrac{2}{a}\pi$ time units, instead of 2π. Compute the curvature of $\overrightarrow{s}(t)$ and compare your answer to problem 1.

3. Compute the curvature of the helix defined by $\overrightarrow{r}(t) = \langle \cos(t), \sin(t), t \rangle$.

4. Compute the unit tangent vector field $\overrightarrow{T}(t)$, unit normal vector field $\overrightarrow{N}(t)$, and the cross product $\overrightarrow{T}(t) \times \overrightarrow{N}(t)$, of the helix $\overrightarrow{r}(t) = \langle \cos(t), \sin(t), t \rangle$.

5. Explain why $\overrightarrow{T}(t) \times \overrightarrow{N}(t)$ is always a unit vector field.

6. (a) Find the formula for the curvature of a standard curve $y = f(x)$ by converting it to parametric form.

   (b) Use this curvature formula to find the curvature function $\kappa(x)$ for $y = \sin(x)$.

   (c) Explain why the only functions $y = f(x)$ with constant 0 curvature function are $f(x) = ax + b$, which are lines.

7. This section could really be described as "*Sliding a ball along a curve*", since we are not really rotating the circle along the curve. A circle is symmetric about its center, therefore it is easy to imagine that the circle really is

rolling along the curve. Devise a method for actually *rolling* the circle along the curve. Do the same for rolling an ellipse along a curve. *Hint: Arclength may come into play here.*

8. Find (look up the *pseudosphere*) a parametric surface whose curvature is a fixed negative real constant, and verify that this is true. This requires a new definition of curvature to allow it to be negative, and so that it will also apply to surfaces.

Maple Problems

1. Graph the curvature of an ellipse given by the position vector field $\vec{r}(t) = \langle 2\cos(t), 3\sin(t)\rangle$ for $t \in [0, 2\pi]$.

2. Graph the curvature of an ellipse given by the position vector field $\vec{r}(t) = \langle 2\cos(2t), 3\sin(2t)\rangle$ for $t \in [0, \pi]$.

3. Plot the unit tangent and unit normal vector fields to the ellipse $\vec{r}(t) = \langle 2\cos(t), 3\sin(t)\rangle$ for $t \in [0, 2\pi]$.

4. Roll a circle of radius $\frac{1}{6}$ in a counterclockwise fashion around the interior of the ellipse $\vec{r}(t) = \langle 2\cos(t), 3\sin(t)\rangle$ for $t \in [0, 2\pi]$.

5. Roll a circle of radius $\frac{1}{5}$ in a clockwise fashion around the interior of the ellipse $\vec{r}(t) = \langle 2\cos(t), 3\sin(t)\rangle$ for $t \in [0, 2\pi]$.

6. Roll a circle of radius $\frac{1}{\kappa}$ in a counterclockwise fashion around the interior of the ellipse $\vec{r}(t) = \langle 2\cos(t), 3\sin(t)\rangle$ for $t \in [0, 2\pi]$.

7. Roll a circle of radius $\frac{1}{\kappa}$ in a clockwise fashion around the interior of the ellipse $\vec{r}(t) = \langle 2\cos(t), 3\sin(t)\rangle$. for $t \in [0, 2\pi]$.

8. Roll a circle of radius $\frac{1}{\kappa}$ around the inside of the helix defined by $\vec{r}(t) = \langle \cos(t), \sin(t), t\rangle$.

9. Use your method devised from *Homework* problem 7 section to actually rotate the circle in problem 4.

10. Plot the original function $y = \sin(x)$ together with its curvature function from *Homework* problem 6 (b).

7.3   The TNB Frame

Since we introduced the tangent and normal vector fields, $\overrightarrow{T}(t)$ and $\overrightarrow{N}(t)$, respectively, to the position vector field $\overrightarrow{r}(t) = \langle x(t), y(t), z(t) \rangle$, in Section 7.2, we might as well introduce one last vector field. It was shown that the tangent and normal vectors fields were orthogonal, therefore by the definition of the cross product, we can find a vector field that is orthogonal to both $\overrightarrow{T}(t)$ and $\overrightarrow{N}(t)$. This is how we define the *binormal vector field* $\overrightarrow{B}(t)$:

$$\overrightarrow{B}(t) = \overrightarrow{T}(t) \times \overrightarrow{N}(t) \tag{7.19}$$

The binormal vector field $\overrightarrow{B}(t)$, together with the tangent and normal vector fields $\overrightarrow{T}(t)$ and $\overrightarrow{N}(t)$, make up what is known as the *TNB frame*.

As a side note, we could have defined $\overrightarrow{B}(t)$ as the cross product of $\overrightarrow{N}(t)$ with $\overrightarrow{T}(t)$. This still would have yielded a third orthogonal vector, however, we wish for the TNB frame to satisfy the *right-hand rule*. The right-hand rule stipulates that if we start with our hand in the direction of the tangent vector and curl our fingers in the direction of the normal vector and then stick our thumb out, we get the direction of the binormal vector. This rule is commonly used in physics.

Note that since $\overrightarrow{T}(t)$ and $\overrightarrow{N}(t)$ are unit vector fields, the binormal vector field is also a unit vector. Remember that we have the formula:

$$\left| \overrightarrow{B}(t) \right| = \left| \overrightarrow{T}(t) \times \overrightarrow{N}(t) \right| = \left| \overrightarrow{T}(t) \right| \left| \overrightarrow{N}(t) \right| \sin(\theta)$$

and by construction, $\left| \overrightarrow{T}(t) \right| = \left| \overrightarrow{N}(t) \right| = 1$ and $\sin(\theta) = 1$. Therefore we can conclude that $\left| \overrightarrow{B}(t) \right| = 1$, and thus is of unit length. Now, we have three orthogonal unit length vectors fields in $\mathbb{R}^3$ associated with any space curve $\overrightarrow{r}(t)$. The set

$$\left\{ \overrightarrow{T}(t), \overrightarrow{N}(t), \overrightarrow{B}(t) \right\} \tag{7.20}$$

forms an *orthonormal basis* for $\mathbb{R}^3$ for any appropriate value of t in the domain of $\overrightarrow{r}(t)$. This means that any vector in $\mathbb{R}^3$ can be expressed as a linear combination of the unit vectors in the TNB frame. We will explore the concept of orthonormal bases in Chapter 8.

The TNB frame is used, for example, in spacecraft navigation, where having an orthogonal set of linearly independent vectors with all three vectors having physical interpretations, comes in very handy. The TNB frame is sometimes referred to as the *Frenet-Serret frame*, and is related to the Frenet-Serret formulas found in differential geometry.

Example 7.3.1. This section is quite short, as very little new information was introduced. However, to use *Maple* to illustrate this concept requires a reasonable amount of coding. As an example, we will compute both the curvature (see Figure 7.13), and the TNB frame for the space curve given parametrically by

$$\overrightarrow{r}(t) = \langle \cos(3t), \sin(2t), \cos(2t) \rangle$$

Pay special attention to the procedure at the end of this section of *Maple* code. Together with the *animate* command immediately proceeding it, the procedure is an exceptionally simple and flexible way to animate the TNB frame along the given curve. Figure 7.14 illustrates the TNB frame at $t = 2$, while Figure 7.15 illustrates the TNB frame for $t = \frac{2}{3}\pi$, $t = \frac{3}{2}\pi$ and $t = \frac{9}{4}\pi$.

```
> with(plots): with(linalg): with(plottools):
> x:= t->cos(3*t):  y:=t->sin(2*t):  z:=t->cos(2*t):
> DX:= D(x):  DY:=D(y):  DZ:=D(z):
> D2X:= (D@@2)(x):  D2Y:=(D@@2)(y):  D2Z:=(D@@2)(z):
> PositionV:= t-> <x(t),y(t),z(t)>:
> TangentVX:= t->DX(t)/sqrt(DX(t)^2+DY(t)^2+DZ(t)^2):
> TangentVY:= t->DY(t)/sqrt(DX(t)^2+DY(t)^2+DZ(t)^2):
> TangentVZ:= t->DZ(t)/sqrt(DX(t)^2+DY(t)^2+DZ(t)^2):
> TangentV:= t-> <TangentVX(t), TangentVY(t), TangentVZ(t)>:
> DTangentVX:= D(TangentVX):
> DTangentVY:= D(TangentVY):
> DTangentVZ:= D(TangentVZ):
> NormalVX:= t-> DTangentVX(t)/sqrt( DTangentVX(t)^2+DTangentVY
(t)^2 + DTangentVZ(t)^2 ):
> NormalVY:= t-> DTangentVY(t)/sqrt( DTangentVX(t)^2+DTangentVY
(t)^2 + DTangentVZ(t)^2 ):
> NormalVZ:= t-> DTangentVZ(t)/sqrt( DTangentVX(t)^2+DTangentVY
(t)^2 + DTangentVZ(t)^2 ):
> NormalV:= t-> < NormalVX(t), NormalVY(t), NormalVZ(t) >:
> BiNormalV:= t-> crossprod(TangentV(t), NormalV(t)):
> kappa:= t-> sqrt((DY(t)*D2Z(t) - D2Y(t)*DZ(t))^2 + (DX(t)*D2Z(t) -
D2X(t)*DZ(t))^2 + (DX(t)*D2Y(t) - D2X(t)*DY(t))^2) / ((DX(t)^2 + DY
(t)^2 + DZ(t)^2)^(3/2)):
```

> plot(kappa(t), t=0..2*Pi);

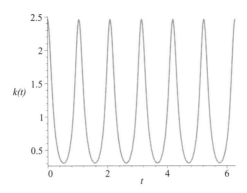

Figure 7.13: Graph of curvature for the space curve $\overrightarrow{r}(t)$.

> TangentArrow0:= arrow(PositionV(2), TangentV(2), .1, .25, .1, cylindri-cal_arrow, color = red):

> NormalArrow0:= arrow(PositionV(2),NormalV(2),.1,.25,.1,cylindrical_arrow, color = tan):

> BiNormalArrow0:= arrow(PositionV(2), BiNormalV(2), .1, .25, .1, cylindri-cal_arrow, color = blue):

> Plot1:= spacecurve([x(t), y(t), z(t), t=0..2*Pi], color = black, thickness = 2, numpoints = 1000):

> display({TangentArrow0, NormalArrow0, BiNormalArrow0, Plot1}, axes = FRAME, scaling = CONSTRAINED, orientation = [44, 70]);

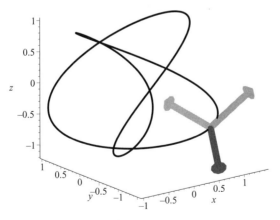

Figure 7.14: TNB frame for the space curve at time $t = 2$.

> F := proc(t)
    plots[display](
        arrow(PositionV(t), TangentV(t), .1, .25, .1, cylindrical_arrow, color = red),
        arrow(PositionV(t), NormalV(t), .1, .25, .1, cylindrical_arrow, color = tan),
        arrow(PositionV(t), BiNormalV(t), .1, .25, .1, cylindrical_arrow, color = blue));
    end:

> animate(F, [t], t = 0..2*Pi, background = Plot1, axes = FRAME, scaling = CONSTRAINED, frames = 100, orientation = [56, 50], view = [-2..2, -2..2, -2..2]);

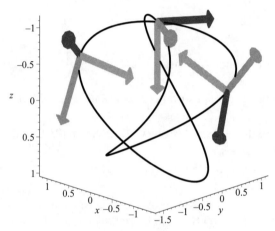

Figure 7.15: Graph of the three frames in the TNB example corresponding to $t = \frac{2}{3}\pi$, $t = \frac{3}{2}\pi$, and $t = \frac{9}{4}\pi$.

Maple Problems

1. Plot the TNB frame for the helix given by $\overrightarrow{r}(t) = \langle \cos(t), \sin(t), t \rangle$ with $0 \le t \le 4\pi$.

2. Plot the TNB frame for the space curve given by $\overrightarrow{r}(t) = \langle t, t\cos(4t), t\sin(4t) \rangle$ with $0 \le t \le 4\pi$.

3. Plot the TNB frame for the space curve

$$\overrightarrow{r}(t) = \langle(2 + \cos(2t))\cos(3t),\ (2 + \cos(2t))\sin(3t),\ \sin(4t)\rangle$$

with $0 \le t \le 4\pi$. This space curve is known as a figure-eight knot.

4. The following space curve lies on a torus and is given by

$$\overrightarrow{r}(t) = \langle(3 + 2\cos(t))\cos(10t),\ (3 + 2\cos(t))\sin(10t),\ 2\sin(t)\rangle$$

with $0 \le t \le 2\pi$. Plot, in the same figure, the following: $\overrightarrow{r}(t)$, the TNB frame for $\overrightarrow{r}(t)$, and the torus on which $\overrightarrow{r}(t)$ lies, which is given parametrically by

$$\overrightarrow{s}(u, v) = \langle(3+2\cos(v))\cos(u),\ (3+2\cos(v))\sin(u),\ 2\sin(v)\rangle,\ \text{for } -\pi \le u, v \le \pi$$

5. The space curve

$$\overrightarrow{r}(t) = \langle(3 + 2\cos(3t))\cos(7t),\ (3 + 2\cos(3t))\sin(7t),\ 2\sin(3t)\rangle$$

for $0 \le t \le 2\pi$, also lies on a torus defined in the previous problem. Plot, in the same figure, $\overrightarrow{r}(t)$, the TNB frame for $\overrightarrow{r}(t)$, and the torus.

6. Plot the curvature function, $\kappa(t)$, for each of the space curves given in problems 1 through 5 above.

7. Look up the Frenet-Serret equations and the *torsion* function $\tau(t)$.

8. Verify the Frenet-Serret equations for the space curves in problems 1 through 5 above.

9. For a parametric surface with position vector field $\overrightarrow{r}(u, v)$, $\dfrac{\partial r}{\partial u}$ and $\dfrac{\partial r}{\partial v}$ are two tangent vector fields for this surface. Then $\dfrac{\partial r}{\partial u} \times \dfrac{\partial r}{\partial v}$ is a normal vector field for this surface. How can we use this information to find the equation of the tangent plane to this surface at the point $P(\overrightarrow{r}(u_0, v_0))$ on this surface?

10. Find the equation of the tangent plane to the torus defined in problem 4 at the point $P\left(\overrightarrow{s}\left(\dfrac{\pi}{3}, \dfrac{\pi}{4}\right)\right)$. Plot both the torus and the tangent plane you found.

Chapter 8

Independence, Basis, and Dimension for Subspaces of $\mathbb{R}^n$

8.1 Subspaces of $\mathbb{R}^n$

If you recall, in Section 6.1 we introduced the definition of $\mathbb{R}^n$ as a vector space for the scalars chosen from $\mathbb{R}$. For completeness, we redefine it here.

Definition 8.1.1. The *vector space* $\mathbb{R}^n$ over the scalar field $\mathbb{R}$ is the set of vectors $\overrightarrow{v}$ represented by arrows starting at the base point of the origin $(0, 0, \dots, 0)$ in the point set $\mathbb{R}^n$ and stopping at the point $P = (p_1, p_2, \dots, p_n)$ in the point set $\mathbb{R}^n$. We write the vector $\overrightarrow{v}$ as $\overrightarrow{v} = \langle p_1, p_2, \dots, p_n \rangle$ for the n real components $p_1, p_2, \dots, p_n$.

$$\mathbb{R}^n = \left\{ \langle p_1, p_2, \dots, p_n \rangle \,\middle|\, (p_1, p_2, \dots, p_n) \in \mathbb{R}^n \right\}$$

The vector space $\mathbb{R}^n$ only allows scalar multiplication to be done when the scalar is a real number, which is what is meant when we say that $\mathbb{R}^n$ is a vector space over the field of scalars $\mathbb{R}$. The way we do arithmetic (addition and scalar multiplication) in the vector space $\mathbb{R}^n$ is as if its elements are real row or column matrices in $\mathbb{R}^{1 \times n}$ or $\mathbb{R}^{n \times 1}$. As well, the context and difference in notation will hopefully be made clear when we are using a vector $\overrightarrow{v}$ from $\mathbb{R}^n$ versus a point P from $\mathbb{R}^n$, as both sets are called $\mathbb{R}^n$, but are very different things mathematically since only in the vector space version of $\mathbb{R}^n$ can we do arithmetic.

The vector space $\mathbb{R}^n$ over the scalar field $\mathbb{R}$ is said to be the *real Euclidean vector space* of dimension n or for short, real Euclidean n-space. The two real

Euclidean n-spaces we are most familiar with are

$$\mathbb{R}^2 = \{\langle p_1, p_2 \rangle \mid p_i \in \mathbb{R}\}$$

which is the vector space of two-dimensional vectors and is visualized as the xy-plane, and

$$\mathbb{R}^3 = \{\langle p_1, p_2, p_3 \rangle \mid p_i \in \mathbb{R}\}$$

which is the vector space of three-dimensional vectors and is visualized as xyz-space. For completeness sake, we will say that $\mathbb{R}^0$ is only the real number zero and it is the real zero-dimensional vector space.

In Definition 8.1.1, we defined the real Euclidean vector space $\mathbb{R}^n$, but it is not always possible to restrict ourselves to merely using real numbers, sometimes out of necessity we must use complex numbers instead, such as when we find the roots of real coefficient polynomials; for instance $x^2 + 1 = 0$ has purely complex roots $x = \pm i$. This will also be true in Chapter 12, when we study eigenvalues and eigenvectors. As such, we need to define the complex Euclidean vector space $\mathbb{C}^n$ of dimension n which is merely the complex version of the real Euclidean vector space $\mathbb{R}^n$ of dimension n.

Definition 8.1.2. The *vector space* $\mathbb{C}^n$ over the scalar field $\mathbb{C}$ (short for complex Euclidean n-space) is the set of vectors written as $\vec{v} = \langle q_1, q_2, \ldots, q_n \rangle$ for the n complex components $q_1, q_2, \ldots, q_n$ with

$$\mathbb{C}^n = \left\{ \langle q_1, q_2, \ldots, q_n \rangle \mid (q_1, q_2, \ldots, q_n) \in \mathbb{C}^n \right\}$$

The vector space $\mathbb{C}^n$ over the field of scalars $\mathbb{C}$ only allows scalar multiplication to be done when the scalar is a complex number. This is what we mean by saying that $\mathbb{C}^n$ is a vector space over the scalar field $\mathbb{C}$. Furthermore, $\mathbb{R} \subset \mathbb{C}$, and so a real number is also a scalar for the vector space $\mathbb{C}^n$ over the scalar field $\mathbb{C}$. The way addition and scalar multiplication are done in the vector space $\mathbb{C}^n$ is as if the elements of $\mathbb{C}^n$ are complex row or column matrices in $\mathbb{C}^{1 \times n}$ or $\mathbb{C}^{n \times 1}$. Remember that if we are doing arithmetic in $\mathbb{C}^n$, then $\mathbb{C}^n$ is the vector space and not the point set.

These two examples of the idea of vector space lead us to the following general definition of a vector space $\mathbb{V}$ over a field of scalars $\mathbf{F}$. A field $\mathbf{F}$ is a set with at least two elements $0_{\mathbf{F}}$ and $1_{\mathbf{F}}$ in which we can do all of the operations of arithmetic in the standard manner, such as in the field of rational numbers $\mathbb{Q}$, the field of real numbers $\mathbb{R}$, and the field of complex numbers $\mathbb{C}$. The integers, $\mathbb{Z}$, are not a field since we cannot do division in $\mathbb{Z}$ and always stay in $\mathbb{Z}$. Note also that $\mathbb{Q} \subset \mathbb{R} \subset \mathbb{C}$, which makes $\mathbb{Q}$ a subfield of $\mathbb{R}$ and in turn $\mathbb{R}$ is a subfield of $\mathbb{C}$.

For simplicity, a field $\mathbf{F}$ in this text will be either $\mathbb{Q}$, $\mathbb{R}$, or $\mathbb{C}$. There are an infinite number of other fields $\mathbf{F}$, but we will not need or study them here.

Definition 8.1.3. A vector space $\mathbb{V}$ over the scalar field $\mathbf{F}$ is a nonempty set on which there are two operations of scalar multiplication (using the scalars from $\mathbf{F}$) and vector addition. The vector space $\mathbb{V}$ must satisfy all of the following properties:

	Closure Properties
Scalar Multiplication	For any $\vec{x} \in \mathbb{V}$ and scalar $a \in \mathbf{F}$, $a\vec{x} \in \mathbb{V}$.
Vector Addition	For all $\vec{x}, \vec{y} \in \mathbb{V}$, $\vec{x} + \vec{y} \in \mathbb{V}$.

	Additive Axioms
Associativity	For all $\vec{w}, \vec{x}, \vec{y} \in \mathbb{V}$, $\vec{w} + (\vec{x} + \vec{y}) = (\vec{w} + \vec{x}) + \vec{y}$.
Commutativity	For all $\vec{x}, \vec{y} \in \mathbb{V}$, $\vec{x} + \vec{y} = \vec{y} + \vec{x}$.
Existence of Vector Identity	There exists $\vec{0} \in \mathbb{V}$ such that $\vec{0} + \vec{x} = \vec{x}$.
Existence of Additive Inverse	For each $\vec{x} \in \mathbb{V}$, there exists a $\vec{w}$ such that $\vec{x} + \vec{w} = \vec{0}$. (The additive inverse $\vec{w}$ is expressed as $-\vec{x}$.)

	Scalar Multiplication Axioms
Distributivity over vector addition	For any scalar $a \in \mathbf{F}$ and all $\vec{x}, \vec{y} \in \mathbb{V}$, $a(\vec{x} + \vec{y}) = a\vec{x} + a\vec{y}$.
Distributivity over scalar addition	For any scalars $a, b \in \mathbf{F}$, and all $\vec{x} \in \mathbb{V}$, $(a + b)\vec{x} = a\vec{x} + b\vec{x}$.
Distributivity over scalar multiplication	For any scalars $a, b \in \mathbf{F}$, and all $\vec{x} \in \mathbb{V}$, $(ab)\vec{x} = a(b\vec{x})$.
Existence of scalar identity	There exists a scalar $1 \in \mathbf{F}$ such that for $\vec{x} \in \mathbb{V}$, $1\vec{x} = \vec{x}$.

Upon inspection of the closure properties and axioms in the definition of a vector space $\mathbb{V}$ over the scalar field $\mathbf{F}$, it should be apparent that $\mathbb{R}^n$ is a vector space over the field $\mathbb{R}$, but it is also a vector space over the field $\mathbb{Q}$ while it is not a vector space over the field $\mathbb{C}$. As well, $\mathbb{C}^n$ is a vector space over the field $\mathbb{C}$, but it is also a vector space over both of the fields $\mathbb{Q}$ and $\mathbb{R}$.

Example 8.1.1. Let $V = \mathbb{Q}^{2 \times 3}$ be the set of all rational 2×3 matrices. Then $\mathbb{V}$ is a vector space over the field of scalars $\mathbb{Q}$, where vector addition and scalar multiplication are the usual matrix addition and matrix multiplication by a scalar. Here $\mathbb{V}$ is not a vector space over either the field $\mathbb{R}$ or $\mathbb{C}$ since in each of these cases the closure property of scalar multiplication would fail.

Example 8.1.2. Let $\mathbb{V} = \left\{ ax^2 + bx + c \mid a, b, c \in \mathbb{R} \right\}$ be the set of all polynomials of degree ≤ 2 with real coefficients. Then $\mathbb{V}$ is a vector space over the field of scalars $\mathbb{R}$, where vector addition and scalar multiplication are the usual polynomial addition and multiplication by a scalar. Here $\mathbb{V}$ is also a vector space over the field $\mathbb{Q}$, but it is not a vector space over the field $\mathbb{C}$ since in this case the closure property of scalar multiplication would fail.

Example 8.1.3. Let $\mathbb{V} = \{ f : [0,1] \to \mathbb{R} \mid f \text{ is a continuous function } \}$. A function f is called *continuous* if its graph can be drawn without raising your pen from the paper, that is, its graph has no holes, jumps, or breaks of any kind. Examples of such functions' rules are $\sin(x)$, $\cos(x)$ and e^x, while $\frac{1}{x} \notin \mathbb{V}$ since it has a jump at $x = 0$ due to its vertical asymptote here. Then $\mathbb{V}$ is a vector space over the field of scalars $\mathbb{R}$ where the vector addition and scalar multiplication are the usual function addition and multiplication by a scalar of the functions' rules. Now $\mathbb{V}$ is also a vector space over the field $\mathbb{Q}$, but it is not a vector space over the field $\mathbb{C}$, since in this case the closure property of scalar multiplication would fail.

Example 8.1.4. Let $\mathbb{V} = \{ a + b \cos(x) + c \cos(2x) \mid a, b, c \in \mathbb{Q} \}$. Then $\mathbb{V}$ is a vector space over the field of scalars $\mathbb{Q}$, where the vector addition and scalar multiplication are the usual function addition and multiplication by a scalar. Here $\mathbb{V}$ is not a vector space over either the field $\mathbb{R}$ or $\mathbb{C}$, since in each of these cases the closure property of scalar multiplication would fail.

Note that in all of these examples of a vector space $\mathbb{V}$ over a scalar field $\mathbf{F}$, the field $\mathbf{F}$ is intimately incorporated into the way the elements of $\mathbb{V}$ are defined, this is not a coincidence, but necessary to have some hope that $\mathbb{V}$ might be a vector space over the field $\mathbf{F}$. As well, these examples point out that if $\mathbb{V}$ is a vector space over the scalar field $\mathbf{F}$, then $\mathbb{V}$ is also a vector space over any *subfield* of **F**, but not over any *superfield* of **F**. Both $\mathbb{Q}$ and $\mathbb{R}$ are subfields of $\mathbb{C}$ while $\mathbb{R}$ is a superfield of $\mathbb{Q}$ and $\mathbb{C}$ is a superfield of both $\mathbb{Q}$ and $\mathbb{R}$.

In the remainder of this section and throughout the rest of this text, we will restrict ourselves mainly to the vector space $\mathbb{R}^n$ over the field of scalars $\mathbb{R}$ and occasionally to the vector space $\mathbb{C}^n$ over the field of scalars $\mathbb{C}$. The reason for this restriction is to simplify our discussion to its basics, but in fact all that we will do in $\mathbb{R}^n$ will also be similarly true in any general finite dimensional vector space $\mathbb{V}$. This is the case since every finite dimensional vector space $\mathbb{V}$ over the scalar field **F** of dimension n is essentially the same vector space as $\mathbb{F}^n$ over the scalar field **F**, which is very much like $\mathbb{R}^n$ and $\mathbb{C}^n$.

We are now interested in when a nonempty subset $\mathbb{S}$ of the vector space $\mathbb{R}^n$ over the field of scalars $\mathbb{R}$ is a vector space over the field of scalars $\mathbb{R}$ in its own right. When this is true, we will say that $\mathbb{S}$ is a *subspace* of $\mathbb{R}^n$. In order to define more clearly what a subspace $\mathbb{S}$ of $\mathbb{R}^n$ looks like, we need first to discuss linear combinations of elements of $\mathbb{R}^n$.

Definition 8.1.4. For any collection of k elements $\vec{v_1}, \vec{v_2}, \ldots, \vec{v_k}$ of $\mathbb{R}^n$, and any real scalars $a_1, a_2, \ldots, a_k$ we say that:

$$\sum_{i=1}^{k} a_i \vec{v_i} = a_1 \vec{v_1} + a_2 \vec{v_2} + \cdots + a_k \vec{v_k} \tag{8.1}$$

is a (*k element*) *linear combination* of the vectors $\vec{v_1}, \vec{v_2}, \ldots, \vec{v_k}$.

Definition 8.1.5. A *subspace* $\mathbb{S}$ of the vector space $\mathbb{R}^n$ over the field of scalars $\mathbb{R}$ is any nonempty subset of $\mathbb{R}^n$, where any two element linear combination of elements from $\mathbb{S}$ is also in $\mathbb{S}$, that is, $\mathbb{S}$ satisfies the closure properties of a general vector space. In other words, $\mathbb{S}$ is a subspace of $\mathbb{R}^n$ if for any two scalars $a, b \in \mathbb{R}$ and any two vectors $\vec{u}, \vec{v} \in \mathbb{S}$ we have

$$a \vec{u} + b \vec{v} \in \mathbb{S}$$

Then $\mathbb{S}$ is itself a vector space over the field $\mathbb{R}$, since it satisfies the closure properties of a vector space and it inherits the addition and scalar multiplication axioms from $\mathbb{R}^n$.

Note that the zero vector, $\vec{0}$, must be in every subspace $\mathbb{S}$. This forces any subspace to be nonempty since we can take the linear combination $0\vec{v} + 0\vec{v} = \vec{0}$. Also, if $\vec{v} \in \mathbb{S}$, then $-\vec{v} \in \mathbb{S}$ since $(-1)\vec{v} + 0\vec{v} = -\vec{v}$ and moreover, all scalar multiples $a\vec{v}$, of $\vec{v}$, are in $\mathbb{S}$ since $a\vec{v} + 0\vec{v} = a\vec{v} \in \mathbb{S}$. Thus, in a subspace $\mathbb{S}$, we can do basic vector arithmetic of addition and scalar multiplication in $\mathbb{S}$ and stay in $\mathbb{S}$. This is an intuitive result that we use when we perform addition, subtraction, and multiplication of integers. By adding, subtracting or multiplying any two integers, you always end up with another integer.

Example 8.1.5. Now, we get to the promised plot, depicted in Figure 8.1, of the two element linear combinations of vectors, which we choose to be $\vec{u} = \langle 5, -7, 11 \rangle$ and $\vec{v} = \langle -3, 15, 9 \rangle$ in $\mathbb{R}^3$. In order to see that they form the plane through the origin parallel to both $\vec{u}$ and $\vec{v}$, we will take the vectors that are linear combinations as $s\vec{u} + t\vec{v}$ for real parameters s and t, and plot them as well.

```
> with(plots): with(linalg): with(plottools):
> u:= vector([5,-7,11]): v:= vector([-3,15,9]):
> plane_uv:= plot3d(convert(evalm(s*u+t*v),list), s=-2..2, t=-2..2, color = red, style = patchnogrid):
> arrow_u:= arrow([0, 0, 0], u, 1, 3.5, .4, cylindrical_arrow, color = blue):
> arrow_v:= arrow([0, 0, 0], v, 1, 3.5, .4, cylindrical_arrow, color = blue):
> plotu:= textplot3d([op(convert(u,list)),"U"], align=ABOVE, color=black):

> plotv:= textplot3d([op(convert(v,list)),"V"], align=ABOVE, color=black):
> display({plane_uv, arrow_u, arrow_v, plotu, plotv}, axes = boxed);
```

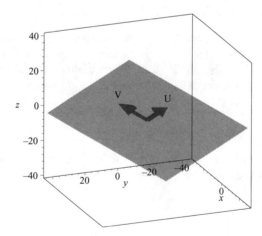

Figure 8.1: Vector Space formed by two nonparallel vectors $\vec{u}$ and $\vec{v}$.

Example 8.1.6. We next consider the following subset of $\mathbb{R}^4$:

$$\mathbb{S} = \{\langle a, 0, b, 0 \rangle \mid a, b \in \mathbb{R}\}$$

Can we show the closure properties hold? Let us check vector addition first:

$$\langle a_1, 0, b_1, 0 \rangle + \langle a_2, 0, b_2, 0 \rangle = \langle a_1 + a_2, 0, b_1 + b_2, 0 \rangle$$

Since the sum of two real numbers is another real number, the resulting vector sum is still of the form $\langle a, 0, b, 0 \rangle$. Next, we check scalar multiplication

$$\alpha \langle a, 0, b, 0 \rangle = \langle \alpha\, a, 0, \alpha\, b, 0 \rangle$$

Since the product of two real numbers is a real number, the resulting vector is also in $\mathbb{S}$. Therefore $\mathbb{S}$ is a vector subspace of $\mathbb{R}^4$, and looks very much like $\mathbb{R}^2$.

Example 8.1.7. The set: $\mathbb{S} = \{ \langle a, b, c, 0, 0 \rangle \mid a, b, c \in \mathbb{R} \}$, is a vector subspace of $\mathbb{R}^5$, and looks like a copy of $\mathbb{R}^3$ inside of $\mathbb{R}^5$. To show this, a similar argument can be used as in Example 8.1.6.

Example 8.1.8. The previous two example were relatively straightforward, let us attempt to discern why the following set is a subspace of $\mathbb{R}^6$:

$$\mathbb{S} = \{ \langle 5a - 3b + 8c, a + 2b - c, -a + 7b - 4c, 0, 4a + b + 3c, 2a + 6b - 9c \rangle \mid a, b, c \in \mathbb{R} \}$$

Upon inspection, notice that we can actually express $\mathbb{S}$ as

$$\mathbb{S} = \{ a\, \vec{u} + b\, \vec{v} + c\, \vec{w} \mid a, b, c \in \mathbb{R} \} \tag{8.2}$$

where $\vec{u} = \langle 5, 1, -1, 0, 4, 2 \rangle$, $\vec{v} = \langle -3, 2, 7, 0, 1, 6 \rangle$, $\vec{w} = \langle 8, -1, -4, 0, 3, -9 \rangle$. This automatically gives that $\mathbb{S}$ is a subspace of $\mathbb{R}^6$, since upon expressing $\mathbb{S}$ in the form of equation (8.2), $\mathbb{S}$ satisfies the closure properties of a vector space.

Example 8.1.9. So far we have seen examples of what subspaces of $\mathbb{R}^n$ can be, the following set is not a subspace of $\mathbb{R}^5$:

$$\mathbb{S} = \{ \langle a, b, c, 7, d \rangle \mid a, b, c, d \in \mathbb{R} \}$$

To see this, add any two vectors from $\mathbb{S}$ together:

$$\langle a_1, b_1, c_1, 7, d_1 \rangle + \langle a_2, b_2, c_2, 7, d_2 \rangle = \langle a_1 + a_2, b_1 + b_2, c_1 + c_2, 14, d_1 + d_2 \rangle$$

and notice that the fourth component is not 7, which implies that the vector addition given above does not result in another vector in $\mathbb{S}$, hence not satisfying the closure properties of a vector subspace. We could have also used the fact that scalar multiplication also does not result in another vector of $\mathbb{S}$, for if $\alpha \neq 1$, then

$$\alpha \langle a_1, b_1, c_1, 7, d_1 \rangle = \langle \alpha\, a_1, \alpha\, b_1, \alpha\, c_1, 7\, \alpha, \alpha\, d_1 \rangle$$

is also not a vector of $\mathbb{S}$.

Example 8.1.10. What if we consider the following subset of $\mathbb{R}^5$:

$$\mathbb{S} = \{\langle 0,0,0,0,0 \rangle\} = \{\vec{0}\}$$

Does the zero vector alone constitute a subspace of $\mathbb{R}^5$? Well, if we check the closure properties, notice that $\vec{0} + \vec{0} = \vec{0}$ and $a\,\vec{0} = \vec{0}$ for any scalar a. One can conclude that the zero vector of $\mathbb{R}^n$ forms a subspace of $\mathbb{R}^n$ always, and is also the smallest subspace of $\mathbb{R}^n$.

Theorem 8.1.1. *Let $\{\vec{u_1}, \vec{u_2}, \ldots, \vec{u_k}\}$ be any k element subset of the vector space $\mathbb{R}^n$. Then the subset $\mathbb{S}$ of $\mathbb{R}^n$ consisting of all the linear combinations of these k elements is a subspace of $\mathbb{R}^n$.*

Proof. The subset $\mathbb{S}$ is defined as

$$\mathbb{S} = \{a_1\vec{u_1} + a_2\vec{u_2} + \cdots + a_k\vec{u_k} \mid a_1, a_2, \ldots, a_k \in \mathbb{R}\}$$

Hopefully, it is clear that if we form a two element linear combination of any two elements of $\mathbb{S}$, then we get another element of $\mathbb{S}$, and so $\mathbb{S}$ is a subspace of $\mathbb{R}^n$. $\square$

In $\mathbb{R}^n$, the *intersection* $\mathbb{U} \cap \mathbb{V}$ and *sum* $\mathbb{U} + \mathbb{V}$ of two subspaces $\mathbb{U}$ and $\mathbb{V}$ are also subspaces. The sum $\mathbb{U} + \mathbb{V}$ is the smallest subspace of $\mathbb{R}^n$, which contains both the subspaces $\mathbb{U}$ and $\mathbb{V}$.

Definition 8.1.6. The *sum* $\mathbb{U} + \mathbb{V}$ of two subspaces of $\mathbb{R}^n$ is defined by

$$\mathbb{U} + \mathbb{V} = \{\vec{u} + \vec{v} \mid \vec{u} \in \mathbb{U} \text{ and } \vec{v} \in \mathbb{V}\} \tag{8.3}$$

Example 8.1.11. If we define $\mathbb{U}$ and $\mathbb{V}$ to be the following two subsets of $\mathbb{R}^7$:

$$\mathbb{U} = \{a\,\vec{u_1} + b\,\vec{u_2} \mid a, b, \in \mathbb{R}\}, \quad \mathbb{V} = \{a\,\vec{v_1} + b\,\vec{v_2} + c\,\vec{v_3} \mid a, b, c \in \mathbb{R}\}$$

where $\vec{u_1}, \vec{u_2}, \vec{v_1}, \vec{v_2}, \vec{v_3}$ are five vectors of $\mathbb{R}^7$, with the added property that $\vec{v_3} \in \mathbb{U}$ as well. Then,

$$\mathbb{U} + \mathbb{V} = \{a\,\vec{u_1} + b\,\vec{u_2} + c\,\vec{v_1} + d\,\vec{v_2} \mid a, b, c, d \in \mathbb{R}\}$$

is a subspace of $\mathbb{R}^7$. Notice that we have excluded the vector $\vec{v_3}$ in the definition of $\mathbb{U} + \mathbb{V}$, as it can be expressed as a linear combination of vectors in $\mathbb{U}$.

Example 8.1.12. Now we do an example involving subspaces from $\mathbb{R}^8$. Let

$$\mathbb{U} = \left\{ \langle a, b, 0, 0, c, d, 0, 0 \rangle \, \middle| \, a, b, c, d \in \mathbb{R} \right\}, \quad \mathbb{V} = \left\{ \langle 0, 0, a, b, 0, c, 0, 0 \rangle \, \middle| \, a, b, c \in \mathbb{R} \right\}$$

which are clearly both vector subspaces of $\mathbb{R}^8$. Then,

$$\mathbb{U} + \mathbb{V} = \left\{ \langle a, b, c, d, e, f, 0, 0 \rangle \mid a, b, c, d, e, f \in \mathbb{R} \right\}$$

is a subspace of $\mathbb{R}^8$ and looks like a copy of $\mathbb{R}^6$ inside $\mathbb{R}^8$.

The *union* of two subspaces $\mathbb{U}$ and $\mathbb{V}$ of $\mathbb{R}^n$, denoted $\mathbb{U} \cup \mathbb{V}$, is not necessarily a subspace of $\mathbb{R}^n$. We can see this since two different planes $\mathbb{U}$ and $\mathbb{V}$, through the origin in $\mathbb{R}^3$, are different subspaces of $\mathbb{R}^3$, but if we add an element $\overrightarrow{p}$ of $\mathbb{U}$ to an element $\overrightarrow{q}$ of $\mathbb{V}$, then $\overrightarrow{p} + \overrightarrow{q}$ is not in either $\mathbb{U}$ or $\mathbb{V}$ and so is not in their union. As an example, consider $\mathbb{U} = \{\langle 0, b \rangle \mid b \in \mathbb{R}\}$ and $\mathbb{V} = \{\langle a, 0 \rangle \mid a \in \mathbb{R}\}$. Clearly $\mathbb{U}$ is a subspace of $\mathbb{R}^2$ and corresponds to the y-axis and similarly, $\mathbb{V}$ corresponds to the x-axis. By taking $\langle 0, 1 \rangle \in \mathbb{U}$ and $\langle 1, 0 \rangle \in V$, note that $\langle 0, 1 \rangle + \langle 1, 0 \rangle = \langle 1, 1 \rangle$ is not in either subspace $\mathbb{U}$ or $\mathbb{V}$.

You should picture a subspace $\mathbb{S}$ of $\mathbb{R}^n$ as any subset of $\mathbb{R}^n$ that contains the origin and is flat or noncurving, that is, like a line or plane in $\mathbb{R}^3$. Although you must keep in mind that a line or plane in $\mathbb{R}^3$ that does not pass through the origin is not a subspace of $\mathbb{R}^3$, since the origin is in every subspace.

Now we return to systems of linear equations. A system of linear equations written as the matrix equation $A\overrightarrow{x} = \overrightarrow{0}$, where the matrix of coefficients A is a $k \times n$ matrix, requires that the column of variables $\overrightarrow{x}$ be an $n \times 1$ matrix, and the zero column $\overrightarrow{0}$ be a $k \times 1$ matrix. The matrix equation (or linear system it represents) $A\overrightarrow{x} = \overrightarrow{0}$ is said to be *homogeneous* because of the zero column vector on the RHS, and *nonhomogeneous* if we have the system $A\overrightarrow{x} = \overrightarrow{c}$ with $\overrightarrow{c} \neq \overrightarrow{0}$. This linear system has n variables making up the column matrix $\overrightarrow{x}$. We will think of the solution column matrix $\overrightarrow{x}$ as a vector in $\mathbb{R}^n$, and the zero column matrix $\overrightarrow{0}$ as the zero vector in $\mathbb{R}^k$.

The set of solution vectors $\overrightarrow{x}$ in $\mathbb{R}^n$ to the matrix equation $A\overrightarrow{x} = \overrightarrow{0}$ is a subspace $\mathbb{S}$ of $\mathbb{R}^n$. In order to see this we do a little algebra. Let $\overrightarrow{y}$ and $\overrightarrow{z}$ be two solutions to $A\overrightarrow{x} = \overrightarrow{0}$ and a and b be two real numbers. Then,

$$A(a\overrightarrow{y} + b\overrightarrow{z}) = a(A\overrightarrow{y}) + b(A\overrightarrow{z}) = a\overrightarrow{0} + b\overrightarrow{0} = \overrightarrow{0} \tag{8.4}$$

and so the linear combination $a\overrightarrow{y} + b\overrightarrow{z}$ is also a solution, thus the solutions form a subspace of $\mathbb{R}^n$.

On the other hand, the set of solution vectors $\overrightarrow{x}$ in $\mathbb{R}^n$ to the matrix equation $A\overrightarrow{x} = \overrightarrow{c}$ is not a subspace of $\mathbb{R}^n$ if $\overrightarrow{c} \neq \overrightarrow{0}$. If we repeat our calculation above for $\vec{y}$ and $\vec{z}$ two solutions to $A\vec{x} = \vec{c}$ and a and b be two real numbers, then

$$A(a\vec{y} + b\vec{z}) = a(A\vec{y}) + b(A\vec{z}) = a\vec{c} + b\vec{c} = (a+b)\vec{c} \qquad (8.5)$$

which is not necessarily $\vec{c}$ and so the linear combination $a\vec{y} + b\vec{z}$ is not a solution.

Upon comparing equations (8.4) and (8.5), we can see that the relationship between solutions to $A\vec{x} = \vec{0}$ and $A\vec{x} = \vec{c}$ is quite intimate. If $\vec{y}$ and $\vec{z}$ are any two solutions to $A\vec{x} = \vec{c}$ for $\vec{c} \neq 0$, then their difference $\vec{y} - \vec{z}$ is a solution to $A\vec{x} = \vec{0}$, since

$$A(\vec{y} - \vec{z}) = A\vec{y} - A\vec{z} = \vec{c} - \vec{c} = \vec{0}$$

This says that given any solution $\vec{y}$ and a fixed particular solution $\vec{y_f}$ to $A\vec{x} = \vec{c}$ for $\vec{c} \neq \vec{0}$, there is a solution $\vec{z}$ to $A\vec{x} = \vec{0}$ so that $\vec{y} = \vec{z} + \vec{y_f}$. In other words, if $\mathbb{T}$ is the set of solutions to $A\vec{x} = \vec{c}$, then

$$\mathbb{T} = \left\{ \vec{w} + \vec{y_f} \mid \vec{w} \text{ satisfies } A\vec{x} = \vec{0} \text{ and } \vec{y_f} \text{ satisfies } A\vec{x} = \vec{c} \right\} \qquad (8.6)$$

Thus $\mathbb{T} = \mathbb{S} + \vec{y_f}$, where $\mathbb{S}$ is the subspace of $\mathbb{R}^n$ consisting of the solutions to $A\vec{x} = \vec{0}$ and $\vec{y_f}$ is any fixed particular solution to $A\vec{x} = \vec{c}$. Here $\mathbb{T}$ is said to be a *translate* of the subspace $\mathbb{S}$ by the vector $\vec{y_f}$. That is, $\mathbb{T}$ is parallel to $\mathbb{S}$ and shifted so that $\mathbb{T}$ goes through $\vec{y_f}$ instead of the origin.

Example 8.1.13. Let us now have *Maple* do an example of this for us. Let our nonhomogeneous system be

$$11x - 8y + 3z = -5$$
$$-7x + y - 6z = 2$$

Its matrix of coefficients A and the column matrix $\vec{c}$ are given respectively by

$$A = \begin{bmatrix} 11 & -8 & 3 \\ -7 & 1 & -6 \end{bmatrix}, \quad \vec{c} = \begin{bmatrix} -5 \\ 2 \end{bmatrix}$$

```
> A:= matrix([[11, -8, 3], [-7, 1, -6]]):
> C:= matrix([[-5], [2]]):
> rref(augment(A,C));
```

$$\begin{bmatrix} 1 & 0 & 1 & -\frac{11}{45} \\ 0 & 1 & 1 & \frac{13}{45} \end{bmatrix}$$

From this *rref* matrix, we see that the solution to this nonhomogeneous linear system is one-dimensional, and given by

$$\left\{ x = -z - \frac{11}{45}, \ y = -z + \frac{13}{45} \right\}$$

This solution can be written using column vectors as

$$\vec{x} = \begin{bmatrix} x \\ y \\ z \end{bmatrix} = \begin{bmatrix} -z - \frac{11}{45} \\ -z + \frac{13}{45} \\ z \end{bmatrix} = z \begin{bmatrix} -1 \\ -1 \\ 1 \end{bmatrix} + \begin{bmatrix} -\frac{11}{45} \\ \frac{13}{45} \\ 0 \end{bmatrix}$$

The column vector $y_f = \langle -\frac{11}{45}, \frac{13}{45}, 0 \rangle = \begin{bmatrix} -\frac{11}{45} & \frac{13}{45} & 0 \end{bmatrix}^T$ is a particular fixed solution to this nonhomogeneous system, corresponding to $z = 0$.

> rref(augment(A,matrix([[0],[0]])));

$$\begin{bmatrix} 1 & 0 & 1 & 0 \\ 0 & 1 & 1 & 0 \end{bmatrix}$$

From this new *rref* matrix, we see that the solution to this homogeneous linear system is $x = -z$ and $y = -z$. This solution can be written using column vectors as

$$\vec{X} = \begin{bmatrix} x \\ y \\ z \end{bmatrix} = \begin{bmatrix} -z \\ -z \\ z \end{bmatrix}$$

Note that if we add to all of these homogeneous solutions the particular solution $\vec{y_f}$, we get all the nonhomogeneous solutions as expected. The solutions to the homogeneous linear system form a line in space through the origin parallel to the vector $\langle -1, -1, 1 \rangle$. The solutions to the nonhomogeneous linear system form a line in space through the point $\left(-\frac{11}{45}, \frac{13}{45}, 0 \right)$ parallel to the vector $\langle -1, -1, 1 \rangle$. In Figure 8.2, we plot both the general solution subspace $\mathbb{S}$, to the homogeneous system, given by

$$\mathbb{S} = \{ \vec{x} \mid \vec{x} = \langle -t, -t, t \rangle, \ t \in \mathbb{R} \}$$

along with the general solution subspace translate $\mathbb{T}$, of the nonhomogeneous system, given by

$$\mathbb{T} = \mathbb{S} + \vec{y_f} = \left\{ \vec{x} \mid \vec{x} = \langle -t, -t, t \rangle + \left\langle -\frac{11}{45}, \frac{13}{45}, 0 \right\rangle, \ t \in \mathbb{R} \right\}$$

In the following *Maple* code, we have changed the variable z to the parameter t to set these solutions up parametrically.

> S:= evalm(t*vector([-1, -1, 1]));

$$S := \begin{bmatrix} -t & t & t \end{bmatrix}$$

> P:= vector([-11/45, 13/45, 0]):
> T:= evalm(S+P);

$$T := \begin{bmatrix} -t - \dfrac{11}{45} & -t + \dfrac{13}{45} & t \end{bmatrix}$$

> plotS:= spacecurve(convert(S,list), t = -3..3, color = blue):
> plotT:= spacecurve(convert(T,list), t = -3..3, color = red):
> display({plotS, plotT}, axes = boxed);

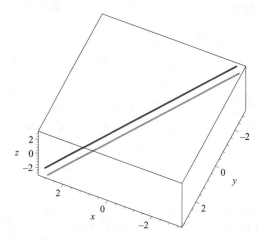

Figure 8.2: The subspace $\mathbb{S}$, and subspace translate $\mathbb{T}$.

◼

Example 8.1.14. Now for one more example of a slightly larger system of linear equations. Let our nonhomogeneous system be

$$11x - 8y + 3z + 2w - u = -5$$
$$-7x + y - 6z + 12w + 4u = 2$$

Its matrix of coefficients, A, and right side column vector, $\vec{c}$, are given by

$$A = \begin{bmatrix} 11 & -8 & 3 & 2 & -1 \\ -7 & 1 & -6 & 12 & 4 \end{bmatrix}, \quad \vec{c} = \begin{bmatrix} -5 \\ 2 \end{bmatrix}$$

> A:= matrix([[11, -8, 3, 2, -1], [-7, 1, -6, 12, 4]]):

> c:= matrix([[-5], [2]]):

> rref(augment(A,c));

$$\begin{bmatrix} 1 & 0 & 1 & -\frac{98}{45} & -\frac{31}{45} & -\frac{11}{45} \\ 0 & 1 & 1 & -\frac{146}{45} & -\frac{37}{45} & \frac{13}{45} \end{bmatrix}$$

From this *rref* matrix, we see that the solution to this nonhomogeneous linear system is given by

$$\left\{ x = -z + \frac{98}{45}w + \frac{31}{45}u - \frac{11}{45}, \; y = -z + \frac{146}{45}w + \frac{37}{45}u + \frac{13}{45} \right\}$$

Notice that the solution is three-dimensional, since the solution can be expressed in terms of the three independent variables, u, w, and z. In column vector notation, we can express the solution as a linear combination of three vectors, multiplied by the arbitrary scalars u, w, and z:

$$\vec{X} = \begin{bmatrix} x \\ y \\ z \\ w \\ u \end{bmatrix} = \begin{bmatrix} -z + \frac{98}{45}w + \frac{31}{45}u - \frac{11}{45} \\ -z + \frac{146}{45}w + \frac{37}{45}u + \frac{13}{45} \\ z \\ w \\ u \end{bmatrix}$$

$$= z\begin{bmatrix} -1 \\ -1 \\ 1 \\ 0 \\ 0 \end{bmatrix} + w\begin{bmatrix} \frac{98}{45} \\ \frac{146}{45} \\ 0 \\ 1 \\ 0 \end{bmatrix} + u\begin{bmatrix} \frac{31}{45} \\ \frac{37}{45} \\ 0 \\ 0 \\ 1 \end{bmatrix} + \begin{bmatrix} -\frac{11}{45} \\ \frac{13}{45} \\ 0 \\ 0 \\ 0 \end{bmatrix}$$

The column vector

$$\vec{y_f} = \left\langle -\frac{11}{45}, \frac{13}{45}, 0, 0, 0 \right\rangle = \left[-\frac{11}{45}, \frac{13}{45}, 0, 0, 0 \right]^T$$

is a particular fixed solution to this nonhomogeneous system corresponding to the choices $z = 0$, $w = 0$, and $u = 0$ in the solution. Notice that the solution $\vec{X}$ is a three-element linear combination in $\mathbb{R}^5$ of the three vectors

$$\left\{ \langle -1, -1, 1, 0, 0 \rangle, \; \left\langle \frac{98}{45}, \frac{146}{45}, 0, 1, 0 \right\rangle, \; \left\langle \frac{31}{45}, \frac{37}{45}, 0, 0, 1 \right\rangle \right\}$$

with the fixed vector $\overrightarrow{y_f}$ added to every linear combination. The general solution to the homogeneous linear system $A\overrightarrow{x} = \overrightarrow{0}$ should now be

$$\overrightarrow{X} = z \begin{bmatrix} -1 \\ -1 \\ 1 \\ 0 \\ 0 \end{bmatrix} + w \begin{bmatrix} \frac{98}{45} \\ \frac{146}{45} \\ 0 \\ 1 \\ 0 \end{bmatrix} + u \begin{bmatrix} \frac{31}{45} \\ \frac{37}{45} \\ 0 \\ 0 \\ 1 \end{bmatrix}$$

Let us check this by using *rref* on its augmented matrix. Note that the solution $\overrightarrow{X} = [-1, -1, 1, 0, 0]^T$ is gotten by setting $z = 1$, $w = 0$ and $u = 0$ in the solution to the homogeneous linear system, the solution $\overrightarrow{X} = \left[\frac{98}{45}, \frac{146}{45}, 0, 1, 0\right]^T$ is gotten by setting $z = 0$, $w = 1$ and $u = 0$ in the solution to the homogeneous linear system, and the solution $\overrightarrow{X} = \left[\frac{31}{45}, \frac{37}{45}, 0, 0, 1\right]^T$ is gotten by setting $z = 0$, $w = 0$ and $u = 1$ in the homogeneous linear system.

> rref(augment(A,matrix([[0],[0]])));

$$\begin{bmatrix} 1 & 0 & 1 & -\frac{98}{45} & -\frac{31}{45} & 0 \\ 0 & 1 & 1 & -\frac{146}{45} & -\frac{37}{45} & 0 \end{bmatrix}$$

One last point to make before we end this section. When considering solutions to linear systems geometrically as subspaces, or subspace translates, of $\mathbb{R}^n$, we must now accept the following fact: A linear system of equations has either no solution, exactly one solution or infinitely many solutions. The reason this is true, is that all solutions to a given linear system form either a subspace, or a subspace translate of $\mathbb{R}^n$. Using the properties of subspaces, we can argue that only nonhomogeneous linear systems can have no solution in which case there is no fixed particular solution $\overrightarrow{y_f}$. If there is a single solution, then in the homogeneous case, this would correspond to the subspace given only by the zero vector, which is the only subspace of $\mathbb{R}^n$ that contains only a single vector. A single solution to a nonhomogeneous system corresponds to a subspace translate of the zero vector solution from the homogeneous case. If there is a nonzero vector $\overrightarrow{v}$ in the subspace corresponding to the solution to the homogeneous system, then there are an infinite number of solutions, since all scalar multiples of $\overrightarrow{v}$ must also be in the subspace, and must therefore be solutions to the homogeneous system. The same can be said about the subspace translate for the nonhomogeneous system. If we now look at this last case algebraically, in terms of matrix equations, we can show through sheer force of algebra that this must be true.

Let $A\vec{x} = \vec{c}$ be our matrix equation for a linear system, and let us assume that the system has two different solutions, $\vec{y}$ and $\vec{z}$, where $\vec{y} \neq \vec{z}$. Then, $\vec{y} - \vec{z} \neq \vec{0}$. If $r \in \mathbb{R}$, then for $\vec{w} = \vec{y} + r(\vec{z} - \vec{y})$, we have

$$
\begin{aligned}
A\vec{w} &= A\vec{y} + rA(\vec{y} - \vec{z}) \\
&= \vec{c} + r(A\vec{y} - A\vec{z}) \\
&= \vec{c} + r(\vec{c} - \vec{c}) \\
&= \vec{c} + r(\vec{0} - \vec{0}) \\
&= \vec{c}
\end{aligned}
$$

Since $\vec{y} - \vec{z} \neq \vec{0}$, $\vec{w} = \vec{y} + r(\vec{y} - \vec{z})$ are solutions to $A\vec{x} = \vec{c}$, as r varies over the real numbers, hence we have generated a method of constructing an infinite number of solutions from a single pair of solutions. This tells us that given any two different solutions $\vec{y}$ and $\vec{z}$ to $A\vec{x} = \vec{c}$, the line $\vec{x} = \vec{y} + r(\vec{y} - \vec{z})$ through $\vec{y}$ parallel to $\vec{y} - \vec{z}$ are also solutions to $A\vec{x} = \vec{c}$.

Homework Problems

1. Determine if each of the following sets define a vector space over some field **F**.

   (a) $\mathbb{V} = \{\langle x, y, 0\rangle \mid x, y \in \mathbb{R}\}$

   (b) $\mathbb{V} = \{\langle x, 1, z\rangle \mid x, z \in \mathbb{Q}\}$

   (c) $\mathbb{V} = \{P(x) \mid P(x) \text{ is a polynomial with real coefficients}\}$

   (d) $\mathbb{V} = \{P(x) \mid P(x) \text{ is a cubic polynomial with complex coefficients}\}$

   (e) $\mathbb{V} = \{\langle x, y, z\rangle \mid x, y, z \geq 0\}$

   (f) $\mathbb{V} = \mathbb{Q}^{4\times 3}$, which are the rational 4×3 matrices.

   (g) $\mathbb{V} = \mathbb{Z}^{3\times 3}$, which are the integer 3×3 matrices.

   (h) $\mathbb{V}$ is the set of all polynomials with integer coefficients.

2. Prove or disprove that the following vector subspace unions, $\mathbb{U} \cup \mathbb{V}$, are themselves vector subspaces.

(a) $U = \{\langle 0, y, 0 \rangle \mid y \in \mathbb{R}\}, V = \{\langle 0, 0, z \rangle \mid z \in \mathbb{R}\}$

(b) $U = \{\langle x, y, 0 \rangle, \mid x, y \in \mathbb{R}\}, V = \{\langle 0, 0, z \rangle \mid z \in \mathbb{R}\}$

(c) $U = \{\langle x, y, 0 \rangle \mid x, y \in \mathbb{R}\}, V = \{\langle x, 0, 0 \rangle \mid x \in \mathbb{R}\}$

3. Express the sum, $U + V$, of the following vector subspaces U and V as a single vector subspace.

(a) $U = \{\langle 0, a, 0 \rangle, \mid a \in \mathbb{R}\}, V = \{\langle 0, 0, b \rangle, \mid b \in \mathbb{R}\}$

(b) $U = \{\langle a, b, 0 \rangle, \mid a, b \in \mathbb{R}\}, V = \{\langle 0, 0, c \rangle, \mid c \in \mathbb{R}\}$

(c) $U = \{\langle a, b, 0 \rangle, \mid a, b \in \mathbb{R}\}, V = \{\langle a, 0, 0 \rangle, \mid a \in \mathbb{R}\}$

(d) $U = \{\langle a, 0, b, 0, c, d, e, 0 \rangle, \mid a, b, c, d, e \in \mathbb{R}\}$,
    $V = \{\langle 0, a, b, c, 0, 0, d, 0 \rangle, \mid a, b, c, d \in \mathbb{R}\}$

(e) $U = \{a\overrightarrow{u_1} + b\overrightarrow{u_2} + c\overrightarrow{u_3}, \mid a, b, c \in \mathbb{R}\}$
    $V = \{a\overrightarrow{v_1} + b\overrightarrow{v_2} + c\overrightarrow{v_3} + d\overrightarrow{v_4}, \mid a, b, c, d \in \mathbb{R}\}$,
    for fixed vectors $\overrightarrow{u_1}, \overrightarrow{u_2}, \overrightarrow{u_3}, \overrightarrow{v_1}, \overrightarrow{v_2}, \overrightarrow{v_3}, \overrightarrow{v_4} \in \mathbb{R}^9$ where $\overrightarrow{u_1}, \overrightarrow{u_2} \in V$.

4. Express the solutions to the following homogeneous systems as vector subspaces of $\mathbb{R}^n$, also state the dimension of each solution.

(a) $3x - 6y + 5z = 0$ (b) $x - 2y - 3z = 0$
           $-x + 3y - 2z = 0$ $-2x + 5y - 7z = 0$

(c) $w + 3x + y + 2z = 0$ (d) $-2w + x + y = 0$
           $2w - 3x + 4y + 6z = 0$ $2w + 5x + y - 2z = 0$
                                       $-w + x + y + z = 0$

5. Express the solutions to the following nonhomogeneous systems as vector subspace translates of $\mathbb{R}^n$, also state the dimension of each solution.

(a) $3x - 6y + 5z = -1$ (b) $x - 2y - 3z = 3$
           $-x + 3y - 2z = 5$ $-2x + 5y - 7z = 2$

(c) $w + 3x + y + 2z = -1$ (d) $-2w + x + y = 1$
           $2w - 3x + 4y + 6z = -2$ $2w + 5x + y - 2z = 3$
                                       $-w + x + y + z = 2$

6. Explain why the following sets $\mathbb{S}$ are subspaces, or not, of the appropriate vector space $\mathbb{R}^n$:

(a) $\mathbb{S} = \{\, \langle a + b + 5, 2a + 7b - 1, a - b + 4 \rangle \mid a, b, \in \mathbb{R} \}$

(b) $\mathbb{S} = \{\, \langle 4a - b, -5a + 7b, a - 3b, -10a + \pi b \rangle \mid a, b, \in \mathbb{R} \}$

(c) $\mathbb{S} = \{\, \langle 5, a + 3b, 0, a - b, \pi \rangle \mid a, b, \in \mathbb{R} \}$

(d) $\mathbb{S} = \{\, \langle a + b, 0, 2a + 7b, 0, a - b, 0, 0 \rangle \mid a, b, \in \mathbb{R} \}$

7. Show that $\mathbb{U} \cap \mathbb{V}$ is a subspace of $\mathbb{R}^n$ if both $\mathbb{U}$ and $\mathbb{V}$ are subspaces of $\mathbb{R}^n$. Does this generalize to the intersection of any finite number of subspaces?

8. Find $\mathbb{U} \cap \mathbb{V}$ for the following vector subspaces.

(a) $\mathbb{U} = \{\langle a, 0, b, 0 \rangle \mid a, b \in \mathbb{R}\}$, $\mathbb{V} = \{\langle 0, a, b, c \rangle \mid a, b, c \in \mathbb{R}\}$

(b) $\mathbb{U} = \{\langle a, 0, b, 0, c, d, e, 0 \rangle \mid a, b, c, d, e \in \mathbb{R}\}$
$\mathbb{V} = \{\langle 0, a, b, c, 0, 0, d, 0 \rangle \mid a, b, c, d \in \mathbb{R}\}$

(c) $\mathbb{U} = \{\langle a, 0, b, 0, c, d, 0, e \rangle \mid a, b, c, d, e \in \mathbb{R}\}$
$\mathbb{V} = \{\langle 0, a, b, c, 0, 0, d, e \rangle \mid a, b, c, d, e \in \mathbb{R}\}$

9. Let $\mathbb{U}$ be the subspace of $\mathbb{R}^n$ which is the solution space to the homogeneous linear system $A\overrightarrow{x} = \overrightarrow{0}$ and $\mathbb{V}$ be the subspace of $\mathbb{R}^n$ which is the solution space to another homogeneous linear system $B\overrightarrow{x} = \overrightarrow{0}$. What homogeneous linear system has the subspace $\mathbb{U} \cap \mathbb{V}$ as its solution space? Is there a homogeneous linear system which has the subspace $\mathbb{U} + \mathbb{V}$ as its solution space?

Maple Problems

1. Parametrize and then plot the vector subspaces corresponding to the solution spaces of the following homogeneous systems:

(a) $3x - 6y = 0$ (b) $x - 2y = 0$

(c) $x + 2y - 3z = 0$ (d) $2x - 5y + 7z = 0$

(e) $x + 2y - 3z = 0$ (f) $2x - 5y + 7z = 0$
       $2x - 5y + 7z = 0$ $x + y + z = 0$

2. Parameterize and then plot the vector subspace translates corresponding to the solutions to the following nonhomogeneous systems:

(a) $3x - 6y = -1$ (b) $x - 2y = 3$

(c) $x + 2y - 3z = 5$ (d) $2x - 5y + 7z = -2$

(e) $x + 2y - 3z = -2$ (f) $2x - 5y + 7z = 9$
        $2x - 5y + 7z = 1$ $x + y + z = -2$

3. Plot the corresponding pairs of vector subspaces and their translates, from problems 1 and 2, respectively, together for each of parts (a) through (f).

4. Solve the following systems of linear equations giving the solutions to both the corresponding homogeneous system and the given system. Write these solutions as linear combinations of vectors and/or the translate of a linear combination. Also, give the dimension of each solution.

(a) $3x - y + 5z - 2w + 9u - v = 2$
        $6x + y - 2z + 8w + 2u - 7v = -9$

(b) $2x - y + 7z - 9w + u - 4v + 2r = 11$
        $-5x + y + 2z - 8w + u - v - 3r = -25$
        $7x + 2y - 2z - w + 3u + v - r = 5$

(c) $3x - y + 5z - 2w + 9u - v + r - s + t = -4$
        $6x + y - 2z + 8w + 2u - 7v - 3r + s - t = 7$
        $-5x + y + 2z - 8w + u - v - r + 5s - 2t = -25$
        $7x + 2y - 2z - w + 3u + v + 6r - 4s + 5t = 5$

(d) $3x - y + 5z = -4$
        $6x + y - 2z = 7$
        $-5x + y + 2z = -25$
        $7x + 2y - z = 5$

5. How are the dimensions of the solutions in problem 4 above related to the number of equations and the number of variables in each system?

6. Verify, with examples using *Maple*, your answers to *Homework* problem 9.

8.2 Independent and Dependent Sets of Vectors in $\mathbb{R}^n$

As we have already seen, the general solution $\mathbb{S}$ to a homogeneous linear system $A\overrightarrow{x} = \overrightarrow{0}$ forms a subspace, and the subspace itself can be expressed as all possible linear combinations of just a few of the solutions to the system. On the other hand, the general solution to a non-homogeneous linear system $A\overrightarrow{x} = \overrightarrow{c}$ does not form a subspace although it can be written (as defined in the last section) as $\mathbb{T} = \mathbb{S} + \overrightarrow{y_f}$, where $\overrightarrow{y_f}$ is any fixed particular solution to the nonhomogeneous system. So $\mathbb{T}$ is a translate of $\mathbb{S}$ by $\overrightarrow{y_f}$, that is, $\mathbb{T}$ is parallel to $\mathbb{S}$ passing through $\overrightarrow{y_f}$ while $\mathbb{S}$ passes through the origin.

Now, we want to answer an important question about how all of this fits together. Can every subspace $\mathbb{S}$ of $\mathbb{R}^n$ be gotten as the general solution to a homogeneous linear system $A\overrightarrow{x} = \overrightarrow{0}$, and if so, how? We already know that the general solution to a homogeneous linear system $A\overrightarrow{x} = \overrightarrow{0}$ is a subspace, and so it is natural to ask if the converse to this statement is also true.

In order to answer this question, we first introduce the definition of a *spanning subset* **K** of a subspace $\mathbb{S}$ of $\mathbb{R}^n$.

Definition 8.2.1. A finite subset $\mathbf{K} = \left\{ \overrightarrow{k_1}, \overrightarrow{k_2}, \ldots, \overrightarrow{k_m} \right\}$, of a subspace $\mathbb{S}$ in $\mathbb{R}^n$, is said to *span* $\mathbb{S}$ (or be a *spanning set of* $\mathbb{S}$) if every vector in $\mathbb{S}$ can be expressed as a linear combination of vectors in **K**. In other words, for each $\overrightarrow{v} \in \mathbb{S}$ there exists scalars a_1 through a_m such that

$$\overrightarrow{v} = a_1 \overrightarrow{k_1} + a_2 \overrightarrow{k_2} + \cdots + a_m \overrightarrow{k_m}$$

K is a spanning set for the subspace $\mathbb{S}$ if and only if

$$\mathbb{S} = \left\{ a_1 \overrightarrow{k_1} + a_2 \overrightarrow{k_2} + \cdots + a_m \overrightarrow{k_m} \;\middle|\; a_1, a_2, \ldots, a_m \in \mathbb{R} \right\}$$

The next question to ask is: Does every subspace $\mathbb{S}$ have a finite spanning set **K**? Here $\mathbb{R}^n$ is a subspace of itself and it has a finite spanning set **U** called the *standard basis* given by

$$\mathbf{U} = \{ \langle 1, 0, 0, \ldots, 0 \rangle, \langle 0, 1, 0, 0, \ldots, 0 \rangle, \ldots, \langle 0, 0, \ldots, 0, 1 \rangle \}$$

U consists of the rows (or columns) that form the $n \times n$ identity matrix. We will introduce the definition of basis in Section 8.3, however, for now a basis should be thought of as a smallest size spanning set. One important fact to keep in mind is that the vector space $\mathbb{R}^n$ has many other possible spanning sets, the standard basis given above is simply the most natural to construct and usually the most convenient to use.

If every general solution (subspace) $\mathbb{S}$ to a homogeneous linear system $A\overrightarrow{x} = \overrightarrow{0}$ has a finite spanning set **K**, we need a method to find this set of vectors.

The most obvious first step in finding this set is to row reduce the matrix A. But where to go from here? The remaining steps are as follows: Once the matrix A has been row reduced, we will remove from the matrix all nonidentity and nonzero columns, followed by negating them and then augmenting each vertically by the appropriate columns from the identity matrix that has square dimension equal to the number of nonidentity columns in the row reduced matrix. This may seem like a complicated and strange set of instructions to follow, however, let us look at a quick example to see this process in action.

Example 8.2.1. Consider the following homogeneous linear system:

$$\begin{bmatrix} 5 & -7 & 1 & 4 & -9 & 3 \\ -2 & -1 & 6 & 0 & -5 & 8 \\ 11 & 9 & -3 & 2 & 5 & 7 \end{bmatrix} \vec{x} = \vec{0}$$

```
> with(linalg):
> A:= matrix([[5, -7, 1, 4, -9, 3], [-2, -1, 6, 0, -5, 8], [11, 9, -3, 2, 5, 7]]):
> R:= rref(A);
```

$$\begin{bmatrix} 1 & 0 & 0 & \frac{13}{31} & -\frac{157}{341} & \frac{268}{341} \\ 0 & 1 & 0 & -\frac{8}{31} & \frac{285}{341} & \frac{126}{341} \\ 0 & 0 & 1 & \frac{3}{31} & -\frac{289}{341} & \frac{565}{341} \end{bmatrix}$$

The command *stackmatrix* from the *linalg* package allows you to augment matrices vertically and we use it below to create $\vec{u}$, $\vec{v}$, and $\vec{w}$.

```
> u = stackmatrix(evalm(-1*delcols(delcols(R, 1..3), 2..3)), matrix([[1], [0], [0]])),
    v = stackmatrix(evalm(-1*delcols(delcols(R, 1..4), 2..2)), matrix([[0], [1], [0]])),
    w = stackmatrix(evalm(-1*delcols(R, 1..5)), matrix([[0], [0], [1]]));
```

$$u = \begin{bmatrix} -\frac{13}{31} \\ \frac{8}{31} \\ -\frac{3}{31} \\ 1 \\ 0 \\ 0 \end{bmatrix}, v = \begin{bmatrix} \frac{157}{341} \\ -\frac{285}{341} \\ \frac{289}{341} \\ 0 \\ 1 \\ 0 \end{bmatrix}, w = \begin{bmatrix} -\frac{268}{341} \\ -\frac{126}{341} \\ -\frac{565}{341} \\ 0 \\ 0 \\ 1 \end{bmatrix}$$

The claim is that the general solution $\mathbb{S}$ to the above given linear system $A\vec{x} = \vec{0}$ consists of all the linear combinations of the 3 vectors $\vec{u}$, $\vec{v}$, and $\vec{w}$

given by:

$$\vec{u} = \left\langle -\frac{13}{31}, \frac{8}{31}, -\frac{3}{31}, 1, 0, 0 \right\rangle, \vec{v} = \left\langle \frac{157}{341}, -\frac{285}{341}, \frac{289}{341}, 0, 1, 0 \right\rangle$$

$$\vec{w} = \left\langle -\frac{268}{341}, -\frac{126}{341}, -\frac{565}{341}, 0, 0, 1 \right\rangle$$

To see this, if $\vec{x} = \langle x_1, x_2, \ldots, x_6 \rangle$ is treated as a column matrix, then we have the following system of equations corresponding to the *rref* matrix:

$$x_1 + \frac{13}{31}x_4 - \frac{157}{341}x_5 + \frac{268}{341}x_6 = 0$$
$$x_2 - \frac{8}{31}x_4 + \frac{285}{341}x_5 + \frac{126}{341}x_6 = 0 \qquad (8.7)$$
$$x_3 + \frac{3}{31}x_4 - \frac{289}{341}x_5 + \frac{565}{341}x_6 = 0$$

Solving for the variables x_1, x_2 and x_3 in the equations given in (8.7), we have

$$x_1 = -\frac{13}{31}x_4 + \frac{157}{341}x_5 - \frac{268}{341}x_6$$
$$x_2 = \frac{8}{31}x_4 - \frac{285}{341}x_5 - \frac{126}{341}x_6 \qquad (8.8)$$
$$x_3 = -\frac{3}{31}x_4 + \frac{289}{341}x_5 - \frac{565}{341}x_6$$

Remember that our solution was of the form $\vec{x} = \langle x_1, x_2, \ldots, x_6 \rangle$, not $\vec{x} = \langle x_1, x_2, x_3 \rangle$. We can express the solution given above in equation (8.8) as follows:

$$
\begin{bmatrix} x_1 \\ x_2 \\ x_3 \\ x_4 \\ x_5 \\ x_6 \end{bmatrix} = x_4 \begin{bmatrix} -\frac{13}{31} \\ \frac{8}{31} \\ -\frac{3}{31} \\ 1 \\ 0 \\ 0 \end{bmatrix} + x_5 \begin{bmatrix} \frac{157}{341} \\ -\frac{285}{341} \\ \frac{289}{341} \\ 0 \\ 1 \\ 0 \end{bmatrix} + x_6 \begin{bmatrix} -\frac{268}{341} \\ -\frac{126}{341} \\ -\frac{565}{341} \\ 0 \\ 0 \\ 1 \end{bmatrix} \qquad (8.9)
$$

But notice this is simply:

$$\vec{x} = x_4 \vec{u} + x_5 \vec{v} + x_6 \vec{w}$$

for the vectors $\vec{u}$, $\vec{v}$, and $\vec{w}$ defined previously. Any arbitrary values of the variables x_4, x_5, and x_6 in the above definition of $\vec{x}$ will satisfy $A\vec{x} = \vec{0}$, and therefore we have shown that solutions to $A\vec{x} = \vec{0}$ can be expressed as linear combinations of the vectors $\vec{u}$, $\vec{v}$, and $\vec{w}$. The subspace $\mathbb{S}$ is therefore spanned by the set $\mathbf{K} = \{\vec{u}, \vec{v}, \vec{w}\}$.

___

Now let us explain why each subspace $\mathbb{S}$ of $\mathbb{R}^n$ has a spanning set $\mathbf{K}$. Consider the following argument: If $\mathbb{S}$ is a subspace consisting of just the zero vector, then we are done as it is its own spanning set, and $\mathbf{K} = \{\vec{0}\}$. If $\mathbb{S} \neq \{\vec{0}\}$, then let $\vec{k_1}$ be any nonzero vector in $\mathbb{S}$. We are again done if $\mathbb{S}$ is all scalar multiples of $\vec{k_1}$ as then $\mathbf{K} = \{\vec{k_1}\}$. Assume this is not the case, and let $\mathbb{S}_1$ be the subspace consisting of all scalar multiples of $\vec{k_1}$. Then $\mathbb{S} \neq \mathbb{S}_1$, and $\mathbb{S}_1$ defines a line through the origin in $\mathbb{R}^n$. Since $\mathbb{S} \neq \mathbb{S}_1$, there is a nonzero vector, $\vec{k_2}$ in $\mathbb{S}$, which is not in $\mathbb{S}_1$. If $\mathbb{S}$ is spanned by $\mathbf{K} = \{\vec{k_1}, \vec{k_2}\}$, then we are done. Assume this is not the case, and let $\mathbb{S}_2$ be the subspace consisting of all linear combinations of $\vec{k_1}$ and $\vec{k_2}$, then $\mathbb{S} \neq \mathbb{S}_2$. We now have that $\mathbb{S}_2$ defines a plane through the origin in $\mathbb{R}^n$, as opposed to the line through the origin given by $\mathbb{S}_1$.

You can continue this process, but not forever. For a second, let us consider the case that we have worked through the process above to the point that you have $n + 1$ vectors $\{\vec{k_1}, \vec{k_2}, \ldots, \vec{k_n}, \vec{k_{n+1}}\}$, where each successive vector in the set cannot be expressed as a linear combination of the previous set of vectors, and all are elements of $\mathbb{S}$. As each one of these $\vec{k_j}$'s are produced, we have also created subspaces, $\mathbb{S}_j$, of $\mathbb{R}^n$, where $\mathbb{S}_j$ is the set of all linear combinations of the vectors $\vec{k_1}, \vec{k_2}, \ldots, \vec{k_j}$. Each time we have done this, we have produced a higher dimensional subspace of $\mathbb{R}^n$ than before. Notice that $\mathbb{S}_1$ is a line that is of dimension 1, while $\mathbb{S}_2$ is a plane and of dimension 2, and then finally $\mathbb{S}_{n+1}$ must have dimension $n + 1$. But $\mathbb{R}^n$ has dimension n, and so how can a subspace of it, such as $\mathbb{S}_{n+1}$, have dimension $n + 1 > n$? This is not possible, and so this process must stop after no more than n steps. Thus, every subspace $\mathbb{S}$ of $\mathbb{R}^n$ has a spanning set $\mathbf{K}$, and it does not need more than n elements if it is chosen by this process. Of course, this process only works if our vector space is of a finite dimension n, but will always work when considering subspaces of $\mathbb{R}^n$ or $\mathbb{C}^n$. In this argument, hopefully you have realized that the subspace $\mathbb{S}_n$ must be $\mathbb{R}^n$ itself which implies that $\vec{k_{n+1}}$ can not exist as constructed.

The process outlined above to explain why every subspace $\mathbb{S}$ of $\mathbb{R}^n$ has a spanning set $\mathbf{K}$ of at most n elements suggests the following definitions of independent and dependent finite subsets of $\mathbb{R}^n$. Basically, we wish to eliminate from a spanning set $\mathbf{K}$, of a subspace $\mathbb{S}$, any element which is a linear combination of the rest of the elements of $\mathbf{K}$ since they are unnecessary in writing out $\mathbb{S}$ as linear combination of $\mathbf{K}$'s elements. If no one element of $\mathbf{K}$ is a linear combination of the rest of $\mathbf{K}$'s elements, then we say $\mathbf{K}$ is *independent* and otherwise we call it *dependent*.

Definition 8.2.2. A finite set of vectors $\mathbf{K}$ in $\mathbb{R}^n$ is called an *independent set* if no single element $\vec{w}$ of $\mathbf{K}$ can be written as a linear combination of the remaining elements of $\mathbf{K}$, that is, for

$$\mathbf{K} = \{\vec{v_1}, \vec{v_2}, \ldots, \vec{v_k}\}$$

$\mathbf{K}$ is independent if $\vec{v_j} \notin \mathbb{S}_j$, where the subspace $\mathbb{S}_j$ of $\mathbb{R}^n$ is spanned by the set

$$\{\vec{v_1}, \vec{v_2}, \ldots, \vec{v_{j-1}}, \vec{v_{j+1}}, \ldots, \vec{v_k}\}$$

for all $1 \leq j \leq k$. Another way to say this is that $\mathbf{K}$ is independent if the subspace $\mathbb{S}$ of $\mathbb{R}^n$ spanned by $\mathbf{K}$ is not spanned by any smaller subset of $\mathbf{K}$.

Definition 8.2.3. A finite set of vectors $\mathbf{K}$ in $\mathbb{R}^n$ is called a *dependent set* if at least one element $\vec{w}$ of $\mathbf{K}$ can be written as a linear combination of the remaining elements of $\mathbf{K}$, that is, for

$$\mathbf{K} = \{\vec{v_1}, \vec{v_2}, \ldots, \vec{v_k}\}$$

$\mathbf{K}$ is dependent if for at least one j, $\vec{v_j} \in \mathbb{S}_j$, where the subspace $\mathbb{S}_j$ of $\mathbb{R}^n$ is spanned by the set

$$\{\vec{v_1}, \vec{v_2}, \ldots, \vec{v_{j-1}}, \vec{v_{j+1}}, \ldots, \vec{v_k}\}$$

Alternatively, $\mathbf{K}$ is dependent if the subspace $\mathbb{S}$ of $\mathbb{R}^n$ spanned by $\mathbf{K}$ is spanned by some smaller subset of $\mathbf{K}$.

Both definitions above can be thought of in the following context. Consider the set of vectors $\{\vec{v_1}, \vec{v_2}, \ldots, \vec{v_k}\}$, then the linear equation

$$a_1\vec{v_1} + a_2\vec{v_2} + \cdots + a_k\vec{v_k} = \vec{0} \tag{8.10}$$

will always be satisfied if $a_j = 0$ for all $1 \leq j \leq k$. If this is the only way in which equation (8.10) can be satisfied, the set $\{\vec{v_1}, \vec{v_2}, \ldots, \vec{v_k}\}$ is linearly independent. If we can find a set of scalars $\{a_1, a_2, \ldots, a_k\}$, for which not all of the a_j's are zero, which satisfy equation (8.10), then the set of vectors $\{\vec{v_1}, \vec{v_2}, \ldots, \vec{v_k}\}$ is linearly dependent.

Any finite set of vectors in $\mathbb{R}^n$ that contains the zero vector is automatically a dependent set as well as any finite set of vectors where one vector is a scalar multiple of another. The fastest way to check if a finite set of vectors $\{\vec{k_1}, \vec{k_2}, \ldots, \vec{k_j}\}$ from $\mathbb{R}^n$ is a dependent or independent set, is to find the *rref* of the matrix with these vectors as its rows. If you get at least one row of all zeros, the set is dependent, otherwise it is independent. The row of all zeros (if there are no row swaps using *rref*) corresponds to a vector that is a linear combination of the previous row vectors; however, this is a somewhat ambiguous statement since any vector involved in the linear combination can be solved for in terms of the others.

Example 8.2.2. Now, let us check out how *rref* can tell us if a set of vectors is dependent or independent. Let $\vec{u} = \langle 5, -7, 3, 1, -2 \rangle, \vec{v} = \langle -9, 0, 8, -4, 6 \rangle$ and $\vec{w} = -3\vec{u} + 10\vec{v}$ in $\mathbb{R}^5$. Note also that by the definition of $\vec{w}$, we have that $\vec{u} = \dfrac{10}{3}\vec{v} - \dfrac{1}{3}\vec{w}$ and $\vec{v} = \dfrac{1}{10}\vec{w} + \dfrac{3}{10}\vec{u}$. Clearly, the set $\{\vec{u}, \vec{v}, \vec{w}\}$ is dependent.

```
> u:= [5, -7, 3, 1, -2]: v:= [-9, 0, 8, -4, 6]:
> w:= convert(evalm(-3*vector(u) + 10*vector(v)),list);
```

$$w := [-105, 21, 71, -43, 66]$$

```
> rref(matrix([u,v,w]));
```

$$\begin{bmatrix} 1 & 0 & -\frac{8}{9} & \frac{4}{9} & -\frac{2}{3} \\ 0 & 1 & -\frac{67}{63} & \frac{11}{63} & -\frac{4}{21} \\ 0 & 0 & 0 & 0 & 0 \end{bmatrix}$$

```
> rref(matrix([v,w,u]));
```

$$\begin{bmatrix} 1 & 0 & -\frac{8}{9} & \frac{4}{9} & -\frac{2}{3} \\ 0 & 1 & -\frac{67}{63} & \frac{11}{63} & -\frac{4}{21} \\ 0 & 0 & 0 & 0 & 0 \end{bmatrix}$$

```
> rref(matrix([w,v,u]));
```

$$\begin{bmatrix} 1 & 0 & -\frac{8}{9} & \frac{4}{9} & -\frac{2}{3} \\ 0 & 1 & -\frac{67}{63} & \frac{11}{63} & -\frac{4}{21} \\ 0 & 0 & 0 & 0 & 0 \end{bmatrix}$$

Notice that the order of the vectors $\vec{u}$, $\vec{v}$, and $\vec{w}$ as rows of the matrix is irrelevant to the *rref* matrix. In either case, we can conclude that this set of three vectors is dependent in $\mathbb{R}^5$. The number of all zero rows in the *rref* matrix indicates how many of the row vectors can be written as linear combinations of the rest. In this case, any one of these three vectors can be written as a linear combination of the remaining two vectors, as described prior to the example.

Example 8.2.3. Now, we do another example by adding a fourth linearly dependent vector to the above set. If we define $\vec{z} = 2\vec{u} - 7\vec{v}$ and apply *rref* to all four vectors $\{\vec{u}, \vec{v}, \vec{w}, \vec{z}\}$, we should get two rows of all zeros at the bottom of the *rref* matrix.

> z:= convert(evalm(2*vector(u) - 7*vector(v)),list);

$$z := [73, -14, -50, 30, -46]$$

> T:= rref(matrix([w,u,z,v]));

$$T := \begin{bmatrix} 1 & 0 & -\frac{8}{9} & \frac{4}{9} & -\frac{2}{3} \\ 0 & 1 & -\frac{67}{63} & \frac{11}{63} & -\frac{4}{21} \\ 0 & 0 & 0 & 0 & 0 \\ 0 & 0 & 0 & 0 & 0 \end{bmatrix}$$

> L:= rref(matrix([u, v]));

$$L := \begin{bmatrix} 1 & 0 & -\frac{8}{9} & \frac{4}{9} & -\frac{2}{3} \\ 0 & 1 & -\frac{67}{63} & \frac{11}{63} & -\frac{4}{21} \end{bmatrix}$$

The set of vectors $\{\vec{u}, \vec{v}, \vec{w}, \vec{z}\}$ above is a dependent set in $\mathbb{R}^5$, and they span the subspace:

$$\mathbb{S} = \{a\,\vec{u} + b\,\vec{v} + c\,\vec{w} + d\,\vec{z} \mid a, b, c, d \in \mathbb{R}\}$$

This subspace $\mathbb{S}$ is the set of all linear combinations of the vectors $\vec{u}$ to $\vec{z}$. $\mathbb{S}$ is also spanned by the set $\{\vec{u}, \vec{v}, \vec{w}\}$ and the set $\{\vec{u}, \vec{v}\}$. Of these three spanning sets of $\mathbb{S}$, $\{\vec{u}, \vec{v}, \vec{w}, \vec{z}\}$ and $\{\vec{u}, \vec{v}, \vec{w}\}$ are dependent sets while $\{\vec{u}, \vec{v}\}$ is an independent set. Note that the independent spanning set $\{\vec{u}, \vec{v}\}$ has the fewest number of elements at two and the dimension of the subspace spanned by $\{\vec{u}, \vec{v}\}$ is 2 since $\vec{u}$ and $\vec{v}$ form a plane in $\mathbb{R}^5$. It should also be noted that the two rows of the matrix L above also span the subspace $\mathbb{S}$, and are an independent set.

The above information should make it fairly clear that, among the spanning sets **K** of a subspace $\mathbb{S}$ of $\mathbb{R}^n$, the spanning sets that contain only linearly independent vectors contain the fewest number of vectors. In fact, the number of elements of an independent spanning set **K** for a subspace $\mathbb{S}$ seems to be its dimension. We will address this idea in greater detail in Section 8.3.

We asked the question in the very beginning of this section if every subspace $\mathbb{S}$ of $\mathbb{R}^n$ is the solution subspace for some homogeneous linear system $A\vec{x} = \vec{0}$. Now, we can answer this question in the affirmative. If $\mathbb{S} = \mathbb{R}^n$, then we can take the matrix A to be any zero matrix with n columns. If $\mathbb{S} = \{\vec{0}\}$ the zero vector, then we can take the matrix A to be the $n \times n$ identity matrix I_n. So what happens if $\mathbb{S}$ is a subspace of $\mathbb{R}^n$ which is neither the zero subspace, or all of $\mathbb{R}^n$?

Definition 8.2.4. Given a vector subspace $\mathbb{S}$ of $\mathbb{R}^n$, the *orthogonal subspace*, (or *perpendicular subspace*), to $\mathbb{S}$, denoted $\mathbb{S}^\perp$, is the subspace of $\mathbb{R}^n$ defined as

$$\mathbb{S}^\perp = \left\{ \vec{u} \in \mathbb{R}^n \mid \vec{u} \cdot \vec{v} = \vec{0}, \ \forall \ \vec{v} \in \mathbb{S} \right\} \tag{8.11}$$

It is left to you to show that $\mathbb{S}^\perp$ is truly a subspace of $\mathbb{R}^n$. Note also that the orthogonal subspace to $\mathbb{S}^\perp$ is $\mathbb{S}$ itself, that is, $(\mathbb{S}^\perp)^\perp = \mathbb{S}$.

Notice that if $\mathbb{S}$ is a planar two-dimensional subspace of $\mathbb{R}^3$, then $\mathbb{S}^\perp$ is the one-dimensional subspace of $\mathbb{R}^3$ that is the line through the origin perpendicular to the plane $\mathbb{S}$. Similarly, if $\mathbb{S}$ is the one-dimensional subspace of $\mathbb{R}^3$ that is a line through the origin, then $\mathbb{S}^\perp$ is the two-dimensional subspace of $\mathbb{R}^3$ which is the plane through the origin perpendicular to the line $\mathbb{S}$. On the other hand, if $\mathbb{S}$ is a planar two-dimensional subspace of $\mathbb{R}^4$, then $\mathbb{S}^\perp$ is the two-dimensional subspace of $\mathbb{R}^4$ that is the plane through the origin perpendicular to the plane $\mathbb{S}$.

Since $\mathbb{S}^\perp$ is a subspace of $\mathbb{R}^n$, it has a finite spanning set $\mathbf{K} = \{\vec{k_1}, \vec{k_2}, \ldots, \vec{k_j}\}$. Let the matrix A be formed by using the vectors in $\mathbf{K}$ as A's rows. Notice that this yields $A\vec{x} = \vec{0}$ for each $\vec{x} \in \mathbb{S}$. As well, if $\vec{y}$ is a solution to $A\vec{x} = \vec{0}$, then $\vec{y}$ is in $\mathbb{S}$ since it is orthogonal to each $\vec{k_i} \in \mathbf{K}$, and so $\vec{y}$ is orthogonal to each element of $\mathbb{S}^\perp$ giving $\vec{y}$ is in $(\mathbb{S}^\perp)^\perp = \mathbb{S}$. Thus, $\mathbb{S}$ is the solution subspace to $A\vec{x} = \vec{0}$. Therefore, to construct a homogeneous linear system corresponding to a particular subspace $\mathbb{S}$ of $\mathbb{R}^n$, we simply have to consider the orthogonal subspace to $\mathbb{S}$.

Example 8.2.4. We will first find the orthogonal subspace to

$$\mathbb{S} = \left\{ \langle a, 0, b, 0, c \rangle \mid a, b, c \in \mathbb{R} \right\}$$

which is a dimension three subspace of $\mathbb{R}^5$. The orthogonal subspace consists of all the vectors in $\mathbb{R}^5$ that are orthogonal to every vector in $\mathbb{S}$, therefore

$$\mathbb{S}^\perp = \left\{ \langle 0, a, 0, b, 0 \rangle \mid a, b \in \mathbb{R} \right\}$$

is the dimension two subspace of $\mathbb{R}^5$ orthogonal to $\mathbb{S}$. To see this, we simply have to show that the following dot product formula is satisfied, for arbitrary real numbers a through e:

$$\langle a, 0, b, 0, c \rangle \cdot \langle 0, d, 0, e, 0 \rangle = 0$$

Note also that the dimensions of the subspace and the dimension of the orthogonal subspace sum to five, the dimension of $\mathbb{R}^5$.

Example 8.2.5. Let us consider a slightly more complex example. The subspace

$$\mathbb{S} = \big\{ a\langle 1, -3, 2, 5, -6, 4 \rangle + b\langle -9, 2, -1, 7, 1, 8 \rangle \,\big|\, a, b \in \mathbb{R} \big\}$$

which has dimension 2 in $\mathbb{R}^6$, has as its orthogonal complement the subspace

$$\mathbb{S}^\perp = \big\{ a\langle 1, 17, 25, 0, 0, 0 \rangle + b\langle 31, 52, 0, 25, 0, 0 \rangle + c\langle -9, -53, 0, 0, 25, 0 \rangle$$
$$+ d\langle 32, 44, 0, 0, 0, 25 \rangle \,\big|\, a, b, c, d \in \mathbb{R} \big\}$$

The subspace $\mathbb{S}^\perp$ has dimension four in $\mathbb{R}^6$, and is a linear combination of the four basic solution vectors:

$$\big\{ \overrightarrow{w_1} = \langle 1, 17, 25, 0, 0, 0 \rangle, \overrightarrow{w_2} = \langle 31, 52, 0, 25, 0, 0 \rangle,$$
$$\overrightarrow{w_3} = \langle -9, -53, 0, 0, 25, 0 \rangle, \overrightarrow{w_4} = \langle 32, 44, 0, 0, 0, 25 \rangle \big\}$$

corresponding to the system of two linear equations given by the two dot products

$$\langle x, y, z, u, v, w \rangle \cdot \langle 1, -3, 2, 5, -6, 4 \rangle = 0$$
$$\langle x, y, z, u, v, w \rangle \cdot \langle -9, 2, -1, 7, 1, 8 \rangle = 0$$

These four basic solutions $\overrightarrow{w_1}$, $\overrightarrow{w_2}$, $\overrightarrow{w_3}$, and $\overrightarrow{w_4}$ to this system are a spanning set for this solution and can be found from the solution to the system written as column matrices (vectors). The solution as a column matrix is

$$\begin{bmatrix} x \\ y \\ z \\ u \\ v \\ w \end{bmatrix} = z \begin{bmatrix} \frac{1}{25} \\ \frac{17}{25} \\ 1 \\ 0 \\ 0 \\ 0 \end{bmatrix} + u \begin{bmatrix} \frac{31}{25} \\ \frac{52}{25} \\ 0 \\ 1 \\ 0 \\ 0 \end{bmatrix} + v \begin{bmatrix} -\frac{9}{25} \\ -\frac{53}{25} \\ 0 \\ 0 \\ 1 \\ 0 \end{bmatrix} + w \begin{bmatrix} \frac{32}{25} \\ \frac{44}{25} \\ 0 \\ 0 \\ 0 \\ 1 \end{bmatrix}$$

where we get the four basic solutions $\overrightarrow{w_1}$, $\overrightarrow{w_2}$, $\overrightarrow{w_3}$, and $\overrightarrow{w_4}$ to this system by successively letting $z = 25$ and $u = v = w = 0$, $u = 25$ and $z = v = w = 0$, $v = 25$ and $z = u = w = 0$, $w = 25$ and $z = u = v = 0$. We use the 25 in order to eliminate fractions from our 4 basic solutions $\overrightarrow{w_1}$, $\overrightarrow{w_2}$, $\overrightarrow{w_3}$, and $\overrightarrow{w_4}$.

> A:= matrix([[1, -3, 2, 5, -6, 4, 0], [-9, 2, -1, 7, 1, 8, 0]]);

$$\begin{bmatrix} 1 & -3 & 2 & 5 & -6 & 4 & 0 \\ -9 & 2 & -1 & 7 & 1 & 8 & 0 \end{bmatrix}$$

> rref(A);

$$
\begin{bmatrix}
1 & 0 & -\frac{1}{25} & -\frac{31}{25} & \frac{9}{25} & -\frac{32}{25} & 0 \\
0 & 1 & -\frac{17}{25} & -\frac{52}{25} & \frac{53}{25} & -\frac{44}{25} & 0
\end{bmatrix}
$$

> X:= vector([x,y,z,u,v,w]):
> U:= vector([1, -3, 2, 5, -6, 4]): V:= vector([-9, 2, -1, 7, 1, 8]):
> solve({dotprod(X, U) = 0, dotprod(X, V) = 0}, {x, y});

$$
\left\{ x = \frac{32}{25}w + \frac{1}{25}z + \frac{31}{25}u - \frac{9}{25}v, y = \frac{44}{25}w + \frac{17}{25}z + \frac{52}{25}u - \frac{53}{25}v \right\}
$$

> w1:= vector([1, 17, 25, 0, 0, 0]); w2:= vector([31, 52, 0, 25, 0, 0]); w3:=
vector([-9, -53, 0, 0, 25, 0]); w4:= vector([32, 44, 0, 0, 0, 25]);

$$
w1 := \begin{bmatrix} 1 & 17 & 25 & 0 & 0 & 0 \end{bmatrix}
$$

$$
w2 := \begin{bmatrix} 31 & 52 & 0 & 25 & 0 & 0 \end{bmatrix}
$$

$$
w3 := \begin{bmatrix} -9 & -53 & 0 & 0 & 25 & 0 \end{bmatrix}
$$

$$
w4 := \begin{bmatrix} 32 & 44 & 0 & 0 & 0 & 25 \end{bmatrix}
$$

> dotprod(U, w1); dotprod(V, w1);

$$0$$
$$0$$

> dotprod(U, w2); dotprod(V, w2);

$$0$$
$$0$$

> dotprod(U, w3); dotprod(V, w3);

$$0$$
$$0$$

> dotprod(U, w4); dotprod(V, w4);

$$0$$
$$0$$

Homework Problems

1. Compute the spanning set **K** for the subspace $\mathbb{S}$ corresponding to the solution of each of the following homogeneous linear systems.

(a) $\begin{bmatrix} 1 & -2 & 1 \\ 4 & -2 & 8 \end{bmatrix} \vec{x} = \vec{0}$

(b) $\begin{bmatrix} 8 & -2 & 2 & 4 \\ -4 & 1 & -1 & -2 \end{bmatrix} \vec{x} = \vec{0}$

(c) $\begin{bmatrix} 0 & 6 & 1 & 4 \\ 2 & -2 & -9 & -1 \\ -1 & 5 & 2 & -1 \end{bmatrix} \vec{x} = \vec{0}$

(d) $\begin{bmatrix} 0 & 6 \\ 2 & 2 \\ -1 & 5 \end{bmatrix} \vec{x} = \vec{0}$

(e) $\begin{bmatrix} -2 & 1 & 4 & 2 \\ 2 & 3 & -5 & 0 \\ 0 & 4 & -2 & 2 \end{bmatrix} \vec{x} = \vec{0}$

(f) $\begin{bmatrix} 2 & 0 & -4 & 2 & 9 \\ 3 & 1 & -7 & 0 & 8 \end{bmatrix} \vec{x} = \vec{0}$

2. For each of the homogeneous systems from problem 1, determine the maximum number of vectors that could possibly be in the spanning set. How does this compare to the actual number of vectors in the spanning set?

3. Determine whether or not each of the following pairs of spanning sets $\mathbf{K}_1$ and $\mathbf{K}_2$ span the same subspace:

(a) $\mathbf{K}_1 = \{\langle 1, 0, 1 \rangle, \langle 0, 1, 1 \rangle\}, \quad \mathbf{K}_2 = \{\langle 2, -1, 1 \rangle, \langle -1, 5, 4 \rangle\}$

(b) $\mathbf{K}_1 = \{\langle 1, 0, 1 \rangle, \langle 0, 1, 1 \rangle\}, \quad \mathbf{K}_2 = \{\langle 2, -1, 2 \rangle, \langle -1, 5, 4 \rangle\}$

(c) $\mathbf{K}_1 = \{\langle 2, 3, -1 \rangle, \langle 3, 3, 1 \rangle\}, \quad \mathbf{K}_2 = \{\langle 1, 0, 2 \rangle, \langle -1, -6, 8 \rangle, \langle 2, 0, 5 \rangle\}$

(d) $\mathbf{K}_1 = \{\langle -1, 0, 1, 0 \rangle, \langle 2, 1, 2, 2 \rangle\}, \quad \mathbf{K}_2 = \{\langle 1, 1, 3, 2 \rangle, \langle 3, 1, 1, 2 \rangle\}$

(e) $\mathbf{K}_1 = \{\langle 2, 3, -4, 1 \rangle, \langle -1, 2, -1, 1 \rangle, \langle 2, 1, -1, 1 \rangle\},$
$\mathbf{K}_2 = \{\langle 1, 3, -2, 2 \rangle, \langle 3, 6, -6, 2 \rangle\}$

(f) $\mathbf{K}_1 = \{\langle 2, 3, -4, 1 \rangle, \langle -1, 2, -1, 1 \rangle, \langle 2, 1, -1, 1 \rangle\},$
$\mathbf{K}_2 = \{\langle 1, 3, -2, 2 \rangle, \langle 3, 6, -6, 3 \rangle, \langle -5, 9, -8, 2 \rangle\}$

4. Prove that given a vector subspace $\mathbb{S}$ of $\mathbb{R}^n$, $\mathbb{S}^\perp$, the set of all vectors orthogonal to every vector of $\mathbb{S}$, is also a vector subspace of $\mathbb{R}^n$.

5. Construct the spanning sets of the orthogonal subspace $\mathbb{S}^\perp$ to the subspaces defined by the following spanning sets.

(a) $\mathbf{K} = \{\langle -2, 3 \rangle\}$

(b) $\mathbf{K} = \{\langle -2, 3, 3 \rangle\}$

(c) $\mathbf{K} = \{\langle -2, 3, 3 \rangle, \langle 0, 3, -7 \rangle\}$

6. What is the orthogonal subspace to $\mathbb{R}^n$?

7. The following two sets, $\mathbf{K}_1$ and $\mathbf{K}_2$, span the same subspace. Explain what this implies about the vectors of $\mathbf{K}_2$.

$$\mathbf{K}_1 = \{\langle 1, 0, 1 \rangle, \langle 0, 1, 1 \rangle\}, \quad \mathbf{K}_2 = \{\langle 2, -1, 1 \rangle, \langle -1, 5, 4 \rangle, \langle 1, 4, 5 \rangle\}$$

8. Let $\mathbb{S}$ be a subspace of $\mathbb{R}^n$. What is $\mathbb{S} + \mathbb{S}^\perp$? Explain your answer.

9. Let $\mathbb{S}$ be a subspace of $\mathbb{R}^n$. What is $\mathbb{S} \cap \mathbb{S}^\perp$? Explain your answer.

10. Let $\mathbb{S}$ be a subspace of $\mathbb{R}^n$. What is the dimension of $\mathbb{S}^\perp$? Explain your answer.

11. Find the orthogonal complement $\mathbb{S}^\perp$ for the following subspaces $\mathbb{S}$, and give their dimensions:

(a) $\mathbb{S} = \{\langle 0, 0, a, b, 0, c, 0 \rangle \mid a, b, c \in \mathbb{R}\}$

(b) $\mathbb{S} = \{\langle a, 0, b, c, d, 0, 0, e \rangle \mid a, b, c, d, e \in \mathbb{R}\}$

Maple Problems

1. For each part of *Homework* problem 2, place all the vectors from the pairs of spanning sets into one matrix. Row reduce the resulting matrix and interpret the results.

2. Compute the spanning set $\mathbf{K}$ for the subspace $\mathbb{S}$ corresponding to the solution of each of the following homogeneous linear systems.

(a) $\begin{bmatrix} 1 & -2 & 1 & 8 & 0 & -1 \\ 4 & -2 & 8 & 3 & 2 & 3 \end{bmatrix} \vec{x} = \vec{0}$

(b) $\begin{bmatrix} 1 & -2 & 1 & 8 & 0 & -1 \\ 4 & -2 & 8 & 3 & 2 & 3 \\ 5 & -4 & 9 & 11 & 2 & 2 \end{bmatrix} \vec{x} = \vec{0}$

(c) $\begin{bmatrix} 1 & -2 & 1 & 8 & 0 & -1 \\ 4 & -2 & 8 & 3 & 2 & 3 \\ 5 & -5 & 9 & 11 & 1 & 2 \\ 0 & -3 & 5 & 23 & 8 & -1 \end{bmatrix} \vec{x} = \vec{0}$

(d) $\begin{bmatrix} 1 & -2 & 1 & 8 & 0 & -1 \\ 4 & -2 & 8 & 3 & 2 & 3 \\ 2 & 1 & 3 & 7 & 8 & 1 \\ 1 & -1 & 4 & -12 & -6 & 3 \end{bmatrix} \vec{x} = \vec{0}$

(e) $\begin{bmatrix} 1 & -2 & 1 & 8 & 0 & -1 \\ 4 & -2 & 8 & 3 & 2 & 3 \\ 2 & 1 & 3 & 7 & 8 & 1 \\ 2 & -1 & 11 & 3 & 10 & 0 \\ 0 & 0 & 14 & 10 & 18 & 1 \end{bmatrix} \vec{x} = \vec{0}$

(f) $\begin{bmatrix} 1 & -2 & 1 & 8 & 0 & -1 \\ 4 & -2 & 8 & 3 & 2 & 3 \\ 2 & 1 & 3 & 7 & 8 & 1 \\ 2 & -1 & 11 & 3 & 10 & 0 \\ 1 & -1 & 4 & -12 & -6 & 3 \end{bmatrix} \vec{x} = \vec{0}$

3. Define $\mathbb{S}$ to be the subset of $\mathbb{R}^7$ given by

$$\mathbb{S} = \big\{ a \langle -1, 3, 10, -5, -6, 4, -9 \rangle + b \langle 9, 2, -4, 6, 1, 8, -3 \rangle$$
$$+ c \langle 11, 7, 2, -8, 0, -1, 5 \rangle \,\big|\, a, b, c \in \mathbb{R} \big\}$$

Find the orthogonal complement $\mathbb{S}^\perp$ and a spanning set for it. What are the likely dimensions of $\mathbb{S}$ and $\mathbb{S}^\perp$?

4. Define $\mathbb{S}$ to be the subset of $\mathbb{R}^9$ given by

$$\mathbb{S} = \big\{ a \langle -1, 3, 10, -5, -6, 4, -9, 2, -7 \rangle$$
$$+ b \langle 9, 2, -4, 6, 1, 8, -3, -5, 11 \rangle \,\big|\, a, b \in \mathbb{R} \big\}$$

Find the orthogonal complement $\mathbb{S}^\perp$ and a spanning set for it. What are the likely dimensions of $\mathbb{S}$ and $\mathbb{S}^\perp$?

8.3 Basis and Dimension for Subspaces of $\mathbb{R}^n$

This section is the culmination of our discussion of subspaces, spanning sets, and the idea of dimension. In Section 8.2, we saw that the best type of spanning set **K** for a subspace $\mathbb{S}$ of $\mathbb{R}^n$ is one that is an independent set since it eliminates unneeded elements of the spanning set, and so gives a smaller spanning set.

Definition 8.3.1. Let $\mathbb{S}$ be a subspace of $\mathbb{R}^n$. Then a *basis* **B** of $\mathbb{S}$ is a minimal length spanning set for $\mathbb{S}$, that is, **B** is a basis for $\mathbb{S}$ if **B** is a finite spanning set for $\mathbb{S}$ and there exists no smaller length spanning set for $\mathbb{S}$.

Theorem 8.3.1. *Every basis* **B** *of a subspace* $\mathbb{S}$ *of* $\mathbb{R}^n$ *has the same length which is called the* dimension *of* $\mathbb{S}$.

Proof. We will argue that this statement is true using the power of *rref*. Let $\mathbb{S}$ be a subspace of $\mathbb{R}^n$ with two bases $\mathbf{K}_1 = \{\vec{u_1}, \vec{u_2}, \ldots, \vec{u_s}\}$ and $\mathbf{K}_2 = \{\vec{v_1}, \vec{v_2}, \ldots, \vec{v_m}\}$, of lengths s and m, respectively, with $s > m$. Now let us form the two matrices P and Q, where P is the matrix whose rows are the elements of $\mathbf{K}_1$ followed by $\mathbf{K}_2$, while the matrix Q is the matrix whose rows are the elements of $\mathbf{K}_2$ followed by $\mathbf{K}_1$:

$$
P = \begin{bmatrix}
\text{---} & \vec{u_1} & \text{---} \\
 & \vdots & \\
\text{---} & \vec{u_s} & \text{---} \\
\text{---} & \vec{v_1} & \text{---} \\
 & \vdots & \\
\text{---} & \vec{v_m} & \text{---}
\end{bmatrix}, \quad
Q = \begin{bmatrix}
\text{---} & \vec{v_1} & \text{---} \\
 & \vdots & \\
\text{---} & \vec{v_m} & \text{---} \\
\text{---} & \vec{u_1} & \text{---} \\
 & \vdots & \\
\text{---} & \vec{u_s} & \text{---}
\end{bmatrix}
$$

The first thing to notice is that both P and Q are of the same size $(s+m) \times n$. If we apply *rref* to P, the matrix $rref(P)$ has exactly m rows of all zeros since every element of $\mathbf{K}_2$ is a linear combination of the elements of $\mathbf{K}_1$, and $\mathbf{K}_1$ is an independent set. On the other hand, if we apply *rref* to Q, the matrix $rref(Q)$ has exactly s rows of all zeros since every element of $\mathbf{K}_1$ is a linear combination of the elements of $\mathbf{K}_2$, and $\mathbf{K}_2$ is an independent set. But we have already seen that *rref* gives the same matrix independent of the ordering of the rows of the matrix, thus $rref(P)$ is equal to $rref(Q)$. We therefore have a contradiction, since these two *rref* matrices have a different number of zero rows. Hence, $s = m$ and so any two bases of a subspace $\mathbb{S}$ must have the same number of elements. $\qquad\square$

We can therefore conclude that a basis **B** for a subspace $\mathbb{S}$ is a minimal length spanning set of the subspace $\mathbb{S}$. As well, the basis also gives the maximum possible number of linearly independent vectors that can be grouped together at any one time from the subspace $\mathbb{S}$. We leave this last fact as something for the reader to verify. It should also be clear that if you have a spanning set **K** for a subspace $\mathbb{S}$, then you can get a basis for $\mathbb{S}$ from **K** by finding *rref* of the matrix whose rows are the elements of $\mathbf{K}$ and then taking all of the nonzero rows as a basis for $\mathbb{S}$.

Definition 8.3.2. The *row rank* of $A \in \mathbb{R}^{m \times n}$ is the number of nonzero rows in the matrix $rref(A)$ and it is the *dimension* of the subspace $\mathbb{S}$ of $\mathbb{R}^n$ spanned by the rows of the matrix A.

Notationally, the dimension of a subspace $\mathbb{S}$ will be denoted $\dim(\mathbb{S})$ when actually performing a computation of the dimension of a subspace. The two statements: The dimension of $\mathbb{S}$ is k, and $\dim(\mathbb{S}) = k$, are equivalent.

Definition 8.3.3. The subspace $\mathbb{S}$ of $\mathbb{R}^n$ spanned by the rows of the matrix A is called the *row space* of A.

In a similar fashion, the *column rank* of a matrix A is the number of non-zero rows in the matrix $rref(A^T)$, and it is the dimension of the subspace $\mathbb{Q}$ of $\mathbb{R}^m$ spanned by the columns of the matrix A. This subspace $\mathbb{Q}$ is called the *column space* of A.

Example 8.3.1. As an example, let us compute the row and column ranks for the matrix A given by

$$A = \begin{bmatrix} -1 & 5 & 2 & 9 & -3 \\ 4 & 1 & -6 & 0 & 7 \\ 3 & 6 & -4 & 9 & 4 \end{bmatrix}$$

The row rank of A is the dimension, m, of the subspace $\mathbb{S}$ of $\mathbb{R}^5$ spanned by the rows of A. For this example, $1 \leq m \leq 3$. The column rank of A is the dimension, n, of the subspace $\mathbb{Q}$ of $\mathbb{R}^3$ spanned by the columns of A and it is the row rank of A^T. For this example, $1 \leq n \leq 5$, however, as we shall see, n cannot be greater than the largest possible value of m (and vice versa).

```
> with(linalg):
> A:= matrix([[-1, 5, 2, 9, -3], [4, 1, -6, 0, 7], [3, 6, -4, 9, 4]]):
> rref(A);
```

$$\begin{bmatrix} 1 & 0 & -\frac{32}{21} & -\frac{3}{7} & \frac{38}{21} \\ 0 & 1 & \frac{2}{21} & \frac{12}{7} & -\frac{5}{21} \\ 0 & 0 & 0 & 0 & 0 \end{bmatrix}$$

```
> rref(transpose(A));
```

$$\begin{bmatrix} 1 & 0 & 1 \\ 0 & 1 & 1 \\ 0 & 0 & 0 \\ 0 & 0 & 0 \\ 0 & 0 & 0 \end{bmatrix}$$

Note that the row rank of the matrix A is 2, and so is its column rank. Is this a coincidence or not? Let us do another example to test things out.

■

Example 8.3.2. This time, we will use *Maple's* random matrix generator from the *linalg* package called *randmatrix*, which will generate a matrix whose entries are random integers between -99 and 99. Note that if you perform the following commands in *Maple*, your matrix A will look different.

> A:= randmatrix(5,9);

$$A := \begin{bmatrix} -85 & -55 & -37 & -35 & 97 & 50 & 79 & 56 & 49 \\ 63 & 57 & -59 & 45 & -8 & -93 & 92 & 43 & -62 \\ 77 & 66 & 54 & -5 & 99 & -61 & -50 & -12 & -18 \\ 31 & -26 & -62 & 1 & -47 & -91 & -47 & -61 & 41 \\ -58 & -90 & 53 & -1 & 94 & 83 & -86 & 23 & -84 \end{bmatrix}$$

> rref(A);

$$\begin{bmatrix} 1 & 0 & 0 & 0 & 0 & -\frac{2433967181}{1740805298} & -\frac{11876764583}{6963221192} & -\frac{7925117287}{6963221192} & \frac{1082661601}{6963221192} \\ 0 & 1 & 0 & 0 & 0 & \frac{517060525}{1740805298} & \frac{12043417875}{6963221192} & \frac{5451069379}{6963221192} & \frac{1645097739}{6963221192} \\ 0 & 0 & 1 & 0 & 0 & \frac{1267131163}{1740805298} & -\frac{7007636545}{6963221192} & -\frac{1506968601}{6963221192} & -\frac{3285345561}{6963221192} \\ 0 & 0 & 0 & 1 & 0 & \frac{785051345}{1740805298} & \frac{6750387327}{6963221192} & \frac{9398652287}{6963221192} & -\frac{17912744681}{6963221192} \\ 0 & 0 & 0 & 0 & 1 & -\frac{87876116}{870402649} & \frac{927517897}{3481610596} & \frac{1491284665}{3481610596} & -\frac{1158759791}{3481610596} \end{bmatrix}$$

> rref(transpose(A));

$$\begin{bmatrix} 1 & 0 & 0 & 0 & 0 \\ 0 & 1 & 0 & 0 & 0 \\ 0 & 0 & 1 & 0 & 0 \\ 0 & 0 & 0 & 1 & 0 \\ 0 & 0 & 0 & 0 & 1 \\ 0 & 0 & 0 & 0 & 0 \\ 0 & 0 & 0 & 0 & 0 \\ 0 & 0 & 0 & 0 & 0 \\ 0 & 0 & 0 & 0 & 0 \end{bmatrix}$$

For the random matrix A in the above example, notice that both A and A^T have the same row ranks, which also means that A has the same row and column rank, both being equal to 5. It seems not to be a coincidence, and it is indeed true that the row and column ranks of any matrix are equal! In fact, many times, the word *"row"* (or *"column"*) is dropped from the front of the term rank.

■

Definition 8.3.4. The *rank* of matrix A is the dimension of the row space or column space of A.

Now, let us see why this works. Let $A \in \mathbb{R}^{m \times n}$ and $\overrightarrow{u_i} = \langle a_{i,1}, a_{i,2}, \ldots, a_{i,n} \rangle$, for $1 \leq i \leq m$, be the rows of A expressed as vectors in $\mathbb{R}^n$. Next, if we let r be the row rank of A, then there is a basis $\mathbf{K} = \{\overrightarrow{v_1}, \overrightarrow{v_2}, \ldots, \overrightarrow{v_r}\}$ for the subspace $\mathbb{S}$ of $\mathbb{R}^n$ spanned by the rows of the matrix A. We can now write each row vector $\overrightarrow{u_i}$ of A as a linear combination of the $\overrightarrow{v_k}$'s of the basis $\mathbf{K}$. So for each row, we get the following expression:

$$\overrightarrow{u_i} = b_{i,1}\overrightarrow{v_1} + b_{i,2}\overrightarrow{v_2} + \cdots + b_{i,r}\overrightarrow{v_r} \tag{8.12}$$

for scalars $b_{i,1}$ through $b_{i,r}$ for $1 \leq i \leq m$. We will call these our row equations, and from these, we get a matrix B whose entries are the $b_{i,j}$'s. Now let each vector $\overrightarrow{v_j}$, from the basis $\mathbf{K}$, be written as $\overrightarrow{v_j} = \langle c_{j,1}, c_{j,2}, \ldots, c_{j,n} \rangle$ for $1 \leq j \leq r$. Then looking at our row equations in terms of components gives us:

$$\begin{aligned} \langle a_{i,1}, a_{i,2}, \ldots, a_{i,n} \rangle &= b_{i,1}\overrightarrow{v_1} + b_{i,2}\overrightarrow{v_2} + \cdots + b_{i,r}\overrightarrow{v_r} \\ &= b_{i,1}\langle c_{1,1}, c_{1,2}, \ldots, c_{1,n} \rangle + \cdots + b_{i,r}\langle c_{r,1}, c_{r,2}, \ldots, c_{r,n} \rangle \end{aligned}$$

for $1 \leq i \leq m$. Now, by equating components for each i, we get a system of equations for each column of A. For $i = 1$, we have

$$\begin{aligned} a_{1,1} &= b_{1,1}c_{1,1} + b_{1,2}c_{2,1} + \cdots + b_{1,r}c_{r,1} \\ a_{2,1} &= b_{2,1}c_{1,1} + b_{2,2}c_{2,1} + \cdots + b_{2,r}c_{r,1} \\ &\vdots \qquad\qquad\qquad\qquad\qquad\quad \vdots \\ a_{m,1} &= b_{m,1}c_{1,1} + b_{m,2}c_{2,1} + \cdots + b_{m,r}c_{r,1} \end{aligned} \tag{8.13}$$

So if $\overrightarrow{w_1}$ is the first column of the matrix A, then our system of equations (8.13) above is

$$\overrightarrow{w_1} = c_{1,1}\overrightarrow{B_1} + c_{2,1}\overrightarrow{B_2} + \cdots + c_{r,1}\overrightarrow{B_r} \tag{8.14}$$

where $\overrightarrow{B_j}$ is the jth column of the matrix B. Similarly, by equating second components for $1 \leq i \leq m$, we have

$$\overrightarrow{w_2} = c_{1,2}\overrightarrow{B_1} + c_{2,2}\overrightarrow{B_2} + \cdots + c_{r,2}\overrightarrow{B_r}$$

where $\overrightarrow{w_2}$ is the second column of the matrix A. Thus, for the columns of A given by the set $\{\overrightarrow{w_1}, \overrightarrow{w_2}, \ldots, \overrightarrow{w_n}\}$, we have

$$\overrightarrow{w_j} = c_{1,j}B_1 + c_{2,j}B_2 + \cdots + c_{r,j}B_r \tag{8.15}$$

for $1 \leq j \leq n$. Now we know that the vector columns of B, given by the set $\{\overrightarrow{B_1}, \overrightarrow{B_2}, \ldots, \overrightarrow{B_r}\}$, are a spanning set for the subspace $\mathbb{Q}$ of $\mathbb{R}^m$ spanned by the columns of the matrix A. So we now know that column rank of A is less than or equal to the row rank of A. If we use A^T instead of A as above, then by a similar argument, we get that the column rank of A^T is less than or equal to the row rank of A^T, or that the row rank of A is less than or equal to the column rank of A. Thus, the two ranks are equal.

Theorem 8.3.2. *Every element of a subspace* $\mathbb{S}$ *can be written in exactly one way as a linear combination of the elements of a basis.*

Proof. Let $\mathbf{B} = \{\overrightarrow{u_1}, \overrightarrow{u_2}, \dots, \overrightarrow{u_s}\}$ be a basis for a subspace $\mathbb{S}$. For an arbitrary $\overrightarrow{w} \in \mathbb{S}$, since $\mathbf{B}$ spans $\mathbb{S}$, we have that:

$$\overrightarrow{w} = a_1\overrightarrow{u_1} + a_2\overrightarrow{u_2} + \cdots + a_s\overrightarrow{u_s} \tag{8.16}$$

for scalars a_k, $1 \le k \le s$. Now, let us assume that, for an instant, the vector $\overrightarrow{w}$ can be written as a different linear combination of the basis vectors:

$$\overrightarrow{w} = b_1\overrightarrow{u_1} + b_2\overrightarrow{u_2} + \cdots + b_s\overrightarrow{u_s} \tag{8.17}$$

for scalars b_k's with $1 \le k \le s$. Now, if we subtract equation (8.17) from (8.16) we get

$$(a_1 - b_1)\,\overrightarrow{u_1} + (a_2 - b_2)\,\overrightarrow{u_2} + \cdots + (a_s - b_s)\,\overrightarrow{u_s} = \overrightarrow{0} \tag{8.18}$$

The independence of the basis $\mathbf{B}$ now tells us that all of these coefficients are 0, and so $a_j = b_j$ for $1 \le j \le s$. Thus, there is exactly one way of writing each element $\overrightarrow{w}$ of a subspace $\mathbb{S}$ as a linear combination in a given basis $\mathbf{B}$ of $\mathbb{S}$. $\square$

Theorem 8.3.2 is sometimes used as the definition of a basis, as it is equivalent to combining the properties of independence and spanning. One other interesting fact about how bases work, is that if $\mathbb{S}$ is a subspace of $\mathbb{R}^n$ and has dimension k, then any linearly independent subset of $\mathbb{S}$ having k elements is automatically a basis of $\mathbb{S}$. This also implies that any spanning subset of $\mathbb{S}$ having k elements is also automatically a basis of $\mathbb{S}$. We now give an example.

Example 8.3.3. Define $\mathbf{B}$ to be the subset of $\mathbb{R}^3$ given by the three vectors $\overrightarrow{u} = \langle -1, 5, -2 \rangle$, $\overrightarrow{v} = \langle 7, -4, 3 \rangle$, and $\overrightarrow{w} = \langle 10, 6, -9 \rangle$. Since there are three vectors in $\mathbf{B}$, if $\mathbf{B}$ is an independent set then it will be a basis of $\mathbb{R}^3$. So by defining A to be the matrix whose rows are the three vectors from $\mathbf{B}$, we can compute $rref(A)$. If the result is I_3, then $\mathbf{B}$ really is a basis. Here, A is explicitly written as

$$A = \begin{bmatrix} -1 & 5 & -2 \\ 7 & -4 & 3 \\ 10 & 6 & -9 \end{bmatrix}$$

> u:= [-1, 5, -2]: v:= [7, -4, 3]: w:= [10, 6, -9]:
> A:= matrix([u,v,w]):

> rref(A);

$$\begin{bmatrix} 1 & 0 & 0 \\ 0 & 1 & 0 \\ 0 & 0 & 1 \end{bmatrix}$$

Since **B** truly is a basis, then we can find the unique scalars a, b, and c so that $\vec{s} = a\vec{u} + b\vec{v} + c\vec{w}$ for $\vec{s} = \langle 8, 0, -5 \rangle$. Solving for the unknown constants, we have

$$\begin{bmatrix} a \\ b \\ c \end{bmatrix} = \left(A^T \right)^{-1} \vec{s} \tag{8.19}$$

where $\vec{s}$ is written as a column.

> s:= vector([8, 0, -5]):
> rref(augment(transpose(A),s));

$$\begin{bmatrix} 1 & 0 & 0 & -\frac{266}{283} \\ 0 & 1 & 0 & -\frac{16}{283} \\ 0 & 0 & 1 & \frac{211}{283} \end{bmatrix}$$

> evalm(inverse(transpose(A))&*s);

$$\begin{bmatrix} -\dfrac{266}{283} & -\dfrac{16}{283} & \dfrac{211}{283} \end{bmatrix}$$

So we now have that $a = -\dfrac{266}{283}$, $b = -\dfrac{16}{283}$ and $c = \dfrac{211}{283}$. Let us verify our solution below.

> evalm(-266/283*vector(u)-16/283*vector(v)+211/283*w);

$$\begin{bmatrix} 8 & 0 & -5 \end{bmatrix}$$

Note that if we write the unknown constants a, b, c as a row vector (matrix of dimension 1×3) instead, we end up with the equation $\vec{s} = \langle a, b, c \rangle A$. We can solve for the unknown constants simply enough:

$$\langle a, b, c \rangle = \vec{s} A^{-1} \tag{8.20}$$

We have *Maple* perform the computations below. Notice that the solution remains the same.

> evalm(s&*inverse(A));

$$\begin{bmatrix} -\dfrac{266}{283} & -\dfrac{16}{283} & \dfrac{211}{283} \end{bmatrix}$$

We conclude this section with two lists of equivalent statements. The first is a list of equivalent statements for *A set* **B** *of k elements to be a basis of a subspace* $\mathbb{S}$ *of* $\mathbb{R}^n$. (Note that if $k = n$, then $\mathbb{S} = \mathbb{R}^n$.)

A set **B** of k elements is a basis of a subspace $\mathbb{S}$ of $\mathbb{R}^n$.

(a) **B** is an independent spanning set for the subspace $\mathbb{S}$ of $\mathbb{R}^n$.

(b) **B** is a spanning set of k elements for the subspace $\mathbb{S}$ of $\mathbb{R}^n$.

(c) **B** is an independent set of k elements in the subspace $\mathbb{S}$ of $\mathbb{R}^n$.

(d) **B** is a minimal size spanning set for the subspace $\mathbb{S}$ of $\mathbb{R}^n$.

(e) **B** is a maximal size independent subset of the subspace $\mathbb{S}$ of $\mathbb{R}^n$.

(f) The $k \times n$ matrix $M_{\mathbf{B}}$, whose rows are the elements of the set **B**, has rank k.

The second list of equivalent statements is for: *The square $n \times n$ linear system $A\vec{x} = \vec{b}$ has a unique solution for $\vec{x}$.*

The square $n \times n$ linear system $A\vec{x} = \vec{b}$ has a unique solution for $\vec{x}$.

(a) The $n \times n$ matrix A has an inverse A^{-1}, which gives the unique solution $\vec{x} = A^{-1}\vec{b}$.

(b) $\det(A) \neq 0$.

(c) $\text{rref}(A) = I_n$.

(d) The rank of A is n.

(e) The rank of A^T is n.

(f) The linear system $A\vec{x} = \vec{0}$ only has the unique solution $\vec{x} = \vec{0}$.

(g) The rows (or columns) of A form an independent subset of $\mathbb{R}^n$.

(h) The rows (or columns) of A form a spanning set of $\mathbb{R}^n$.

Homework Problems

1. Determine which of the following sets are bases for $\mathbb{R}^2$:

(a) $\{\langle 2, 3 \rangle, \langle 2, 1 \rangle\}$ (b) $\{\langle -2, 3 \rangle, \langle -3, 1 \rangle\}$ (c) $\{\langle 1, -2 \rangle, \langle -3, 6 \rangle\}$

(d) $\{\langle 0, 2 \rangle, \langle 1, 4 \rangle\}$

2. Determine which of the following sets are bases for $\mathbb{R}^3$:

(a) $\{\langle 2, 3, 1\rangle, \langle 0, 2, 1\rangle, \langle -1, 2, 1\rangle\}$ (b) $\{\langle -2, 0, 1\rangle, \langle 0, 2, 0\rangle, \langle 0, 0, 5\rangle\}$

(c) $\{\langle 1, 1, 0\rangle, \langle 0, 1, 1\rangle, \langle 1, 0, 1\rangle\}$ (d) $\{\langle 3, -2, 2\rangle, \langle 1, -1, 0\rangle, \langle -5, 3, -4\rangle\}$

3. Compute the row rank of the following matrices:

(a) $\begin{bmatrix} 1 & 3 & 1 \\ -2 & 1 & -2 \\ 0 & 8 & 1 \end{bmatrix}$ (b) $\begin{bmatrix} -3 & -4 & 3 \\ 4 & 1 & 2 \\ 5 & -2 & 7 \end{bmatrix}$ (c) $\begin{bmatrix} -3 & -4 & 3 \\ 4 & 1 & 2 \\ 5 & -2 & 7 \\ 3 & -2 & 1 \end{bmatrix}$

(d) $\begin{bmatrix} -3 & -4 & 3 \\ 2 & 2 & 10 \\ 5 & -2 & 7 \\ 8 & -6 & 4 \end{bmatrix}$ (e) $\begin{bmatrix} -3 & -4 \\ 4 & 1 \\ 5 & -2 \end{bmatrix}$ (f) $\begin{bmatrix} -3 & -4 & 3 & 0 \\ 4 & 1 & 2 & -1 \\ 5 & -2 & 7 & 3 \end{bmatrix}$

4. Determine the maximum possible row rank for each of the matrices from problem 3.

5. Construct a basis for each of the following subspaces of $\mathbb{R}^n$.

(a) the set of all vectors in $\mathbb{R}^3$ of the form $\langle a, b, a\rangle$

(b) the set of all vectors in $\mathbb{R}^4$ of the form $\langle a, b, -a, -b\rangle$

(c) the set of all vectors in $\mathbb{R}^3$ of the form $\langle a, b, a - b\rangle$

(d) the set of all vectors in $\mathbb{R}^4$ of the form $\langle a, 2b, a - 3b, 2a + 3b + c\rangle$

6. Let $C = (A \,|\, 0)$ be the augmented matrix for the homogeneous linear system $A\overrightarrow{x} = \overrightarrow{0}$, where $A \in \mathbb{R}^{m \times n}$. Now apply $rref$ to this matrix C in order to read off the solutions $\overrightarrow{x}$ to this system. You will get from rref(C) that the solutions $\overrightarrow{x}$ are of the form

$$\overrightarrow{x} = x_{k_1}\overrightarrow{u_1} + x_{k_2}\overrightarrow{u_2} + \cdots + x_{k_p}\overrightarrow{u_p}$$

where $x_{k_1}, x_{k_1}, \ldots, x_{k_p}$ are arbitrary solution variables from $\overrightarrow{x}$, and $\overrightarrow{u_1}, \overrightarrow{u_2}, \ldots,$ $\overrightarrow{u_p} \in \mathbb{R}^n$ are p fixed solutions. Are the column vectors $\overrightarrow{u_1}, \overrightarrow{u_2}, \ldots, \overrightarrow{u_p}$ automatically a basis of the subspace $\mathbb{S}$ of $\mathbb{R}^n$ consisting of all the solutions $\overrightarrow{x}$ to the homogeneous system $A\overrightarrow{x} = \overrightarrow{0}$? Explain your answer in detail.

7. (a) Let $\mathbb{S}$ be a subspace of $\mathbb{R}^n$ with a basis $\mathbf{B} = \{\vec{u_1}, \vec{u_2}, \ldots, \vec{u_p}\}$. Explain in detail how you can find a basis for the orthogonal complement $\mathbb{S}^{\perp}$.

(b) What is the dimension of $\mathbb{S}^{\perp}$ in terms of n and the dimension p of $\mathbb{S}$ Explain why this is true.

8. For the following subspaces $\mathbb{S}$ of $\mathbb{R}^n$, find a basis for both $\mathbb{S}$ and its orthogonal complement $\mathbb{S}^{\perp}$, giving the dimension of each.

(a) $\mathbb{S} = \{\langle 0, 0, a, b, 0, c, 0 \rangle \,|\, a, b, c \in \mathbb{R}^7 \}$

(b) $\mathbb{S} = \{\langle a, 0, b, c, d, 0, 0, e \rangle \,|\, a, b, c, d, e \in \mathbb{R}^7 \}$

Maple Problems

1. Determine if the following sets of vectors are linearly independent:

(a) $\{\langle -1, 2, 1 \rangle, \langle -3, 4, 0 \rangle, \langle -1, 1, 1 \rangle\}$

(b) $\{\langle -1, 2, 0, 1 \rangle, \langle -3, 4, 0, 2 \rangle, \langle -1, 1, 1, 0 \rangle, \langle 0, 2, 1, 0 \rangle\}$

(c) $\{\langle -1, 2, 0, 1 \rangle, \langle -3, 4, 0, 2 \rangle, \langle -1, 1, 1, 0 \rangle, \langle 3, -2, 1, -1 \rangle, \langle -2, -1, -4, 2 \rangle\}$

(d) $\{\langle -1, 2, 0, 1, 0 \rangle, \langle -3, 1, 4, 0, 2 \rangle, \langle -1, 1, 3, 1, 0 \rangle, \langle 2, 1, 2, 1, 0 \rangle\}$

(e) $\{\langle -1, 2, 0, 1, 0 \rangle, \langle -3, 1, 4, 0, 2 \rangle, \langle -1, 1, 3, 1, 0 \rangle, \langle 2, 1, 2, 1, 0 \rangle, \langle -3, 5, 9, 3, 2 \rangle\}$

2. Compute the row rank of the following matrices:

(a) $\begin{bmatrix} 1 & 3 & 1 & -2 \\ -2 & 1 & -2 & 2 \\ 0 & 8 & 1 & 3 \\ 2 & -7 & 5 & 1 \end{bmatrix}$

(b) $\begin{bmatrix} -3 & -4 & 3 \\ 4 & 1 & 2 \\ -4 & 6 & 7 \\ 2 & -5 & 8 \\ 8 & 2 & -2 \\ 0 & 3 & 3 \end{bmatrix}$

(c) $\begin{bmatrix} -3 & -4 & 3 & 3 \\ 4 & 1 & 2 & -2 \\ 1 & -3 & 5 & 1 \\ -2 & -7 & 8 & 4 \\ 0 & -13 & 18 & 6 \\ -1 & -10 & 13 & 5 \end{bmatrix}$

(d) $\begin{bmatrix} -3 & -4 & 3 & 3 & 5 & 1 \\ -2 & 3 & 0 & 3 & -2 & 3 \\ 1 & -10 & 3 & -3 & 9 & -5 \end{bmatrix}$

(e) $\begin{bmatrix} -3 & -4 & 3 & 3 & 5 & 1 \\ -2 & 3 & 0 & 3 & -2 & 3 \\ -4 & 6 & 0 & 6 & -4 & 6 \\ 1 & -10 & 3 & -3 & 9 & -5 \end{bmatrix}$

(f) $\begin{bmatrix} -3 & -4 \\ 4 & 1 \\ 5 & -2 \\ -2 & 1 \\ 0 & -2 \\ 3 & 0 \end{bmatrix}$

3. Compute the row rank of the transpose of each of the matrices from problem 2 and compare your answers to those of problem 2.

4. Let

$$\mathbb{S} = \big\{ a \, \langle 2, -3, 7, -5, -6, 4, -9 \rangle + b \, \langle -1, 8, -4, 6, 0, 8, -3 \rangle$$
$$+ \, c \, \langle 9, -7, 31, -19, -30, 28, -48 \rangle \, \big| \, a, b, c \in \mathbb{R} \big\}$$

Find a basis and the dimension for both $\mathbb{S}$ and its orthogonal complement $\mathbb{S}^{\perp}$.

5. Let
$$\mathbb{S} = \big\{ a \, \langle -1, 3, 10, -5, 7 \rangle + b \, \langle 9, 2, -4, 6, -8 \rangle \, \big| \, a, b \in \mathbb{R} \big\}$$
Find a basis and the dimension for both $\mathbb{S}$ and its orthogonal complement $\mathbb{S}^{\perp}$.

8.4   Vector Projection onto a Subspace of $\mathbb{R}^n$

Since we now know about subspaces, bases, and projections, we can combine all these concepts and will now generalize the process of vector projection of one vector $\vec{v}$ onto another vector $\vec{w}$ to that of projecting one vector $\vec{v}$ onto an entire subspace $\mathbb{S}$ of $\mathbb{R}^n$. Projecting $\vec{v}$ onto $\vec{w}$ resulted in a vector in the direction of $\vec{w}$. The vector projection $\text{proj}_{\mathbb{S}}(\vec{v})$ should be the vector of the subspace $\mathbb{S}$, where $\vec{v} - \text{proj}_{\mathbb{S}}(\vec{v})$ is orthogonal to the subspace $\mathbb{S}$, that is, every element $\vec{w}$ of $\mathbb{S}$ is orthogonal to $\vec{v} - \text{proj}_{\mathbb{S}}(\vec{v})$. In terms of dot products, this implies that:
$$\vec{w} \cdot (\vec{v} - \text{proj}_{\mathbb{S}}(\vec{v})) = 0 \qquad (8.21)$$

for all $\vec{w} \in \mathbb{S}$. By the linearity of the dot product, it is enough that the above formula hold for all $\vec{w} \in \mathbf{B}$, where $\mathbf{B}$ is any basis of the subspace $\mathbb{S}$. It may not seem like this condition will determine the vector projection $\text{proj}_{\mathbb{S}}(\vec{v})$ uniquely, but as we shall see next, it does indeed. Furthermore, we should also get that $\text{proj}_{\mathbb{S}}(\vec{v}) = \vec{v}$, whenever $\vec{v} \in \mathbb{S}$.

We now derive the process for projecting a vector onto a subspace of $\mathbb{R}^n$. From the previous section, we are guaranteed that the subspace $\mathbb{S}$ of $\mathbb{R}^n$ has a basis $\mathbf{B} = \{\vec{w_1}, \vec{w_2}, \ldots, \vec{w_k}\}$. If we choose a fixed $\vec{v} \in \mathbb{R}^n$, then

$$\vec{w_i} \cdot (\vec{v} - \text{proj}_{\mathbb{S}}(\vec{v})) = 0 \tag{8.22}$$

Using the fact that $\vec{u} \cdot (\vec{v} - \vec{w}) = \vec{u} \cdot \vec{v} - \vec{u} \cdot \vec{w}$, equation (8.22) can be rewritten as

$$\vec{w_i} \cdot \text{proj}_{\mathbb{S}}(\vec{v}) = \vec{w_i} \cdot \vec{v} \tag{8.23}$$

for all $\vec{w_i} \in \mathbf{B}$. Since $\text{proj}_{\mathbb{S}}(\vec{v}) \in \mathbb{S}$, it can be written as a linear combination of the vectors of the basis $\mathbf{B}$. So we can define $\text{proj}_{\mathbb{S}}(\vec{v})$ as

$$\text{proj}_{\mathbb{S}}(\vec{v}) = a_1\vec{w_1} + a_2\vec{w_2} + \cdots + a_k\vec{w_k} = \sum_{j=1}^{k} a_j\vec{w_j} \tag{8.24}$$

for unique, and so far unknown, scalars $a_1, a_2, \ldots, a_k$. Plugging this into equation (8.23), and using the linearity of the dot product, gives the equation:

$$\sum_{j=1}^{k} a_j\vec{w_i} \cdot \vec{w_j} = \vec{w_i} \cdot \vec{v} \tag{8.25}$$

for $1 \leq i \leq k$. We now need to determine the values of the a_j's that will satisfy these equations. To simplify matters, we will express the above k equations in matrix form. First, we explicitly write out the equations:

$$\begin{aligned} a_1\vec{w_1} \cdot \vec{w_1} + a_2\vec{w_1} \cdot \vec{w_2} + \cdots + a_k\vec{w_1} \cdot \vec{w_k} = \vec{w_1} \cdot \vec{v} \\ a_1\vec{w_2} \cdot \vec{w_1} + a_2\vec{w_2} \cdot \vec{w_2} + \cdots + a_k\vec{w_2} \cdot \vec{w_k} = \vec{w_2} \cdot \vec{v} \\ \vdots \qquad\qquad\qquad \vdots \\ a_1\vec{w_k} \cdot \vec{w_1} + a_2\vec{w_k} \cdot \vec{w_2} + \cdots + a_k\vec{w_k} \cdot \vec{w_k} = \vec{w_k} \cdot \vec{v} \end{aligned} \tag{8.26}$$

Next, we recognize that the LHS of the system given in equation (8.26) can be expressed as a matrix multiplication, so we now have the following expression:

$$\begin{bmatrix} \vec{w_1} \cdot \vec{w_1} & \vec{w_1} \cdot \vec{w_2} & \cdots & \vec{w_1} \cdot \vec{w_k} \\ \vec{w_2} \cdot \vec{w_1} & \vec{w_2} \cdot \vec{w_2} & \cdots & \vec{w_2} \cdot \vec{w_k} \\ \vdots & \vdots & \ddots & \vdots \\ \vec{w_k} \cdot \vec{w_1} & \vec{w_k} \cdot \vec{w_2} & \cdots & \vec{w_k} \cdot \vec{w_k} \end{bmatrix} \begin{bmatrix} a_1 \\ a_2 \\ \vdots \\ a_k \end{bmatrix} = \begin{bmatrix} \vec{w_1} \cdot \vec{v} \\ \vec{w_2} \cdot \vec{v} \\ \vdots \\ \vec{w_k} \cdot \vec{v} \end{bmatrix} \tag{8.27}$$

So, now we see that if A is the $k \times 1$ column (vector) of the scalar a_j's, then $WA = V$ where the $k \times k$ matrix W is defined entry wise by $W_{i,j} = \overrightarrow{w_i} \cdot \overrightarrow{w_j}$, and the column matrix (vector) V is defined similarly by $V_i = \overrightarrow{w_i} \cdot \overrightarrow{v}$. This matrix equation is now simple to solve via matrix arithmetic. By using the inverse of W, we have $A = W^{-1}V$.

We will now perform some matrix manipulations to simplify the process of computing $\mathrm{proj}_\mathbb{S}(\overrightarrow{v})$. First, we note that we can express the projection as

$$\mathrm{proj}_\mathbb{S}(\overrightarrow{v}) = \langle a_1, a_2, \ldots, a_k \rangle \cdot \langle \overrightarrow{w_1}, \overrightarrow{w_2}, \ldots, \overrightarrow{w_k} \rangle \tag{8.28}$$

Next, we will use the following fact: If $\overrightarrow{u}, \overrightarrow{v} \in \mathbb{R}^n$ are column vectors, the dot product $\overrightarrow{u} \cdot \overrightarrow{v}$ can be expressed as the matrix multiplication $U^T V$, where U and V are the column matrices corresponding to $\overrightarrow{u}$ and $\overrightarrow{v}$, respectively. Thus, the formula given in equation (8.28) can be expressed in terms of the matrix multiplication:

$$\mathrm{proj}_\mathbb{S}(\overrightarrow{v}) = \begin{bmatrix} a_1 & a_2 & \cdots & a_k \end{bmatrix} \begin{bmatrix} \overrightarrow{w_1} \\ \overrightarrow{w_2} \\ \vdots \\ \overrightarrow{w_k} \end{bmatrix} = A^T M_\mathbf{B} \tag{8.29}$$

where A is the column matrix corresponding to the column vector whose components are the a_j's, and $M_\mathbf{B}$ is the matrix whose rows are the $\overrightarrow{w_j}$ vectors. Now remember, we are given the $\overrightarrow{w_j}$ vectors, so if we can use these vectors, along with the vector $\overrightarrow{v}$ exclusively, in our formula, it would make the process of computing $\mathrm{proj}_\mathbb{S}(\overrightarrow{v})$ programmatically simple. To this end we define $C = M_\mathbf{B}^T$, explicitly given by

$$C = \begin{bmatrix} \overrightarrow{w_1} & \overrightarrow{w_2} & \cdots & \overrightarrow{w_k} \end{bmatrix} \tag{8.30}$$

As a result of (8.27), we know that $A = W^{-1}V$. Furthermore, since W is the dot product matrix, we can express W as $W = C^T C$, and in a similar fashion, $V = C^T \overrightarrow{v}$. Putting all of these facts together gives

$$\begin{aligned} \mathrm{proj}_\mathbb{S}(\overrightarrow{v}) &= A^T M_\mathbf{B} \\ &= \left(W^{-1}V\right)^T C^T \\ &= \left(C\left(W^{-1}V\right)\right)^T \\ &= \left(C\left(C^T C\right)^{-1} C^T \overrightarrow{v}\right)^T \end{aligned} \tag{8.31}$$

Notice that the above formula involves only the matrix C, and the vector $\overrightarrow{v}$. C, as defined in (8.30), is simply the matrix whose columns are the basis vectors

$\overrightarrow{w_j}$, for $1 \leq j \leq k$. If we wish to express $\text{proj}_\mathbb{S}(\overrightarrow{v})$ as a row vector, notice that we can remove the outer transpose in the last line of equation (8.31):

$$\text{proj}_\mathbb{S}(\overrightarrow{v})^T = C\left(C^T C\right)^{-1} C^T \overrightarrow{v} \tag{8.32}$$

In the following examples, we will use the row form of the projection vector given in (8.32).

Example 8.4.1. As a first example, we define $\mathbb{S}$ to be the subspace of $\mathbb{R}^3$ with basis $\mathbf{B} = \{\overrightarrow{w_1} = \langle 5, -2, 9 \rangle, \overrightarrow{w_2} = \langle -3, 7, 1 \rangle\}$. Let us find and plot $\text{proj}_\mathbb{S}(\overrightarrow{v})$ for $\overrightarrow{v} = \langle 4, 8, -15 \rangle$. We will use *rref* to see that $\overrightarrow{v}$ is not in $\mathbb{S}$. Note from the work below that the dot product works equally well on lists as on vectors. Later on in this section, the same is shown to be true for cross products.

```
> with(linalg):
> w[1]:= [5, -2, 9]: w[2]:= [-3, 7, 1]:
> w_ row:= matrix([w[1], w[2]]);
```

$$\begin{bmatrix} 5 & -2 & 9 \\ -3 & 7 & 1 \end{bmatrix}$$

```
> v:= [4, 8, -15]:
> rref(matrix([w[1],w[2],v]));
```

$$\begin{bmatrix} 1 & 0 & 0 \\ 0 & 1 & 0 \\ 0 & 0 & 1 \end{bmatrix}$$

```
> W:= matrix(2,2,0);
```

$$\begin{bmatrix} 0 & 0 \\ 0 & 0 \end{bmatrix}$$

```
> for i from 1 to 2 do
      for j from 1 to 2 do
         W[i,j]:= dotprod(w[i],w[j])
      od:
   od:
> print(W);
```

$$\begin{bmatrix} 110 & -20 \\ -20 & 59 \end{bmatrix}$$

```
> V:= matrix(2,1,0);
```

$$V := \begin{bmatrix} 0 \\ 0 \end{bmatrix}$$

> for i from 1 to 2 do
  V[i,1]:= dotprod(w[i],v)
  od:
> print(V);

$$V := \begin{bmatrix} -131 \\ 29 \end{bmatrix}$$

> A:= evalm(inverse(W)&*V);

$$A := \begin{bmatrix} -\frac{2383}{2030} \\ \frac{19}{203} \end{bmatrix}$$

> Proj_vontoS:= add(A[i,1]*w[i], i=1..2);

$$Proj\_vontoS := \begin{bmatrix} -\frac{2497}{406}, & \frac{3048}{1015}, & -\frac{733}{70} \end{bmatrix}$$

> dotprod(v-Proj_vontoS,w[1]);

$$0$$

> dotprod(v-Proj_vontoS,w[2]);

$$0$$

Let us now see that the formula (8.31) gives the same solution as the previous method.

> C:= transpose(w_row);

$$C := \begin{bmatrix} 5 & -3 \\ -2 & 7 \\ 9 & 1 \end{bmatrix}$$

> NewFormula_Proj_vontoS:= evalm(C&*inverse(transpose(C)&*C)&*transpose(C)&*v);

$$NewFormula\_Proj\_vontoS := \begin{bmatrix} -\frac{2497}{406} & \frac{3048}{1015} & -\frac{733}{70} \end{bmatrix}$$

Everything checks out and $\vec{v} - \text{proj}_\mathbb{S}(\vec{v})$ is orthogonal to the subspace $\mathbb{S}$, which in this case is the plane through the origin spanned by the basis **B** as defined previously. Now we should plot all of this in $\mathbb{R}^3$ to see it geometrically.

> with(plots): with(plottools):
> CP:= crossprod(w[1],w[2]);

$$CP := \begin{bmatrix} -65 & -32 & 29 \end{bmatrix}$$

> Plane_S:= CP[1]*x + CP[2]*y + CP[3]*z = 0;

$$Plane\_S := -65x - 32y + 29z = 0$$

> plot_S:= plot3d((65*x+32*y)/29, x = -10..10, y = -10..10, color = red):

> arrow_w[1]:= arrow([0, 0, 0], w[1], .4, 1.2, .2, cylindrical_arrow, color = black):

> arrow_w[2]:= arrow([0, 0, 0], w[2], .4, 1.2, .2, cylindrical_arrow, color = black):

> arrow_v:= arrow([0, 0, 0], v, .4, 1.2, .2, cylindrical_arrow, color = black):

> arrow_proj:= arrow([0, 0, 0], Proj_vontoS, .4, 1.2, .2, cylindrical_arrow, color = blue):

> arrow_vminusproj:= arrow(Proj_vontoS, v, .4, 1.2, .2, cylindrical_arrow, color = blue):

> text_w[1]:= textplot3d([op(w[1]),"w[1]"], align = BELOW, color = black):

> text_w[2]:= textplot3d([op(w[2]),"w[2]"], align = BELOW, color = black):

> text_v:= textplot3d([op(.5*v),"V"], align = {ABOVE, RIGHT}, color = black):

> text_proj:= textplot3d([op(Proj_vontoS),"Proj"], align = ABOVE, color = black):

> text_vminusproj:=textplot3d([op(.5*(v+Proj_vontoS)),"V-Proj"], align = LEFT, color = black):

> display({plot_S, arrow_w[1], arrow_w[2], arrow_v, arrow_proj, arrow_vminusproj, text_w[1], text_w[2], text_v, text_proj, text_vminusproj}, axes = boxed, style = patchnogrid, view = [-15..15, -15..15, -15..15], scaling = constrained);

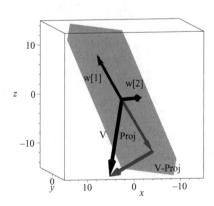

Figure 8.3: The projection of a vector $\vec{v}$ onto a plane in $\mathbb{R}^3$

In Figure 8.3, the two vectors $\overrightarrow{w_1}$ and $\overrightarrow{w_2}$ form the basis for the plane. The triangle that has one edge on the plane has hypotenuse $\overrightarrow{v}$, the edge in the plane is the projection onto the plane of $\overrightarrow{v}$, while the edge normal to the plane is the difference between $\overrightarrow{v}$ and the projection. The distance from the terminal point of $\overrightarrow{v}$ to the plane defined by $\mathbb{S}$ previously is the length of the vector $\overrightarrow{v} - \text{proj}_\mathbb{S}(\overrightarrow{v})$, since it is the shortest distance from the point $\overrightarrow{v}$ to any point of the plane $\mathbb{S}$. This also holds true for any more generalized subspace $\mathbb{S}$ of $\mathbb{R}^n$ and any point $\overrightarrow{v}$ of $\mathbb{R}^n$.

> dist_vtoS:= evalf(sqrt(dotprod(v-Proj_vontoS, v-Proj_vontoS)));

$$dist\_vtoS := 12.18630015$$

What we have done in the command above is compute the shortest distance from the point to the plane. If you recall, this idea was discussed in Section 6.4. Remember that any plane in $\mathbb{R}^3$ has equation $R : ax + by + cz = d$, where the vector $\overrightarrow{n} = \langle a, b, c \rangle$ is the normal vector to the plane. The shortest distance from a point $P(x_0, y_0, z_0)$ to this plane should have the formula:

$$D(P, R) = \frac{|ax_0 + by_0 + cz_0 - d|}{\sqrt{a^2 + b^2 + c^2}}$$

Let us check this formula's value for the plane $\mathbb{S}$ through the origin and parallel to both $\overrightarrow{w_1}$ and $\overrightarrow{w_2}$ and the terminal point of $\overrightarrow{v}$ of the last example. We can take the normal vector $\overrightarrow{n}$ to this plane to be $\overrightarrow{n} = \overrightarrow{w_1} \times \overrightarrow{w_2}$, which is called CP in the *Maple* code above and below.

> dist:= evalf(abs(dotprod(CP,v))/sqrt(dotprod(CP,CP)));

$$dist := 12.18630015$$

In doing this vector projection, we computed the vector projection $\text{proj}_\mathbb{S}(\overrightarrow{v})$ of $\overrightarrow{v}$ onto the subspace $\mathbb{S}$ using any basis **B** of $\mathbb{S}$. One may wonder if some bases are "better" than others for a vector space. If so, what properties would be desired for vectors in a "better" basis?

Definition 8.4.1. A basis $\mathbf{B} = \{\overrightarrow{w_1}, \overrightarrow{w_2}, \ldots, \overrightarrow{w_k}\}$ of a subspace $\mathbb{S}$ is said to be *orthogonal* if $\overrightarrow{w_i} \cdot \overrightarrow{w_j} = 0$ for all $1 \leq i \neq j \leq k$.

Definition 8.4.2. A basis $\mathbf{B} = \{\overrightarrow{w_1}, \overrightarrow{w_2}, \ldots, \overrightarrow{w_k}\}$ of a subspace $\mathbb{S}$ is said to be *orthonormal* if it is an orthogonal basis with the added property that each vector has unit length, that is, $|\overrightarrow{w_i}|^2 = \overrightarrow{w_i} \cdot \overrightarrow{w_i} = 1$ for $1 \leq i \leq k$.

As an obvious example, the standard basis for $\mathbb{R}^n$ is an orthonormal basis, and is given by the set:

$$\{\langle 1,0,0,\ldots,0\rangle, \langle 0,1,0,\ldots,0\rangle, \ldots, \langle 0,0,\ldots,0,1\rangle\}$$

Now one must ask the question: Why is this a "better", or perhaps even "best", basis to use? Relating this basis to our discussion of projecting vectors onto subspaces, notice that the $k \times k$ matrix W with $W_{i,j} = \overrightarrow{w_i} \cdot \overrightarrow{w_j}$ is the $k \times k$ identity matrix I_k if the basis $\mathbf{B}$ is orthonormal, and as a result, $C^T C = I_k$. This holds true not just for vectors chosen from the standard basis, but also for vectors chosen from any orthonormal basis. The vector projection of $\overrightarrow{v}$ onto $\mathbb{S}$ simplifies quite a bit to

$$\text{proj}_{\mathbb{S}}(\overrightarrow{v}) = (\overrightarrow{w_1} \cdot \overrightarrow{v})\,\overrightarrow{w_1} + (\overrightarrow{w_2} \cdot \overrightarrow{v})\,\overrightarrow{w_2} + \cdots + (\overrightarrow{w_k} \cdot \overrightarrow{v})\,\overrightarrow{w_k}$$

$$= \sum_{j=1}^{k} (\overrightarrow{w_j} \cdot \overrightarrow{v})\,\overrightarrow{w_j} \tag{8.33}$$

Referring back to the matrix expression of the projection formula, we see that under the assumption that $C^T C = I_k$, equation (8.32) simplifies to

$$\text{proj}_{\mathbb{S}}(\overrightarrow{v}) = \left(CC^T \overrightarrow{v}\right)^T. \tag{8.34}$$

Both of the above formulas are very elegant and computationally simple as long as we can always find an orthonormal basis for a subspace $\mathbb{S}$ from any given basis $\mathbf{B}$ of $\mathbb{S}$. In effect, what this states is that given an orthonormal basis, any vector $\overrightarrow{v}$ in the subspace can always be expressed as a linear combination of those basis vectors with the coefficients in front of each basis vector being the dot product of the basis vector with $\overrightarrow{v}$. So the question now becomes: Is there a systematic process that takes a basis $\mathbf{B}$ and from it, generates an orthonormal basis? A procedure called the *Gram-Schmidt orthonormalization process* exists to do precisely this. We conclude this section with a theorem, whose proof will be left as an exercise.

Theorem 8.4.1. *Let $\mathbb{S}$ be a subspace of $\mathbb{R}^n$. Then a basis $\boldsymbol{B} = \{\overrightarrow{w_1}, \overrightarrow{w_2}, \ldots, \overrightarrow{w_k}\}$ is an orthonormal basis of $\mathbb{S}$ if and only if for all vectors $\overrightarrow{v} \in \mathbb{S}$*

$$\overrightarrow{v} = (\overrightarrow{w_1} \cdot \overrightarrow{v})\,\overrightarrow{w_1} + (\overrightarrow{w_2} \cdot \overrightarrow{v})\,\overrightarrow{w_2} + \cdots + (\overrightarrow{w_k} \cdot \overrightarrow{v})\,\overrightarrow{w_k}$$

Homework Problems

1. Verify that if $\mathbb{S}$ is a one-dimensional subspace of $\mathbb{R}^n$ spanned by the single vector $\overrightarrow{w}$, then the formula for $\text{proj}_{\mathbb{S}}(\overrightarrow{v})$ reduces to the formula given by $\text{proj}_{\overrightarrow{w}}(\overrightarrow{v})$.

2. Given the basis $\mathbf{B} = \{\overrightarrow{w_1}, \overrightarrow{w_2}, \ldots, \overrightarrow{w_k}\}$ for a subspace of $\mathbb{R}^n$, prove that the matrix $W \in \mathbb{R}^{k \times k}$, defined by $W_{i,j} = \overrightarrow{w_i} \cdot \overrightarrow{w_j}$, satisfies $W^T = W$.

3. Given the basis $\mathbf{B} = \{\overrightarrow{w_1}, \overrightarrow{w_2}, \ldots, \overrightarrow{w_k}\}$ for a subspace of $\mathbb{C}^n$, and $W \in \mathbb{C}^{k \times k}$ defined by $W_{i,j} = \overrightarrow{w_i} \cdot \overrightarrow{w_j}$, determine a relationship between W and W^T.

4. Project the following vectors on the subspace of $\mathbb{R}^3$ generated by the basis $\{\langle 1, 1, 0 \rangle, \langle 0, 0, 1 \rangle\}$.

    (a) $\langle -1, 2, 1 \rangle$ (b) $\langle -1, 2, 0 \rangle$ (c) $\langle 0, 2, 3 \rangle$

5. Modify the vector projection formula given in equation (8.31) for when the vectors are complex valued.

6. Project the vector $\overrightarrow{v} = \langle -2 + i, 6 - i, 3 + 2i \rangle$ onto the subspace of $\mathbb{C}^3$ generated by the basis $\{\langle i, 1, 1 + i \rangle, \langle 0, i, 1 - i \rangle\}$.

7. Prove that the dot product matrix W, when considering a set of orthogonal vectors, is diagonal. Furthermore, what do the values on the diagonal correspond to?

8. Determine if each of the following sets of vectors constitute an orthogonal set:

    (a) $\{\langle 1, 0, -1 \rangle, \langle 0, 1, 0 \rangle, \langle 1, 0, 1 \rangle\}$

    (b) $\{\langle 1, 1, 1 \rangle, \langle 2, -2, 2 \rangle, \langle 6, 0, -6 \rangle\}$

    (c) $\{\langle 1, 2, 1 \rangle, \langle 2, -2, 2 \rangle, \langle 6, 0, -6 \rangle\}$

    (d) $\{\langle 1, 1, -1, 1 \rangle, \langle 0, 1, 0, -1 \rangle, \langle -2, 1, 2, 2 \rangle\}$

    (e) $\{\langle 1, 1, -1, 1 \rangle, \langle 0, 1, 0, -1 \rangle, \langle -3, 2, 1, 2 \rangle\}$

9. Convert each set from problem 8 that was determined to be an orthogonal set into an orthonormal basis.

10. Compute the distance from the plane spanned by the vectors $\{\langle 1, 0, 1 \rangle, \langle 1, 2, -1 \rangle\}$ to the point $(8, 8, 1)$.

11. Prove Theorem 8.4.1.

12. Is vector projection linear? That is, for a subspace $\mathbb{S}$ of $\mathbb{R}^n$ and two vectors $\overrightarrow{u}, \overrightarrow{v} \in \mathbb{R}^n$ and scalar c, determine if the following two properties hold. If the

properties do not hold, give an example.

(a) $\text{proj}_{\mathbb{S}}(c\vec{v}) = c\,\text{proj}_{\mathbb{S}}(\vec{v})$

(b) $\text{proj}_{\mathbb{S}}(\vec{u} + \vec{v}) = \text{proj}_{\mathbb{S}}(\vec{u}) + \text{proj}_{\mathbb{S}}(\vec{v})$

13. Let $\mathbb{S}$ be a subspace of $\mathbb{R}^n$ with basis $\mathbf{B} = \{\vec{w_1}, \vec{w_2}, \ldots, \vec{w_k}\}$ and let $\vec{v} \in \mathbb{R}^n$. Determine the conditions on $\vec{v}$ under which the following condition holds:

$$\text{proj}_{\mathbb{S}}(\vec{v}) = \text{proj}_{\vec{w_1}}(\vec{v}) + \text{proj}_{\vec{w_2}}(\vec{v}) + \cdots + \text{proj}_{\vec{w_k}}(\vec{v})$$

14. What is $\text{proj}_{\mathbb{S}}(\text{proj}_{\mathbb{S}}(\vec{v}))$?

15. Let $\mathbb{S}$ be a subspace of $\mathbb{R}^n$ where $\mathbf{B} = \{\vec{w_1}, \vec{w_2}, \ldots, \vec{w_k}\}$ is a basis of $\mathbb{S}$. Can we extend $\mathbf{B}$ into a full basis of all of $\mathbb{R}^n$? *Hint: Consider a basis of the orthogonal complement* $\mathbb{S}^{\perp}$.

16. (a) Let $\mathbb{S}$ be a subspace of $\mathbb{R}^n$ with basis $\mathbf{B} = \{\vec{w_1}, \vec{w_2}, \vec{w_3}\}$. Show that the set $\mathbf{B_1} = \{\vec{q_1}, \vec{q_2}, \vec{q_3}\}$ is an orthogonal basis of $\mathbb{S}$, where

$$\vec{q_1} = \vec{w_1}, \quad \vec{q_2} = \vec{w_2} - \text{proj}_{\vec{q_1}}(\vec{w_2}), \quad \vec{q_3} = \vec{w_3} - \text{proj}_{\vec{q_1}}(\vec{w_3}) - \text{proj}_{\vec{q_2}}(\vec{w_3})$$

(b) Generalize part (a) to a subspace $\mathbb{S}$ of any dimension.

17. Let $\mathbb{U}$ and $\mathbb{V}$ be two subspaces of $\mathbb{R}^n$. For $\vec{w} \in \mathbb{R}^n$, find conditions on $\mathbb{U}$ and $\mathbb{V}$ such that

$$\text{proj}_{\mathbb{U}+\mathbb{V}}(\vec{w}) = \text{proj}_{\mathbb{U}}(\vec{w}) + \text{proj}_{\mathbb{V}}(\vec{w})$$

18. (*Fourier Series*) Let $\mathbb{V}$ be the real vector space of all continuous functions $f : [0, 2\pi] \to \mathbb{R}$. Define the the dot product of two elements $f(x)$ and $g(x)$ of $\mathbb{V}$ by

$$f(x) \cdot g(x) = \int_0^{2\pi} f(x)g(x)\,dx$$

Compute the vector projection $\text{proj}_{\mathbb{S}}(\vec{v})$, for $\vec{v} = e^x \in \mathbb{V}$, and $\mathbb{S}$, the subspace of $\mathbb{V}$, having basis

$$\mathbf{B} = \left\{ \frac{1}{\sqrt{2\pi}}, \frac{1}{\sqrt{\pi}}\cos(x), \frac{1}{\sqrt{\pi}}\sin(x), \frac{1}{\sqrt{\pi}}\cos(2x), \frac{1}{\sqrt{\pi}}\sin(2x) \right\}$$

(You should first check if $\mathbf{B}$ is orthonormal.) Now graph together both $\vec{v}$ and $\text{proj}_{\mathbb{S}}(\vec{v})$. Is $\text{proj}_{\mathbb{S}}(\vec{v})$ a reasonable approximation of $\vec{v}$, and how can you make $\text{proj}_{\mathbb{S}}(\vec{v})$ into a better approximation of $\vec{v}$?

Maple Problems

1. For each of the following sets of vectors, compute the dot product matrix W, where $W_{i,j} = \overrightarrow{w_i} \cdot \overrightarrow{w_j}$.

  (a) $\{\langle -1, 0, 2, 1 \rangle, \langle 1, -5, 6, 1 \rangle, \langle 2, 1, 0, -3 \rangle\}$

  (b) $\{\langle -1, 0, 2, 1, 0 \rangle, \langle 1, -5, 6, 1, -1 \rangle, \langle 2, 1, 0, -3, -2 \rangle\}$

  (c) $\{\langle -1, 0, 2, 1, 0 \rangle, \langle 1, -5, 6, 1, -1 \rangle, \langle 2, 1, 0, -3, -2 \rangle, \langle 0, 0, 2, 1, 0 \rangle\}$

  (d) $\{\langle 1 + i, 1 - i, i \rangle, \langle 1 - 2i, 0, 1 \rangle, \langle 0, 1 + i, 1 - 2i \rangle\}$

  (e) $\{\langle 1 + i, 1 - i, i \rangle, \langle 1 - 2i, 0, 1 \rangle\}$

2. Verify that $W^T = W$ for parts (a) - (c) of problem 1. For parts (d) and (e) of problem 1, verify that W satisfies the property given as the solution to *Homework* problem 3.

3. For each part of *Homework* problem 4, construct a graph similar to that depicted in Figure 8.4.1. Be sure to include the plane, the two basis vectors for the plane, the projection of the vector $\overrightarrow{v}$ into the plane, along with the original vector $\overrightarrow{v}$.

4. Determine whether or not each of the following sets of vectors constitute an orthogonal set of vectors by computing the dot product matrix W.

  (a) $\{\langle 1, 0, -1 \rangle, \langle 0, 1, 0 \rangle, \langle 1, 0, 1 \rangle\}$

  (b) $\{\langle 1, 1, 1 \rangle, \langle 2, -2, 2 \rangle, \langle 6, 0, -6 \rangle\}$

  (c) $\{\langle 1, 2, 1 \rangle, \langle 2, -2, 2 \rangle, \langle 6, 0, -6 \rangle\}$

  (d) $\{\langle 1, 1, -1, 1 \rangle, \langle 0, 1, 0, -1 \rangle, \langle -2, 1, 2, 2 \rangle\}$

  (e) $\{\langle 1, 1, -1, 1 \rangle, \langle 0, 1, 0, -1 \rangle, \langle 1, 0, 1, 0 \rangle\}$

5. Project the following vectors onto the subspaces generated by the given bases from problem 1.

  (a) $\langle -2, 3, 1, 2 \rangle$ onto the basis from 1 (a)

  (b) $\langle 3, 0, -6, -3 \rangle$ onto the basis from 1 (a)

(c) $\langle 2, -4, -8, -1, -3 \rangle$ onto the basis from 1 (b)

(d) $\langle 2, -4, -8, -1, -3 \rangle$ onto the basis from 1 (c)

(e) $\langle 2, -4, 8, -1, 2 \rangle$ onto the basis from 1 (b)

(f) $\langle 2, -4, 8, -1, 2 \rangle$ onto the basis from 1 (c)

(g) $\langle 1, 0, 1 \rangle$ onto the basis from 1 (d)

(h) $\langle 0, i, 1 \rangle$ onto the basis from 1 (e)

(i) $\langle 1 - 2i, 1 + i, 2 - 2i \rangle$ onto the basis from 1 (e)

6. Using *Homework* problem 16, find an orthogonal basis $\mathbf{B}_1$ for the subspace $\mathbb{S}$ of $\mathbb{R}^5$ having basis:

$$\mathbf{B} = \{\langle -2, 7, 4, 1, 9 \rangle, \langle 6, 1, -3, -1, 7 \rangle, \langle 13, 0, -8, 2, 9 \rangle\}$$

First, check that this set $\mathbf{B}$ is independent, and next check that $\mathbf{B}_1$ is orthogonal.

8.5   The Gram-Schmidt Orthonormalization Process

From the end of Section 8.4, it was pointed out that if we have an orthonormal set of vectors, vector projections onto subspaces becomes computationally simple. If $\mathbf{B} = \{\vec{w_1}, \vec{w_2}, \ldots, \vec{w_k}\}$ is basis for a subspace $\mathbb{S}$ of $\mathbb{R}^n$, our goal in this section is to convert $\mathbf{B}$ to an *orthogonal basis* $\mathbf{Q} = \{\vec{q_1}, \vec{q_2}, \ldots, \vec{q_k}\}$ of $\mathbb{S}$, where $\vec{q_i} \cdot \vec{q_j} = 0$ for all $i \neq j$. Once we have a set of orthogonal vectors, to make them orthonormal, we simply divide each vector by its magnitude. The result is an *orthonormal basis* for $\mathbb{S}$ of the form

$$\mathbf{P} = \left\{ \frac{\vec{q_1}}{|\vec{q_1}|}, \frac{\vec{q_2}}{|\vec{q_2}|}, \ldots, \frac{\vec{q_k}}{|\vec{q_k}|} \right\}$$

In order to start the procedure for getting the $\vec{q}$'s, let $\vec{q_1} = \vec{w_1}$. Now to get $\vec{q_2}$, we start by first assuming that $\vec{q_2} = a\vec{q_1} + \vec{w_2}$, for some unknown scalar a. In other words, we wish to express $\vec{q_2}$ as a sum of two vectors, the first being in the direction of $\vec{q_1}$. So now we find the value of the scalar a so that $\vec{q_1}$ and

$\vec{q_2}$ are orthogonal. We arrive at the following string of equations:

$$\vec{q_1} \cdot \vec{q_2} = 0$$
$$\vec{q_1} \cdot (a\vec{q_1} + \vec{w_2}) = 0$$
$$a\vec{q_1} \cdot \vec{q_1} + \vec{q_1} \cdot \vec{w_2} = 0$$

Solving for a gives:

$$a = -\frac{\vec{q_1} \cdot \vec{w_2}}{\vec{q_1} \cdot \vec{q_1}} = -\frac{\vec{q_1} \cdot \vec{w_2}}{|\vec{q_1}|^2}$$

Therefore,

$$\vec{q_2} = \vec{w_2} - \frac{\vec{q_1} \cdot \vec{w_2}}{|\vec{q_1}|^2}\vec{q_1}$$

$$= \vec{w_2} - \text{proj}_{\vec{q_1}}(\vec{w_2}) \tag{8.35}$$

Now we have two orthogonal vectors $\{\vec{q_1}, \vec{q_2}\}$, which span the same subspace as $\{\vec{w_1}, \vec{w_2}\}$. The idea that these two different bases span the same subspace is very important, so be sure that you understand why this is so.

Now to get $\vec{q_3}$, we assume that $\vec{q_3} = a\vec{q_1} + b\vec{q_2} + \vec{w_3}$, and then find the values of the scalars a and b so that $\vec{q_1} \cdot \vec{q_3} = 0$ and $\vec{q_2} \cdot \vec{q_3} = 0$. To solve for a we use the first:

$$\vec{q_1} \cdot \vec{q_3} = 0$$
$$\vec{q_1} \cdot (a\vec{q_1} + b\vec{q_2} + \vec{w_3}) = 0$$
$$a\vec{q_1} \cdot \vec{q_1} + b\vec{q_1} \cdot \vec{q_2} + \vec{q_1} \cdot \vec{w_3} = 0$$
$$a\vec{q_1} \cdot \vec{q_1} + \vec{q_1} \cdot \vec{w_3} = 0$$

Here we explicitly used the fact that $\vec{q_1}$ and $\vec{q_2}$ are orthogonal to simplify the above expression. Solving for a gives

$$a = -\frac{\vec{q_1} \cdot \vec{w_3}}{\vec{q_1} \cdot \vec{q_1}} = -\frac{\vec{q_1} \cdot \vec{w_3}}{|\vec{q_1}|^2}$$

In a similar fashion, using the expression $\vec{q_2} \cdot \vec{q_3} = 0$ allows us to solve for b:

$$\vec{q_2} \cdot \vec{q_3} = 0$$
$$\vec{q_2} \cdot (a\vec{q_1} + b\vec{q_2} + \vec{w_3}) = 0$$
$$a\vec{q_2} \cdot \vec{q_1} + b\vec{q_2} \cdot \vec{q_2} + \vec{q_2} \cdot \vec{w_3} = 0$$
$$b\vec{q_2} \cdot \vec{q_2} + \vec{q_2} \cdot \vec{w_3} = 0$$

therefore:

$$b = -\frac{\vec{q_2} \cdot \vec{w_3}}{\vec{q_2} \cdot \vec{q_2}} = -\frac{\vec{q_2} \cdot \vec{w_3}}{|\vec{q_2}|^2}$$

Now we have the values of the scalars a and b, and thus a complete expression for $\vec{q_3}$:

$$
\begin{aligned}
\vec{q_3} &= \vec{w_3} - \frac{\vec{w_3} \cdot \vec{q_1}}{|\vec{q_1}|^2}\vec{q_1} - \frac{\vec{w_3} \cdot \vec{q_2}}{|\vec{q_2}|^2}\vec{q_2} \\
&= \vec{w_3} - \text{proj}_{\vec{q_1}}(\vec{w_3}) - \text{proj}_{\vec{q_2}}(\vec{w_3})
\end{aligned}
\tag{8.36}
$$

This process clearly follows a pattern, so we now have a systematic way of computing the jth vector in the orthogonal set. The formula is as follows:

$$
\begin{aligned}
\vec{q_j} &= \vec{w_j} - \frac{\vec{w_j} \cdot \vec{q_1}}{|\vec{q_1}|^2}\vec{q_1} - \frac{\vec{w_j} \cdot \vec{q_2}}{|\vec{q_2}|^2}\vec{q_2} - \cdots - \frac{\vec{w_j} \cdot \vec{q_{j-1}}}{|\vec{q_{j-1}}|^2}\vec{q_{j-1}} \\
&= \vec{w_j} - \sum_{i=1}^{j-1} \frac{\vec{w_j} \cdot \vec{q_i}}{|\vec{q_i}|^2}\vec{q_i}
\end{aligned}
\tag{8.37}
$$

for $2 \le j \le k$, where $\vec{q_1} = \vec{w_1}$. We can rewrite the above expression in terms of projections, similar to the definitions of $\vec{q_2}$ and $\vec{q_3}$:

$$
\vec{q_j} = \vec{w_j} - \sum_{i=1}^{j-1} \text{proj}_{\vec{q_i}}(\vec{w_j})
\tag{8.38}
$$

It should be clear that the subspace spanned by $\{\vec{q_1}, \vec{q_2}, \ldots, \vec{q_j}\}$ is the same as that spanned by $\{\vec{w_1}, \vec{w_2}, \ldots, \vec{w_j}\}$, for each $1 \le j \le k$. Normalizing each $\vec{q_j}$ to unit length will give us our orthonormal basis $\mathbf{P}$ of the subspace $\mathbb{S}$.

Example 8.5.1. To illustrate this, we will redo the subspace projection example 8.4.1 from Section 8.4. The basis for the subspace $\mathbb{S}$ was given by the set $\mathbf{B} = \{\langle 5, -2, 9\rangle, \langle -3, 7, 1\rangle\}$, and the vector $\vec{v} = \langle 4, 8, -15\rangle$ was chosen to be projected onto the subspace. We should end up with the same result, $\text{proj}_{\mathbb{S}}(\vec{v}) = \langle -\frac{2497}{406}, \frac{3048}{1015}, -\frac{733}{70}\rangle$. However, we first construct an orthonormal basis from our original pair of basis vectors.

```
> with(linalg):
> w[1]:= [5, -2, 9]: w[2]:= [-3, 7, 1]:
> v:= [4, 8, -15]:
> q[1]:= w[1]:
> q[2]:= w[2] - dotprod(w[2],q[1])/dotprod(q[1], q[1])*q[1];
```

$$
q2 := \left[-\frac{23}{11}, \frac{73}{11}, \frac{29}{11} \right]
$$

```
> dotprod(q[1], q[2]);
```

$$
0
$$

The last calculation above shows that the set $\{\langle 5, -2, 9 \rangle, \langle -\frac{23}{11}, \frac{73}{11}, \frac{29}{11} \rangle\}$ is an orthogonal basis for the plane $\mathbb{S}$. Now, we can use the much simpler formulation of projection given in equation (8.33), assuming that we first divide both of the vectors in the above given set by their magnitude to create unit length vectors.

> p[1]:= evalm(q[1]/sqrt(dotprod(q[1],q[1])));

$$ p_1 := \left[\begin{array}{ccc} \dfrac{1}{22} \sqrt{110} & -\dfrac{1}{55} \sqrt{110} & \dfrac{9}{110} \sqrt{110} \end{array} \right] $$

> p[2]:= evalm(q[2]/sqrt(dotprod(q[2],q[2])));

$$ p_2 := \left[\begin{array}{ccc} -\dfrac{23}{6699} \sqrt{6699} & \dfrac{73}{6699} \sqrt{6699} & \dfrac{1}{231} \sqrt{6699} \end{array} \right] $$

> proj_vontoS:= evalm(dotprod(v,p[1])*p[1] + dotprod(v,p[2])*p[2]);

$$ proj\_vontoS := \left[\begin{array}{ccc} -\dfrac{2497}{406} & \dfrac{3048}{1015} & -\dfrac{733}{70} \end{array} \right] $$

The vector projection of $\vec{v}$ onto $\mathbb{S}$ is the same, independent of the basis used to represent $\mathbb{S}$. ∎

Example 8.5.2. Now we will do one more example, this time with vectors in $\mathbb{R}^5$, starting with the basis:

$$ \mathbf{B} = \{ \vec{w_1} = \langle 1, -2, 1, 0, 1 \rangle, \vec{w_2} = \langle -2, -2, 0, 2, 7 \rangle, $$
$$ \vec{w_3} = \langle 4, 1, 2, 0, 3 \rangle, w_4 = \langle -1, 0, 0, 0, 1 \rangle \} $$

We will implement the Gram-Schmidt orthonormalization to create an orthonormal basis $\mathbf{P}$ of the subspace $\mathbb{S}$ of $\mathbb{R}^5$ whose basis is $\mathbf{B}$. Pay special attention to the way in which we have *Maple* perform this.

> w[1]:= [1, -2, 1, 0, 1]: w[2]:= [-2, -2, 0, 2, 7]:
> w[3]:= [4, 1, 2, 0, 3]: w[4]:= [-1, 0, 0, 0, 1]:
> rref(matrix([w[1],w[2],w[3],w[4]]));

$$ \left[\begin{array}{ccccc} 1 & 0 & 0 & 0 & -1 \\ 0 & 1 & 0 & 0 & \frac{3}{5} \\ 0 & 0 & 1 & 0 & \frac{16}{5} \\ 0 & 0 & 0 & 1 & \frac{31}{10} \end{array} \right] $$

Notice that since the *rref* command gives no rows of all zeroes, the set **B** is a set of linearly independent vectors.

```
> q:= [0,0,0,0]:
> q[1]:= w[1]:
> for j from 2 to 4 do
     q[j]:= evalm(w[j] - add(dotprod(w[j], q[i])/dotprod(q[i], q[i])*q[i], i = 1..(j-1)));
  od:
> seq(print(q[i]),i=1..4);
```

$$[1, -2, 1, 0, 1]$$

$$\left[-\frac{23}{7} \quad \frac{4}{7} \quad -\frac{9}{7} \quad 2 \quad \frac{40}{7} \right]$$

$$\left[\frac{542}{173} \quad \frac{515}{173} \quad \frac{182}{173} \quad -\frac{14}{173} \quad \frac{306}{173} \right]$$

$$\left[-\frac{1707}{7930} \quad \frac{58}{793} \quad \frac{181}{610} \quad -\frac{1463}{3965} \quad \frac{257}{3965} \right]$$

```
> P:= [0,0,0,0]:
> for j from 1 to 4 do
    P[j]:= evalm(q[j]/sqrt(dotprod(q[j],q[j])));
  od:
> seq(print(P[i]),i=1..4);
```

$$\left[\frac{1}{7}\sqrt{7} \quad -\frac{2}{7}\sqrt{7} \quad \frac{1}{7}\sqrt{7} \quad 0 \quad \frac{1}{7}\sqrt{7} \right]$$

$$\left[-\frac{23}{2422}\sqrt{2422} \quad \frac{2}{1211}\sqrt{2422} \quad -\frac{9}{2422}\sqrt{2422} \quad \frac{1}{173}\sqrt{2422} \quad \frac{20}{1211}\sqrt{2422} \right]$$

$$\left[\frac{542}{685945}\sqrt{685945} \quad \frac{103}{137189}\sqrt{685945} \quad \frac{14}{52765}\sqrt{685945} \quad -\frac{14}{685945}\sqrt{685945} \quad \frac{306}{685945}\sqrt{685945} \right]$$

$$\left[-\frac{1707}{17612530}\sqrt{17612530} \quad \frac{58}{1761253}\sqrt{17612530} \quad \frac{181}{1354810}\sqrt{17612530} \quad -\frac{1463}{8806265}\sqrt{17612530} \quad \frac{257}{8806265}\sqrt{17612530} \right]$$

```
> W:= matrix(4,4,0):
> for i from 1 to 4 do
    for j from 1 to 4 do
        W[i,j]:= dotprod(P[i],P[j]);
  od:
  od:
```

> print(W);

$$\begin{bmatrix} 1 & 0 & 0 & 0 \\ 0 & 1 & 0 & 0 \\ 0 & 0 & 1 & 0 \\ 0 & 0 & 0 & 1 \end{bmatrix}$$

So **P** is an orthonormal basis of the subspace of $\mathbb{R}^5$ corresponding to the basis **B**.

As a side note, one can combine the two steps of the Gram-Schmidt orthonormalization process in a more integrated fashion. Instead of computing the $\vec{q_j}$'s first and then making them all unit length, we can make $\vec{q_1}$ a unit vector first, by replacing $\vec{q_1}$ with $\dfrac{\vec{w_1}}{|\vec{w_1}|}$, and then continuing on to $\vec{q_2}$ from equation (8.35), in which case we get

$$\vec{q_2} = \vec{w_2} - \frac{\vec{q_1} \cdot \vec{w_2}}{|\vec{q_1}|^2} \vec{q_1}$$
$$= \vec{w_2} - (\vec{q_1} \cdot \vec{w_2}) \vec{q_1}$$

Now, we do an intermediate step before computing $\vec{q_3}$, which is making $\vec{q_2}$ unit length as well. So we perform the reassignment $\dfrac{\vec{q_2}}{|\vec{q_2}|} \to \vec{q_2}$. We get that $\vec{q_3}$ from equation (8.36) is now given by

$$\vec{q_3} = \vec{w_3} - (\vec{w_3} \cdot \vec{q_1}) \vec{q_1} - (\vec{w_3} \cdot \vec{q_2}) \vec{q_2}$$

This alternative method of the Gram-Schmidt orthonormalization process can be expressed as follows for each step in the process:

$$\text{part 1)} \quad \frac{\vec{q}_{j-1}}{|\vec{q_{j-1}}|} \to \vec{q_{j-1}}$$

$$\text{part 2)} \quad \vec{q_j} = \vec{w_j} - \sum_{i=1}^{j-1} (\vec{w_j} \cdot \vec{q_i}) \vec{q_i}$$

for $2 \leq j \leq k$, where $\vec{q_1} = \dfrac{\vec{w_1}}{|\vec{w_1}|}$. Finally, as one last step we must perform a normalization of $\vec{q_k}$. As a complete process, we have

$$\text{step 1)} \quad \vec{q_1} = \vec{w_1}$$

$$\text{step 2)} \quad \text{for } 2 \leq j \leq k:$$

$$\frac{\vec{q_{j-1}}}{|\vec{q_{j-1}}|} \rightarrow \vec{q_{j-1}}$$

$$\vec{q_j} = \vec{w_j} - \sum_{i=1}^{j-1} (\vec{w_j} \cdot \vec{q_i}) \, \vec{q_i}$$

$$\text{step 3)} \quad \frac{\vec{q_k}}{|\vec{q_k}|} \rightarrow \vec{q_k}$$

Now we take the previous basis **B** and use the new method to produce the orthonormal basis **P**.

```
> qn:= [0,0,0,0]:
> qn[1]:= w[1]:
> for j from 2 to 4 do
    qn[j-1]:= evalm(qn[j-1]/sqrt(dotprod(qn[j-1],qn[j-1]))):
    qn[j]:= evalm(w[j] - add(dotprod(w[j],qn[i])*qn[i], i = 1..(j- 1))):
  od:
> qn[4]:= evalm(qn[4]/sqrt(dotprod(qn[4],qn[4]))):
> seq(print(evalm(qn[k]-P[k])),k=1..4);
```

$$\begin{bmatrix} 0 & 0 & 0 & 0 \end{bmatrix}$$
$$\begin{bmatrix} 0 & 0 & 0 & 0 \end{bmatrix}$$
$$\begin{bmatrix} 0 & 0 & 0 & 0 \end{bmatrix}$$
$$\begin{bmatrix} 0 & 0 & 0 & 0 \end{bmatrix}$$

The last command shows that either method for computing the orthonormal basis results in the same set of unit length, mutually orthogonal vectors.

Homework Problems

1. Explain what happens if one attempts to apply the Gram-Schmidt orthonormalization process to a set of vectors that is linearly dependent. It may be easiest to assume that $\mathbf{K} = \{\vec{v_1}, \vec{v_2}, \ldots, \vec{v_k}, \vec{v_{k+1}}\}$ and that

$$\vec{v}_{k+1} = \sum_{j=1}^{k} a_k \vec{v_k}$$

for some scalars a_k, $1 \leq j \leq k$, of which at least one is nonzero.

2. What happens when one applies the Gram-Schmidt orthonormalization process to a set of vectors that are already mutually orthogonal?

3. Convert each of the following sets of vectors to an orthonormal set of vectors:

    (a) $\mathbf{K}_1 = \{\langle -2, 3 \rangle, \langle 6, 1 \rangle\}$

    (b) $\mathbf{K}_2 = \{\langle 1, 0, 1 \rangle, \langle 0, -1, 1 \rangle\}$

    (c) $\mathbf{K}_3 = \{\langle 1, 0, 1, 1 \rangle, \langle 0, -1, 2, 1 \rangle, \langle 3, 1, 0, -2 \rangle\}$

    (d) $\mathbf{K}_4 = \{\langle 1, 1, 0, 1 \rangle, \langle 2, 1, -1, 1 \rangle, \langle -2, -1, 1, 0 \rangle\}$

4. Project each of the following vectors onto the corresponding orthonormal basis found in problem 3.

(a) $\langle 1, 1, 1 \rangle$ onto $\mathbf{K}_2$ (b) $\langle 3, 4, -2 \rangle$ onto $\mathbf{K}_2$ (c) $\langle 4, -5, -3 \rangle$ onto $\mathbf{K}_2$

(d) $\langle 2, 1, 2, 3 \rangle$ onto $\mathbf{K}_3$ (e) $\langle 3, 5, -5, 7 \rangle$ onto $\mathbf{K}_3$ (f) $\langle 3, 5, -5, 7 \rangle$ onto $\mathbf{K}_4$

5. A square matrix $P \in \mathbb{R}^{n \times n}$ whose columns form an orthonormal basis of $\mathbb{R}^n$ is called an *orthogonal matrix*. Prove the following identities. *Hint: Consider the matrix multiplication PP^T.*

    (a) $P^{-1} = P^T$ (b) $\det(P) = \pm 1$

6. Use problem 5 to show that any real matrix A of the form

$$A = \frac{1}{a^2 + b^2} \begin{bmatrix} a & b \\ -b & a \end{bmatrix}$$

is orthogonal if $a^2 + b^2 \neq 0$.

7. Use problems 5 and 6 to show that every 2×2 orthogonal matrix can be written as

$$\begin{bmatrix} \cos(\theta) & \sin(\theta) \\ -\sin(\theta) & \cos(\theta) \end{bmatrix}$$

for some angle θ.

8. Use problem 5 to show that if both P and Q are orthogonal matrices of the same size, then their two products PQ and QP are also orthogonal.

9. Let $\mathbf{B} = \{\overrightarrow{w_1}, \overrightarrow{w_2}, \ldots, \overrightarrow{w_n}\}$ be any orthogonal basis of $\mathbb{R}^n$. Let $\mathbb{S}$ be the subspace of $\mathbb{R}^n$ with basis $\mathbf{B}_1 = \{\overrightarrow{w_1}, \overrightarrow{w_2}, \ldots, \overrightarrow{w_k}\}$ for $k < n$. What is a basis of $\mathbb{S}^\perp$?

10. Let $\mathbf{B}$ be any finite orthonormal subset of $\mathbb{R}^n$. Prove that $\mathbf{B}$ is an independent set. Is this also true if $\mathbf{B}$ is merely orthogonal? Is any n-element orthonormal subset of $\mathbb{R}^n$ automatically a basis of $\mathbb{R}^n$?

Maple Problems

1. Determine which of the following are orthonormal sets of vectors:

(a) $\left\{\left\langle \dfrac{1}{\sqrt{17}}, \dfrac{4}{\sqrt{17}} \right\rangle, \left\langle -\dfrac{4}{\sqrt{17}}, \dfrac{1}{\sqrt{17}} \right\rangle \right\}$

(b) $\left\{\left\langle \dfrac{1}{\sqrt{10}}, \dfrac{3}{\sqrt{10}} \right\rangle, \left\langle -\dfrac{4}{\sqrt{10}}, \dfrac{1}{\sqrt{10}} \right\rangle \right\}$

(c) $\left\{\left\langle \dfrac{1}{\sqrt{10}}, 0, \dfrac{3}{\sqrt{10}} \right\rangle, \left\langle -\dfrac{9}{\sqrt{190}}, \dfrac{1}{19\sqrt{190}}, \dfrac{3}{\sqrt{190}} \right\rangle, \left\langle \dfrac{3}{\sqrt{19}}, \dfrac{3}{\sqrt{19}}, -\dfrac{1}{\sqrt{19}} \right\rangle, \right\}$

(d) $\left\{\left\langle 0, \dfrac{3}{\sqrt{26}}, -\dfrac{4}{\sqrt{26}}, \dfrac{1}{\sqrt{26}} \right\rangle, \left\langle -\dfrac{26}{\sqrt{16302}}, -\dfrac{99}{\sqrt{16302}}, -\dfrac{76}{\sqrt{16302}}, -\dfrac{7}{\sqrt{16302}} \right\rangle, \right.$
$\left. \left\langle \dfrac{928}{\sqrt{914793}}, -\dfrac{132}{\sqrt{914793}}, -\dfrac{133}{\sqrt{914793}}, -\dfrac{136}{\sqrt{914793}} \right\rangle, \right\}$

2. Each of the following sets of vectors forms a basis for $\mathbb{R}^n$. Construct the orthogonal matrix for each set, and verify the properties of *Homework* problem 5.

(a) $\mathbf{K} = \{\langle -2, 5\rangle, \langle 7, 9\rangle\}$

(b) $\mathbf{K} = \{\langle 1, 2, 1\rangle, \langle -1, 1, 1\rangle, \langle 1, 0, 1\rangle\}$

(c) $\mathbf{K} = \{\langle 1, -2, 1, 3\rangle, \langle 1, -1, 1, 1\rangle, \langle 1, 0, 1, 1\rangle, \langle 2, 1, 1, -3\rangle\}$

(d) $\mathbf{K} = \{\langle 2, -2, 1, -2, 3\rangle, \langle 1, -2, 2, 1, -1\rangle, \langle -2, 3, 1, 1, 0\rangle,$
$\langle 0, 2, 1, -3, 2\rangle, \langle 2, 1, -1, -1, -2\rangle\}$

3. A *unitary matrix* is the complex version of an orthogonal matrix. A matrix P is unitary if $P^{-1} = \overline{P}^T$. Construct the orthonormal basis given the following set of vectors and show that the resulting matrix, whose columns are these vectors, is unitary:

$$\mathbf{K} = \{\langle 2 + i, 1 - 2i, 2 + 3i\rangle, \langle 4 - i, 3, -3i\rangle, \langle 2 - 3i, 2 + 4i, 3\rangle\}$$

4. Write the vector $\overrightarrow{v} = \langle 7, 1, -2, 9\rangle$ as a linear combination of the possible orthonormal sets in problem 1(d) or (e), whichever is orthonormal if both are not.

5. Write the vector $\vec{v} = \langle 7, 1, -2, 9, -5 \rangle$ as a linear combination of the orthonormal basis of $\mathbb{R}^5$ from problem 2(d).

6. Let $f(x)$ and $g(x)$ be two continuous functions for x in the interval $[0, 2\pi]$. We can define the dot product of these two functions as

$$f(x) \cdot g(x) = \int_0^{2\pi} f(x)\, g(x)\, dx$$

Show that the set of trigonometric functions

$$\left\{ \frac{1}{\sqrt{2\pi}}, \frac{1}{\sqrt{\pi}} \cos(x), \frac{1}{\sqrt{\pi}} \sin(x), \frac{1}{\sqrt{\pi}} \cos(2x), \frac{1}{\sqrt{\pi}} \sin(2x) \right\}$$

is an orthonormal set in this dot product.

7. Let $f(x)$ and $g(x)$ be two continuous functions for x in the interval $[-1, 1]$. We can define the dot product of these two functions as

$$f(x) \cdot g(x) = \int_{-1}^{1} f(x)\, g(x)\, dx$$

Apply the Gram-Schmidt orthonormalization process to the set of functions

$$\left\{ 1, x, x^2, x^3 \right\}$$

Chapter 9

Linear Maps from $\mathbb{R}^n$ to $\mathbb{R}^m$

9.1 Basics About Linear Maps

This section will discuss the idea of linear maps, sometimes referred to as linear transformations. We begin with a definition.

Definition 9.1.1. A *linear map* T is a function from $\mathbb{R}^n$ to $\mathbb{R}^m$ that preserves linear combinations, and is denoted $T : \mathbb{R}^n \to \mathbb{R}^m$. Thus, for $\vec{u}, \vec{v} \in \mathbb{R}^n$, and real scalars a and b, the linear map T has the property that:

$$T(a\vec{u} + b\vec{v}) = aT(\vec{u}) + bT(\vec{v}) \tag{9.1}$$

First, we explore properties common to all linear maps. One simple fact to recognize is that every linear map T takes the zero vector $\vec{0}_n$ of $\mathbb{R}^n$ to the zero vector $\vec{0}_m$ of $\mathbb{R}^m$. To see this, notice that:

$$\begin{aligned}
T\left(\vec{0}_n\right) &= T\left(\vec{0}_n + \vec{0}_n\right) \\
&= T\left(\vec{0}_n\right) + T\left(\vec{0}_n\right) \\
&= 2T\left(\vec{0}_n\right),
\end{aligned}$$

which forces us to conclude that $T\left(\vec{0}_n\right) = \vec{0}_m$.

Now we need some examples of linear and nonlinear maps in order to see what patterns might appear in the rules for such types of functions.

Example 9.1.1. Let $T : \mathbb{R}^3 \to \mathbb{R}$ have the rule $T(\langle x, y, z \rangle) = ax + by + cz$ for some fixed real scalars a, b, c. Then T is a linear map since for any real scalars

α and β, and any two elements $\langle x_1, y_1, z_1 \rangle$ and $\langle x_2, y_2, z_2 \rangle$ of $\mathbb{R}^3$, we have

$$
\begin{aligned}
T(\alpha \langle x_1, y_1, z_1 \rangle + \beta \langle x_2, y_2, z_2 \rangle) &= T(\langle \alpha x_1 + \beta x_2, \alpha y_1 + \beta y_2, \alpha z_1 + \beta z_2 \rangle) \\
&= a\,(\alpha x_1 + \beta x_2) + b\,(\alpha y_1 + \beta y_2) + c\,(\alpha z_1 + \beta z_2) \\
&= \alpha\,(a x_1 + b y_1 + c z_1) + \beta\,(a x_2 + b y_2 + c z_2) \\
&= \alpha\, T(\langle x_1, y_1, z_1 \rangle) + \beta\, T(\langle x_2, y_2, z_2 \rangle).
\end{aligned}
$$

Consider what happens if we change the rule for T slightly by adding a constant to it, $T(\langle x, y, z \rangle) = ax + by + cz + d$ for some fixed nonzero real scalar d. Then the new T is not a linear map since

$$
T(\alpha \langle x_1, y_1, z_1 \rangle + \beta \langle x_2, y_2, z_2 \rangle) \neq \alpha\, T(\langle x_1, y_1, z_1 \rangle) + \beta\, T(\langle x_2, y_2, z_2 \rangle)
$$

which is left for you to verify. This addition of d translates the original linear map so that it no longer sends $\overrightarrow{0_3}$ to 0. This new version of T is called an *affine map* since it is a linear map (the original T) plus a nonzero real scalar d.

Example 9.1.2. Let $T : \mathbb{R}^2 \to \mathbb{R}^3$ have the rule $T(\langle x, y \rangle) = x \langle 1, -5, 2 \rangle + y \langle -7, 3, 9 \rangle$. Then T is a linear map since for any real scalars α and β, and any two elements $\langle x_1, y_1 \rangle, \langle x_2, y_2 \rangle \in \mathbb{R}^2$, we have

$$
\begin{aligned}
T(\alpha \langle x_1, y_1 \rangle + \beta \langle x_2, y_2 \rangle) &= T(\langle \alpha x_1 + \beta x_2, \ \alpha y_1 + \beta y_2 \rangle) \\
&= (\alpha\, x_1 + \beta\, x_2) \langle 1, -5, 2 \rangle + (\alpha\, y_1 + \beta\, y_2) \langle -7, 3, 9 \rangle \\
&= \alpha\ (x_1 \langle 1, -5, 2 \rangle + y_1 \langle -7, 3, 9 \rangle) \\
&\quad + \beta\ (x_2 \langle 1, -5, 2 \rangle + y_2 \langle -7, 3, 9 \rangle) \\
&= \alpha\, T(\langle x_1, y_1 \rangle) + \beta\, T(\langle x_2, y_2 \rangle)
\end{aligned}
$$

If we change the rule for T to $T(\langle x, y \rangle) = x \langle 1, -5, 2 \rangle + y \langle -7, 3, 9 \rangle + \langle 4, 0, 6 \rangle$, then the new T is not a linear map, but as in Example 9.1.1, the new T is an affine map since it is a translate of a linear map by the constant vector $\langle 4, 0, 6 \rangle$.

Example 9.1.3. Let $T : \mathbb{R}^3 \to \mathbb{R}^2$ have the rule $T(\langle x, y, z \rangle) = \langle 5x^2 - y + z^2, x + y + z \rangle$. Then T is not a linear map due to the squaring of the variables x and z in the first component of the rule for T. If this squaring was not present, then T would be a linear map. In fact, all linear maps must have as their rules something similar to Example 9.1.2 or equivalently, Example 9.1.4.

Example 9.1.4. Let $T : \mathbb{R}^3 \to \mathbb{R}^4$ have the rule

$$
\begin{aligned}
T(\langle x, y, z \rangle) &= \langle 5x - y + z, \ x + 2y + z, \ 7x - z, \ y + 4z \rangle \\
&= x \langle 5, 1, 7, 0 \rangle + y \langle -1, 2, 0, 1 \rangle + z \langle 1, 1, -1, 4 \rangle \\
&= x\, T(\langle 1, 0, 0 \rangle) + y\, T(\langle 0, 1, 0 \rangle) + z\, T(\langle 0, 0, 1 \rangle)
\end{aligned}
$$

If we treat these vectors as column matrices, then the rule for T can also be rewritten as

$$
T\left(\begin{bmatrix} x \\ y \\ z \end{bmatrix}\right) = \begin{bmatrix} 5x - y + z \\ x + 2y + z \\ 7x - z \\ y + 4z \end{bmatrix} = \begin{bmatrix} 5 & -1 & 1 \\ 1 & 2 & 1 \\ 7 & 0 & -1 \\ 0 & 1 & 4 \end{bmatrix} \begin{bmatrix} x \\ y \\ z \end{bmatrix}
$$

Since matrix multiplication is distributive, this T is a linear map. ∎

From these examples, we can come to a couple of immediate conclusions, which we now present as theorems.

Theorem 9.1.1. $T : \mathbb{R}^n \to \mathbb{R}^m$ *is a linear map if and only if*

$$
T(\langle x_1, x_2, \ldots, x_n \rangle) = x_1\, T(\langle 1, 0, \ldots, 0 \rangle) + x_2\, T(\langle 0, 1, 0, \ldots, 0 \rangle)
$$
$$
+ \cdots + x_n\, T(\langle 0, 0, \ldots, 1 \rangle)
$$

Second and equivalently, if we write our vectors as columns, we have the following theorem:

Theorem 9.1.2. $T : \mathbb{R}^n \to \mathbb{R}^m$ *is a linear map if and only if*

$$
T(\langle x_1, x_2, \ldots, x_n \rangle) = A \begin{bmatrix} x_1 \\ x_2 \\ \vdots \\ x_n \end{bmatrix}
$$

where the $m \times n$ matrix A has as its columns the n vectors in $\mathbb{R}^m$:

$$
T(\langle 1, 0, \ldots, 0 \rangle), T(\langle 0, 1, 0, \ldots, 0 \rangle), \ldots, T(\langle 0, 0, \ldots, 1 \rangle)
$$

and we are writing the output of T as a column vector in $\mathbb{R}^m$.

There are two very important ways we can generate new subspaces of $\mathbb{R}^n$ and $\mathbb{R}^m$ from original subspaces of each using a linear map $T : \mathbb{R}^n \to \mathbb{R}^m$ by taking the *image* and *inverse image* of the original subspaces by T. The process called image will use T to take a subspace $\mathbb{S}$ of $\mathbb{R}^n$ and returns a subspace $T(\mathbb{S})$ of $\mathbb{R}^m$ while the one called inverse image reverses things and uses T to take a subspace $\mathbb{K}$ of $\mathbb{R}^m$ and returns a subspace $T^{-1}(\mathbb{K})$ of $\mathbb{R}^n$.

Definition 9.1.2. Let $T : \mathbb{R}^n \to \mathbb{R}^m$ be a linear map with $\mathbb{S}$ a subspace of $\mathbb{R}^n$. Then the *image $T(\mathbb{S})$ of the subspace $\mathbb{S}$ under T* is given by

$$
T(\mathbb{S}) = \{ T(\vec{v}) \mid \vec{v} \in \mathbb{S} \}
$$

which is a subspace of $\mathbb{R}^m$.

In order to see that $T(\mathbb{S})$ is truly a subspace of $\mathbb{R}^m$, let $T\left(\overrightarrow{u}\right), T\left(\overrightarrow{v}\right) \in T(\mathbb{S})$ for $\overrightarrow{u}, \overrightarrow{v} \in \mathbb{S}$, and a, b be two real scalars. Then we have $aT\left(\overrightarrow{u}\right) + bT\left(\overrightarrow{v}\right) = T\left(a\overrightarrow{u} + b\overrightarrow{v}\right) \in T(\mathbb{S})$, since $a\overrightarrow{u} + b\overrightarrow{v}$ is an element of $\mathbb{S}$, and T is linear. Thus, $T(\mathbb{S})$ is a subspace of $\mathbb{R}^m$.

Definition 9.1.3. If $\mathbb{K}$ is a subspace of $\mathbb{R}^m$, then

$$T^{-1}(\mathbb{K}) = \{\overrightarrow{v} \in \mathbb{R}^n | T\left(\overrightarrow{v}\right) \in \mathbb{K}\}$$

is a subspace of $\mathbb{R}^n$, called the *inverse image of the subspace* $\mathbb{K}$ under the linear map $T : \mathbb{R}^n \to \mathbb{R}^m$.

In order to see that $T^{-1}(\mathbb{K})$ is a subspace of $\mathbb{R}^n$, let $\overrightarrow{u}, \overrightarrow{v} \in T^{-1}(\mathbb{K})$, and a, b be two real scalars. Then we have

$$T\left(a\overrightarrow{u} + b\overrightarrow{v}\right) = aT\left(\overrightarrow{u}\right) + bT\left(\overrightarrow{v}\right)$$

is an element of $\mathbb{K}$ since $T\left(\overrightarrow{u}\right), T\left(\overrightarrow{v}\right) \in \mathbb{K}$ and $\mathbb{K}$ is a subspace of $\mathbb{R}^m$. Thus, $T^{-1}(\mathbb{K})$ is a subspace of $\mathbb{R}^n$.

Example 9.1.5. We will now find the image $T(\mathbb{S})$ of the subspace

$$\mathbb{S} = \{a\langle 1, 2, -3\rangle \mid a \in \mathbb{R}\}$$

and inverse image $T^{-1}(\mathbb{K})$ of the subspace

$$\mathbb{K} = \{a\langle -1, 4, -2, 1, -3\rangle + b\langle 4, -1, -6, -5, 4\rangle | a, b \in \mathbb{R}\} \tag{9.2}$$

using the linear map $T : \mathbb{R}^3 \to \mathbb{R}^5$ given by

$$T(\langle x, y, z\rangle) = \langle 5x - 3y + 2z, x - y - z, -7x + z, 6y - 11z, 8x + y - 5z\rangle \tag{9.3}$$

First, note that

$$\begin{aligned} T(\mathbb{S}) &= \{T(a\langle 1, 2, -3\rangle) \mid a \in \mathbb{R}\} \\ &= \{aT(\langle 1, 2, -3\rangle) \mid a \in \mathbb{R}\} \\ &= \{a\langle -7, 2, -10, 45, 25\rangle \mid a \in \mathbb{R}\} \end{aligned}$$

so the vector $\langle -7, 2, -10, 45, 25\rangle$ spans the subspace $T(\mathbb{S})$ and since it is an independent set it is also a basis of $T(\mathbb{S})$, which says that $T(\mathbb{S})$ has dimension one.

In order to find $T^{-1}(\mathbb{K})$, we need to solve the equation:

$$T\left(\begin{bmatrix} x \\ y \\ z \end{bmatrix}\right) = \begin{bmatrix} -a + 4b \\ 4a - b \\ -2a - 6b \\ a - 5b \\ -3a + 4b \end{bmatrix} \tag{9.4}$$

for x, y and z in terms of a and b. To solve equation (9.4), we simply have to solve the following set of equations, which equates corresponding components of the left and right sides of the equation:

$$5x - 3y + 2z = -a + 4b$$
$$x - y - z = 4a - b$$
$$-7x + z = -2a - 6b \quad (9.5)$$
$$6y - 11z = a - 5b$$
$$8x + y - 5z = -3a + 4b$$

In augmented matrix form, with the first three columns corresponding to the LHS of the above system, and last two columns to the RHS, we have

$$\begin{bmatrix} 5 & -3 & 2 & -1 & 4 \\ 1 & -1 & -1 & 4 & -1 \\ -7 & 0 & 1 & -2 & -6 \\ 0 & 6 & -11 & 1 & -5 \\ 8 & 1 & -5 & -3 & 4 \end{bmatrix}$$

which row reduces to

$$\begin{bmatrix} 1 & 0 & 0 & 0 & 1 \\ 0 & 1 & 0 & 0 & 1 \\ 0 & 0 & 1 & 0 & 1 \\ 0 & 0 & 0 & 1 & 0 \\ 0 & 0 & 0 & 0 & 0 \end{bmatrix}$$

Do not forget that there are two columns in the above *rref* matrix representing the RHS of equation (9.5). We get the solution is given by $x = b, y = b, z = b, a = 0$. Therefore, we have that any vector of the form

$$\langle x, y, z \rangle = b \langle 1, 1, 1 \rangle$$

will satisfy equation (9.4). We can therefore conclude that $T^{-1}(\mathbb{K})$ has dimension one and is given by

$$T^{-1}(\mathbb{K}) = \{ b \langle 1, 1, 1 \rangle \mid b \in \mathbb{R} \}$$

and $\{ \langle 1, 1, 1 \rangle \}$ is the basis for $T^{-1}(\mathbb{K})$. Remember that $\mathbb{K}$, defined in equation (9.2), has basis $\{ \langle -1, 4, -2, 1, -3 \rangle, \langle 4, -1, -6, -5, 4 \rangle \}$. An interesting item to notice in this example is that $T(\langle 1, 1, 1 \rangle) = \langle 4, -1, -6, -5, 4 \rangle$, and since $\langle 1, 1, 1 \rangle$ was the only element in the basis of $T^{-1}(\mathbb{K})$, the equation $T(\langle x, y, z \rangle) = \langle -1, 4, -2, 1, -3 \rangle$ has no solution. Therefore, if we defined

$$\mathbb{K}_2 = \{ a \langle -1, 4, -2, 1, -3 \rangle \mid a \in \mathbb{R} \}$$

then $T^{-1}(\mathbb{K}_2) = \{ \overrightarrow{0}_3 \}$. Note that for most subspaces $\mathbb{K}$, $T^{-1}(\mathbb{K}) = \{ \overrightarrow{0} \}$.

It turns out that every linear map $T : \mathbb{R}^n \to \mathbb{R}^m$ can be represented as the multiplication of some $m \times n$ matrix A by vectors in $\mathbb{R}^n$. The question is, How do we find the matrix A? The answer is actually hinted at in the previous examples. To determine A, we start by taking the standard basis vectors $\mathbf{S}_n = \{\vec{s_1}, \vec{s_2}, \ldots, \vec{s_n}\}$ of $\mathbb{R}^n$, which consists of the rows of the $n \times n$ identity matrix I_n, and plug them into T. The vectors $\{T(\vec{s_1}), T(\vec{s_2}), \ldots, T(\vec{s_n})\}$ of $\mathbb{R}^m$ are the columns of the matrix A.

We can verify that this matrix A works to give us the linear map T by first taking an element $\vec{x} \in \mathbb{R}^n$ expressed as

$$\vec{x} = \langle x_1, x_2, \ldots, x_n \rangle = x_1 \vec{s_1} + x_2 \vec{s_2} + \cdots + x_n \vec{s_n}$$

and plugging it into T. We get that

$$T(\vec{x}) = x_1 T(\vec{s_1}) + x_2 T(\vec{s_2}) + \cdots + x_n T(\vec{s_n})$$

and so writing $\vec{x}, T(\vec{x}), T(\vec{s_1}), \ldots, T(\vec{s_n})$ as columns, we arrive at the matrix equation:

$$T(\vec{x}) = A\vec{x}$$

where

$$A = \left[T(\vec{s_1}) \,\middle|\, T(\vec{s_2}) \,\middle|\, \cdots \,\middle|\, T(\vec{s_n}) \right] \tag{9.6}$$

Of course, the argument requires us to use the standard basis to determine A, which in turn requires us to know the image of each standard basis vector under the map T. The question becomes, what if we have a different basis for $\mathbb{R}^n$ for which we know the image instead? Clearly, the above argument does not hold. However, the claim is that given a basis $\mathbf{B} = \{\vec{w_1}, \vec{w_2}, \ldots, \vec{w_n}\}$ of $\mathbb{R}^n$, if we know the values of $T(\vec{w_k})$, for $1 \le k \le n$, then we can determine A. Before we go through the process in an abstract fashion, we will consider a concrete example. Pay attention to the ideas though, as they will be used in the general situation.

Example 9.1.6. We will define $T : \mathbb{R}^3 \to \mathbb{R}^2$ to be the linear map such that:

$$T(\langle 2, 5, -1 \rangle) = \langle 4, -9 \rangle$$
$$T(\langle -7, 1, 3 \rangle) = \langle 1, 6 \rangle$$
$$T(\langle 4, -8, 0 \rangle) = \langle -2, 10 \rangle$$

We now wish to find the 2×3 matrix A such that $T(\vec{x}) = A\vec{x}$ for all vectors $\vec{x} \in \mathbb{R}^3$. The columns of the matrix A are the vectors $T(\langle 1, 0, 0 \rangle)$, $T(\langle 0, 1, 0 \rangle)$ and $T(\langle 0, 0, 1 \rangle)$, in order. We must find out how to write the standard basis,

$$\left\{ \vec{i}, \vec{j}, \vec{k} \right\} = \{\langle 1, 0, 0 \rangle, \langle 0, 1, 0 \rangle, \langle 0, 0, 1 \rangle\} \tag{9.7}$$

of $\mathbb{R}^3$, in terms of the new basis

$$\{\overrightarrow{u}, \overrightarrow{v}, \overrightarrow{w}\} = \{\langle 2, 5, -1 \rangle, \langle -7, 1, 3 \rangle, \langle 4, -8, 0 \rangle\} \tag{9.8}$$

Now let us write down our problem in matrix notation. First, in most general terms, we wish to find the matrix A such that:

$$T\left(\begin{bmatrix} x \\ y \\ z \end{bmatrix}\right) = \begin{bmatrix} A_{11} & A_{12} & A_{13} \\ A_{21} & A_{22} & A_{23} \end{bmatrix} \begin{bmatrix} x \\ y \\ z \end{bmatrix} = \begin{bmatrix} \alpha \\ \beta \end{bmatrix}$$

where $\langle x, y, z \rangle \in \mathbb{R}^3$ and $\langle \alpha, \beta \rangle \in \mathbb{R}^2$. We will also require that:

$$\begin{bmatrix} A_{11} & A_{12} & A_{13} \\ A_{21} & A_{22} & A_{23} \end{bmatrix} \begin{bmatrix} 2 \\ 5 \\ -1 \end{bmatrix} = \begin{bmatrix} 4 \\ -9 \end{bmatrix}$$

$$\begin{bmatrix} A_{11} & A_{12} & A_{13} \\ A_{21} & A_{22} & A_{23} \end{bmatrix} \begin{bmatrix} -7 \\ 1 \\ 3 \end{bmatrix} = \begin{bmatrix} 1 \\ 6 \end{bmatrix} \tag{9.9}$$

$$\begin{bmatrix} A_{11} & A_{12} & A_{13} \\ A_{21} & A_{22} & A_{23} \end{bmatrix} \begin{bmatrix} 4 \\ -8 \\ 0 \end{bmatrix} = \begin{bmatrix} -2 \\ 10 \end{bmatrix}$$

Furthermore, we also get the following equations as a result of the definition of the standard basis $\{\overrightarrow{i}, \overrightarrow{j}, \overrightarrow{k}\}$ for $\mathbb{R}^3$:

$$T\left(\overrightarrow{i}\right) = \begin{bmatrix} A_{11} \\ A_{21} \end{bmatrix}, \; T\left(\overrightarrow{j}\right) = \begin{bmatrix} A_{12} \\ A_{22} \end{bmatrix}, \; T\left(\overrightarrow{k}\right) = \begin{bmatrix} A_{13} \\ A_{23} \end{bmatrix} \tag{9.10}$$

Also, the following expressions involving the standard basis will prove useful:

$$\langle 2, 5, -1 \rangle = 2\overrightarrow{i} + 5\overrightarrow{j} - \overrightarrow{k}, \langle -7, 1, 3 \rangle = -7\overrightarrow{i} + \overrightarrow{j} + 3\overrightarrow{k},$$
$$\langle 4, -8, 0 \rangle = 4\overrightarrow{i} - 8\overrightarrow{j} \tag{9.11}$$

Combining equations (9.9), (9.10), and (9.11) gives

$$T\left(\langle 2, 5, -1 \rangle\right) = 2T\left(\overrightarrow{i}\right) + 5T\left(\overrightarrow{j}\right) - T\left(\overrightarrow{k}\right) = \langle 4, -9 \rangle$$
$$T\left(\langle -7, 1, 3 \rangle\right) = -7T\left(\overrightarrow{i}\right) + T\left(\overrightarrow{j}\right) + 3T\left(\overrightarrow{k}\right) = \langle 1, 6 \rangle$$
$$T\left(\langle 4, -8, 0 \rangle\right) = 4T\left(\overrightarrow{i}\right) - 8T\left(\overrightarrow{j}\right) + 0T\left(\overrightarrow{k}\right) = \langle -2, 10 \rangle$$

Using the definitions of $T\left(\overrightarrow{i}\right)$, $T\left(\overrightarrow{j}\right)$, and $T\left(\overrightarrow{k}\right)$ described above, the previous system of three equations now becomes:

$$T\left(\langle 2,5,-1\rangle\right) = 2\begin{bmatrix} A_{11} \\ A_{21} \end{bmatrix} + 5\begin{bmatrix} A_{12} \\ A_{22} \end{bmatrix} - \begin{bmatrix} A_{13} \\ A_{23} \end{bmatrix} = \begin{bmatrix} 4 \\ -9 \end{bmatrix}$$

$$T\left(\langle -7,1,3\rangle\right) = -7\begin{bmatrix} A_{11} \\ A_{21} \end{bmatrix} + \begin{bmatrix} A_{12} \\ A_{22} \end{bmatrix} - 3\begin{bmatrix} A_{13} \\ A_{23} \end{bmatrix} = \begin{bmatrix} 1 \\ 6 \end{bmatrix} \quad (9.12)$$

$$T\left(\langle 4,-8,0\rangle\right) = 4\begin{bmatrix} A_{11} \\ A_{21} \end{bmatrix} - 8\begin{bmatrix} A_{12} \\ A_{22} \end{bmatrix} + 0\begin{bmatrix} A_{13} \\ A_{23} \end{bmatrix} = \begin{bmatrix} -2 \\ 10 \end{bmatrix}$$

Note that we have used the linearity of the map T to arrive at the above equations. The system of equations can now be written in matrix form:

$$\begin{bmatrix} A_{11} & A_{12} & A_{13} \\ A_{21} & A_{22} & A_{23} \end{bmatrix} \begin{bmatrix} 2 & -7 & 4 \\ 5 & 1 & -8 \\ -1 & 3 & 0 \end{bmatrix} = \begin{bmatrix} 4 & 1 & -2 \\ -9 & 6 & 10 \end{bmatrix} \quad (9.13)$$

We write this symbolically as $AM^T = B$, where

$$M = \begin{bmatrix} 2 & 5 & -1 \\ -7 & 1 & 3 \\ 4 & -8 & 0 \end{bmatrix}$$

seems a natural way to define the matrix that right multiplies A when looking at equation (9.12), hence the requirement of using M^T in the equation $AM^T = B$. The columns of B are the vectors in $\mathbb{R}^2$ corresponding to the image of the columns of the matrix M^T, or the rows of M, under the map T. So now that we have our equation involving the unknown matrix A, with the known matrices M and B, it is simple enough to solve for A:

$$A = B\left(M^T\right)^{-1} \quad (9.14)$$

We will now perform the necessary calculations in *Maple*, also making sure to verify the independence of the vectors in the domain of T, hence making them a basis for $\mathbb{R}^3$.

```
> with(linalg):
> u:= [2, 5, -1]: v:= [-7, 1, 3]: w:= [4, -8, 0]:
> M:= matrix([u,v,w]);
```

$$M := \begin{bmatrix} 2 & 5 & -1 \\ -7 & 1 & 3 \\ 4 & -8 & 0 \end{bmatrix}$$

```
> rref(M);
```

$$\begin{bmatrix} 1 & 0 & 0 \\ 0 & 1 & 0 \\ 0 & 0 & 1 \end{bmatrix}$$

> B:= matrix(2,3,[4,1,-2,-9,6,10]);

$$B := \begin{bmatrix} 4 & 1 & -2 \\ -9 & 6 & 10 \end{bmatrix}$$

> A:= multiply(B,inverse(transpose(M)));

$$A := \begin{bmatrix} \frac{9}{7} & \frac{25}{28} & \frac{85}{28} \\ -\frac{1}{7} & -\frac{37}{28} & \frac{59}{28} \end{bmatrix}$$

Now, let us check that this 2×3 matrix A really is our linear map T through multiplication by A.

> multiply(A, transpose(M));

$$\begin{bmatrix} 4 & 1 & -2 \\ -9 & 6 & 10 \end{bmatrix}$$

Next consider the more abstract case, where $\mathbf{B} = \{\overrightarrow{w_1}, \overrightarrow{w_2}, \ldots, \overrightarrow{w_n}\}$ is a basis of $\mathbb{R}^n$, with known values for $\overrightarrow{v_j} = T(\overrightarrow{w_j}) \in \mathbb{R}^m$ for $1 \leq j \leq n$. Since we wish to express T as multiplication by a matrix $A \in \mathbb{R}^{m \times n}$, the following n equations must be satisfied:

$$A\overrightarrow{w_j} = \overrightarrow{v_j}, \ 1 \leq j \leq n \tag{9.15}$$

We can put all n of these equations into the form $AW = V$, where

$$W = \begin{bmatrix} \overrightarrow{w_1} & \overrightarrow{w_2} & \cdots & \overrightarrow{w_n} \end{bmatrix}, \ V = \begin{bmatrix} \overrightarrow{v_1} & \overrightarrow{v_2} & \cdots & \overrightarrow{v_n} \end{bmatrix} \tag{9.16}$$

We must make sure that the dimensions of the matrices allow us to perform the above matrix multiplication. As previously stated, $A \in \mathbb{R}^{m \times n}$, and W consists of n vectors treated as columns, each of length n, therefore $W \in \mathbb{R}^{n \times n}$. Similarly, V consists of n vectors of dimension m represented as column vectors, so $V \in \mathbb{R}^{m \times n}$. Since W is a square matrix whose columns form a basis for $\mathbb{R}^n$, it is invertible. Therefore, we can conclude that $A = VW^{-1}$. It is quite easy to verify that this general argument applied to the previous problem once again yields equation (9.13). We use *Maple* to solve Example 9.1.6 with the more general approach.

> W:= matrix(3,3,[2,-7,4,5,1,-8,-1,3,0]);

$$W := \begin{bmatrix} 2 & -7 & 4 \\ 5 & 1 & -8 \\ -1 & 3 & 0 \end{bmatrix}$$

> V:= matrix(2,3,[4, 1, -2, -9, 6, 10]);

$$V := \begin{bmatrix} 4 & 1 & -2 \\ -9 & 6 & 10 \end{bmatrix}$$

> multiply(V, inverse(W));

$$\begin{bmatrix} \frac{9}{7} & \frac{25}{28} & \frac{85}{28} \\ -\frac{1}{7} & -\frac{37}{28} & \frac{59}{28} \end{bmatrix}$$

Homework Problems

1. Given a linear map $T : \mathbb{R}^m \to \mathbb{R}^m$, expressed in terms of matrix multiplication by $A\vec{x} = \vec{y}$, where A is invertible, does $A^{-1}y = x$ correspond to the inverse map $T^{-1} : \mathbb{R}^m \to \mathbb{R}^m$?

2. For each of the given linear maps, determine the matrix A such that $T(\vec{x}) = A\vec{x}$.

(a) $T(\langle 2,1 \rangle) = \langle 0,1 \rangle$ (b) $T(\langle 2,1 \rangle) = \langle 0,1,1 \rangle$
        $T(\langle 1,2 \rangle) = \langle 1,1 \rangle$ $T(\langle 1,2 \rangle) = \langle 1,1,0 \rangle$

(c) $T(\langle 1,0,1 \rangle) = \langle 0,1 \rangle$ (d) $T(\langle 1,1,1 \rangle) = \langle 0,1,2 \rangle$
        $T(\langle 0,1,0 \rangle) = \langle 1,1 \rangle$ $T(\langle 1,2,0 \rangle) = \langle 0,1,0 \rangle$
        $T(\langle 1,1,0 \rangle) = \langle 1,0 \rangle$ $T(\langle 0,0,-1 \rangle) = \langle 1,1,1 \rangle$

3. Compute the inverse map, T^{-1}, to the maps from problem 2 parts (a) and (d).

4. So far, we have considered linear maps from $\mathbb{R}^n$ to $\mathbb{R}^m$ where we know the image of a basis of $\mathbb{R}^n$. Consider the following map:

$$T(\langle 1,2,0,1 \rangle) = \langle -5,-1 \rangle,$$
$$T(\langle -1,0,3,2 \rangle) = \langle 8,14 \rangle$$

Explain why this map cannot be defined by a unique matrix $A \in \mathbb{R}^{2 \times 4}$?

5. Construct two 2×4 matrices that satisfy the map given in problem 4.

6. Define A_b and A_c to be the matrices corresponding to the linear maps from parts (b) and (c) of problem 2. Verify the following:

(a) $A_c A_b \langle 2, 1 \rangle^T = \langle 1, 3 \rangle^T$ (b) $A_c A_b \langle 1, 2 \rangle^T = \langle 1, 0 \rangle^T$

7. Construct a linear map from $\mathbb{R}^2$ to the subspace of $\mathbb{R}^3$ given by

$$\mathbb{S} = \{ \langle x, 0, z \rangle \,|\, x, z, \in \mathbb{R} \}$$

8. For the linear map T in Example 9.1.4, find the image $T(\mathbb{S})$ of the subspace

$$\mathbb{S} = \{ a \langle 1, 2, -3 \rangle + b \langle -4, 5, 1 \rangle \,|\, a, b \in \mathbb{R} \}$$

and inverse image $T^{-1}(\mathbb{K})$ of the subspace

$$\mathbb{K} = \{ a \langle 5, 4, 6, 5 \rangle + b \langle 6, 5, 5, 9 \rangle + c \langle -1, 2, -3, 7 \rangle \,|\, a, b, c \in \mathbb{R} \}$$

Give a basis and the dimension for both the image $T(\mathbb{S})$ and the inverse image $T^{-1}(\mathbb{K})$.

9. Let $T : \mathbb{R}^2 \to \mathbb{R}^2$ have the rule

$$T(\langle x, y \rangle) = x \, \langle \cos(\theta), -\sin(\theta) \rangle + y \, \langle \sin(\theta), \cos(\theta) \rangle$$

Show that T is a linear map and explain what it does geometrically.

10. Let $\mathbb{S}$ be a subspace of $\mathbb{R}^n$, and $T : \mathbb{R}^n \to \mathbb{R}^n$ be defined by $T(\overrightarrow{v}) = \text{proj}_{\mathbb{S}}(\overrightarrow{v})$, for $\overrightarrow{v} \in \mathbb{R}$.

(a) Is the function T a linear map? Explain why if it is, but if it is not give an example to illustrate why not.

(b) If the function T in part a) is a linear map, what is the image $T(\mathbb{R}^n)$?

(c) If the function T in part a) is a linear map, what is $T^{-1}(\overrightarrow{0}_n)$?

11. (a) Let $T : \mathbb{R}^n \to \mathbb{R}^m$ and $T : \mathbb{R}^m \to \mathbb{R}^k$ be two linear maps. Show that their *composite*, $S \circ T : \mathbb{R}^n \to \mathbb{R}^k$, is also a linear map. Recall that the composite function $S \circ T : \mathbb{R}^n \to \mathbb{R}^k$ has the rule $(S \circ T)(\overrightarrow{v}) = S\,(T\,(\overrightarrow{v}))$, for all $\overrightarrow{v} \in \mathbb{R}^n$.

(b) Let $\mathbb{K}$ be a subspace of $\mathbb{R}^k$. Explain why the inverse image is

$$(S \circ T)^{-1}(\mathbb{K}) = T^{-1}\left(S^{-1}(\mathbb{K})\right).$$

12. Let $T : \mathbb{R}^n \to \mathbb{R}^m$ be a linear map that has an inverse function $T^{-1} : \mathbb{R}^m \to \mathbb{R}^n$. Prove that T^{-1} is also a linear map and $n = m$.

13. Let $T : \mathbb{R}^n \to \mathbb{R}^n$ be a linear map which has an inverse linear map $T^{-1} : \mathbb{R}^n \to \mathbb{R}^n$. Prove that if T can be written as the matrix multiplication $T(\vec{x}) = A\vec{x}$, then T^{-1} can be written as the matrix multiplication $T^{-1}(\vec{y}) = A^{-1}\vec{y}$.

14. Let $T : \mathbb{R}^n \to \mathbb{R}^n$ be a linear map and $\mathbf{B}$ be a basis of $\mathbb{R}^n$. Show that T has an inverse linear map T^{-1} if and only if $T(\mathbf{B})$ is also a basis of $\mathbb{R}^n$. This says that T is invertible if and only if T sends a basis to a basis.

15. How can you define a linear map $T : \mathbb{V} \to \mathbb{W}$, for $\mathbb{V}$ a subspace of $\mathbb{R}^n$ and $\mathbb{W}$ a subspace of $\mathbb{R}^m$? Explain how the material of this section can be altered for this new more general linear map. Does everything about linear maps also work if we replace $\mathbb{R}$ by $\mathbb{C}$ and even mix $\mathbb{R}$ and $\mathbb{C}$.

Maple Problems

1. For each of the given linear maps, determine the matrix A such that $T(\vec{x}) = A\vec{x}$.

(a) $T(\langle 1,0,1,1 \rangle) = \langle 1,0,1,1 \rangle$
$T(\langle 1,2,1,0 \rangle) = \langle 0,1,1,1 \rangle$
$T(\langle 1,-2,1,-1 \rangle) = \langle 1,2,1,0 \rangle$
$T(\langle 0,1,1,1 \rangle) = \langle 1,-2,1,-1 \rangle$

(b) $T(\langle 1,0,-1,0 \rangle) = \langle 1,-3,1,9,0 \rangle$
$T(\langle -3,2,1,5 \rangle) = \langle 4,1,1,8,0 \rangle$
$T(\langle 0,-2,1,0 \rangle) = \langle 1,-7,1,0,3 \rangle$
$T(\langle -3,0,0,-7 \rangle) = \langle 1,-2,1,-1,2 \rangle$

(c) $T(\langle 1,0,1,1 \rangle) = \langle 1,-3,1 \rangle$
$T(\langle 1,2,1,0 \rangle) = \langle 7,1,0 \rangle$
$T(\langle 1,-2,1,-1 \rangle) = \langle 3,2,-1 \rangle$
$T(\langle 0,1,1,1 \rangle) = \langle 1,-2,6 \rangle$

(d) $T(\langle 1,0,1,1,0 \rangle) = \langle 1,0,1,1,0 \rangle$
$T(\langle 0,1,1,0,1 \rangle) = \langle 0,1,1,1,-1 \rangle$
$T(\langle 1,1,1,0,1 \rangle) = \langle 1,2,1,0,1 \rangle$
$T(\langle -1,1,-1,-1,0 \rangle) = \langle 1,-2,1,-1,0 \rangle$
$T(\langle 0,0,1,0,0 \rangle) = \langle 1,-2,1,-1,1 \rangle$

2. Compute the inverse map to those maps that are invertible from problem 1.

3. Consider the following two maps, T and S, defined as follows:

$S(\langle -2,0,1,1 \rangle) = \langle 1,0,2,1,3 \rangle$
$S(\langle 1,2,1,0 \rangle) = \langle 0,1,-3,1,4 \rangle$
$S(\langle 3,-2,1,-1 \rangle) = \langle 1,2,1,-5,6 \rangle$
$S(\langle 0,-4,1,5 \rangle) = \langle 7,-2,1,-1,3 \rangle$

$T(\langle 1,0,2,1,3 \rangle) = \langle 1,1,0,1,1,1 \rangle$
$T(\langle 0,1,-3,1,4 \rangle) = \langle -1,2,1,0,0,1 \rangle$
$T(\langle 1,2,1,-5,6 \rangle) = \langle 1,-1,1,2,-1,-1 \rangle$
$T(\langle 7,-2,1,-1,3 \rangle) = \langle -3,2,-1,2,-2,2 \rangle$
$T(\langle 4,-2,2,0,3 \rangle) = \langle 1,5,-2,1,9,-6 \rangle$

(a) Compute the matrices A_S and A_T corresponding to the maps S and T, respectively.

(b) Verify that the composition of the maps S and T, given by $A_T A_S$ takes each vector of $\mathbb{R}^4$ in the definition of S to its corresponding counterpart vector in $\mathbb{R}^6$ found in the range of the definition of T.

4. Verify your answer to *Homework* problem 8.

5. Let $T : \mathbb{R}^5 \rightarrow \mathbb{R}^8$ have the rule:

$$T((x, y, z, u, v)) = (5x - y + z + 2u - v, 3x + 2y + z - u + v,$$
$$7x + y + z - 6u + 9v, x - 4y + 10z + 13u - 11v,$$
$$- x + 3y - 9z + u + v, -3x + 6y + 8z - 7u + 6v,$$
$$- 2x + 15y + z - 10u + 2v, 17x - 3y + 14z - 5u + 8v)$$

Find the image, $T(\mathbb{S})$, of the subspace

$$\mathbb{S} = \{a\,(1, 2, -3, -1, -6) + b\,(-1, 5, -12, 17, -9)$$
$$+ c\,(1, -1, 2, -5, -1) + d\,(1, 0, 1, 0, -1) \mid a, b, c, d \in \mathbb{R}\}$$

and the inverse image, $T^{-1}(\mathbb{K})$, of the subspace

$$\mathbb{K} = \{a\,(5, 6, 9, 7, -7, 11, 14, 28) + b\,(2, 1, 4, 12, -7, 7, -7, 17)$$
$$+ c\,(-4, 2, 1, 6, -1, 3, -5, 9) \mid a, b, c \in \mathbb{R}\}$$

Give a basis and the dimension for both the image, $T(\mathbb{S})$, and the inverse image, $T^{-1}(\mathbb{K})$.

6. For the linear maps given in problem 1, determine which of them have inverse functions T^{-1} and find for them the matrix B that represents them as matrix multiplication. Show that this matrix B is the inverse to the matrix which represents T.

9.2   The Kernel and Image Subspaces of a Linear Map

From Section 9.1, we know that linear maps take subspaces of $\mathbb{R}^n$ to subspaces of $\mathbb{R}^m$. For a linear map $T : \mathbb{R}^n \rightarrow \mathbb{R}^m$ the most important two subspaces of $\mathbb{R}^n$ and $\mathbb{R}^m$ are the *kernel* of T, denoted $\mathrm{Ker}(T)$, and the *image* of T, denoted $\mathrm{Im}(T)$ corresponding to $T^{-1}(\overrightarrow{0}_m)$ and $T(\mathbb{R}^n)$, respectively.

Definition 9.2.1. The *kernel* of a linear map $T : \mathbb{R}^n \to \mathbb{R}^m$, denoted $\text{Ker}(T)$, is defined as

$$\text{Ker}(T) = \left\{ \vec{v} \in \mathbb{R}^n \mid T(\vec{v}) = \vec{0}_m \right\} = T^{-1}(\vec{0}_m)$$

Definition 9.2.2. The *image* of a linear map $T : \mathbb{R}^n \to \mathbb{R}^m$, denoted $\text{Im}(T)$, is defined as

$$\text{Im}(T) = \{T(\vec{v}) \mid \vec{v} \in \mathbb{R}^n\} = T(\mathbb{R}^n)$$

The first thing to notice is that the kernel of T is a subspace of $\mathbb{R}^n$, while the image of T is a subspace of $\mathbb{R}^m$. These two subspaces contain important information about the nature of T.

Example 9.2.1. Let us do an example of computing the kernel and image of the linear map $T : \mathbb{R}^4 \to \mathbb{R}^7$ given by

$$T((x, y, z, w)) = \langle x - y + z + w, 3x + 5y + 8w, x + 7z + 8w,$$
$$x + y + 2z + 4w, 6y + 6w, x + 2z + 3w, -8x + y + 11z + 4w \rangle$$

We first find the 7×4 matrix A that represents T as a multiplication, then we solve the system $A\vec{v} = \vec{0}_7$ to get the kernel of T. The image of T is the subspace of $\mathbb{R}^7$ spanned by the columns of the matrix A. We will find a basis for the image of T by applying *rref* to A^T and taking its nonzero rows.

> with(linalg):

> T:= (x, y, z, w) - > [x - y + z + w, 3*x + 5*y + 8*w, x + 7*z + 8*w, x + y + 2*z + 4*w, 6*y +6*w, x + 2*z + 3*w, -8*x + y + 11*z + 4*w]:

> T(-3,5,1,-8);

$$[-15, -48, -60, -28, -18, -25, 8]$$

> A:= transpose(matrix([T(1,0,0,0), T(0,1,0,0), T(0,0,1,0), T(0,0,0,1)]));

$$A := \begin{bmatrix} 1 & -1 & 1 & 1 \\ 3 & 5 & 0 & 8 \\ 1 & 0 & 7 & 8 \\ 1 & 1 & 2 & 4 \\ 0 & 6 & 0 & 6 \\ 1 & 0 & 2 & 3 \\ -8 & 1 & 11 & 4 \end{bmatrix}$$

> rref(augment(A, matrix(7,1,0)));

$$\begin{bmatrix} 1 & 0 & 0 & 1 & 0 \\ 0 & 1 & 0 & 1 & 0 \\ 0 & 0 & 1 & 1 & 0 \\ 0 & 0 & 0 & 0 & 0 \\ 0 & 0 & 0 & 0 & 0 \\ 0 & 0 & 0 & 0 & 0 \\ 0 & 0 & 0 & 0 & 0 \end{bmatrix}$$

The *rref* matrix given above tells us that the kernel of T can be expressed by the equation:

$$\begin{bmatrix} x \\ y \\ z \\ w \end{bmatrix} = w \begin{bmatrix} -1 \\ -1 \\ -1 \\ 1 \end{bmatrix}$$

for w arbitrary. So $\mathrm{Ker}(T)$ has dimension one, with basis vector $\langle -1, -1, -1, 1 \rangle$. Now let us find a basis for the image of T. Before we do this, however, we should have a strategy. It was easy to compute $\mathrm{Ker}(T)$, since we simply determine the spanning set for the homogeneous solution. To determine the image of T, we first write the map as follows:

$$T(\langle x, y, z, w \rangle) = x \langle 1, 3, 1, 1, 0, 1, -8 \rangle + y \langle -1, 5, 0, 1, 6, 0, 1 \rangle$$
$$+ z \langle 1, 0, 7, 2, 0, 2, 11 \rangle + w \langle 1, 8, 8, 4, 6, 3, 4 \rangle \qquad (9.17)$$
$$= x \overrightarrow{r_1} + y \overrightarrow{r_2} + z \overrightarrow{r_3} + w \overrightarrow{r_4}$$

Then the matrix A that turns T into multiplication by A is the 7×4 matrix with columns $\overrightarrow{r_1}, \overrightarrow{r_2}, \overrightarrow{r_3}, \overrightarrow{r_4}$ or

$$A = \begin{bmatrix} \overrightarrow{r_1} & \Big| & \overrightarrow{r_2} & \Big| & \overrightarrow{r_3} & \Big| & \overrightarrow{r_4} \end{bmatrix}$$

As well, $\overrightarrow{r_j} \in T\left(\mathbb{R}^4\right)$ for $1 \le j \le 4$ since $\overrightarrow{r_1} = T(\langle 1, 0, 0, 0 \rangle)$, $\overrightarrow{r_2} = T(\langle 0, 1, 0, 0 \rangle)$, $\overrightarrow{r_3} = T(\langle 0, 0, 1, 0 \rangle)$, and $\overrightarrow{r_4} = T(\langle 0, 0, 0, 1 \rangle)$. The set $\{\overrightarrow{r_1}, \overrightarrow{r_2}, \overrightarrow{r_3}, \overrightarrow{r_4}\}$ also spans $T\left(\mathbb{R}^4\right)$ since we can write the rule of T as a linear combination of the $\overrightarrow{r_j}$'s. To determine the spanning set corresponding to the set of $\overrightarrow{r_j}$'s, we simply transpose the matrix A, whose columns are the $\overrightarrow{r_j}$'s, and row reduce. The result is given below.

> ImT:= rref(transpose(A));

$$ImT := \begin{bmatrix} 1 & 0 & 0 & \frac{4}{51} & -\frac{42}{17} & \frac{25}{51} & -\frac{356}{51} \\ 0 & 1 & 0 & \frac{11}{51} & \frac{12}{17} & \frac{5}{51} & -\frac{61}{51} \\ 0 & 0 & 1 & \frac{14}{51} & \frac{6}{17} & \frac{11}{51} & \frac{131}{51} \\ 0 & 0 & 0 & 0 & 0 & 0 & 0 \end{bmatrix}$$

From the matrix ImT above, we see that the subspace $\text{Im}(T)$ has dimension three, with basis the first three rows of the matrix ImT. Thus, any vector in $\text{Im}(T)$ can be expressed as a linear combination of the three vectors:

$$\left\{ \left\langle 1,0,0,\frac{4}{51},-\frac{42}{17},\frac{25}{51},-\frac{356}{51} \right\rangle, \left\langle 0,1,0,\frac{11}{51},\frac{12}{17},\frac{5}{51},-\frac{61}{51} \right\rangle, \left\langle 0,0,1,\frac{14}{51},\frac{6}{17},\frac{11}{51},\frac{131}{51} \right\rangle \right\}$$

Note that the dimension of the kernel of T added together with the dimension of the image of T is four, which is the dimension of $\mathbb{R}^4$, the domain of T. Is this a coincidence or a general fact? If it is a general fact, then why is it true?

Happily, it is true in general. If $T : \mathbb{R}^n \to \mathbb{R}^m$ is a linear map that can be represented in matrix form by $T(\overrightarrow{x}) = A\overrightarrow{x}$, then the dimension of $\text{Im}(T)$ is equal to the column rank of A, which is equivalent to the number of dependent variables in solving $A\overrightarrow{x} = \overrightarrow{0}$, which is how we compute $\text{Ker}(T)$ in the first place. This leaves us to assume that the number of independent variables must be the dimension of $\text{Ker}(T)$, since the number of independent and dependent variables must sum to n.

Theorem 9.2.1. *Given a linear map $T : \mathbb{R}^n \to \mathbb{R}^m$, $\text{Im}(T)$ and $\text{Ker}(T)$ satisfy the property:*

$$\dim(\text{Im}(T)) + \dim(\text{Ker}(T)) = n$$

Proof. Let $T : \mathbb{R}^n \to \mathbb{R}^m$ be a linear map. Then $\text{Ker}(T) = T^{-1}(\overrightarrow{0}_m)$ is a subspace of $\mathbb{R}^n$. Its orthogonal complement, $\text{Ker}(T)^\perp$, is also a subspace of $\mathbb{R}^n$, where the sum and intersection of the two subspaces satisfy:

$$\text{Ker}(T) + \text{Ker}(T)^\perp = \mathbb{R}^n, \quad \text{Ker}(T) \cap \text{Ker}(T)^\perp = \left\{ \overrightarrow{0}_n \right\} \qquad (9.18)$$

Hence, if we have any basis $\{\overrightarrow{v_1}, \overrightarrow{v_2}, \ldots, \overrightarrow{v_k}\}$ of $\text{Ker}(T)$, and any basis $\{\overrightarrow{w_1}, \overrightarrow{w_2}, \ldots, \overrightarrow{w_p}\}$ of $\text{Ker}(T)^\perp$, then the union, $\{\overrightarrow{v_1}, \overrightarrow{v_2}, \ldots, \overrightarrow{v_k}, \overrightarrow{w_1}, \overrightarrow{w_2}, \ldots, \overrightarrow{w_p}\}$, is a basis for $\mathbb{R}^n$ with $k + p = n$.

Now, if we take any element $\overrightarrow{u} \in \mathbb{R}^n$, it can be expressed as

$$\overrightarrow{u} = a_1\overrightarrow{v_1} + a_2\overrightarrow{v_2} + \cdots + a_k\overrightarrow{v_k} + b_1\overrightarrow{w_1} + b_2\overrightarrow{w_2} + \cdots + b_p\overrightarrow{w_p}$$

for real scalars a_i's and b_j's, $1 \le i \le k$ and $1 \le j \le p$. Then plugging $\overrightarrow{u}$ into T we get

$$\begin{aligned} T(\overrightarrow{u}) &= a_1T(\overrightarrow{v_1}) + a_2T(\overrightarrow{v_2}) + \cdots + a_kT(\overrightarrow{v_k}) \\ &\quad + b_1T(\overrightarrow{w_1}) + b_2T(\overrightarrow{w_2}) + \cdots + b_pT(\overrightarrow{w_p}) \\ &= b_1T(\overrightarrow{w_1}) + b_2T(\overrightarrow{w_2}) + \cdots + b_pT(\overrightarrow{w_p}) \end{aligned}$$

since $\{\overrightarrow{v_1}, \overrightarrow{v_2}, \ldots, \overrightarrow{v_k}\} \subseteq \text{Ker}(T)$ says that $T(\overrightarrow{v_j}) = \overrightarrow{0}_m$, for $1 \le j \le k$. So the set $\{T(\overrightarrow{w_1}), T(\overrightarrow{w_2}), \ldots, T(\overrightarrow{w_p})\}$ spans $\text{Im}(T(\mathbb{R}^n))$.

We will be done if we can show that this set is also independent, implying that $\{T(\overrightarrow{w_1}), T(\overrightarrow{w_2}), \ldots, T(\overrightarrow{w_p})\}$ is a basis of $\text{Im}(T(\mathbb{R}^n))$ and giving $\text{Im}(T(\mathbb{R}^n))$ dimension p. Let

$$b_1 T(\overrightarrow{w_1}) + b_2 T(\overrightarrow{w_2}) + \cdots + b_p T(\overrightarrow{w_p}) = \overrightarrow{0}_m \tag{9.19}$$

for scalars the b_j's, for $1 \le j \le p$. The set $\{T(\overrightarrow{w_1}), T(\overrightarrow{w_2}), \ldots, T(\overrightarrow{w_p})\}$ is independent if all the b_j's must be 0 in order to satisfy (9.19). We know that

$$b_1 T(\overrightarrow{w_1}) + b_2 T(\overrightarrow{w_2}) + \cdots + b_p T(\overrightarrow{w_p}) = T(b_1\overrightarrow{w_1} + b_2\overrightarrow{w_2} + \cdots + b_p\overrightarrow{w_p}) \tag{9.20}$$

and substituting this into equation (9.19) gives that $b_1\overrightarrow{w_1} + b_2\overrightarrow{w_2} + \cdots + b_p\overrightarrow{w_p} \in \text{Ker}(T)$, thus $b_1\overrightarrow{w_1} + b_2\overrightarrow{w_2} + \cdots + b_p\overrightarrow{w_p} \in \text{Ker}(T) \cap \text{Ker}(T)^{\perp}$, which, by the second identity given in equation (9.18), implies that the linear combination is the zero vector of $\mathbb{R}^n$:

$$b_1\overrightarrow{w_1} + b_2\overrightarrow{w_2} + \cdots + b_p\overrightarrow{w_p} = \overrightarrow{0}_n$$

Since $\{\overrightarrow{w_1}, \overrightarrow{w_2}, \ldots, \overrightarrow{w_p}\}$ is an independent set, $b_j = 0$ for $1 \le j \le p$. We can now conclude that $\dim(\text{Im}(T)) + \dim(\text{Ker}(T)) = n$. $\qquad\square$

In the last part of this section, we will relate the kernel and image of a linear map T to whether the linear map is either a one-to-one or onto function. This will be very important in deciding if a linear map T has an inverse linear map T^{-1} since a function only possesses an inverse function when it is both one-to-one and onto.

Definition 9.2.3. A function $F : D \to R$, with domain set D and range set R, is said to be *one-to-one* if whenever $F(x) = F(y)$ for $x, y \in D$, then $x = y$.

Definition 9.2.4. A function $F : D \to R$ is said to be *onto* if $F(D) = R$, that is, $R = \{F(x) \mid x \in D\}$.

Definition 9.2.5. A function $F : D \to R$ is said to be *bijective* if F is both one-to-one and onto.

A one-to-one function F is one that never repeats its range values, while an onto function F is one which attains all of its range values. If we consider real valued functions, with $D, R \subseteq \mathbb{R}$, then $F : D \to R$ is one-to-one if any horizontal line crosses its graph at most once, and onto if each horizontal line $y = r$, for $r \in R$, crosses the graph of F at least once. See Figure 9.1 below. As examples, $\sin : [0, 2\pi] \to [-1, 1]$ is not one-to-one, since $\sin(0) = \sin(2\pi)$ but it is onto since for each $y \in [-1, 1]$, there is an angle $\theta \in [0, 2\pi]$ with $\sin(\theta) = y$. Now if we alter this to $\sin : \left[0, \frac{\pi}{2}\right] \to [-1, 1]$, then it is one-to-one but not onto. Finally, if we define $\sin : \left[-\frac{\pi}{2}, \frac{\pi}{2}\right] \to [-1, 1]$, then our function is bijective. If you are familiar with the definition of arcsin, then you will notice that the domain and range in the final example are the principle range and

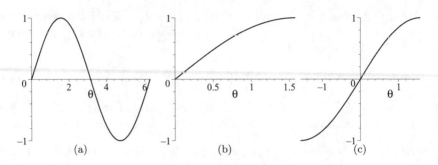

Figure 9.1: $(a)\sin(\theta)$ is onto, (b) one-to-one, and (c) bijective.

domain, respectively, for arcsin. Also, note that in the second definition, if we had limited the range to $[0, 1]$, the function would have been bijective. The graphs in Figure 9.1 illustrate the three different situations of onto, one-to-one and bijective. For a linear map $T : \mathbb{R}^n \to \mathbb{R}^m$, one-to-one and onto are related to its kernel and image subspaces.

Theorem 9.2.2. *A linear map* $T : \mathbb{R}^n \to \mathbb{R}^m$ *is one-to-one, if* $\text{Ker}(T) = \{\vec{0}_n\}$, *and it is onto if* $\text{Im}(T) = \mathbb{R}^m$.

Proof. Another way of saying this is that T is one-to-one if $\dim(\text{Ker}(T)) = 0$, and onto if $\dim(\text{Im}(T)) = m$. Let us look at one-to-one first. If $T(\vec{x}) = T(\vec{y})$ for $\vec{x}, \vec{y} \in \mathbb{R}^n$, Then $T(\vec{x}) - T(\vec{y}) = \vec{0}_m$. However, since T is linear, this gives $T(\vec{x} - \vec{y}) = \vec{0}_m$. We can therefore conclude that $\vec{x} - \vec{y} \in \text{Ker}(T)$. So T is one-to-one exactly when $\text{Ker}(T) = \{\vec{0}_n\}$. The onto condition is automatic from the definition of onto and $\text{Im}(T)$. $\square$

Note that the linear map $T : \mathbb{R}^4 \to \mathbb{R}^7$ given in Example 9.2.1 is neither one-to-one nor onto.

Theorem 9.2.3. *A linear map* $T : \mathbb{R}^n \to \mathbb{R}^m$ *is both one-to-one and onto if and only if* $\dim(\text{Ker}(T)) = 0$ *and* $\dim(\text{Im}(T)) = m$.

From Theorem 9.2.1, if T is both one-to-one and onto, then $m = n$. It does not work the other way around since if $m = n$, then we do not know anything about whether T is one-to-one or onto. It does follow though, that if $m = n$ and T is one-to-one, then T is onto. Similarly, if $m = n$ and T is onto, then T is one-to-one.

Theorem 9.2.4. *A general linear map* $T : \mathbb{R}^n \to \mathbb{R}^m$ *is both one-to-one and onto if, and only if,* $m = n$ *and the matrix* $A \in \mathbb{R}^{n \times n}$ *representing* T *is invertible.*

This follows from our discussion above and the fact that $T(\vec{x}) = A\vec{x}$ allows us to do some simple algebra: Solving $A\vec{x} = \vec{0}_m$ gives $\text{Ker}(T) = \{\vec{0}_n\}$, for a square matrix A, exactly when A has an inverse. Also, $\vec{y} = A\vec{x}$ has exactly one solution for each $y \in \mathbb{R}^m$, for a square matrix A, exactly when A has an inverse, which gives the image of T is $\mathbb{R}^m$ exactly when A has an inverse.

Corollary 9.2.5. *A linear map $T : \mathbb{R}^n \to \mathbb{R}^m$ is one-to-one if, and only if, for any independent set of vectors $\boldsymbol{K} = \{\vec{w}_1, \ldots, \vec{w}_k\}$ of $\mathbb{R}^n$ we have that $T(\boldsymbol{K})$ is an independent set of vectors in $\mathbb{R}^m$.*

If $\boldsymbol{K}$ is a set that spans all of $\mathbb{R}^n$, then the above corollary becomes:

Corollary 9.2.6. *A linear map $T : \mathbb{R}^n \to \mathbb{R}^m$ is one-to-one if and only if T sends a basis of $\mathbb{R}^n$ to a basis of the image $T(\mathbb{R}^n)$.*

Next, a corollary involving the onto property:

Corollary 9.2.7. *A linear map $T : \mathbb{R}^n \to \mathbb{R}^n$ is onto if and only if for any spanning subset $\mathbb{S}$ of $\mathbb{R}^n$, we have that $T(\mathbb{S})$ is a spanning subset of $\mathbb{R}^n$.*

Combining the last two corollaries yields the following:

Corollary 9.2.8. *A linear map $T : \mathbb{R}^n \to \mathbb{R}^n$ is bijective if and only if T sends a basis of the domain $\mathbb{R}^n$ to a basis of the range $\mathbb{R}^n$.*

In terms of matrix representation, the above corollaries yield the following:

Corollary 9.2.9. *If $A \in \mathbb{R}^{m \times n}$ is the matrix representation of the linear map $T : \mathbb{R}^n \to \mathbb{R}^m$, then T is onto if $\text{rank}(A) = m$, and one-to-one if $\text{rank}(A) = n$.*

Homework Problems

1. Each of the following matrices represent a linear map $T : \mathbb{R}^n \to \mathbb{R}^m$. Compute both the $\text{Im}(T)$ and $\text{Ker}(T)$ for each map, expressing your answer in terms of basis vectors.

(a) $\begin{bmatrix} 2 & -3 \\ 8 & -4 \end{bmatrix}$
(b) $\begin{bmatrix} 2 & -2 \\ -1 & 1 \end{bmatrix}$
(c) $\begin{bmatrix} 2 & -8 & -2 \\ -4 & 16 & 1 \end{bmatrix}$

(d) $\begin{bmatrix} 2 & -8 & -2 \\ -4 & 16 & 1 \\ -2 & 8 & -1 \end{bmatrix}$
(e) $\begin{bmatrix} 5 & -2 & -1 \\ -4 & 2 & 1 \\ -2 & 1 & -4 \end{bmatrix}$
(f) $\begin{bmatrix} 1 & 3 & 0 & -2 \\ 0 & 1 & 3 & 3 \end{bmatrix}$

2. For each of the maps from problem 1, verify that Theorem 9.2.1 holds.

3. Classify each map from problem 1 as one-to-one, onto or bijective; if the map does not satisfy any of the given properties, then state so.

4. Compute both the $\text{Im}(T)$ and $\text{Ker}(T)$ for each of the following maps.

(a) $T(\langle x, y \rangle) = \langle x, -x, x \rangle$

(b) $T(\langle x, y \rangle) = \langle y, x, y, x \rangle$

(c) $T(\langle x, y, z \rangle) = \langle x + y, y - z, x - y \rangle$

(d) $T(\langle x, y, z \rangle) = x + y + z$

5. Classify each map from problem 4 as one-to-one, onto or bijective; if the map does not satisfy any of the given properties, then state so.

6. In problems 1 and 4, find a basis of $\text{Ker}(T)^\perp$ and show that:

$$\dim\left(\text{Ker}(T)^\perp\right) = \dim(\text{Im}(T))$$

7. Prove that a linear map $T : \mathbb{R}^n \to \mathbb{R}^m$ is not one-to-one if $m < n$.

8. Prove that a linear map $T : \mathbb{R}^n \to \mathbb{R}^m$ is not onto if $m > n$.

9. (See *Homework* problem 11 of Section 9.1). Let $T : \mathbb{R}^n \to \mathbb{R}^m$ and $S : \mathbb{R}^m \to \mathbb{R}^k$ be two linear maps. Their composite $S \circ T : \mathbb{R}^n \to \mathbb{R}^k$ is also a linear map.

(a) Let $S \circ T$ be one-to-one. Then must both S and T be one-to-one? If yes, explain why. If no, then give an example of why not.

(b) Let $S \circ T$ be onto. Then must both S and T be onto? If yes, explain why. If no, then give an example of why not.

10. Let A be the $n \times n$ matrix representing the linear map $T : \mathbb{R}^n \to \mathbb{R}^n$ through multiplication by A.

(a) Explain how the columns of the matrix A are found and what they are in the range $\mathbb{R}^n$.

(b) Explain why $\det(A) \neq 0$ if and only if T sends a basis to a basis.

Maple Problems

1. Each of the following matrices represent a linear map $T : \mathbb{R}^n \to \mathbb{R}^m$. Compute both the Im($T$) and Ker($T$) for each map. Also, find a basis and the dimension for both Im(T) and Ker(T).

(a) $\begin{bmatrix} 1 & 3 & 0 & -2 \\ -4 & 2 & 1 & 3 \\ 3 & -2 & 1 & 4 \end{bmatrix}$
(b) $\begin{bmatrix} 1 & 3 & 0 & -2 \\ -4 & 2 & 1 & 3 \\ 0 & -2 & 1 & 4 \\ 1 & -2 & 1 & 3 \end{bmatrix}$
(c) $\begin{bmatrix} 1 & -3 & -5 \\ 5 & 1 & 3 \\ -1 & 7 & 2 \\ 3 & 3 & -1 \\ -1 & 3 & 5 \end{bmatrix}$

(d) $\begin{bmatrix} 1 & 3 & 0 \\ -4 & 2 & 1 \\ 2 & -5 & 6 \\ 3 & -2 & 1 \end{bmatrix}$
(e) $\begin{bmatrix} 1 & 3 & 0 & -2 \\ -4 & 2 & 1 & 3 \\ 0 & -2 & 1 & 4 \\ -4 & 7 & -7 & 3 \\ 1 & -2 & 1 & 3 \end{bmatrix}$
(f) $\begin{bmatrix} 1 & -3 & -5 \\ 0 & 1 & 3 \\ -1 & 7 & 2 \\ 3 & 3 & -1 \\ -1 & 0 & 5 \\ -1 & 8 & 3 \\ 6 & 3 & -1 \\ -1 & -9 & 0 \end{bmatrix}$

2. Compute the rank of each of the following matrices to determine if the maps defined by them are one-to-one:

(a) $\begin{bmatrix} 1 & 1 \\ -4 & 2 \end{bmatrix}$
(b) $\begin{bmatrix} 1 & 3 & -2 \\ 5 & 0 & 3 \\ -4 & -3 & -5 \end{bmatrix}$

(c) $\begin{bmatrix} 1 & 3 & -2 \\ 5 & 0 & 3 \\ -4 & 3 & -5 \end{bmatrix}$
(d) $\begin{bmatrix} 3 & -2 & 2 & -2 \\ 9 & -2 & 6 & -14 \\ 3 & -6 & 2 & 6 \\ 0 & -2 & 0 & 4 \end{bmatrix}$

(e) $\begin{bmatrix} 3 & -8 & 2 & 10 \\ 0 & -2 & 0 & 4 \\ 1 & 3 & 3 & -7 \\ 3 & -2 & 2 & -2 \end{bmatrix}$
(f) $\begin{bmatrix} 1 & -3 & -5 & 3 & 1 \\ 0 & 1 & 3 & -4 & 0 \\ 1 & -1 & 1 & -5 & 1 \\ -2 & 7 & 13 & -10 & -2 \\ 3 & -7 & -9 & 1 & 3 \end{bmatrix}$

3. For the following matrix A, which represents a linear map $T : \mathbb{C}^4 \to \mathbb{C}^5$ by

multiplication by A, find a basis and the dimension for both $\text{Im}(T)$ and $\text{Ker}(T)$.

$$A = \begin{bmatrix} 3-i & 5i & 2+i & 8 \\ -7+i & 3 & i & -6i \\ 1 & i & 1+i & 0 \\ i & -i & 0 & 2i \\ 5+i & -i & 1 & i \end{bmatrix}$$

9.3   Composites of Two Linear Maps and Inverses

Next, we will focus our attention on the composition of two linear maps, $T : \mathbb{R}^n \to \mathbb{R}^m$ and $S : \mathbb{R}^m \to \mathbb{R}^l$, denoted $S \circ T$. By definition of the composition of functions, $S \circ T : \mathbb{R}^n \to \mathbb{R}^l$, where $(S \circ T)(\vec{x}) = S(T(\vec{x}))$, for all $\vec{x} \in \mathbb{R}^n$. The composition $S \circ T$ is a linear map, since:

$$\begin{aligned} (S \circ T)(a\vec{u} + b\vec{v}) &= S(T(a\vec{u} + b\vec{v})) \\ &= S(aT(\vec{u}) + bT(\vec{v})) \\ &= aS(T(\vec{u})) + bS(T(\vec{v})) \end{aligned}$$

for all $\vec{u}, \vec{v} \in \mathbb{R}^n$ and scalars a, b.

If the $m \times n$ matrix A represents T and the $l \times m$ matrix B represents S, then BA is the $l \times n$ matrix that represents the composite linear map $S \circ T$. In order to see this, let $T(\vec{x}) = A\vec{x}$ and $S(\vec{y}) = B\vec{y}$ for $\vec{x} \in \mathbb{R}^n$ and $\vec{y} \in \mathbb{R}^m$. Then,

$$\begin{aligned} (S \circ T)(\vec{x}) &= S(T(\vec{x})) \\ &= S(A\vec{x}) \\ &= B(A\vec{x}) \\ &= (BA)(\vec{x}) \end{aligned}$$

Example 9.3.1. Let us do an example to see that this is correct. Let $T : \mathbb{R}^4 \to \mathbb{R}^7$ and $S : \mathbb{R}^7 \to \mathbb{R}^3$ be given, respectively, by

$$T((\langle x, y, z, w \rangle)) = \langle x - y + z + w, 3x + 5y - 2w, x + 7z + 3w,$$
$$x + y + 4z, 6y - 9w, x + 2z, -8x + y + 11z \rangle$$
$$S((\langle a, b, c, d, e, f, g \rangle)) = \langle -6a + 3c - 2d + f - 9g, a - 5b + c - d + e + f - g,$$
$$7a + 3b - 4c + 5d - e - f + 8g \rangle$$

We will have *Maple* verify that $S \circ T$ can be represented as the 3×4 matrix C, with $C = BA$ for $B \in \mathbb{R}^{3 \times 7}$ representing S, and $A \in \mathbb{R}^{7 \times 4}$ representing T. The *Maple* notation for the composition of two functions is @. Pay special attention to how the matrix C is constructed, as the *transpose* command must be used since the maps T and S are defined as row vectors using *Maple*. This same situation was encountered in the previous section.

```
> with(linalg):
> T:= (x, y, z, w) -> (x - y + z + w, 3*x + 5*y - 2*w, x + 7*z + 3*w, x + y
+ 4*z, 6*y - 9*w, x + 2*z, -8*x + y + 11*z):
> S:= (a, b, c, d, e, f, g) -> (-6*a + 3*c - 2*d + f - 9*g, a - 5*b + c - d + e +
f - g, 7*a + 3*b - 4*c + 5*d - e - f + 8*g):
> (S@T)(x,y,z,w);
```

$$68x - 5y - 90z + 3w, -5x - 22y - 5z + 5w, -48x + 15y + 85z - 2w$$

```
> C:= transpose(matrix([[(S@T)(1,0,0,0)], [(S@T)(0,1,0,0)], [(S@T)(0,0,1,0)],
[(S@T)(0,0,0,1)]]));
```

$$C := \begin{bmatrix} 68 & -5 & -90 & 3 \\ -5 & -22 & -5 & 5 \\ -48 & 15 & 85 & -2 \end{bmatrix}$$

So we first calculated C using *Maple*'s composition command @. We now construct C through matrix multiplication and show that indeed $C = BA$.

```
> A:= transpose(matrix([[T(1,0,0,0)], [T(0,1,0,0)], [T(0,0,1,0)], [T(0,0,0,1)]]));
```

$$A := \begin{bmatrix} 1 & -1 & 1 & 1 \\ 3 & 5 & 0 & -2 \\ 1 & 0 & 7 & 3 \\ 1 & 1 & 4 & 0 \\ 0 & 6 & 0 & -9 \\ 1 & 0 & 2 & 0 \\ -8 & 1 & 11 & 0 \end{bmatrix}$$

```
> B:= transpose(matrix([[S(1,0,0,0,0,0,0)], [S(0,1,0,0,0,0,0)], [S(0,0,1,0,0,0,0)],
[S(0,0,0,1,0,0,0)], [S(0,0,0,0,1,0,0)], [S(0,0,0,0,0,1,0)], [S(0,0,0,0,0,0,1)]]));
```

$$B := \begin{bmatrix} -6 & 0 & 3 & -2 & 0 & 1 & -9 \\ 1 & -5 & 1 & -1 & 1 & 1 & -1 \\ 7 & 3 & -4 & 5 & -1 & -1 & 8 \end{bmatrix}$$

```
> evalm(B&*A);
```

$$\begin{bmatrix} 68 & -5 & -90 & 3 \\ -5 & -22 & -5 & 5 \\ -48 & 15 & 85 & -2 \end{bmatrix}$$

Once the concept of compositions of linear maps has been introduced, it is only natural to consider inverse maps. We begin with the definition of a function being invertible.

Definition 9.3.1. A function $F : D \to R$ is said to be *invertible*, or have an *inverse function*, if there exists a function $G : R \to D$ such that both $(F \circ G) : R \to R$ and $(G \circ F) : D \to D$ are the identity functions, I_R and I_D, on R and D, respectively.

This means that both $(F \circ G)(r) = r$ and $(G \circ F)(d) = d$, for all $r \in R$ and $d \in D$. G, and F are said to be inverse functions of each other and both are said to be invertible functions.

Here are a few examples of functions, their inverses along with corresponding domains and ranges:

$$\begin{cases} F(x) &= x^2 : (-\infty, 0] \to [0, \infty) \\ G(y) &= -\sqrt{y} : [0, \infty) \to (-\infty, 0] \end{cases}$$

$$\begin{cases} F(x) &= e^x : \mathbb{R} \to (0, \infty) \\ G(y) &= \ln y : (0, \infty) \to \mathbb{R} \end{cases}$$

$$\begin{cases} F(x) &= \tan(x) : \left(-\frac{\pi}{2}, \frac{\pi}{2}\right) \to \mathbb{R} \\ G(y) &= \arctan(y) : \mathbb{R} \to \left(-\frac{\pi}{2}, \frac{\pi}{2}\right) \end{cases}$$

For each of the above examples, with the specified domains, it can be shown that $(G \circ F)(x) = x$ and $(F \circ G)(y) = y$, for all x and y in the domain and range of F, respectively. We now connect the concept of invertibility to the concepts of one-to-one and onto.

Theorem 9.3.1. *A function $F : D \to R$ has an inverse function $G : R \to D$, if, and only if, F (as well as G) is both one-to-one and onto, that is, F is bijective or a one-to-one correspondence.*

Proof. To prove this theorem, we consider the following argument: Let F and G be an inverse function pair. Then $G \circ F = I_D$, and for $d \neq c$ two distinct elements of D, we have $(G \circ F)(d) = G(F(d)) = d$, while $(G \circ F)(c) = G(F(c)) = c$. So $F(d) \neq F(c)$, and therefore we can conclude that F is one-to-one.

Now, let $r \in R$, then $F(G(r)) = (F \circ G)(r) = I_R(r) = r$, and so F is onto. If we now switch and let F be both one-to-one and onto, then for any $r \in R$ we define $G(r) = d$ in D, where $F(d) = r$. We know d exists and is unique since F is one-to-one and onto. Therefore both $F \circ G = I_R$ and $G \circ F = I_D$, giving G is the inverse to F. $\qquad\square$

From the previous discussion, we know that a linear map $T : \mathbb{R}^n \to \mathbb{R}^m$ has an inverse function $S : \mathbb{R}^m \to \mathbb{R}^n$, if, and only if, $n = m$ and $\dim(\text{Ker}(T)) = 0$, or $\dim(\text{Im}(T)) = n$. Equivalently, a linear map $T : \mathbb{R}^n \to \mathbb{R}^m$, with $m \times n$ matrix A representing it, has an inverse linear map $S : \mathbb{R}^m \to \mathbb{R}^n$, if, and only if, A^{-1} exists, in which case the matrix representing the inverse linear map S is A^{-1}. To see this, we would require that $S \circ T(\overrightarrow{x}) = \overrightarrow{x}$, but this is represented in matrix form as $A^{-1}A\overrightarrow{x}$, which is of course, just $\overrightarrow{x}$. The same argument holds for $T \circ S(\overrightarrow{y}) = \overrightarrow{y}$, since $AA^{-1}\overrightarrow{y} = \overrightarrow{y}$.

One very useful way to uniquely define a linear map $T : \mathbb{R}^n \to \mathbb{R}^m$ is by taking a basis $\mathbf{B} = \{\overrightarrow{v_1}, \overrightarrow{v_2}, \ldots, \overrightarrow{v_n}\}$ of its domain $\mathbb{R}^n$, and then providing the values

$$T(\mathbf{B}) = \{T(\overrightarrow{v_1}), T(\overrightarrow{v_2}), \ldots, T(\overrightarrow{v_n})\}$$

The question then arises as to what do we know about the linear map T if we know something about the set $T(\mathbf{B})$. Theorem 9.3.2 is one possible answer to this question.

Theorem 9.3.2. *A linear map $T : \mathbb{R}^n \to \mathbb{R}^n$ is invertible, if, and only if, for any basis $\mathbf{B}$ of $\mathbb{R}^n$, we have that $T(\mathbf{B})$ is also a basis of $\mathbb{R}^n$.*

Example 9.3.2. Now, let us do an example of finding the matrix A representing an invertible linear map T, where T is defined by sending a basis $\mathbf{B}_V$ to a basis $\mathbf{B}_W$. We also will show that A has an inverse matrix, which corresponds to the inverse map S. Let the linear map $T : \mathbb{R}^4 \to \mathbb{R}^4$ be defined by

$$T((-1, 3, 1, 5)) = (2, -4, 1, 0)$$
$$T((6, -2, 0, -4)) = (1, 3, -7, 2)$$
$$T((9, 11, -2, 8)) = (-5, 6, -2, 1)$$
$$T((-4, -9, 3, 1)) = (0, 7, -2, -10)$$

From Section 9.1, we learned that the matrix representing A is given by $A = WV^{-1}$, where V is the matrix whose columns are the elements of the basis $\mathbf{B}_V$ and W is the matrix whose columns are the elements of the basis $\mathbf{B}_W$.

> v[1]:= [-1,3,1,5]: v[2]:= [6,-2,0,-4]: v[3]:= [9,11,-2,8]: v[4]:= [-4,-9,3,1]:
> V_row:= matrix([v[1],v[2],v[3],v[4]]);

$$V\_row := \begin{bmatrix} -1 & 3 & 1 & 5 \\ 6 & -2 & 0 & -4 \\ 9 & 11 & -2 & 8 \\ -4 & -9 & 3 & 1 \end{bmatrix}$$

> rref(V_row);

$$\begin{bmatrix} 1 & 0 & 0 & 0 \\ 0 & 1 & 0 & 0 \\ 0 & 0 & 1 & 0 \\ 0 & 0 & 0 & 1 \end{bmatrix}$$

> w[1]:= [2,-4,1,0]: w[2]:= [1,3,-7,2]: w[3]:= [-5,6,-2,1]: w[4]:= [0,7,-2,-10]:
> W_row:= matrix([w[1],w[2],w[3],w[4]]);

$$W\_row := \begin{bmatrix} 2 & -4 & 1 & 0 \\ 1 & 3 & -7 & 2 \\ -5 & 6 & -2 & 1 \\ 0 & 7 & -2 & -10 \end{bmatrix}$$

> rref(W_row);

$$\begin{bmatrix} 1 & 0 & 0 & 0 \\ 0 & 1 & 0 & 0 \\ 0 & 0 & 1 & 0 \\ 0 & 0 & 0 & 1 \end{bmatrix}$$

> A:= evalm(transpose(W_row)&*inverse(transpose(V_row)));

$$A := \begin{bmatrix} -\frac{25}{132} & \frac{41}{33} & \frac{1019}{264} & -\frac{305}{264} \\ \frac{389}{396} & -\frac{271}{99} & -\frac{4171}{792} & \frac{1657}{792} \\ -\frac{127}{132} & \frac{5}{33} & -\frac{415}{264} & \frac{61}{264} \\ -\frac{13}{99} & \frac{179}{99} & \frac{485}{198} & -\frac{317}{198} \end{bmatrix}$$

Now that we have A, we check to see if it is invertible and that it satisfies the four vector maps used to define T.

> det(A);

$$\frac{161}{264}$$

> evalm(A&*vector(v[1]));

$$\begin{bmatrix} 2 & -4 & 1 & 0 \end{bmatrix}$$

> evalm(A&*vector(v[2]));

$$\begin{bmatrix} 1 & 3 & -7 & 2 \end{bmatrix}$$

> evalm(A&*vector(v[3]));

$$\begin{bmatrix} -5 & 6 & -2 & 1 \end{bmatrix}$$

> evalm(A&*vector(v[4]));

$$\begin{bmatrix} 0 & 7 & -2 & -10 \end{bmatrix}$$

As one can see, the vectors above do agree with the columns of the matrix W_row below.

> print(W_row);

$$\begin{bmatrix} 2 & -4 & 1 & 0 \\ 1 & 3 & -7 & 2 \\ -5 & 6 & -2 & 1 \\ 0 & 7 & -2 & -10 \end{bmatrix}$$

$\blacksquare$

Homework Problems

1. For each of the following pairs of maps, determine which order, if possible, S and T can be composed in.

(a) $T : \mathbb{R}^3 \to \mathbb{R}^5$
     $S : \mathbb{R}^3 \to \mathbb{R}^3$

(b) $T : \mathbb{R}^3 \to \mathbb{R}^5$
     $S : \mathbb{R}^5 \to \mathbb{R}^3$

(c) $T : \mathbb{R}^2 \to \mathbb{R}$
     $S : \mathbb{R} \to \mathbb{R}^3$

(d) $T : \mathbb{R}^5 \to \mathbb{R}^3$
     $S : \mathbb{R}^4 \to \mathbb{R}^3$

(e) $T : \mathbb{R} \to \mathbb{R}^2$
     $S : \mathbb{R}^4 \to \mathbb{R}$

(f) $T : \mathbb{R}^2 \to \mathbb{R}^3$
     $S : \mathbb{R}^3 \to \mathbb{R}^3$

2. For each of the following pairs of maps, compute $S \circ T$ without using matrices.

(a) $T(\langle x, y \rangle) = x + y$
     $S(a) = \langle a, -a \rangle$

(b) $T(x) = \langle x, -x, 2x \rangle$
     $S(\langle a, b, c \rangle) = \langle a + b, a - c \rangle$

(c) $T(\langle x, y \rangle) = \langle -x, y, x + y \rangle$
     $S(\langle a, b, c \rangle) = \langle -a, b \rangle$

(d) $T(x) = \langle 3x, 2x, -4x \rangle$
     $S(\langle a, b, c \rangle) = a + b + c$

3. The compositions from problem 2 should have yielded two such that $(S \circ T)(\overrightarrow{v}) = \overrightarrow{v}$, for all $\overrightarrow{v}$ in the domain of T. Which two are they?

4. The two compositions found in problem 3 highlight an important sticking point in Definition 9.3.1 and Theorem 9.3.1. Verify that $(T \circ S)(\overrightarrow{v}) \neq \overrightarrow{v}$, for all $\overrightarrow{v}$ in the domain of S. Why is this the case?

5. Determine the linear map given by each of the following matrices.

(a) $\begin{bmatrix} 1 & -1 & 2 \end{bmatrix}$

(b) $\begin{bmatrix} 1 & 1 & 0 \\ 1 & 0 & -1 \end{bmatrix}$

(c) $\begin{bmatrix} -1 & 0 \\ 0 & 1 \\ 1 & 1 \end{bmatrix}$

(d) $\begin{bmatrix} -1 & 0 \\ 0 & 1 \\ 0 & 0 \end{bmatrix}$

(e) $\begin{bmatrix} 1 \\ -1 \end{bmatrix}$

(f) $\begin{bmatrix} 1 \\ -1 \\ 2 \end{bmatrix}$

6. Four of the matrices from problem 5 correspond to linear maps, S or T, from problem 2. Find them and state which maps they correspond to.

7. If the matrix C, which represents a composite of two linear maps through multiplication by C, has nonzero determinant, then must the same be true for each of the two matrices A and B, which represent the individual linear maps? If yes, then explain why. If not, then give an example to verify it.

8. Let $T : \mathbb{R}^n \to \mathbb{R}^n$ be a linear map that is invertible as a function. Show directly that its inverse function, $T^{-1} : \mathbb{R}^n \to \mathbb{R}^n$, is also a linear map.

Maple Problems

1. Compute the rank of each of the following matrices to determine if the corresponding linear map has an inverse:

(a) $\begin{bmatrix} 1 & 2 \\ -3 & 4 \end{bmatrix}$
(b) $\begin{bmatrix} 1 & 1 & 1 \\ 2 & 3 & 1 \\ 4 & 5 & 1 \end{bmatrix}$

(c) $\begin{bmatrix} 1 & 2 & -1 & 2 \\ 2 & 1 & -1 & 2 \\ 0 & -3 & 1 & -2 \\ 1 & 8 & -3 & 6 \end{bmatrix}$
(d) $\begin{bmatrix} 1 & 2 & -1 & 2 \\ 0 & -3 & 1 & -2 \\ 4 & -4 & -4 & 1 \\ -1 & 0 & 6 & -8 \end{bmatrix}$

(e) $\begin{bmatrix} -3 & 5 & -2 & 7 \\ -1 & 0 & 6 & -8 \\ 4 & -4 & -4 & 1 \\ 0 & 1 & 0 & 0 \end{bmatrix}$
(f) $\begin{bmatrix} 1 & -2 & 3 & -4 \\ -5 & 6 & -7 & 8 \\ 9 & -10 & 11 & -12 \\ -13 & 14 & -15 & 16 \end{bmatrix}$

2. For each of the linear maps found to be invertible from problem 1, compute the inverse map.

3. Graph each of the following pairs of functions on an appropriate domain, along with the graph of $y = x$. As discussed in this section, these pairs of functions are inverses of each other on the specified domain. Explain geometrically the relationship between the functions and the line $y = x$.

(a) $\begin{cases} F(x) & = x^2 \\ G(x) & = -\sqrt{x} \end{cases}$
(b) $\begin{cases} F(x) & = e^x \\ G(x) & = \ln x \end{cases}$
(c) $\begin{cases} F(x) & = \tan(x) \\ G(x) & = \arctan(x) \end{cases}$

4. A curve $y = f(x)$ can be thought of as the collection of points $(x, f(x))$ for all x in the domain of f. Using this information, can you geometrically

construct the inverse to a bijective function $y = f(x)$ through matrix multiplication? *Hint: There are two reasonable approaches, however, if you consider the curve to lie in the xy-plane in $\mathbb{R}^3$, then you can simply rotate the curve about a line that passes through the origin in $\mathbb{R}^3$ by a fixed angle.*

5. For the matrix A below that represents a linear map $T : \mathbb{C}^4 \to \mathbb{C}^4$ by multiplication by A, find the matrix B that represents its inverse map. Also check, on the level of the respective two linear maps, that these are inverse functions.

$$A = \begin{bmatrix} 3 - i & 5i & 2 + i & 8 \\ -7 + i & 3 & i & -6i \\ 1 & i & 1 + i & 0 \\ i & -i & 0 & 2i \end{bmatrix}$$

9.4   Change of Bases for the Matrix Representation of a Linear Map

A linear map $T : \mathbb{R}^n \to \mathbb{R}^m$ is usually represented by the $m \times n$ matrix A where the standard bases $\mathbf{S}_n$ of $\mathbb{R}^n$ and $\mathbf{S}_m$ of $\mathbb{R}^m$ are used to write out the column vectors of $\mathbb{R}^n$ and $\mathbb{R}^m$, respectively. This means that $\overrightarrow{y} = T(\overrightarrow{x}) = A\overrightarrow{x}$, where the column vectors $\overrightarrow{x}$ of $\mathbb{R}^n$, and $\overrightarrow{y}$ of $\mathbb{R}^m$, are written out as linear combinations in the standard bases. That is, if

$$\overrightarrow{x} = \langle x_1, x_2, \ldots, x_n \rangle$$

as a row, then

$$\overrightarrow{x} = x_1 \langle 1, 0, \ldots, 0 \rangle + x_2 \langle 0, 1, 0, \ldots, 0 \rangle + \cdots + x_n \langle 0, 0, \ldots, 0, 1 \rangle$$

What if we want to perform linear maps, not using the standard bases $\mathbf{S}_n$ of $\mathbb{R}^n$ and $\mathbf{S}_m$ of $\mathbb{R}^m$, but instead with a pair of different bases $\mathbf{B}_n = \{\overrightarrow{v_1}, \overrightarrow{v_2}, \ldots, \overrightarrow{v_n}\}$ of $\mathbb{R}^n$ and $\mathbf{B}_m = \{\overrightarrow{w_1}, \overrightarrow{w_2}, \ldots, \overrightarrow{w_m}\}$ of $\mathbb{R}^m$? What we seek is an $m \times n$ matrix A' where $\overrightarrow{y} = T(\overrightarrow{x}) = A'\overrightarrow{x}$ for column vectors $\overrightarrow{x}$ of $\mathbb{R}^n$ and $\overrightarrow{y}$ of $\mathbb{R}^m$ written out as linear combinations in the new bases.

Before we can fully understand what this means, we will need to introduce some notation. When we express a vector $\overrightarrow{x}$ as a linear combination of vectors in a basis $\mathbf{B} = \{\overrightarrow{v_1}, \overrightarrow{v_2}, \ldots, \overrightarrow{v_n}\}$, we mean that

$$\overrightarrow{x} = x_1\overrightarrow{v_1} + x_2\overrightarrow{v_2} + \cdots + x_n\overrightarrow{v_n} \tag{9.21}$$

If we are dealing with more than one basis, we now have a way to recognize the different potential representations of a vector using different bases. For instance, as long as we have n linearly independent vectors $\mathbf{C} = \{\overrightarrow{z_1}, \overrightarrow{z_2}, \ldots, \overrightarrow{z_n}\}$ in $\mathbb{R}^n$, we know that they form a basis for $\mathbb{R}^n$. Hence, any vector $\overrightarrow{x}$ can be written as a linear combination of vectors in the basis $\mathbf{C}$. So for an arbitrary $\vec{x}$,

$$\vec{x} = c_1\vec{z_1} + c_2\vec{z_2} + \cdots + c_n\vec{z_n} \tag{9.22}$$

for some scalars c_1 through c_n. It should be fairly obvious that the coefficients in the expansions of a vector using two different bases will not be the same. Clearly if $\vec{z_k} \neq \vec{v_j}$ for all $1 \leq j, k \leq n$ when comparing two bases, then one may assume that most likely, $c_k \neq x_k$ for all k as well. But we know that

$$x_1\vec{v_1} + x_2\vec{v_2} + \cdots x_n\vec{v_n} = c_1\vec{z_1} + c_2\vec{z_2} + \cdots + c_n\vec{z_n} \tag{9.23}$$

Thus, to avoid confusion, if we are not using the standard basis $\mathbf{S}_n$, we will place the name of the basis as a subscript on our vectors to denote which basis the vectors have been expanded in. For instance, equation 9.22 can be expressed in the following notation:

$$\vec{x}_\mathbf{C} = c_1\vec{z_1} + c_2\vec{z_2} + \cdots c_n\vec{z_n} = \langle c_1, c_2, \ldots, c_n \rangle_\mathbf{C} \tag{9.24}$$

If there is no subscript on the vector, we will assume that the vector is represented as a linear combination of the vectors in the standard basis.

Example 9.4.1. To clarify this idea, we now consider an example involving the following basis for $\mathbb{R}^3$:

$$\mathbf{C} = \{\vec{z_1} = \langle -1, 5, 2 \rangle, \ \vec{z_2} = \langle 3, -4, 7 \rangle, \ \vec{z_3} = \langle 8, 1, -9 \rangle\}$$

It is easy to show that:

$$\langle 10, 2, 0 \rangle = \vec{z_1} + \vec{z_2} + \vec{z_3} = \langle 1, 1, 1 \rangle_\mathbf{C}$$

We now turn our attention back to finding the matrix A'. If we have the following:

$$\vec{x}_{\mathbf{B}_n} = \langle x_1, x_2, \ldots, x_n \rangle_{\mathbf{B}_n}, \quad \vec{y}_{\mathbf{B}_m} = \langle y_1, y_2, \ldots, y_m \rangle_{\mathbf{B}_m}$$

then we now know, respectively, that

$$\vec{x}_{\mathbf{B}_n} = x_1\vec{v_1} + x_2\vec{v_2} + \cdots + x_n\vec{v_n}, \quad \vec{y}_{\mathbf{B}_m} = y_1\vec{w_1} + y_2\vec{w_2} + \cdots + y_m\vec{w_m}$$

Since T is a linear map, we have that

$$T(\vec{x}_{\mathbf{B}_n}) = x_1 T(\vec{v_1}) + x_2 T(\vec{v_2}) + \cdots + x_n T(\vec{v_n}) \tag{9.25}$$

Furthermore, since $T(\vec{v_j}) \in \mathbb{R}^m$, for $1 \leq j \leq n$, we can express each vector as a linear combination of the basis vectors in $\mathbf{B}_m$:

$$T(\vec{v_j}) = c_{1,j}\vec{w_1} + c_{2,j}\vec{w_2} + \cdots + c_{m,j}\vec{w_m} \tag{9.26}$$

for each $1 \leq j \leq n$, and scalar $c_{k,j}$'s with $1 \leq j \leq n$ and $1 \leq k \leq m$. The claim now is that $A'_{i,j} = c_{i,j}$, where the $c_{i,j}$'s are defined as in (9.26). To see this, we start at the beginning:

$$
\begin{aligned}
T\left(\overrightarrow{x}_{\mathbf{B}_n}\right) = T &\left(\sum_{j=1}^{n} x_j \overrightarrow{v_j}\right) \\
&= \sum_{j=1}^{n} x_j T\left(\overrightarrow{v_j}\right) \\
&= \sum_{j=1}^{n} x_j \left(\sum_{k=1}^{m} c_{k,j} \overrightarrow{w_k}\right) \\
&= \sum_{j=1}^{n} \sum_{k=1}^{m} x_j c_{k,j} \overrightarrow{w_k} \\
&= \sum_{k=1}^{m} \left(\sum_{j=1}^{n} x_j c_{k,j}\right) \overrightarrow{w_k} \\
&= \sum_{k=1}^{m} \alpha_k \overrightarrow{w_k}
\end{aligned}
$$

In the last step of the above chain of equalities, we set $\alpha_k = \displaystyle\sum_{j=1}^{n} x_j c_{k,j}$ to make it more clear that we have a linear combination of basis elements of $\mathbf{B}_m$. We therefore now have the following:

$$
\begin{aligned}
\overrightarrow{y}_{\mathbf{B}_m} &= \langle \alpha_1, \dots, \alpha_m \rangle_{\mathbf{B}_m} \\
&= \begin{bmatrix} \displaystyle\sum_{j=1}^{n} x_j c_{1,j} \\ \displaystyle\sum_{j=1}^{n} x_j c_{2,j} \\ \vdots \\ \displaystyle\sum_{j=1}^{n} x_j c_{m,j} \end{bmatrix}_{\mathbf{B}_m} \\
&= \begin{bmatrix} c_{1,1} & c_{1,2} & \cdots & c_{1,n} \\ c_{2,1} & \ddots & & c_{2,n} \\ \vdots & & \ddots & \vdots \\ c_{m,1} & c_{m,2} & \cdots & c_{m,n} \end{bmatrix} \begin{bmatrix} x_1 \\ x_2 \\ \vdots \\ x_n \end{bmatrix}_{\mathbf{B}_n} \\
&= T\left(\overrightarrow{x}_{\mathbf{B}_n}\right)
\end{aligned}
$$

(9.27)

Now, unfortunately this tells us what form the matrix A' takes, but we still do not know the specific values of the $c_{i,j}$'s. To do this, refer to equation (9.26). Notice that for all $1 \leq k \leq n$,

$$
T(\overrightarrow{v_k}) = c_{1,k}\overrightarrow{w_1} + c_{2,k}\overrightarrow{w_2} + \cdots + c_{m,k}\overrightarrow{w_m}
$$

$$
= \left[\begin{array}{c|c|c|c} \overrightarrow{w_1} & \overrightarrow{w_2} & \cdots & \overrightarrow{w_m} \end{array} \right] \begin{bmatrix} c_{1,k} \\ c_{2,k} \\ \vdots \\ c_{m,k} \end{bmatrix} \tag{9.28}
$$

Applying this to all the vectors allows us to construct the following matrix equation:

$$
\left[\begin{array}{c|c|c|c} T(\overrightarrow{v_1}) & T(\overrightarrow{v_2}) & \cdots & T(\overrightarrow{v_n}) \end{array} \right] = \left[\begin{array}{c|c|c|c} \overrightarrow{w_1} & \overrightarrow{w_2} & \cdots & \overrightarrow{w_m} \end{array} \right] A' \tag{9.29}
$$

This says that the matrix $A' = W^{-1}T(V)$, where W is the matrix whose columns are the basis elements of $\mathbf{B}_m$ and $T(V)$ is the matrix whose columns are results of applying T to the basis elements of $\mathbf{B}_n$. We also have $A' = W^{-1}AV$ where V is the matrix whose columns are the elements of the basis $\mathbf{B}_n$ since the matrix $T(V) = AV$ for A the matrix representing T in the standard bases.

Example 9.4.2. Of course, now we must do an example to verify our reasoning above. Let $T : \mathbb{R}^3 \longrightarrow \mathbb{R}^5$ be the map given as follows:

$$
T(\langle x, y, z \rangle) = \langle 5x - 3y + 2z, x - y - z, -7x + z, 6y - 11z, 8x + y - 5z \rangle
$$

whose corresponding matrix form is given by

$$
A\overrightarrow{v} = \begin{bmatrix} 5 & -3 & 2 \\ 1 & -1 & -1 \\ -7 & 0 & 1 \\ 0 & 6 & -11 \\ 8 & 1 & -5 \end{bmatrix} \begin{bmatrix} x \\ y \\ z \end{bmatrix}
$$

In our case, we wish to find the 5×3 matrix A' representing T in the two bases

$$
\mathbf{B}_3 = \left\{ \overrightarrow{v_1} = \langle -1, 5, 2 \rangle, \overrightarrow{v_2} = \langle 3, -4, 7 \rangle, \overrightarrow{v_3} = \langle 8, 1, -9 \rangle \right\}
$$

of $\mathbb{R}^3$ and

$$
\mathbf{B}_5 = \left\{ \overrightarrow{w_1} = \langle 1, -1, 3, -2, 5 \rangle, \overrightarrow{w_2} = \langle -4, 2, 0, 1, -3 \rangle, \overrightarrow{w_3} = \langle -2, 1, 5, -7, 2 \rangle, \right.
$$
$$
\left. \overrightarrow{w_4} = \langle 0, 9, 1, -2, 1 \rangle, \overrightarrow{w_5} = \langle 6, 8, 1, -1, -3 \rangle \right\}
$$

of $\mathbb{R}^5$. Our matrix A' must satisfy

$$T\left(\overrightarrow{v_1}\right) \hat{=} A' \begin{bmatrix} 1 \\ 0 \\ 0 \end{bmatrix}, \quad T\left(\overrightarrow{v_2}\right) \hat{=} A' \begin{bmatrix} 0 \\ 1 \\ 0 \end{bmatrix}, \quad T\left(\overrightarrow{v_3}\right) \hat{=} A' \begin{bmatrix} 0 \\ 0 \\ 1 \end{bmatrix}$$

where $\hat{=}$ means when each $T\left(\overrightarrow{v_j}\right)$ is written as a linear combination in the $\mathbf{B}_5$ basis of $\overrightarrow{w_j}$'s.

> with(linalg):
> T:= (x, y, z) - > [5*x-3*y+2*z, x-y-z, -7*x+z, 6*y-11*z, 8*x+y-5*z]:
> v[1]:= (-1, 5, 2): v[2]:= (3, -4, 7): v[3]:= (8, 1, -9):
> T(v[1]);

$$[-16, -8, 9, 8, -13]$$

> w[1]:= [1, -1, 3, -2, 5]: w[2]:= [-4, 2, 0, 1, -3]: w[3]:= [-2, 1, 5, -7, 2]: w[4]:= [0, 9, 1, -2, 1]: w[5]:= [6, 8, 1, -1, -3]:
> W:= transpose(matrix([w[1], w[2], w[3], w[4], w[5]]));

$$W := \begin{bmatrix} 1 & -4 & -2 & 0 & 6 \\ -1 & 2 & 1 & 9 & 8 \\ 3 & 0 & 5 & 1 & 1 \\ -2 & 1 & -7 & -2 & -1 \\ 5 & -3 & 2 & 1 & -3 \end{bmatrix}$$

> TV:= transpose(matrix([T(v[1]), T(v[2]), T(v[3])]));

$$TV := \begin{bmatrix} -16 & 41 & 19 \\ -8 & 0 & 16 \\ 9 & -14 & -65 \\ 8 & -101 & 105 \\ -13 & -15 & 110 \end{bmatrix}$$

> A_Prime:= evalm(inverse(W)&*TV);

$$A\_Prime := \begin{bmatrix} \dfrac{11217}{2608} & -\dfrac{46945}{1304,} & \dfrac{43531}{2608} \\[2mm] \dfrac{42179}{5216} & -\dfrac{93259}{2608} & -\dfrac{34815}{5216} \\[2mm] -\dfrac{2031}{5216} & \dfrac{48327}{2608} & -\dfrac{126789}{5216} \\[2mm] -\dfrac{19997}{5216} & \dfrac{16117}{2608} & \dfrac{97185}{5216} \\[2mm] \dfrac{4897}{2608} & -\dfrac{6297}{1304} & -\dfrac{31733}{2608} \end{bmatrix}$$

> C1:= evalm(A_Prime&*matrix([[1], [0], [0]]));

$$C1 := \begin{bmatrix} \dfrac{11217}{2608} \\[2mm] \dfrac{42179}{5216} \\[2mm] -\dfrac{2031}{5216} \\[2mm] -\dfrac{19997}{5216} \\[2mm] \dfrac{4897}{2608} \end{bmatrix}$$

> evalm(C1[1,1]*w[1] + C1[2,1]*w[2] + C1[3,1]*w[3] + C1[4,1]*w[4]
+ C1[5,1]*w[5]);

$$\begin{bmatrix} -16 & -8 & 9 & 8 & -13 \end{bmatrix}$$

> T(v[1]);

$$[-16, -8, 9, 8, -13]$$

> C2:= evalm(A_Prime&*matrix([[0], [1], [0]]));

$$C2 := \begin{bmatrix} -\dfrac{46945}{1304} \\[2mm] -\dfrac{93259}{2608} \\[2mm] \dfrac{48327}{2608} \\[2mm] \dfrac{16117}{2608} \\[2mm] -\dfrac{6297}{1304} \end{bmatrix}$$

> evalm(C2[1,1]*w[1] + C2[2,1]*w[2] + C2[3,1]*w[3] + C2[4,1]*w[4]
+ C2[5,1]*w[5]);

$$\begin{bmatrix} 41 & 0 & -14 & -101 & -15 \end{bmatrix}$$

> T(v[2]);

$$[41, 0, -14, -101, -15]$$

> C3:= evalm(A_Prime&*matrix([[0], [0], [1]]));

$$C3 := \begin{bmatrix} \dfrac{43531}{2608} \\[2mm] -\dfrac{34815}{5216} \\[2mm] -\dfrac{126789}{5216} \\[2mm] \dfrac{97185}{5216} \\[2mm] -\dfrac{31733}{2608} \end{bmatrix}$$

> evalm(C3[1,1]*w[1] + C3[2,1]*w[2] + C3[3,1]*w[3] + C3[4,1]*w[4]
+ C3[5,1]*w[5]);

$$\begin{bmatrix} 19 & 16 & -65 & 105 & 110 \end{bmatrix}$$

> T(v[3]);

$$[19, 16, -65, 105, 110]$$

So $A' = W^{-1}T(V)$, where W is the matrix whose columns are the basis elements of $\mathbf{B}_5$ and $T(V)$ is the matrix whose columns are T of the basis elements of $\mathbf{B}_3$.

Now, we also can write the matrix A' in terms of the matrix A that represents the linear map T in the standard bases and the two matrices whose columns consist of the new basis elements. We have $A' = W^{-1}AV$ where V is the matrix whose columns are the elements of the basis $\mathbf{B}_3$ since the matrix $T(V) = AV$.

> A:= transpose(matrix([T(1,0,0),T(0,1,0),T(0,0,1)]));

$$A := \begin{bmatrix} 5 & -3 & 2 \\ 1 & -1 & -1 \\ -7 & 0 & 1 \\ 0 & 6 & -11 \\ 8 & 1 & -5 \end{bmatrix}$$

> V:= transpose(matrix([[v[1]],[v[2]],[v[3]]]));

$$V := \begin{bmatrix} -1 & 3 & 8 \\ 5 & -4 & 1 \\ 2 & 7 & -9 \end{bmatrix}$$

> evalm(A&*V);

$$\begin{bmatrix} -16 & 41 & 19 \\ -8 & 0 & 16 \\ 9 & -14 & -65 \\ 8 & -101 & 105 \\ -13 & -15 & 110 \end{bmatrix}$$

> evalm(TV - A&*V);

$$\begin{bmatrix} 0 & 0 & 0 \\ 0 & 0 & 0 \\ 0 & 0 & 0 \\ 0 & 0 & 0 \\ 0 & 0 & 0 \end{bmatrix}$$

> New_A_Prime:= evalm(inverse(W)&*A&*V);

$$
New\_A\_Prime := \begin{bmatrix}
\dfrac{11217}{2608} & -\dfrac{46945}{1304} & \dfrac{43531}{2608} \\[2mm]
\dfrac{42179}{5216} & -\dfrac{93259}{2608} & -\dfrac{34815}{5216} \\[2mm]
-\dfrac{2031}{5216} & \dfrac{48327}{2608} & -\dfrac{126789}{5216} \\[2mm]
-\dfrac{19997}{5216} & \dfrac{16117}{2608} & \dfrac{97185}{5216} \\[2mm]
\dfrac{4897}{2608} & -\dfrac{6297}{1304} & -\dfrac{31733}{2608}
\end{bmatrix}
$$

> evalm(New_A_Prime - A_Prime);

$$
\begin{bmatrix}
0 & 0 & 0 \\
0 & 0 & 0 \\
0 & 0 & 0 \\
0 & 0 & 0 \\
0 & 0 & 0
\end{bmatrix}
$$

Clearly, this method works. Next, we will try to illustrate the derivation of the formula $A' = W^{-1}AV$ in a much simpler way. Mathematicians have come up with a very nice tool to help visualize this derivation process. The following is an example of a *commutative diagram*:

$$
\begin{array}{ccc}
\vec{x}_{\mathbf{B}_n} & \xrightarrow{\;\;A'\;\;} & \vec{y}_{\mathbf{B}_m} \\[2mm]
{\scriptstyle V}\downarrow & & \uparrow{\scriptstyle W^{-1}} \\[2mm]
\vec{x} & \xrightarrow{\;\;A\;\;} & \vec{y}
\end{array}
$$

We will now attempt to interpret this diagram. Remember that the matrix A' corresponds to the linear map T with the property that $T(\vec{x}_{\mathbf{B}_n}) = \vec{y}_{\mathbf{B}_m}$, and uses the basis $\mathbf{B}_n$ for $\mathbb{R}^n$ and $\mathbf{B}_m$ for $\mathbb{R}^m$. Starting in the upper-left corner of this diagram at $\vec{x}_{\mathbf{B}_n}$ and following the arrow to the right, which means applying A', we end up with $\vec{y}_{\mathbf{B}_m}$, hence $A'\vec{x}_{\mathbf{B}_n} = \vec{y}_{\mathbf{B}_m}$. This in itself does us no good, however, notice that instead of following the arrow to the right, we could have just as easily applied the matrix V to $\vec{x}_{\mathbf{B}_n}$ and ended up with the vector $\vec{x}$ in the lower-left corner of the diagram. To see why this is true, remember that V was the matrix whose columns were the elements of the basis $\mathbf{B}_n$, which yields the expansion,

$$
\vec{x}_{\mathbf{B}_n} = \sum_{j=1}^{n} x_j \vec{v_j} = V \cdot [x_1, x_2, \ldots, x_n]^T \tag{9.30}
$$

which is then interpreted in the standard basis $\mathbf{S}_n$ as

$$
\sum_{j=1}^{n} x_j \overrightarrow{v_j} = \sum_{j=1}^{n} x_j \left(\sum_{k=1}^{n} v_{k,j} \overrightarrow{e_j} \right)
$$
$$
= \sum_{k=1}^{n} \left(\sum_{j=1}^{n} x_j v_{k,j} \right) \overrightarrow{e_j}
$$

(9.31)

Now that we have the vector $\overrightarrow{x}$ after applying V by left multiplication to $\overrightarrow{x}_{\mathbf{B}_n}$, we simply left multiply by A, which is the map matrix under the standard basis. Thus, we are now in the lower-right corner of the diagram. At this point, we know that $AV\overrightarrow{x}_{\mathbf{B}_n} = \overrightarrow{y}$. In a similar argument to that for the case of V, we can go from $\overrightarrow{y}$ to $\overrightarrow{y}_{\mathbf{B}_m}$ by left multiplication of W^{-1}, where $W \in \mathbb{R}^{m \times m}$ is the matrix whose columns are the basis vectors of $\mathbf{B}_n$. Hence, we are now in the upper-right corner of the diagram. As a result, we see that $A' = W^{-1}AV$ for all vectors expressed in terms of the basis $\mathbf{B}_n$. The following is the commutation diagram expressed in terms of vector spaces and respective bases.

$$
\begin{array}{ccc}
\mathbb{R}^n \text{ with basis } \mathbf{B}_n & \xrightarrow{\ A'\ } & \mathbb{R}^m \text{ with basis } \mathbf{B}_m \\
\Big\downarrow{\scriptstyle V} & & \Big\uparrow{\scriptstyle W^{-1}} \\
\mathbb{R}^n \text{ with basis } \mathbf{S}_n & \xrightarrow{\ A\ } & \mathbb{R}^m \text{ with basis } \mathbf{S}_m
\end{array}
$$

The power of a commutative diagram lies in the fact that it allows you to perform function mappings that you do not know, in terms of functions that you do. In our case, we do not know A', however, we do know V, W and can readily determine A. Therefore, it is much easier to use mappings that we already know and express our unknown map A' in terms of them. In this case, we end up with $A' = W^{-1}AV$.

Homework Problems

1. Represent each vector of the standard basis $\mathbf{S}_2$ of $\mathbb{R}^2$ as a linear combination of vectors in the basis $\mathbf{B} = \{\langle 1, 1 \rangle, \langle 1, -1 \rangle\}$. Also, write your answer in vector form, using the correct notation.

2. Express each of the following vectors in $\mathbb{R}^2$ in vector form using the basis $\mathbf{B}$ from problem 1:

(a)	$\langle 2, 0 \rangle$	(b)	$\langle 2, -3 \rangle$	(c)	$\langle 9, -1 \rangle$
(d)	$\langle -2, 0 \rangle$	(e)	$\langle -\frac{1}{2}, \frac{3}{2} \rangle$	(f)	$\langle 7, -6 \rangle$

3. Verify that the following equation has only the trivial solution:

$$a\vec{e_1} + b\vec{e_2} = a\langle 1, 1 \rangle + b\langle 1, -1 \rangle$$

Here, $\vec{e_1}$ and $\vec{e_2}$ are the standard basis vectors for $\mathbb{R}^2$. What does this imply in regards to representation of vectors with the standard basis and the basis **B** from problem 1?

4. Represent each vector of the standard basis $\mathbf{S}_3$ of $\mathbb{R}^3$ as a linear combination of vectors in the basis $\mathbf{B} = \{\langle 1, 0, 1 \rangle, \langle 0, 1, 1 \rangle, \langle 0, -1, 0 \rangle\}$. Also, write your answer in vector form, using the correct notation.

5. Express each of the following vectors in $\mathbb{R}^3$ in vector form using the basis **B** from problem 4.

     (a) $\langle 1, 1, 0 \rangle$ (b) $\langle 2, 2, -2 \rangle$ (c) $\langle -5, 6, 6 \rangle$

     (d) $\langle 2, -2, 3 \rangle$ (e) $\langle 1, 1, 1 \rangle$ (f) $\langle 3, -4, -10 \rangle$

6. Construct a matrix A' that takes vectors in $\mathbb{R}^2$ expressed in terms of the basis $\mathbf{B}_1 = \{\langle 1, 1 \rangle, \langle 1, -1, \rangle\}$, and expresses them in terms of the basis $\mathbf{B}_2 = \{\langle 2, -1 \rangle, \langle 3, 2, \rangle\}$, i.e., $A'\vec{x}_{\mathbf{B}_1} = \vec{x}_{\mathbf{B}_2}$.

7. Construct a matrix A' that takes vectors in $\mathbb{R}^3$ expressed in terms of the basis $\mathbf{B}_1 = \{\langle 1, 0, 1 \rangle, \langle 0, 1, 1 \rangle, \langle 0, -1, 0 \rangle\}$, and expresses them in terms of the basis $\mathbf{B}_2 = \{\langle 1, 1, -1 \rangle, \langle 1, -1, 1 \rangle, \langle 1, -1, -1 \rangle\}$.

8. Given the map $T : \mathbb{R}^2 \rightarrow \mathbb{R}^3$ defined by $T(\langle x, y \rangle) = \langle 0, y, x \rangle$, find the matrix A' corresponding to T under the two bases $\mathbf{B}_2 = \{\langle 1, 1 \rangle, \langle 1, -1 \rangle\}$ and $\mathbf{B}_3 = \{\langle 0, 1, 1 \rangle, \langle 1, 0, 0 \rangle, \langle 0, 1, 0 \rangle\}$.

9. Construct the commutation diagram for the map from problem 8.

10. Given the map $S : \mathbb{R}^3 \rightarrow \mathbb{R}^4$ defined by $S(\langle x, y, z \rangle) = \langle 0, z, y, x \rangle$, find the matrix A' corresponding to S under the two bases

$$\mathbf{B}_3 = \{\langle 0, 1, 1 \rangle, \langle 1, 0, 0 \rangle, \langle 0, 1, 0 \rangle\}$$

and

$$\mathbf{B}_4 = \{\langle 0, 1, 1, 1 \rangle, \langle 1, 0, 0, 1 \rangle, \langle 0, 1, 1, 0 \rangle, \langle 0, 0, 1, 1 \rangle\}$$

11. Construct the commutation diagram for the map from problem 10.

12. Construct the commutation diagram for $S \circ T$, where T and S are the maps from problems 8 and 10, respectively. Use the diagram to find the matrix

corresponding to the map $S \circ T$.

13. Consider the case of a linear map whose domain is represented by a non-standard basis $\mathbf{B_n}$, and whose image is also represented by a non-standard basis $\mathbf{B}_m$. Hence, we already have $T'(\vec{x}_{\mathbf{B}_n}) = \vec{y}_{\mathbf{B}_m}$. How can you recover the original maps's matrix A in the standard bases, given the matrix A' that represents T' for the pair of nonstandard bases? *Hint: Drawing a commutation diagram can help.*

14. In problems 6 and 7, verify directly that $(A')^{-1}$ reverses the order of the two bases.

15. Let $T : \mathbb{R}^n \to \mathbb{R}^m$ be a linear map where we have two (different) pairs of bases, bases $\mathbf{B}$ and $\mathbf{C}$ for $\mathbb{R}^n$ and bases $\mathbf{D}$ and $\mathbf{E}$ for $\mathbb{R}^m$. Let $T_{\mathbf{D}}^{\mathbf{B}}$ be the matrix that represents the linear map T in the two bases $\mathbf{B}$ on the domain $\mathbb{R}^n$ and $\mathbf{D}$ on the range $\mathbb{R}^m$. Similarly, $T_{\mathbf{E}}^{\mathbf{C}}$ be the matrix that represents the linear map T in the two bases $\mathbf{C}$ on the domain $\mathbb{R}^n$ and $\mathbf{E}$ on the range $\mathbb{R}^m$. Also, for I the identity linear map from $\mathbb{R}^n$ to $\mathbb{R}^n$, we have the matrix $I_{\mathbf{C}}^{\mathbf{B}}$ that represents the linear map I in the two bases $\mathbf{B}$ and $\mathbf{C}$ while similarly, the matrix $I_{\mathbf{D}}^{\mathbf{E}}$ represents the identity linear map I from $\mathbb{R}^m$ to $\mathbb{R}^m$ in the two bases $\mathbf{D}$ and $\mathbf{E}$.

(a) Explain the meaning of, and discuss the validity of, the following commutative diagram:

$$\begin{array}{ccc} \mathbb{R}_{\mathbf{B}}^n & \xrightarrow{\ T_{\mathbf{D}}^{\mathbf{B}}\ } & \mathbb{R}_{\mathbf{D}}^m \\ {\scriptstyle I_{\mathbf{C}}^{\mathbf{B}}}\downarrow & & \uparrow{\scriptstyle I_{\mathbf{D}}^{\mathbf{E}}} \\ \mathbb{R}_{\mathbf{C}}^n & \xrightarrow{\ T_{\mathbf{E}}^{\mathbf{C}}\ } & \mathbb{R}_{\mathbf{E}}^m \end{array}$$

(b) Is the matrix equation $T_{\mathbf{D}}^{\mathbf{B}} = I_{\mathbf{D}}^{\mathbf{E}} T_{\mathbf{E}}^{\mathbf{C}} I_{\mathbf{C}}^{\mathbf{B}}$ correct? Explain your reasoning.

(c) Verify with an example the matrix equation in part (b) when only one pair $\mathbf{B}$ and $\mathbf{D}$ of bases are the standard bases of $\mathbb{R}^2$.

(d) Verify with an example the matrix equation in part (b) when all four bases are not the standard bases of $\mathbb{R}^2$.

16. (Continuation of problem 15.) Let I be the identity linear map from $\mathbb{R}^n$ to $\mathbb{R}^n$ and $\mathbf{B}$, $\mathbf{C}$ be two bases of $\mathbb{R}^n$.

(a) Explain why $I_{\mathbf{B}}^{\mathbf{B}} = I_{\mathbf{C}}^{\mathbf{C}} = I_n$, where I_n is the $n \times n$ identity matrix.

(b) Explain why $I_{\mathbf{C}}^{\mathbf{B}} = \left(I_{\mathbf{B}}^{\mathbf{C}}\right)^{-1}$.

(c) Let $T : \mathbb{R}^n \to \mathbb{R}^n$ be a linear map. Explain why

$$T_{\mathbf{B}}^{\mathbf{B}} = I_{\mathbf{B}}^{\mathbf{C}} T_{\mathbf{C}}^{\mathbf{C}} I_{\mathbf{C}}^{\mathbf{B}}$$

is correct, and thus

$$T_{\mathbf{B}}^{\mathbf{B}} = I_{\mathbf{B}}^{\mathbf{C}} T_{\mathbf{C}}^{\mathbf{C}} \left(I_{\mathbf{B}}^{\mathbf{C}} \right)^{-1}$$

This last equation says that the two $n \times n$ matrices $T_{\mathbf{B}}^{\mathbf{B}}$ and $T_{\mathbf{C}}^{\mathbf{C}}$ are *similar* matrices.

(d) Explain why $\left(T^{-1} \right)_{\mathbf{C}}^{\mathbf{B}} = \left(T_{\mathbf{B}}^{\mathbf{C}} \right)^{-1}$.

(e) Why is $\det \left(T_{\mathbf{C}}^{\mathbf{C}} \right) = \det \left(T_{\mathbf{B}}^{\mathbf{B}} \right)$?

(f) Let $n = 2$, and do examples to illustrate parts (a)–(e) above.

Maple Problems

1. (a)Represent each vector of the first basis **B** as a linear combination of vectors from the second basis **C**, and vice versa.

$$\mathbf{B} = \{ \langle 1, 1, -1 \rangle, \langle 1, -1, 1 \rangle, \langle 1, -1, -1 \rangle \}$$
$$\mathbf{C} = \{ \langle 0, 1, -1 \rangle, \langle 1, 0, 1 \rangle, \langle 1, -1, 0 \rangle \}$$

(b) Construct the matrix A' that takes vectors in $\mathbb{R}^3$ expressed in terms of the basis **B**, and expresses them in terms of the basis **C**, where the two bases are defined as in the previous problem. As in the *Homework Problems* section, A' should satisfy $A' \vec{x}_{\mathbf{B}} = \vec{x}_{\mathbf{C}}$. Now do this in reverse and relate the two matrices A' which result.

2. (a) Represent each vector of the first basis **B** as a linear combination of vectors from the second basis **C**, and vice versa.

$$\mathbf{B} = \{ \langle 1, 1, -1, 0 \rangle, \langle 0, 1, -1, 1 \rangle, \langle 1, 0, -1, -1 \rangle, \langle 1, 0, 1, 0 \rangle \}$$
$$\mathbf{C} = \{ \langle 1, 0, 1, 1 \rangle, \langle 0, 1, 1, 1 \rangle, \langle 0, 0, 1, 0 \rangle, \langle -1, -1, 1, 0 \rangle \}$$

(b) Construct the matrix A' that takes vectors in $\mathbb{R}^4$ expressed in terms of the basis **B**, and expresses them in terms of the basis **C**, where the two bases are defined as in the previous problem. Now do this in reverse and relate the two matrices A' which result.

3. Consider the maps $S : \mathbb{R}^4 \to \mathbb{R}^5$, $T : \mathbb{R}^5 \to \mathbb{R}^2$ and $U : \mathbb{R}^2 \to \mathbb{R}^4$ defined by

$$S(\langle w, x, y, z \rangle) = \langle w + x, x + y, y + z, z + w, w + y \rangle$$
$$T(\langle a, b, c, d, e \rangle) = \langle a + b - c + d - e, a + c + e \rangle$$
$$U(\langle j, k \rangle) = \langle j, j + k, j - k, k \rangle$$

and bases $\mathbf{B}_2$, $\mathbf{B}_4$ and $\mathbf{B}_5$, given by

$\mathbf{B}_2 = \{\langle 1, 1 \rangle, \langle -1, 1, \rangle\}$
$\mathbf{B}_4 = \{\langle 1, 1, -1, 0 \rangle, \langle 0, 1, -1, 1 \rangle, \langle 1, 0, -1, -1 \rangle, \langle 1, 0, 1, 0 \rangle\}$
$\mathbf{B}_5 = \{\langle 1, 1, 0, 0, 0 \rangle, \langle 0, 1, 1, 0, 0 \rangle, \langle 1, 0, 1, 1, 0 \rangle, \langle 0, 0, 1, 1, 0 \rangle, \langle 0, 1, 0, 1, 1 \rangle\}$

Compute the matrices corresponding to the following maps with the given bases for domains and ranges:

(a) S with $\mathbf{B}_4$ and $\mathbf{B}_5$ (b) T with $\mathbf{B}_5$ and $\mathbf{B}_2$

(c) U with $\mathbf{B}_2$ and $\mathbf{B}_4$ (d) $T \circ S$ with $\mathbf{B}_4$ and $\mathbf{B}_2$

(e) $U \circ T$ with $\mathbf{B}_5$ and $\mathbf{B}_4$ (f) $S \circ U$ with $\mathbf{B}_2$ and $\mathbf{B}_5$

(g) $U \circ T \circ S$ with $\mathbf{B}_4$ and $\mathbf{B}_4$ (h) $T \circ S \circ U$ with $\mathbf{B}_2$ and $\mathbf{B}_2$

(i) $S \circ U \circ T$ with $\mathbf{B}_5$ and $\mathbf{B}_5$

4. Use *Homework* problems 15 and 16 to

(a) Find the following matrices:

(i) I_D^B (ii) I_B^D (iii) I_E^C (iv) I_B^B

(v) T_D^B (vi) T_B^D (vii) T_B^B (viii) T_D^D

(ix) $(T^{-1})_B^D$ (x) $(T^{-1})_D^B$ (xi) $(T^{-1})_B^B$ (xii) $(T^{-1})_D^D$

(xiii) T_E^C (xiv) T_C^E (xv) T_C^C (xvi) T_E^E

(xvii) $(T^{-1})_E^C$ (xviii) $(T^{-1})_C^E$ (xix) $(T^{-1})_C^C$ (xx) $(T^{-1})_E^E$

for the linear map $T : \mathbb{R}^3 \to \mathbb{R}^3$ given by

$$T(\langle x, y, z \rangle) = \langle 7x - y + 2z, -6x + 3y - 5z, x + y + 4z \rangle$$

in the four bases:

$$\mathbf{B} = \{\langle -2, 1, 4 \rangle, \langle 5, -9, 6 \rangle, \langle 7, 3, -11 \rangle\}$$
$$\mathbf{C} = \{\langle -17, 4, 8 \rangle, \langle 10, -5, 13 \rangle, \langle 15, 6, -7 \rangle\}$$
$$\mathbf{D} = \{\langle -13, 2, -10 \rangle, \langle 1, 2, 9 \rangle, \langle 5, -1, 3 \rangle\}$$
$$\mathbf{E} = \{\langle -5, -4, -1 \rangle, \langle 2, 7, 4 \rangle, \langle 7, -3, 6 \rangle\}$$

(b) Find the determinants of all of these matrices.

(c) Explain all possible relationships between these matrices and their determinants.

5. Use *Homework* problems 15 and 16 to

(a) Find the following matrices:

(i)	I_D^B	(ii)	I_B^D	(iii)	I_E^C	(iv)	I_B^B
(v)	T_D^B	(vi)	T_B^D	(vii)	T_B^B	(viii)	T_D^D
(ix)	$(T^{-1})_B^D$	(x)	$(T^{-1})_D^B$	(xi)	$(T^{-1})_B^B$	(xii)	$(T^{-1})_D^D$
(xiii)	T_E^C	(xiv)	T_C^E	(xv)	T_C^C	(xvi)	T_E^E
(xvii)	$(T^{-1})_E^C$	(xviii)	$(T^{-1})_C^E$	(xix)	$(T^{-1})_C^C$	(xx)	$(T^{-1})_E^E$,

for the linear map $T : \mathbb{C}^2 \to \mathbb{C}^2$ given by

$$T(\langle x, y \rangle) = \langle ix - y, 6x + (2 + 5i)y \rangle$$

in the four bases:

$$\mathbf{B} = \{\langle i, -i \rangle, \langle 1 + i, 1 \rangle\}$$
$$\mathbf{C} = \{\langle 5 - i, i \rangle, \langle 2 + i, -i \rangle\}$$
$$\mathbf{D} = \{\langle 1, i \rangle, \langle i, -1 \rangle\}$$
$$\mathbf{E} = \{\langle i, -3 \rangle, \langle -1, -i \rangle\}$$

(b) Find the determinants of all of these matrices.

(c) Explain all possible relationships between these matrices and their determinants.

Chapter 10

The Geometry of Linear and Affine Maps

10.1 The Effect of a Linear Map on Area and Arclength in Two Dimensions

As discussed in Chapter 9, a linear map T can be represented by matrix multiplication. In particular, $T : \mathbb{R}^2 \to \mathbb{R}^2$ can be represented by a 2×2 matrix A. Now we will investigate how such functions transform standard geometric objects, such as line segments, types of quadrilaterals, polygons, and circles, as well as general simple closed curves in $\mathbb{R}^2$. Also, we want to see how the determinant of the matrix A affects the geometric behavior of the linear map T.

We shall begin with a line segment S joining two points, expressed in vector form as $\overrightarrow{P}\langle x_1, y_1 \rangle$ and $\overrightarrow{Q}\langle x_2, y_2 \rangle$. All the points of this line segment can be parametrically written as

$$\overrightarrow{S} = t\overrightarrow{P} + (1-t)\overrightarrow{Q}$$
$$= \langle tx_1 + (1-t)x_2, ty_1 + (1-t)y_2 \rangle$$

for the parameter $t \in [0, 1]$. The point $\overrightarrow{Q}$ corresponds to $t = 0$ and $\overrightarrow{P}$ corresponds to $t = 1$. Plugging the above equation for the line segment into the linear map T we have

$$T(\overrightarrow{S}) = A\overrightarrow{S}$$
$$= A(t\overrightarrow{P} + (1-t)\overrightarrow{Q})$$
$$= tA\overrightarrow{P} + (1-t)A\overrightarrow{Q}$$
$$= tT(\overrightarrow{P}) + (1-t)T(\overrightarrow{Q})$$

which is the line segment joining the point $T(\vec{P})$ to the point $T(\vec{Q})$.

A linear map $T : \mathbb{R}^2 \to \mathbb{R}^2$ is uniquely determined by sending a line segment $L_1 = \{\langle x_1, y_1 \rangle, \langle x_2, y_2 \rangle \}$ in its domain, where L_1 does not determine a line through the origin, to another line segment $L_2 = \{\langle z_1, w_1 \rangle, \langle z_2, w_2 \rangle\}$ in its range. This means that $T(\langle x_j, y_j \rangle) = \langle z_j, w_j \rangle$, for $j = 1, 2$. Equivalently, if we set $A = \begin{bmatrix} a & b \\ c & d \end{bmatrix}$, then

$$\begin{bmatrix} a & b \\ c & d \end{bmatrix} \begin{bmatrix} x_j \\ y_j \end{bmatrix} = \begin{bmatrix} z_j \\ w_j \end{bmatrix},$$

for $j = 1, 2$ and the four unknowns a, b, c, and d. This matrix equation gives us the two pairs of linear systems:

$$\begin{cases} ax_j + by_j = z_j, & \text{for } j = 1, 2 \\ cx_j + dy_j = w_j, & \text{for } j = 1, 2 \end{cases} \tag{10.1}$$

The first equation in system (10.1) gives rise to the matrix system:

$$\begin{bmatrix} x_1 & y_1 \\ x_2 & y_2 \end{bmatrix} \begin{bmatrix} a \\ b \end{bmatrix} = \begin{bmatrix} z_1 \\ z_2 \end{bmatrix}$$

and similarly for the second equation in system (10.1), we get:

$$\begin{bmatrix} x_1 & y_1 \\ x_2 & y_2 \end{bmatrix} \begin{bmatrix} c \\ d \end{bmatrix} = \begin{bmatrix} w_1 \\ w_2 \end{bmatrix}$$

We can combine the above two matrix equations into the single equation:

$$\begin{bmatrix} x_1 & y_1 \\ x_2 & y_2 \end{bmatrix} \begin{bmatrix} a & c \\ b & d \end{bmatrix} = \begin{bmatrix} z_1 & w_1 \\ z_2 & w_2 \end{bmatrix} \tag{10.2}$$

We can now solve this for A, although we have to be careful, since it is not A that appears in the above formula, but A^T. So equation (10.2) can be expressed in terms of A, L_1 and L_2 as

$$L_1 A^T = L_2 \tag{10.3}$$

where

$$L_1 = \begin{bmatrix} x_1 & y_1 \\ x_2 & y_2 \end{bmatrix}, \quad L_2 = \begin{bmatrix} z_1 & w_1 \\ z_2 & w_2 \end{bmatrix}$$

First, we get

$$A^T = (L_1)^{-1} L_2,$$

and then we transpose both sides to get

$$A = \left(L_1^{-1} L_2 \right)^T = L_2^T \left(L_1^{-1} \right)^T$$

Here we used the fact that $(BC)^T = C^T B^T$ for any matrices B and C of the correct dimensions. Now, note that we could have started solving equation (10.3) for A by first taking a transpose. This would have given

$$A L_1^T = L_2^T$$

Then solving for A gives

$$A = L_2^T \left(L_1^T\right)^{-1}$$

which shows that $\left(L_1^{-1}\right)^T = \left(L_1^T\right)^{-1}$, thus reminding us that the transpose of the inverse of a matrix is equivalent to the inverse of the transpose of a matrix. Now to solve for A, we prefer to use the second expression, which in full matrix form is

$$A = \left[\begin{array}{cc} z_1 & z_2 \\ w_1 & w_2 \end{array} \right] \left[\begin{array}{cc} x_1 & x_2 \\ y_1 & y_2 \end{array} \right]^{-1}$$

Note that $\left[\begin{array}{cc} x_1 & x_2 \\ y_2 & y_2 \end{array} \right]^{-1}$ exists if and only if the line segment L_1 does not determine a line through the origin.

Example 10.1.1. Now, let us see if this really works by finding the matrix A which represents the linear map T where $T(\langle 5, 13\rangle) = \langle -2, -9\rangle$ and $T(\langle -7, 3\rangle) = \langle 4, -1\rangle$.

```
> with(linalg): with(plots): with(plottools):
> P:= [5,13]: Q:= [-7,3]:
> L1:= transpose(matrix([P,Q]));
```

$$L1 := \left[\begin{array}{cc} 5 & -7 \\ 13 & 3 \end{array} \right]$$

```
> TP:= [-2,-9]: TQ:= [4,-1]:
> L2:= transpose(matrix([TP,TQ]));
```

$$L2 := \left[\begin{array}{cc} -2 & 4 \\ -9 & -1 \end{array} \right]$$

```
> A:= evalm(L2&*inverse(L1));
```

$$A := \left[\begin{array}{cc} -\frac{29}{53} & \frac{3}{53} \\ -\frac{7}{53} & -\frac{34}{53} \end{array} \right]$$

```
> evalm(A&*P);
```

$$\left[\begin{array}{cc} -2 & -9 \end{array} \right]$$

```
> evalm(A&*Q);
```

$$\left[\begin{array}{cc} 4 & -1 \end{array} \right]$$

Example 10.1.2. In the first example, we computed the matrix A corresponding to a given map of two points in $\mathbb{R}^2$. For our next example, we will instead define the matrix A as:

$$A = \begin{bmatrix} -2 & 7 \\ 5 & 3 \end{bmatrix}$$

and explore the geometric ramifications of this map to lines and squares in $\mathbb{R}^2$. First, we start with lines, and will plot the line segment joining $\overrightarrow{P}\langle -4, -9 \rangle$ to $\overrightarrow{Q}\langle 8, 3 \rangle$, along with its transformed version, as depicted in Figure 10.1.

```
> A:= matrix([[-2, 7], [5, 3]]):
> P:= [-4, -9]:  Q:= [8, 3]:
> dist_PQ:= sqrt(dotprod(P-Q,P-Q));
```

$$dist\_PQ := 12\sqrt{2}$$

```
> T_P:= evalm(A&*P);
```

$$T\_P := \begin{bmatrix} -55 & -47 \end{bmatrix}$$

```
> T_Q:= evalm(A&*Q);
```

$$T\_Q := \begin{bmatrix} 5 & 49 \end{bmatrix}$$

```
> dist_TPtoTQ:= sqrt(dotprod(T_P-T_Q, T_P-T_Q));
```

$$dist\_TPtoTQ := 12\sqrt{89}$$

```
> det(A);
```

$$-41$$

```
> evalf(dist_TPtoTQ/dist_PQ); evalf(sqrt(41));
```

$$6.670832030$$
$$6.403124237$$

```
> segment_PQ:= line(P,Q, color = blue, thickness = 3):
> segment_TPtoTQ:= line(convert(T_P,list),convert(T_Q,list), color = red,
thickness = 3):
> plotP:= textplot([op(P), "P"],align = BELOW, color = black):
> plotQ:= textplot([op(Q), "Q"],align = ABOVE, color = black):
> plotT_P:= textplot([op(convert(T_P,list)), "T(P)"],align = BELOW, color =
black):
> plotT_Q:= textplot([op(convert(T_Q,list)), "T(Q)"],align = ABOVE, color
= black):
```

> display({segment_PQ, segment_TPtoTQ, plotP, plotQ, plotT_P, plotT_Q},
scaling = constrained);

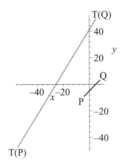

Figure 10.1: The line segment $\overrightarrow{P}\ \overrightarrow{Q}$ is transformed to $T(\overrightarrow{P})\ T(\overrightarrow{Q})$.

Note that the length of the line segment from $T(\overrightarrow{P})$ to $T(\overrightarrow{Q})$ is approximately $\sqrt{|\det(A)|}$ times the length of the line segment from $\overrightarrow{P}$ to $\overrightarrow{Q}$:

$$D(T(\overrightarrow{P}), T(\overrightarrow{Q})) \approx \sqrt{|\det(A)|} D(\overrightarrow{P}, \overrightarrow{Q}) \tag{10.4}$$

> evalf(dist_TPtoTQ);

$$113.2077736$$

> evalf(sqrt(abs(det(A)))*dist_PQ);

$$108.6646216$$

Is this always approximately true or not? We will attempt to answer this shortly. Let us next move on to the simplest geometric figure with an area: the unit square. The unit square has vertices at the points $\big\{\langle 0,0\rangle, \langle 1,0\rangle, \langle 1,1\rangle,$ $\langle 0,1\rangle\big\}$, as we move around its perimeter in a counterclockwise fashion. Every point of the square, and its interior, can be written as $\langle s,t\rangle = s\langle 1,0\rangle + t\langle 0,1\rangle$ for $0 \leq s,t \leq 1$. This should be immediately obvious, since the vectors $\{\langle 0,1\rangle, \langle 1,0\rangle\}$ correspond to the standard basis for $\mathbb{R}^2$, and if $0 \leq s,t \leq 1$, then the resulting point will always have coordinates between 0 and 1 in value. If we wish to apply a linear map T to an arbitrary point $\langle s,t\rangle$ on, or in, the interior of the unit square, notice what happens due to the properties of T being a linear map:

$$\begin{aligned} T(\langle s,t\rangle) &= T\left(s\langle 1,0\rangle + t\langle 0,1\rangle\right) \\ &= s\,T(\langle 1,0\rangle) + t\,T(\langle 0,1\rangle) \end{aligned} \tag{10.5}$$

It appears that given the location of the transformed points corresponding to $\langle 1,0\rangle$ and $\langle 0,1\rangle$, we can determine the location of any other point on, and in, the square. In reality, we also know the locations of two other points as well. In particular, it is easy to show that $T(\langle 0, 0 \rangle) = \langle 0, 0 \rangle$ and $T(\langle 1, 1 \rangle) = T(\langle 1, 0 \rangle) + T(\langle 0, 1 \rangle)$. So the transform of our square is a parallelogram with corners $\langle 0, 0 \rangle, T(\langle 1, 0 \rangle), T(\langle 0, 1 \rangle)$, and $T(\langle 1, 0 \rangle) + T(\langle 0, 1 \rangle)$. Also, if we represent T by a 2×2 matrix A, then notice that $T(\langle 1, 0 \rangle)$ is the first column of A, while $T(\langle 0, 1 \rangle)$ is the second column of A.

Example 10.1.3. Using the matrix A defined in the previous example, we will transform the unit square and its interior, as depicted in Figure 10.2. We also want to see if the areas of the unit square and the resulting parallelogram are related. Recall that the area of a parallelogram with two adjacent sides being the vectors $\overrightarrow{v}$ and $\overrightarrow{w}$ is given by $|\overrightarrow{v} \times \overrightarrow{w}|$. Also, remember that the cross product can only be applied to vectors in $\mathbb{R}^3$. Since we are dealing with points in $\mathbb{R}^2$, we will tack on an extra coordinate to make them points in $\mathbb{R}^3$, with the third coordinate always being 0. We can choose the last coordinate of each of the points to be any number we so choose, as long as it is consistently done for both points. Many formulas set the last coordinate to 1 instead of 0.

```
> UnitSqr:= polygon([[0,0],[1,0],[1,1],[0,1]], color = blue):
> T:= (x,y) -> convert(evalm(A&*[x,y]),list):
> TofUnitSqr:= polygon([[0,0],T(1,0),T(1,1),T(0,1)], color = red):
> plot_T(1,0):= textplot([op(T(1,0)),"T(1,0)"], align = {LEFT,ABOVE},
color = black):
> plot_T(0,1):= textplot([op(T(0,1)),"T(0,1)"], align = {RIGHT,ABOVE},
color = black):
> plot_T(1,1):= textplot([op(T(1,1)),"T(1,1)"], align = {RIGHT,ABOVE},
color = black):
> display([UnitSqr, TofUnitSqr, plot_T(1,0), plot_T(0,1), plot_T(1,1)]);
```

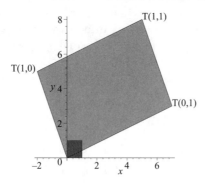

Figure 10.2: The unit square and its transform parallelogram.

> T_3dvector:= (x,y) -> [op(convert(evalm(A&*[x,y]),list)),0]:
> T_3dvector(1,0);
$$[-2, 5, 0]$$
> norm(crossprod(T_3dvector(0,1), T_3dvector(1,0)),frobenius);
$$41$$
> det(A);
$$-41$$

In this example, the linear map T has turned the unit square into a parallelogram with area precisely $|\det(A)| = 41$.

It appears at first glance that linear maps preserve the overall geometric properties of an object in the plane, with area changed by the multiplicative factor $|\det(A)|$. Let us see if this is true for a generic parallelogram. We will define our parallelogram to have vertices $\{\langle 0,0\rangle, \langle a, b\rangle, \langle a+c, b+d\rangle, \langle c, d\rangle\}$, and the matrix A to be $A = \begin{bmatrix} \alpha & \beta \\ \delta & \theta \end{bmatrix}$. We will show, with the help of *Maple*, that the area of transformed parallelograms does satisfy the property seen in the previous example.

> v:= [a, b]: w:= [c, d]:
> A:= matrix([[alpha, beta], [delta, theta]]):
> det(A);
$$\alpha\theta - \beta\delta$$
> T_3dvector:= (x,y) -> [op(convert(evalm(A&*[x,y]),list)),0]:
> norm(crossprod([op(v),0], [op(w),0]),frobenius);
$$|ad - bc|$$
> norm(crossprod(T_3dvector(op(v)), T_3dvector(op(w))),frobenius);
$$|(\alpha a + \beta b)(\delta c + \theta d) - (\delta a + \theta b)(\alpha c + \beta d)|$$
> factor(simplify(%));
$$|(\alpha\theta - \beta\delta)(ad - bc)|$$
So the area of the transformed parallelogram using the linear map T is

$$\begin{aligned}
\text{Transformed Area} &= |(\alpha\theta - \beta\delta)(ad - bc)| \\
&= |(\alpha\theta - \beta\delta)|\,|(ad - bc)| \\
&= |\det(A)|\,|ad - bc| \\
&= |\det(A)| \cdot (\text{the area of the original parallelogram}).
\end{aligned}$$

Now we might ask if this is also the case for general *simple polygons* in the xy-plane. By use of the term simple, we mean that the polygon is not self-intersecting. Let P be a simple polygon of k sides with consecutive (clockwise or counterclockwise) k vertices $\{\langle x_1, y_1\rangle, \langle x_2, y_2\rangle, \ldots, \langle x_k, y_k\rangle\}$. Then the area of the polygon P is given by the formula:

$$\text{Area of } P = \frac{1}{2}\left(\det\left(\begin{bmatrix} x_1 & x_2 \\ y_1 & y_2 \end{bmatrix}\right) + \det\left(\begin{bmatrix} x_2 & x_3 \\ y_2 & y_3 \end{bmatrix}\right)\right.$$
$$\left. + \cdots + \det\left(\begin{bmatrix} x_{k-1} & x_k \\ y_{k-1} & y_k \end{bmatrix}\right) + \det\left(\begin{bmatrix} x_k & x_1 \\ y_k & y_1 \end{bmatrix}\right)\right)$$

(10.6)

Let us apply the general linear map T above to this polygon to get a new polygon Q with k consecutive vertices $\{T(\langle x_1, y_1\rangle), T(\langle x_2, y_2\rangle), \ldots, T(\langle x_k, y_k\rangle)\}$. Are their areas related as they are for parallelograms? We will check this for an arbitrary polygon with $k = 7$ sides.

> P:= matrix([[x[1], y[1]],[x[2], y[2]], [x[3], y[3]], [x[4], y[4]], [x[5], y[5]], [x[6], y[6]], [x[7], y[7]]]);

$$P := \begin{bmatrix} x_1 & y_1 \\ x_2 & y_2 \\ x_3 & y_3 \\ x_4 & y_4 \\ x_5 & y_5 \\ x_6 & y_6 \\ x_7 & y_7 \end{bmatrix}$$

> T:= (x,y) -> convert(evalm(A&*[x,y]),list):
> TP:= matrix([T(op([x[1],y[1]])), T(op([x[2],y[2]])), T(op([x[3],y[3]])), T(op([x[4],y[4]])), T(op([x[5],y[5]])), T(op([x[6],y[6]])), T(op([x[7],y[7]]))]);

$$TP := \begin{bmatrix} \alpha x_1 + \beta y_1 & \delta x_1 + \theta y_1 \\ \alpha x_2 + \beta y_2 & \delta x_2 + \theta y_2 \\ \alpha x_3 + \beta y_3 & \delta x_3 + \theta y_3 \\ \alpha x_4 + \beta y_4 & \delta x_4 + \theta y_4 \\ \alpha x_5 + \beta y_5 & \delta x_5 + \theta y_5 \\ \alpha x_6 + \beta y_6 & \delta x_6 + \theta y_6 \\ \alpha x_7 + \beta y_7 & \delta x_7 + \theta y_7 \end{bmatrix}$$

> AreaofTP:= .5*simplify(TP[7,1]*TP[1,2] + add(TP[i,1]*TP[i+1,2], i=1..6) - TP[7,2]*TP[1,1] - add(TP[i,2]*TP[i+1,1], i=1..6));

$AreaofTP := 0.5\,\alpha\,x_7\theta\,y_1 + 0.5\,\beta\,y_7\delta\,x_1 + 0.5\,\alpha\,x_1\theta\,y_2 + 0.5\,\beta\,y_1\delta\,x_2$
$\qquad + 0.5\,\alpha\,x_2\theta\,y_3 + 0.5\,\beta\,y_2\delta\,x_3 + 0.5\,\alpha\,x_3\theta\,y_4 + 0.5\,\beta\,y_3\delta\,x_4 + 0.5\,\alpha\,x_4\theta\,y_5$
$\qquad + 0.5\,\beta\,y_4\delta\,x_5 + 0.5\,\alpha\,x_5\theta\,y_6 + 0.5\,\beta\,y_5\delta\,x_6 + 0.5\,\alpha\,x_6\theta\,y_7 + 0.5\,\beta\,y_6\delta\,x_7$

$- 0.5\,\delta\,x_7\beta\,y_1 - 0.5\,\theta\,y_7\alpha\,x_1 - 0.5\,\delta\,x_1\beta\,y_2 - 0.5\,\theta\,y_1\alpha\,x_2 - 0.5\,\delta\,x_2\beta\,y_3$
$- 0.5\,\theta\,y_2\alpha\,x_3 - 0.5\,\delta\,x_3\beta\,y_4 - 0.5\,\theta\,y_3\alpha\,x_4 - 0.5\,\delta\,x_4\beta\,y_5 - 0.5\,\theta\,y_4\alpha\,x_5$
$- 0.5\,\delta\,x_5\beta\,y_6 - 0.5\,\theta\,y_5\alpha\,x_6 - 0.5\,\delta\,x_6\beta\,y_7 - 0.5\,\theta\,y_6\alpha\,x_7$

> AreaofP:= .5*(P[7,1]*P[1,2] + add(P[i,1]*P[i+1,2], i=1..6) - P[7,2]*P[1,1] - add(P[i,2]*P[i+1,1], i=1..6));

$AreaofP := 0.5x_7y_1 + 0.5x_1y_2 + 0.5x_2y_3 + 0.5x_3y_4 + 0.5x_4y_5 + 0.5x_5y_6$
$+ 0.5x_6y7 - 0.5y_7x_1 - 0.5y_1x_2 - 0.5y_2x_3 - 0.5y_3x_4 - 0.5y_4x_5 - 0.5y_5x_6$
$- 0.5y_6x_7$

> simplify(AreaofTP - det(A)*AreaofP);

$$0$$

Now we see that the linear map T transforms a polygon P, of k vertices, into a new polygon TP, of k vertices, with

$$\text{Area of } TP = |\det(A)| \cdot (\text{Area of } P) \tag{10.7}$$

at least for the $k = 2$ and $k = 7$ cases. If this still does not convince you that the transformed area does follow the given formula, consider the following argument. First, we refer back to the equation for the area of a polygon, given in equation (10.6). Due to the way in which it is written, it would be enough to show the following:

$$\det\left(\begin{bmatrix} \alpha x_j + \beta y_j & \alpha x_{j+1} + \beta y_{j+1} \\ \delta x_j + \theta y_j & \delta x_{j+1} + \theta y_{j+1} \end{bmatrix}\right) = \det(A) \cdot \det\left(\begin{bmatrix} x_j & x_{j+1} \\ y_j & y_{j+1} \end{bmatrix}\right)$$

for $1 \le j \le k$, with $j = k$ giving $j + 1 = 1$. Notice that the matrix on the LHS of the above equation is simply the product of A and the matrix whose columns are the points $\langle x_j, y_j \rangle$ and $\langle x_{j+1}, y_{j+1} \rangle$. From Section 5.1, we know that given any two square matrices A and B, $\det(AB) = \det(A)\det(B)$. Hence, the above expression automatically holds. This allows a factor of $\det(A)$ to be pulled out of each term in equation (10.6) to give us equation (10.7) instead.

Example 10.1.4. As another example, let us see the effects of the linear map T from Example 10.1.1 on the circle with center $(3, 8)$ and radius 6. The standard equation for this circle is given by $(x - 3)^2 + (y - 8)^2 = 36$, however it can also be written parametrically as

$$C : \langle x(t), y(t) \rangle = \langle 3 + 6\cos(t), 8 + 6\sin(t) \rangle, \ t \in [0, 2\pi]$$

> A:= matrix([[-2, 7], [5, 3]]):
> T:= (x,y) -> convert(evalm(A&*[x,y]),list):
> C:= [3 + 6*cos(t), 8 + 6*sin(t)]:
> TC:= T(op(C));

$$TC := [50 - 12\cos(t) + 42\sin(t), 39 + 30\cos(t) + 18\sin(t)]$$

> plotC:= plot([op(C), t=0..2*Pi], color = blue):
> plotTC:= plot([op(TC), t=0..2*Pi], color = red):
> plot_centers:= pointplot({[3,8], T(3,8)}, symbol = cross, symbolsize = 20, color = black):
> display({plotC, plotTC, plot_centers}, scaling = constrained);

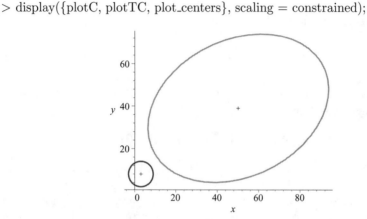

Figure 10.3: Circle is transformed to an ellipse by T.

Figure 10.3 illustrates the fact that T transforms a circle into an ellipse with T taking the circle's center to the ellipse's center. This also implies that T takes the interior of the circle to the interior of the ellipse. To visualize which points on the circle correspond to which points on the ellipse, we will have *Maple* construct an animation. We will plot two vectors, one for the circle and one for the ellipse. The bases of the vectors will be at the corresponding centers, while the tips of the vectors will be on the circle and ellipse. Since both objects have been previously defined parametrically, we can pick $N + 1$ evenly spaced points in the interval $[0, 2\pi]$, with the first point being $t = 0$ and last being $t = 2\pi$ to help with the animation. A frame of this animation can be seen in Figure 10.4

> N:= 30:
> arrows_circle:= [seq(PLOT(arrow([3,8], [3 + 6*cos(2*Pi*j/N), 8 + 6*sin(2*Pi*j/N)], .5, 2, .5, color = blue)), j = 0..N)]:

> arrows_ellipse:= [seq(PLOT(arrow(T(3,8), T(3 + 6*cos(2*Pi*j/N), 8 + 6*sin(2*Pi*j/N)), 1, 5, .4, color = red)), j = 0..N)]:

> display([plotC, plotTC, display(arrows_ellipse, insequence = true), display(arrows_circle, insequence = true)], scaling = constrained);

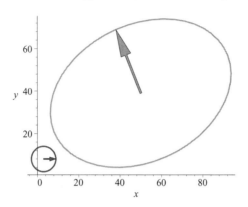

Figure 10.4: The point on the circle at the tip of the short arrow corresponds to the point on the ellipse at the tip of the long arrow. The long arrow rotates clockwise while the short rotates counter-clockwise.

Now, let us compute the arclengths and areas for both curves to see if they are related. To do this, we need to introduce a definition and two formulas.

We now generalize the circle and ellipse to a general type of curve called a simple closed curve. A simple closed curve should be continuously differentiable except possibly for a finite number of corner points and enclose a simply connected region that is a region in one connected piece with no holes in it. Besides the circle and ellipse, a square, rectangle or parallelogram are also simple closed curves, as well as all simple closed polygons

Definition 10.1.1. A parametric curve $C : \langle x, y \rangle = \langle x(t), y(t) \rangle$, $t \in [a, b]$ is said to be a *simple closed* curve if

closed: Its starting point at $t = a$ is also its stopping point at $t = b$,

simple: It intersects itself at only its endpoints which are the same point,

piecewise C^1 curve: It is a continuously differentiable curve except for at most a finite number of corner points.

If $C : \langle x, y \rangle = \langle x(t), y(t) \rangle$ for $t \in [a, b]$ is a parametric curve, then the *arclength* L is given by the integral:

$$L = \int_a^b \sqrt{\left(\frac{dx}{dt}\right)^2 + \left(\frac{dy}{dt}\right)^2}\, dt \tag{10.8}$$

For the circle, the arclength is 12π and the area is 36π.

From Green's theorem of multivariable calculus, the area of the region inside a simple closed curve $C : \langle x, y \rangle = \langle x(t), y(t) \rangle$ for $t \in [a, b]$ is

$$
\begin{aligned}
\text{Area of } C &= \int_a^b x(t)\frac{dy}{dt}\,dt \\
&= -\int_a^b y(t)\frac{dx}{dt}\,dt \\
&= \frac{1}{2}\int_a^b \left[x(t)\frac{dy}{dt} - y(t)\frac{dx}{dt} \right] dt
\end{aligned}
\tag{10.9}
$$

if C is traversed in the counterclockwise direction. The area is the negative of these integrals if C is traversed in the clockwise direction. The ellipse is traversed clockwise and so we take the negative of these integrals for its area.

> Arclength_ellipse:= evalf(Int(sqrt(diff(TC[1], t)^2 + diff(TC[2], t)^2), t = 0..2*Pi));

$$Arclength\_ellipse := 246.8600608$$

> evalf(Arclength_ellipse/(12*Pi)); evalf(sqrt(abs(det(A))));

$$6.548166485$$
$$6.403124237$$

> Area_ellipse:= evalf(-Int(TC[1]*diff(TC[2],t),t=0..2*Pi));

$$Area\_ellipse := 4636.990757$$

> evalf(Area_ellipse/(36*Pi)); abs(det(A));

$$40.99999999$$
$$41$$

Again, we see that the linear map T multiplies arclength by approximately $\sqrt{|\det(A)|}$, while it multiplies area by exactly $|\det(A)|$. Should we expect the same relationships involving area and arclength to hold if the simple closed curve C is more complicated than a circle?

Example 10.1.5. Let us investigate the effects of the linear map T from Example 10.1.1 on the simple closed curve given by

$$C : \langle x(t), y(t) \rangle = \langle (3 - \cos(12t))\cos(t) + 7, (3 - \cos(12t))\sin(t) + 11 \rangle, \ t \in [0, 2\pi].$$

The curve and its transform can be seen in Figures 10.5 and 10.6.

> C:= [(3-cos(12*t))*cos(t)+7, (3-cos(12*t))*sin(t)+11]:
> TC:= T(op(C));

$$TC := [-2(3 - \cos(12t)) \cos(t) + 63 + 7(3 - \cos(12t)) \sin(t),$$
$$5(3 - \cos(12t)) \cos(t) + 68 + 3(3 - \cos(12t)) \sin(t)]$$

> plotC:= plot([op(C), t=0..2*Pi], color = blue):
> plotTC:= plot([op(TC), t=0..2*Pi], color = red):
> plot_centers:= pointplot({[7,11], T(7,11)}, symbol = cross, color = black):
> display({plotC, plotTC, plot_centers}, scaling = constrained);

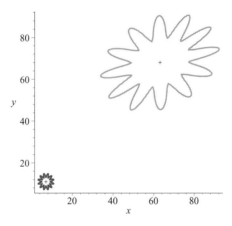

Figure 10.5: Simple closed curve C with center at $(7, 11)$, and transformed closed curve with center at $(63, 68)$.

> Arclength_original:= evalf(Int(sqrt(diff(C[1], t)^2 + diff(C[2], t)^2), t = 0..2*Pi));

$$Arclength\_original := 53.03488065$$

> Arclength_new:= evalf(Int(sqrt(diff(TC[1], t)^2 + diff(TC[2], t)^2), t = 0..2*Pi));

$$Arclength\_new := 347.2812478$$

> evalf(Arclength_new/Arclength_original); evalf(sqrt(abs(det(A))));

$$6.548166858$$
$$6.403124237$$

> Area_original:= evalf(Int(C[1]*diff(C[2],t),t=0..2*Pi));

$$Area\_original := 29.84513021$$

> Area_new:= evalf(-Int(TC[1]*diff(TC[2],t),t=0..2*Pi));

$$Area\_new := 1223.650339$$

> evalf(Area_new/Area_original); abs(det(A));

41.00000001
41

> N:= 100:
> arrows_original:= [seq(PLOT(arrow([7,11],subs(t=2*Pi*j/N,C),.5,2,.5, color = blue)),j = 0..N)]:
> arrows_new:= [seq(PLOT(arrow(T(7,11),T(op(subs(t=2*Pi*j/N,C))),1,5,.4, color = red)),j = 0..N)]:
> display([plotC, plotTC, display(arrows_original, insequence = true), display(arrows_new, insequence = true)], scaling = constrained);

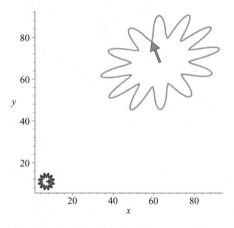

Figure 10.6: The point on the simple closed curve C at the tip of the short arrow corresponds to the point on the transformed simple closed curve at the tip of the long arrow. The long arrow rotates clockwise around the transformed curve, while the short arrow rotates counterclockwise around the original.

Once again, we have that T multiplies arclength by $\sqrt{|\det(A)|}$, approximately, while it multiplies area by exactly $|\det(A)|$. To prove this relation holds in general, we employ a method similar to the argument used for simple closed polygons. Letting $T : \mathbb{R}^2 \to \mathbb{R}^2$ be represented by the 2×2 matrix $A = \begin{bmatrix} \alpha & \beta \\ \delta & \theta \end{bmatrix}$, and $C : \langle x, y \rangle = \langle x(t), y(t) \rangle$ be an arbitrary simple closed curve for $t \in [a, b]$ with counterclockwise orientation, we wish to prove the following:

$$\text{Area of } T(C) = |\det(A)| \cdot (\text{Area of } C) \tag{10.10}$$

The first thing to note is that we have

$$T(C) : \langle x, y \rangle = \langle \alpha x(t) + \beta y(t), \delta x(t) + \theta y(t) \rangle, \quad t \in [a, b]$$

If the curve $T(C)$ is also counterclockwise, then

$$\text{Area of } T(C) = \int_a^b (\alpha x(t) + \beta y(t)) \frac{d}{dt} (\delta x(t) + \theta y(t)) \, dt$$

$$= \int_a^b (\alpha x(t) + \beta y(t)) \left(\delta \frac{dx}{dt} + \theta \frac{dy}{dt} \right) dt$$

$$= \alpha \delta \int_a^b x(t) \frac{dx}{dt} \, dt + \beta \theta \int_a^b y(t) \frac{dy}{dt} \, dt$$

$$+ \alpha \theta \int_a^b x(t) \frac{dy}{dt} \, dt + \beta \delta \int_a^b y(t) \frac{dx}{dt} \, dt$$

$$= \alpha \delta \left. \frac{1}{2} x^2(t) \right|_a^b + \beta \theta \left. \frac{1}{2} y^2(t) \right|_a^b$$

$$+ \alpha \theta \int_a^b x(t) \frac{dy}{dt} \, dt - \beta \delta \int_a^b x(t) \frac{dy}{dt} \, dt$$

$$= (\alpha \theta - \beta \delta) \int_a^b x(t) \frac{dy}{dt} \, dt$$

$$= \det(A) \, (\text{Area of } C)$$

In the above string of equalities, we made use of the two facts

$$\frac{1}{2} \left(x^2(b) - x^2(a) \right) = 0 \quad \frac{1}{2} \left(y^2(b) - y^2(a) \right) = 0$$

due to C being a closed curve. Furthermore, in the above string of equalities, we also used the fact that

$$\int_a^b y(t) \frac{dx}{dt} \, dt = - \int_a^b x(t) \frac{dy}{dt} \, dt$$

which is Green's theorem as stated previously in equation (10.9).

If the curve $T(C)$ is clockwise, then a similar calculation gives

$$\text{Area of } T(C) = -\det(A) \, (\text{Area of } C)$$

Thus, the linear map T maintains the same orientation between the two simple closed curves C and $T(C)$ if $\det(A) > 0$, and switches orientation if $\det(A) < 0$. If $\det(A) = 0$, then the curve $T(C)$ is not a simple closed curve.

Homework Problems

1. Find a linear map that maps the unit square to the parallelogram with vertices at $\langle 0,0 \rangle$, $\langle 2,1 \rangle$, $\langle 1,3 \rangle$, and $\langle 3,4 \rangle$. Express your answer in matrix form.

2. Using the map from problem 1, verify that the area of the transformed parallelogram is equal to the determinant of the matrix A corresponding to the linear map.

3. Find a linear map that maps the unit square to the parallelogram with vertices at $\langle 0,0 \rangle$, $\langle -3,3 \rangle$, $\langle -5,-2 \rangle$, and $\langle -8,1 \rangle$. You can express your answer in matrix form.

4. Using the map from problem 2, verify that the area of the transformed parallelogram is equal to the determinant of the matrix A corresponding to the linear map.

5. Find a linear map that maps the parallelogram with vertices at $\langle 0,0 \rangle$, $\langle 2,1 \rangle$, $\langle 1,3 \rangle$, and $\langle 3,4 \rangle$, to the parallelogram with vertices at $\langle 0,0 \rangle$, $\langle -3,3 \rangle$, $\langle -5,-2 \rangle$, and $\langle -8,1 \rangle$.

6. Explain why there does not exist a linear map that maps the unit square to the parallelogram with vertices at $\langle 1,1 \rangle$, $\langle 3,1 \rangle$, $\langle 2,4 \rangle$, and $\langle 4,4 \rangle$.

7. Find the matrix corresponding to the linear map that rescales the unit square by a factor $a > 0$ in both the x- and y-directions.

8. Find the matrix A_x corresponding to the linear map that flips the unit square about the x-axis.

9. Find the matrix A_y corresponding to the linear map that flips the unit square about the y-axis.

10. Is it true that the matrix $A_x A_y$ corresponds to the linear map that flips the unit square about both the x- and y-axes?

11. (a) Find the arclength of the *spiral parametric curve*

$$C : \langle x(t), y(t) \rangle = \langle t \cos(t),\ t \sin(t) \rangle,\ t \in [0, 2\pi]$$

You will need that

$$\int_0^{2\pi} \sqrt{1+t^2}\, dt = \pi\sqrt{1+4\pi^2} - \frac{1}{2} \ln\left(-2\pi + \sqrt{1+4\pi^2}\right)$$

(b) Find the arclength of the new parametric curve $T(C)$ gotten by applying to C the linear map $T : \mathbb{R}^2 \to \mathbb{R}^2$ with rule $T(\langle x, y \rangle) = \langle 3x - 5y, 5x + 3y \rangle$. Are the two arclengths related as expected?

(c) Find the arclength of the new parametric curve $S(C)$ gotten by applying to C the linear map $S : \mathbb{R}^2 \to \mathbb{R}^2$ with rule $S(\langle x, y \rangle) = \langle ax - by, bx + ay \rangle$, $a, b \in \mathbb{R}$. Is this arclength related to that of C as expected, and why might you expect it to be true from the form of the rule for this linear map S?

12. Find the arclength and the area inside of the astroid parametric curve from *Maple* problem 3.

Maple Problems

1. Consider the *figure-eight curve* given parametrically by

$$C : \langle x(t), y(t) \rangle = \langle 2\cos(t), 2\sin(2t) \rangle, \quad t \in [0, 2\pi]$$

and the linear map matrix $A = \begin{bmatrix} -1 & 1 \\ 1 & 1 \end{bmatrix}$. Graph the parametric function and its corresponding transform. Compute the area enclosed by the original curve and the transformed curve, then determine if the relation Area of $T(C) = |\det(A)| \cdot (\text{Area of } C)$ holds.

2. Repeat problem 1 with except with $A = \dfrac{1}{\sqrt{2}} \begin{bmatrix} -1 & 1 \\ 1 & 1 \end{bmatrix}$.

3. An *astroid* is defined parametrically by

$$\langle x(t), y(t) \rangle = \langle \cos^3(t), \sin^3(t) \rangle, \quad t \in [0, 2\pi]$$

Transform the astroid using the matrix $A = \begin{bmatrix} -1 & 1 \\ 2 & 1 \end{bmatrix}$, then graph both the parametric function and its corresponding transform. Compute the area enclosed by the original curve and the transformed curve and determine if the relation Area of $T(C) = |\det(A)|$ (Area of C) holds.

4. Repeat problem 3 with $A = \dfrac{1}{\sqrt{3}} \begin{bmatrix} -1 & 1 \\ 1 & 1 \end{bmatrix}$.

5. Construct a map matrix A that rotates the astroid from problem 3 through an angle of $\dfrac{\pi}{4}$, with the resulting astroid having five times the area of the original.

6. Redo *Homework* problem 11, plotting the original C and the new $T(C)$.

7. Using the spiral parametric curve C of *Homework* problem 11, find the approximate arclength of the new parametric curve $T(C)$ gotten by applying to C the linear map $T : \mathbb{R}^2 \to \mathbb{R}^2$ with rule $T((x,y)) = (3x + 5y, 7x - 18y)$. Are the two arclengths related as expected? Plot both C and $T(C)$.

8. Repeat problem 7 for the astroid parametric curve of problem 3.

9. Repeat problem 7 for the *bat parametric curve*

$$C : (x(t), y(t)) = (r(t)\ \cos(t), r(t)\ \sin(3t)), \ t \in [0, 200\pi]$$

where

$$r(t) = e^{\sin(t)} - 2\cos(4t) + \cos\left(\frac{t}{6}\right) - \frac{1}{2}\sin\left(\frac{t}{12}\right) + \sin^5\left(\frac{t}{24}\right) - \frac{1}{2}\cos^3\left(\frac{t}{48}\right)$$

Plot this bat using the plot option "numpoints = 1000". (Note that *Maple* may have a great deal of trouble computing the arclength of this full curve, good luck!) Also, see what curve C you get if the coefficient 3 is changed to 2 or 1 in $\sin(3t)$.

10. Let C be the simple closed curve given as the regular pentagon P inscribed in the unit circle.

(a) Find P's perimeter and enclosed area.

(b) Now transform this pentagon P by the linear map T given by multiplication by the matrix

$$A = \begin{bmatrix} 3 & -1 \\ 5 & 7 \end{bmatrix}$$

into the new pentagon $T(P)$.

(c) Find $T(P)$'s perimeter and enclosed area. How are these quantities related to those of P?

(d) Plot both the pentagons P and $T(P)$ in different colors.

10.2 The Decomposition of Linear Maps into Rotations, Reflections, and Rescalings in $\mathbb{R}^2$

In Section 4.4, we discussed how matrix multiplication, and thus linear maps, can be used to rotate points in the plane. Recall that the 2×2 map matrix A corresponding to a rotation about the origin from the positive x-axis by an angle of θ is given by

$$A_\theta = \begin{bmatrix} \cos(\theta) & -\sin(\theta) \\ \sin(\theta) & \cos(\theta) \end{bmatrix} \qquad (10.11)$$

Linear maps can also be used to rescale objects, hence we will call a map a *rescaling* if it does exactly that. For instance, if we wish to rescale objects horizontally by a factor $a > 0$ and vertically by a factor $b > 0$, then $T(\langle x, y \rangle) = \langle ax, by \rangle$. The matrix A corresponding to this map is much easier to determine, and is given by

$$A_{(a,b)} = \begin{bmatrix} a & 0 \\ 0 & b \end{bmatrix} \qquad (10.12)$$

Finally, linear maps can *reflect* objects across the x-axis by $T(\langle x, y \rangle) = \langle x, -y \rangle$ or about the y-axis by $T(\langle x, y \rangle) = \langle -x, y \rangle$ with matrices given, respectively, by

$$A_x = \begin{bmatrix} 1 & 0 \\ 0 & -1 \end{bmatrix} \text{ and } A_y = \begin{bmatrix} -1 & 0 \\ 0 & 1 \end{bmatrix} \qquad (10.13)$$

The amazing fact, is that all linear maps are simply composites of these three basic types of linear maps: rotations T_θ about the origin through the angle θ, general rescalings $T_{(a,b)}$ by factors of a horizontally and b vertically, and reflections such as T_x about the x-axis.

Our first question is: Do these three basic types of linear maps commute? Let us test and see.

```
> with(linalg):
> R:= theta -> matrix([[cos(theta), -sin(theta)], [sin(theta), cos(theta)]]):
> R(theta);
```

$$\begin{bmatrix} \cos(\theta) & -\sin(\theta) \\ \sin(\theta) & \cos(\theta) \end{bmatrix}$$

```
> S:= (a,b) -> matrix([[a, 0], [0, b]]):
> S(a,b);
```

$$\begin{bmatrix} a & 0 \\ 0 & b \end{bmatrix}$$

> Reflectx:= matrix([[1, 0], [0, -1]]);

$$Reflectx := \begin{bmatrix} 1 & 0 \\ 0 & -1 \end{bmatrix}$$

> evalm(R(theta)&*S(a,b));

$$\begin{bmatrix} \cos(\theta)\, a & \sin(\theta)\, b \\ \sin(\theta)\, a & -\cos(\theta)\, b \end{bmatrix}$$

> evalm(S(a,b)&*R(theta));

$$\begin{bmatrix} \cos(\theta)\, a & -\sin(\theta)\, a \\ \sin(\theta)\, b & \cos(\theta)\, b \end{bmatrix}$$

From the above two matrix multiplications, we see that rotations and rescalings do not commute.

> evalm(R(theta)&*Reflectx);

$$\begin{bmatrix} \cos(\theta) & -\sin(\theta) \\ \sin(\theta) & \cos(\theta) \end{bmatrix}$$

> evalm(Reflectx&*R(theta));

$$\begin{bmatrix} \cos(\theta) & -\sin(\theta) \\ -\sin(\theta) & -\cos(\theta) \end{bmatrix}$$

Also rotations and reflection about the x-axis do not commute.

> evalm(S(a,b)&*Reflectx);

$$\begin{bmatrix} a & 0 \\ 0 & -b \end{bmatrix}$$

> evalm(Reflectx &*S(a,b));

$$\begin{bmatrix} a & 0 \\ 0 & -b \end{bmatrix}$$

Now we see that reflections about x-axis and rescalings commute. Next, we want to know if two of the same type commute.

> map(combine,evalm(R(theta)&*R(phi)));

$$\begin{bmatrix} \cos(\theta + \phi) & -\sin(\theta + \phi) \\ \sin(\theta + \phi) & \cos(\theta + \phi) \end{bmatrix}$$

> map(combine,evalm(R(phi)&*R(theta)));

$$\begin{bmatrix} \cos(\theta + \phi) & -\sin(\theta + \phi) \\ \sin(\theta + \phi) & \cos(\theta + \phi) \end{bmatrix}$$

> evalm(S(a,b)&*S(c,d));

$$\begin{bmatrix} ac & 0 \\ 0 & bd \end{bmatrix}$$

> evalm(S(c,d)&*S(a,b));

$$\begin{bmatrix} ac & 0 \\ 0 & bd \end{bmatrix}$$

We have verified with *Maple* that the composite of two rotations is another rotation and they commute, which you were to prove in *Homework* problem 3 of Section 4.4. The same holds true for two rescalings. Now we want to see how we could write an arbitrary linear map T as a composite of these three basic types of linear map.

First of all, we shall look at this on the level of matrix multiplication instead of the composite of linear maps since they are equivalent operations to each other. Second, our rescaling matrices $A_{(a,b)}$ for positive scalars a and b must be generalized to have entries a and b that can be any real numbers. This is not really much of a generalization except to allow the entries a and b to be either, or both, zero since:

$$\begin{bmatrix} a & 0 \\ 0 & -b \end{bmatrix} = \begin{bmatrix} a & 0 \\ 0 & b \end{bmatrix}\begin{bmatrix} 1 & 0 \\ 0 & -1 \end{bmatrix}$$

$$\begin{bmatrix} -a & 0 \\ 0 & b \end{bmatrix} = \begin{bmatrix} a & 0 \\ 0 & b \end{bmatrix}\begin{bmatrix} -1 & 0 \\ 0 & 1 \end{bmatrix}$$

$$\begin{bmatrix} -a & 0 \\ 0 & -b \end{bmatrix} = \begin{bmatrix} a & 0 \\ 0 & b \end{bmatrix}\begin{bmatrix} -1 & 0 \\ 0 & -1 \end{bmatrix}$$

where $\begin{bmatrix} -1 & 0 \\ 0 & 1 \end{bmatrix}$ represents the reflection about the y-axis and $\begin{bmatrix} -1 & 0 \\ 0 & -1 \end{bmatrix}$ represents the reflection about the origin.

Using all of the information we have gathered thus far, we now attempt to decompose an arbitrary linear map given in matrix form by $A = \begin{bmatrix} a & b \\ c & d \end{bmatrix}$. The first step is to recognize that the matrix A can be expressed as

$$\begin{bmatrix} a & b \\ c & d \end{bmatrix} = \begin{bmatrix} R\cos(\theta) & Q\cos(\phi) \\ R\sin(\theta) & Q\sin(\phi) \end{bmatrix} \tag{10.14}$$

if each column of A is treated as a point in the xy-plane at distances R and Q from the origin with angles θ and ϕ measured from the positive x-axis, respectively, where

$$R = \sqrt{a^2 + c^2}, \ Q = \sqrt{b^2 + d^2}, \ \theta = \tan^{-1}\left(\frac{c}{a}\right), \ \phi = \tan^{-1}\left(\frac{d}{b}\right) \qquad (10.15)$$

provided $a, b \neq 0$. So it appears that we have rewritten our original matrix A in a more complex form, however, the point was to decompose the matrix, of which we now have the first step:

$$\begin{bmatrix} R\cos(\theta) & Q\cos(\phi) \\ R\sin(\theta) & Q\sin(\phi) \end{bmatrix} = \begin{bmatrix} \cos(\theta) & \cos(\phi) \\ \sin(\theta) & \sin(\phi) \end{bmatrix} \begin{bmatrix} R & 0 \\ 0 & Q \end{bmatrix} \qquad (10.16)$$

The second matrix on the RHS of the above equation is a rescaling, but the first matrix is not a rotation. So we need to rewrite the first matrix as a product of rotations. To do this, we use the substitution $B = \frac{1}{2}(\theta + \phi)$ and $C = \frac{1}{2}(\theta - \phi)$, which in terms of θ and ϕ is $\theta = B + C$ and $\phi = B - C$. So we can write

$$\begin{bmatrix} \cos(\theta) & \cos(\phi) \\ \sin(\theta) & \sin(\phi) \end{bmatrix} = \begin{bmatrix} \cos(B + C) & \cos(B - C) \\ \sin(B + C) & \sin(B - C) \end{bmatrix} \qquad (10.17)$$

and with the usual trig identities:

$$\begin{aligned} &= \begin{bmatrix} \cos(B)\cos(C) - \sin(B)\sin(C) & \cos(B)\cos(C) + \sin(B)\sin(C) \\ \sin(B)\cos(C) + \cos(B)\sin(C) & \sin(B)\cos(C) - \cos(B)\sin(C) \end{bmatrix} \\ &= \begin{bmatrix} \cos(B) & -\sin(B) \\ \sin(B) & \cos(B) \end{bmatrix} \begin{bmatrix} \cos(C) & \cos(C) \\ \sin(C) & -\sin(C) \end{bmatrix} \end{aligned} \qquad (10.18)$$

The first matrix on the last line is just a simple rotation, however, the second is not in the standard form of a rotation. So we now focus our attention on the second matrix above. Note that

$$\begin{bmatrix} \cos(C) & \cos(C) \\ \sin(C) & -\sin(C) \end{bmatrix} = \begin{bmatrix} \cos(C) & 0 \\ 0 & \sin(C) \end{bmatrix} \begin{bmatrix} 1 & 1 \\ 1 & -1 \end{bmatrix} \qquad (10.19)$$

where the first matrix on the RHS of equation (10.19) is simply a rescaling by the value of $\cos(C)$ in the x-direction, and $\sin(C)$ in the y-direction. The right matrix, however, is not in the form of one of the three types, but we can decompose it as follows:

$$\begin{aligned} \begin{bmatrix} 1 & 1 \\ 1 & -1 \end{bmatrix} &= \begin{bmatrix} 1 & -1 \\ 1 & 1 \end{bmatrix} \begin{bmatrix} 1 & 0 \\ 0 & -1 \end{bmatrix} \\ &= \begin{bmatrix} \sqrt{2} & 0 \\ 0 & \sqrt{2} \end{bmatrix} \begin{bmatrix} \frac{1}{\sqrt{2}} & -\frac{1}{\sqrt{2}} \\ \frac{1}{\sqrt{2}} & \frac{1}{\sqrt{2}} \end{bmatrix} \begin{bmatrix} 1 & 0 \\ 0 & -1 \end{bmatrix} \end{aligned} \qquad (10.20)$$

where

$$\begin{bmatrix} \frac{1}{\sqrt{2}} & -\frac{1}{\sqrt{2}} \\ \frac{1}{\sqrt{2}} & \frac{1}{\sqrt{2}} \end{bmatrix} = \begin{bmatrix} \cos\left(\frac{\pi}{4}\right) & -\sin\left(\frac{\pi}{4}\right) \\ \sin\left(\frac{\pi}{4}\right) & \cos\left(\frac{\pi}{4}\right) \end{bmatrix}$$

We can now represent our original matrix A in terms of rotations and rescalings:

$$\begin{bmatrix} a & b \\ c & d \end{bmatrix} = \begin{bmatrix} \cos(B) & -\sin(B) \\ \sin(B) & \cos(B) \end{bmatrix} \begin{bmatrix} \sqrt{2}\cos(C) & 0 \\ 0 & \sqrt{2}\sin(C) \end{bmatrix} \times$$
$$\begin{bmatrix} \cos\left(\frac{\pi}{4}\right) & -\sin\left(\frac{\pi}{4}\right) \\ \sin\left(\frac{\pi}{4}\right) & \cos\left(\frac{\pi}{4}\right) \end{bmatrix} \begin{bmatrix} R & 0 \\ 0 & -Q \end{bmatrix} \qquad (10.21)$$

where

$$R = \sqrt{a^2 + c^2}, \ Q = \sqrt{b^2 + d^2}, \ B = \frac{\theta + \phi}{2}, \ C = \frac{\theta - \phi}{2} \qquad (10.22)$$

$$a = R\cos(\theta), \ c = R\sin(\theta), \ b = Q\cos(\phi), \ d = Q\sin(\phi)$$

So the matrix A can be written as a product of two rotations and two general rescalings, where general rescalings include the reflections about the x-axis, y-axis, and the origin, as well as normal positive rescalings. In particular, when we perform the following matrix multiplication:

$$\begin{bmatrix} a & b \\ c & d \end{bmatrix} \begin{bmatrix} x \\ y \end{bmatrix}$$

from a geometric point of view, we are first rescaling the point (x, y) by a factor of R in the x-direction, and $-Q$ in the y-direction, then rotating by an angle of $\frac{\pi}{4}$, then rescaling again by a factor of $\sqrt{2}\cos(C)$ in the x-direction and $\sqrt{2}\sin(C)$ in the y-direction, after which we finally rotate through an angle of B to end up at the point $(ax + by, cx + dy)$. To find the values of R, Q, B, and C, we use the formulas from (10.22).

Example 10.2.1. Now, let us do an example of this decomposition with

$$A = \begin{bmatrix} -8 & 5 \\ 3 & -1 \end{bmatrix}$$

> col_A1:= [-8,3]: col_A2:= [5,-1]:
> A:= transpose(matrix([col_A1, col_A2])):
> R:= norm(col_A1, frobenius);

$$R := \sqrt{73}$$

> Q:= norm(col_A2, frobenius);

$$Q := \sqrt{26}$$

> theta:= evalf(Pi + arctan(-3/8));

$$\theta := 2.782821984$$

> evalf(R*cos(theta)); evalf(R*sin(theta));

$$-8.000000002$$
$$2.999999995$$

> phi:= evalf(arctan(-1/5));

$$\phi := -0.1973955598$$

> evalf(Q*cos(phi)); evalf(Q*sin(phi));

$$5.000000000$$
$$-0.9999999999$$

> B:= (theta+phi)/2;

$$B := 1.292713212$$

> C:= (theta-phi)/2;

$$C := 1.490108772$$

> evalf(evalm(matrix([[cos(B), -sin(B)], [sin(B), cos(B)]]) &* matrix([[sqrt(2)* cos(C), 0], [0, sqrt(2)* sin(C)]]) &* matrix([[cos(Pi/4), -sin(Pi/4)], [sin(Pi/4), cos(Pi/4)]]) &* matrix([[R, 0], [0, -Q]])));

$$\begin{bmatrix} -8.000000003 & 5.000000000 \\ 2.999999993 & -1.000000000 \end{bmatrix}$$

Now we see directly that our decomposition of a linear map T into a product of rotations and general rescalings works. In particular, we can write T as $T = T_1 \circ T_2 \circ T_3 \circ T_4$, or in matrix form as $A = A_1 A_2 A_3 A_4$ with

$$A_1 \approx \begin{bmatrix} \cos(1.292713212) & -\sin(1.292713212) \\ \sin(1.292713212) & \cos(1.292713212) \end{bmatrix}$$

$$A_2 \approx \begin{bmatrix} \sqrt{2}\cos(1.490108772) & 0 \\ 0 & \sqrt{2}\sin(1.490108772) \end{bmatrix}$$

$$A_3 = \begin{bmatrix} \cos\left(\frac{\pi}{4}\right) & -\sin\left(\frac{\pi}{4}\right) \\ \sin\left(\frac{\pi}{4}\right) & \cos\left(\frac{\pi}{4}\right) \end{bmatrix}$$

$$A_4 = \begin{bmatrix} \sqrt{73} & 0 \\ 0 & -\sqrt{26} \end{bmatrix}$$

This decomposition cannot be reordered as we check below.

> evalf(evalm(matrix([[cos(B), -sin(B)], [sin(B), cos(B)]])&*matrix([[sqrt(2)*
cos(C), 0], [0, sqrt(2)* sin(C)]])&*matrix([[R, 0], [0, -Q]])&*matrix([[cos(Pi/4),
-sin(Pi/4)], [sin(Pi/4), cos(Pi/4)]])));

$$\begin{bmatrix} 5.076222857 & 4.698137893 \\ -0.7330011379 & -2.057384138 \end{bmatrix}$$

Homework Problems

1. Express each of the following linear map matrices as the product given in (10.21).

(a) $\begin{bmatrix} -\sqrt{3} & 0 \\ 1 & 3 \end{bmatrix}$

(b) $\begin{bmatrix} -\frac{3}{\sqrt{2}} & \frac{1}{\sqrt{2}} \\ \frac{3}{\sqrt{2}} & -\frac{1}{\sqrt{2}} \end{bmatrix}$

(c) $\begin{bmatrix} 0 & -\frac{\sqrt{3}}{6} \\ \frac{1}{2} & -\frac{1}{6} \end{bmatrix}$

(d) $\begin{bmatrix} -1 & -\frac{1}{3} \\ -\sqrt{3} & 0 \end{bmatrix}$

(e) $\begin{bmatrix} -5\sqrt{2} & \frac{11}{\sqrt{2}} \\ -5\sqrt{2} & -\frac{11}{\sqrt{2}} \end{bmatrix}$

(f) $\begin{bmatrix} 0 & 7\sqrt{3} \\ -7 & 7 \end{bmatrix}$

2. In example 10.2.1, it was shown that the decomposition could not be re-ordered in the fashion $A_1 A_2 A_4 A_3$. Determine if any of the following reorderings agree with the decomposition $A_1 A_2 A_3 A_4$.

(a) $A_2 A_1 A_3 A_4$ (b) $A_4 A_3 A_2 A_1$ (c) $A_3 A_2 A_1 A_4$ (d) $A_2 A_3 A_1 A_4$

3. Consider the map that takes a point $\langle x, y \rangle$ to the point $\langle 0, x - y \rangle$. Express the map as the product $A_1 A_2 A_3 A_4$.

4. If $A \in \mathbb{R}^{2\times 2}$ is not invertible, can it still be decomposed into the product of the four matrices?

5. Explain geometrically what the map given by $A = \begin{bmatrix} 0 & 0 \\ 0 & 1 \end{bmatrix}$ does to a point $\langle x, y \rangle$, and then decompose A.

6. Let T be the linear map that is first the rotation about the origin through the angle $\alpha = \frac{7}{6}\pi$ followed by the reflection about the y-axis, and then the

rescaling by a factor of seven in the x-direction. Find the matrix A that represents this linear map T and then decompose it according to formula (10.21). Now decide what this decomposition does geometrically.

7. (a) Explain if the inverse of a rotation, rescaling or reflection is also of the same type, and if it is, give a formula for this inverse which clearly preserves the type.

(b) Explain what happens if the inverse of a rotation, rescaling or reflection does not exist or is not of the same type.

8. Decompose the 2×2 off diagonal matrix $A = \begin{bmatrix} 0 & a \\ b & 0 \end{bmatrix}$

9. If you know the decomposition according to formula (10.21) for a matrix A, then is there any simple way to get the decomposition for A^2? Explain in detail.

10. If we switch to using complex numbers from $\mathbb{C}$ instead of real numbers in $\mathbb{R}$, then does a linear map $T : \mathbb{C}^2 \to \mathbb{C}^2$ and its corresponding 2×2 complex matrix $A \in \mathbb{C}^{2 \times 2}$ representing it have a similar decomposition to the real case discussed in this section? In particular, are there rotations, reflections, and rescalings in the complex case similar to the real case?

11. Let $T : \mathbb{R}^3 \to \mathbb{R}^3$ be a linear map with the matrix $A = \begin{bmatrix} 1 & 0 & 0 \\ 0 & a & b \\ 0 & c & d \end{bmatrix}$

representing it for $a, b, c, d \in \mathbb{R}$.

(a) Decompose A using the methodology of this section and explain your results geometrically in $\mathbb{R}^3$.

(b) Apply the results of part (a) to the matrix $A = \begin{bmatrix} 1 & 0 & 0 \\ 0 & -1 & 3 \\ 0 & 2 & 5 \end{bmatrix}$.

Maple Problems

1. Express each of the following linear map matrices as the product given in (10.21).

(a) $\begin{bmatrix} 2 & -1 \\ 1 & 3 \end{bmatrix}$ (b) $\begin{bmatrix} -1 & 3 \\ 5 & 2 \end{bmatrix}$ (c) $\begin{bmatrix} 0 & 4 \\ -4 & 0 \end{bmatrix}$

(d) $\begin{bmatrix} \frac{1}{3} & -\frac{1}{2} \\ \frac{1}{5} & \frac{1}{4} \end{bmatrix}$ (e) $\begin{bmatrix} 9 & 2 \\ -7 & -8 \end{bmatrix}$ (f) $\begin{bmatrix} -\frac{3}{7} & \frac{4}{5} \\ \frac{7}{3} & -\frac{5}{4} \end{bmatrix}$

2. Let $A = \begin{bmatrix} 5 - 2i & -1 + i \\ 3 + 8i & 2 - 5i \end{bmatrix}$

(a) Write A as $A = B + Ci$, where B and C are the real 2×2 matrices that are the real and imaginary parts of A. Now decompose B and C according to formula (10.21).

(b) Does this tell us anything about the geometric nature of the linear map $T : \mathbb{C}^2 \to \mathbb{C}^2$, which has the matrix A representing it? Explain.

10.3 The Effect of Linear Maps on Volume, Area, and Arclength in $\mathbb{R}^3$

Now we switch from two dimensions to three, and linear maps $T : \mathbb{R}^3 \to \mathbb{R}^3$ with corresponding 3×3 matrix representations. Naturally, we want to see how things stay the same, or change, as we move to a higher dimension. As with the case of maps in $\mathbb{R}^2$, T must take a line segment to another line segment, since this is independent of dimension. But does this determine T uniquely as in two dimensions? At this point you should take a guess before moving on to the next paragraph.

It seems like it should not. This call is made based on the fact that the same formula in two dimensions will no longer work, since we do not have a square matrix that we can invert. Thus, we need three points, and so a triangle in $\mathbb{R}^3$ is needed to replace the line segment of $\mathbb{R}^2$. As with the $\mathbb{R}^2$ case before, we need that no line segment making up the triangle lies on a line through the origin. So if T takes the triangle $T_1 = \{\langle x_1, y_1, z_1 \rangle, \langle x_2, y_2, z_2 \rangle, \langle x_3, y_3, z_3 \rangle\}$ in its domain to another triangle $T_2 = \{\langle u_1, v_1, w_1 \rangle, \langle u_2, v_2, w_2 \rangle, \langle u_3, v_3, w_3 \rangle\}$ in its range, then our two-dimensional formula suggests that the standard matrix A representing the linear map T in three dimensions is

$$A = \begin{bmatrix} u_1 & u_2 & u_3 \\ v_1 & v_2 & v_3 \\ w_1 & w_2 & w_3 \end{bmatrix} \begin{bmatrix} x_1 & x_2 & x_3 \\ y_1 & y_2 & y_3 \\ z_1 & z_2 & z_3 \end{bmatrix}^{-1} \qquad (10.23)$$

Example 10.3.1. We now verify this formula with an example in which we find the standard matrix A representing the linear map T, where

$$T(\langle -4, 7, 3 \rangle) = \langle 5, 2, -9 \rangle, T(\langle 11, -6, 2 \rangle) = \langle -1, 3, 8 \rangle, T(\langle 9, 13, -5 \rangle) = \langle 3, -6, 7 \rangle$$

The triangle and its transform are depicted in Figure 10.7.

> restart: with(linalg): with(plots): with(plottools):

> P:= [-4,7,3]: Q:= [11,-6,2]: R:= [9,13,-5]:

> PQR:= transpose(matrix([P,Q,R]));

$$PQR := \begin{bmatrix} -4 & 11 & 9 \\ 7 & -6 & 13 \\ 3 & 2 & -5 \end{bmatrix}$$

> TP:= [5,2,-9]: TQ:= [-1,3,8]: TR:= [3,-6,7]:

> TP_TQ_TR:= transpose(matrix([TP,TQ,TR]));

$$TP\_TQ\_TR := \begin{bmatrix} 5 & -1 & 3 \\ 2 & 3 & -6 \\ -9 & 8 & 7 \end{bmatrix}$$

> A:= evalm(TP_TQ_TR&*inverse(PQR));

$$A := \begin{bmatrix} \frac{7}{181} & \frac{165}{362} & \frac{237}{362} \\ \frac{19}{543} & -\frac{121}{1086} & \frac{1057}{1086} \\ \frac{130}{181} & -\frac{71}{181} & -\frac{204}{181} \end{bmatrix}$$

> evalm(A&*P);

$$\begin{bmatrix} 5 & 2 & -9 \end{bmatrix}$$

> evalm(A&*Q);

$$\begin{bmatrix} -1 & 3 & 8 \end{bmatrix}$$

> evalm(A&*R);

$$\begin{bmatrix} 3 & -6 & 7 \end{bmatrix}$$

> plot_PQR:= polygon([P,Q,R], color=blue, thickness=2):

> plot_TPTQTR:= polygon([TP,TQ,TR], color=red, thickness=2):

> plotP:= textplot3d([op(P),"P"], color = black):

> plotQ:= textplot3d([op(Q),"Q"], color = black):

> plotR:= textplot3d([op(R),"R"], color = black):

> plotTP:= textplot3d([op(TP),"TP"], color = black):

> plotTQ:= textplot3d([op(TQ),"TQ"], color = black):

> plotTR:= textplot3d([op(TR),"TR"], color = black):

> display({plot_PQR, plot_TPTQTR, plotP, plotQ, plotR, plotTP, plotTQ, plotTR}, axes = boxed);

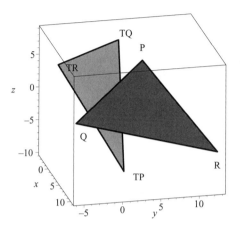

Figure 10.7:   Triangle PQR, pointing left, is transformed to triangle $T(P)T(Q)T(R)$, pointing down, by T.

This example justifies our belief that a linear map T on Euclidean three-space will be uniquely determined by taking a triangle to another triangle in three-space.

Example 10.3.2. Now, let us see how this linear map T transforms the length of these three line segments and the areas of these two triangles. How do you think T will behave? Also, we shall see how T affects the volume of a parallelepiped.

> lengthPQ:= evalf(norm(Q-P,frobenius));

$$lengthPQ := 19.87460691$$

> lengthTPTQ:= evalf(norm(TQ-TP,frobenius));

$$lengthTPTQ := 18.05547009$$

> evalf(lengthTPTQ/lengthPQ);

$$0.9084692931$$

> evalf(det(A));

$$0.3996316759$$

> evalf(abs(det(A))^(1/2));
$$0.6321642791$$

> evalf(abs(det(A))^(1/3));
$$0.7365800773$$

> lengthPR:= evalf(norm(R-P,frobenius));
$$lengthPR := 16.40121947$$

> lengthTPTR:= evalf(norm(TR-TP,frobenius));
$$lengthTPTR := 18.$$

> evalf(lengthTPTR/lengthPR);
$$1.097479369$$

> lengthQR:= evalf(norm(R-Q,frobenius));
$$lengthQR := 20.34698995$$

> lengthTQTR:= evalf(norm(TR-TQ,frobenius));
$$lengthTQTR := 9.899494934$$

> evalf(lengthTQTR/lengthQR);
$$0.4865336327$$

> (.9084692931+1.097479369+.4865336327)/3;
$$0.8308274317$$

It seems that the average effect of T on these line segment lengths is 0.8308274317, which is approximately the cube root of $|\det(A)|$. Now, let us check on the areas of these two triangles.

> area_PQR:= evalf(0.5*norm(crossprod(Q-P,R-P),frobenius));
$$area\_PQR := 150.5240844$$

> area_TPTQTR:= evalf(.5*norm(crossprod(TQ-TP,TR-TP),frobenius));
$$area\_TPTQTR := 85.80209787$$

> area_TPTQTR/area_PQR; evalf(abs(det(A))^(1/2));
$$0.5700223869$$
$$0.6321642791$$

The linear map T seems to change area by approximately the factor the square root of $|\det(A)|$. Now we want to check how T changes the volume of the paralelopiped with three adjacent sides the position vectors for the points P, Q, and R. The volume of the parallelopiped with adjacent sides the vectors $\overrightarrow{P}$, $\overrightarrow{Q}$, and $\overrightarrow{R}$ is

$$\left| \overrightarrow{P} \cdot \left(\overrightarrow{Q} \times \overrightarrow{R} \right) \right| = |\det(PQR)|$$

where the points P, Q, and R make up the rows of the matrix PQR. Figure 10.8 is a depiction of the parallelopiped and its transform.

> parallelopipedPQR:= polygon([[0,0,0], P, P+Q, Q], color = blue), polygon([[0,0,0], P, P+R, R], color = blue), polygon([[0,0,0], R, R+Q, Q], color = blue), polygon([P, P+R, P+Q+R, P+Q], color = blue), polygon([R, P+R, P+Q+R, R+Q], color = blue), polygon([Q, P+Q, P+Q+R, R+Q] ,color = blue):

> parallelopipedTPTQTR:= polygon([[0,0,0],TP,TP+TQ,TQ], color = red), polygon([[0,0,0],TP,TP+TR,TR], color = red), polygon([[0,0,0], TR, TR+TQ, TQ], color = red), polygon([TP,TP+TR, TP+TQ+TR, TP+TQ], color = red), polygon([TR,TP+TR, TP+TQ+TR, TR+TQ], color = red), polygon([TQ, TP + TQ, TP+TQ+TR, TR+TQ], color = red):

> display({parallelopipedPQR, parallelopipedTPTQTR, plotP, plotQ, plotR, plotTP, plotTQ, plotTR}, axes = boxed);

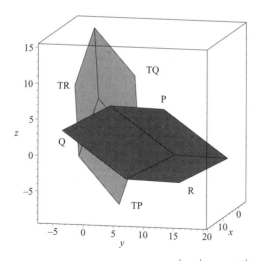

Figure 10.8: Parallelepiped defined by vectors $\overrightarrow{P}$, $\overrightarrow{Q}$, and $\overrightarrow{R}$ is transformed to the parallelepiped defined by vectors $\overrightarrow{TP}$, $\overrightarrow{TQ}$, and $\overrightarrow{TR}$.

> volume_PQR:= abs(det(matrix([P,Q,R])));

$$volume\_PQR := 1086$$

> volume_TPTQTR:= abs(det(matrix([TP,TQ,TR])));

$$volume\_TPTQTR := 434$$

> evalf(volume_TPTQTR/volume_PQR); evalf(det(A));

$$0.3996316759$$
$$0.3996316759$$

The linear map T changes volume by the factor $|\det(A)|$. Now, let us see if this is also true for the volume of a solid inside a simple closed surface S. A *simple closed surface* S is one that does not intersect itself and has no boundary curve, that is, it is a surface that has inside it a simply connected (one connected piece without any holes) solid. A sphere, ellipsoid, infinite (in both directions) cylinder, hyperboloid of one sheet, paraboloid, and infinite cone are all examples of simple closed surfaces.

From the Divergence theorem of multivariable calculus, the volume V of the solid inside a simple closed surface S is given by

$$V = \frac{1}{3} \left| \int_c^d \int_a^b \vec{r}(u,v) \cdot (\vec{r_u} \times \vec{r_v})\, du\, dv \right| \tag{10.24}$$

where the surface S is given by its position vector field $\vec{r}(u,v)$ for $a \le u \le b$ and $c \le v \le d$. Also, $\vec{r}_u = \frac{\partial}{\partial u}\vec{r}(u,v)$ and $\vec{r}_v = \frac{\partial}{\partial v}\vec{r}(u,v)$ are the u- and v-direction tangent vector fields to $\vec{r}(u,v)$, respectively. We also have that

$$\vec{r}(u,v) \cdot (\vec{r}_u \times \vec{r}_v) = \det\left(\begin{bmatrix} \vec{r} \\ \vec{r_u} \\ \vec{r_v} \end{bmatrix} \right) \tag{10.25}$$

Example 10.3.3. Let us test this formula out on the following example. Let the linear map $T : \mathbb{R}^3 \to \mathbb{R}^3$ be the one with standard 3×3 matrix A representing T given by $A = \begin{bmatrix} -8 & 5 & -3 \\ 7 & 2 & -9 \\ 1 & 6 & 4 \end{bmatrix}$. Let T send the sphere S with center at $(8, 2, 4)$ and radius 5 to the surface $T(S)$. S has a position vector field defined by

$$\vec{r}(u,v) = \langle 5\cos(u)\sin(v) + 8, 5\sin(u)\sin(v) + 2, 5\cos(v) + 4 \rangle$$

for $u \in [0, 2\pi]$ and $v \in [0, \pi]$. This parametric surface definition of a sphere can be found in most multivariable calculus books.

```
> A:= matrix([[-8, 5, -3], [7, 2, -9], [1, 6, 4]]):
> r:= [5*cos(u)*sin(v) + 8, 5*sin(u)*sin(v) + 2, 5*cos(v) + 4]:
> ru:= map(diff,r,u);
```

$$ru := [-5\sin(u)\sin(v), 5\cos(u)\sin(v), 0]$$

```
> rv:= map(diff,r,v);
```

$$rv := [5\cos(u)\cos(v), 5\sin(u)\cos(v), -5\sin(v)]$$

```
> Integ:= simplify(det(matrix([r,ru,rv])));
```

$$Integ := -25\sin(v)(8\cos(u)\sin(v) + 5 + 2\sin(u)\sin(v) + 4\cos(v))$$

```
> volume_r:= abs(evalf(1/3*int(int(Integ, u=0..2*Pi), v=0..Pi)));
```

$$volume\_r := 523.5987758$$

```
> evalf(4/3*Pi*5^3);
```

$$523.5987758$$

```
> TS:= evalm(A&*r):
> TSu:= map(diff,TS,u);
```

$$TSu := \big[40\sin(u)\sin(v) + 25\cos(u)\sin(v), -35\sin(u)\sin(v)$$
$$+ 10\cos(u)\sin(v), -5\sin(u)\sin(v) + 30\cos(u)\sin(v)\big]$$

```
> TSv:= map(diff,TS,v);
```

$$TSv := \big[-40\cos(u)\cos(v) + 25\sin(u)\cos(v) + 15\sin(v), 35\cos(u)\cos(v)$$
$$+ 10\sin(u)\cos(v) + 45\sin(v), 5\cos(u)\cos(v) + 30\sin(u)\cos(v) - 20\sin(v)\big]$$

```
> Integ_TS:= simplify(det(matrix([TS,TSu,TSv])));
```

$$Integ\_TS := 20025\sin(v)(8\cos(u)\sin(v) + 5 + 2\sin(u)\sin(v) + 4\cos(v))$$

```
> volume_TS:= abs(evalf(1/3*int(int(Integ_TS, u=0..2*Pi), v=0..Pi)));
```

$$volume\_TS := 419402.6193$$

```
> abs(evalf(det(A)*volume_r)); abs(det(A));
```

$$419402.6194$$
$$801$$

We have the volume relationship as expected, with the linear map T causing a change in volume by the factor of $|\det(A)|$. Now, let us plot both of these solids and see that $T(S)$ is an ellipsoid as we might have expected. This is depicted in Figure 10.9.

```
> plot_r:= plot3d(r, u=0..2*Pi, v=0..Pi, color = blue):
> plot_TS:= plot3d(TS, u=0..2*Pi, v=0..Pi, color = red):
> display({plot_r, plot_TS}, axes = boxed, scaling = constrained, style = patchnogrid);
```

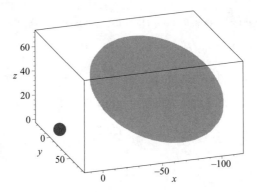

Figure 10.9: The sphere is transformed by T to the ellipsoid.

From multivariable calculus, the surface area A_S of a simple closed surface S can be computed by the formula

$$A_S = \int_c^d \int_a^b |\vec{r_u} \times \vec{r_v}| \; du \; dv$$

where the surface S is given by its position vector field $\vec{r}(u, v)$ for $u \in [a, b]$ and $v \in [c, d]$. Let us see how this works on our two surfaces. We hope once again to get that the surface areas of the two are related by approximately the factor of the square root of $|\det(A)|$.

```
> integr:= simplify(norm(crossprod(ru,rv),frobenius));
```

$$integr := 25\sqrt{|\cos(u)^2 \sin(v)^4| + |\sin(u)^2 \sin(v)^4| + |\sin(v)^2 \cos(v)^2|}$$

```
> surfarea_r:= evalf(Int(Int(integr, u=0..2*Pi), v=0..Pi));
```

$$surfarea\_r := 314.1592654$$

```
> evalf(4*Pi*5^2);
```

$$314.1592654$$

> integr_TS:= simplify(norm(crossprod(TSu,TSv), frobenius));

$$integr\_TS := 25 \left(\left| \sin(v)^2(-40\cos(v) + 37\sin(u)\sin(v) - 62\cos(u)\sin(v))^2 \right| \right.$$
$$+ \left| \sin(v)^2(-53\cos(v) + 29\sin(u)\sin(v) + 38\cos(u)\sin(v))^2 \right|$$
$$\left. + 9 \left| \sin(v)^2(17\cos(v) + 31\sin(u)\sin(v) + 13\cos(u)\sin(v))^2 \right| \right)^{\frac{1}{2}}$$

> surfarea_TS:= evalf(Int(Int(integr_TS, u=0..2*Pi), v=0..Pi));

$$surfarea\_TS := 28211.64213$$

> surfarea_TS/surfarea_r; evalf(sqrt(abs(det(A))));

$$89.80044594$$
$$28.30194340$$

As can be seen by the last two outputs above, the ratio of the surface area of the transformed surface to that of the original is not even close to the value of $\sqrt{|\det(A)|}$ in this example.∎

Homework Problems

1. Compute the matrix A corresponding to the linear map that takes triangle T_1 to triangle T_2 defined by the following corner points.

(a) $T_1 = \{\langle 1, 1, 1 \rangle, \langle 1, -1, -2 \rangle, \langle 1, 1, -1 \rangle\}$
      $T_2 = \{\langle 0, 1, 0 \rangle, \langle 2, 1, -2 \rangle, \langle -2, 0, 1 \rangle\}$

(b) $T_1 = \{\langle 1, 2, 3 \rangle, \langle -3, 2, -1 \rangle, \langle 1, 1, -2 \rangle\}$
      $T_2 = \{\langle -2, 0, 2 \rangle, \langle 0, 3, 0 \rangle, \langle -1, 2, -3 \rangle\}$

(c) $T_1 = \{\langle 4, -6, 2 \rangle, \langle -1, -1, -1 \rangle, \langle 1, 1, -1 \rangle\}$
      $T_2 = \{\langle 3, 3, 2 \rangle, \langle -1, -3, 6 \rangle, \langle 3, 8, -2 \rangle\}$

(d) $T_1 = \{\langle -8, 3, 7 \rangle, \langle -8, 3, 6 \rangle, \langle -7, 3, 6 \rangle\}$
      $T_2 = \{\langle -2, 3, 4 \rangle, \langle 4, 3, -9 \rangle, \langle -1, 3, 3 \rangle\}$

2. Verify equation (10.25).

3. If $A = \begin{bmatrix} 1 & 1 & -2 \\ -2 & 0 & 1 \\ 3 & 3 & 1 \end{bmatrix}$ and $T_2 = \{\langle 3, 3, 2 \rangle, \langle -1, -3, 6 \rangle, \langle 3, 8, -2 \rangle\}$, find the three points which define the corners of the triangle T_1 such that T_2 is the map

of T_1 by the matrix A.

4. Prove that the map defined by the matrix A given below preserves the length of the sides of any triangle. Geometrically, what does the matrix A represent?

$$A = \begin{bmatrix} \frac{1}{\sqrt{2}} & -\frac{1}{\sqrt{2}} & 0 \\ \frac{1}{\sqrt{2}} & \frac{1}{\sqrt{2}} & 0 \\ 0 & 0 & 1 \end{bmatrix}$$

5. Prove that the map defined by the matrix A given below scales line segments in $\mathbb{R}^3$ by a factor of $\frac{5}{3}$, i.e. if $\overrightarrow{v} \in \mathbb{R}^3$, $|A\overrightarrow{v}| = \frac{5}{3}|\overrightarrow{v}|$.

$$A = \begin{bmatrix} \frac{2}{\sqrt{3}} & 0 & \frac{1}{\sqrt{3}} \\ 0 & \sqrt{\frac{5}{3}} & 0 \\ \frac{1}{\sqrt{3}} & 0 & -\frac{2}{\sqrt{3}} \end{bmatrix}$$

6. Prove that the map defined by the matrix A given below preserves the length of the sides of any triangle.

$$A = \begin{bmatrix} \frac{1}{\sqrt{3}} & \frac{1}{\sqrt{3}} & \frac{1}{\sqrt{3}} \\ -\frac{1}{\sqrt{6}} & \sqrt{\frac{2}{3}} & -\frac{1}{\sqrt{6}} \\ -\frac{1}{\sqrt{2}} & 0 & \frac{1}{\sqrt{2}} \end{bmatrix}$$

7. Show that the rows of the matrix from problem 6 form an orthonormal basis for $\mathbb{R}^3$.

8. Prove or disprove the following statement: If the rows of $A \in \mathbb{R}^{3\times3}$ form an orthonormal basis, then the lengths of the sides of the transformed triangle under the map A are the same as that of the original.

9. Let $C : \langle x, y, z \rangle = \langle x(t), y(t), z(t) \rangle$ for $t \in [a, b]$ be a parametric spacecurve. The arclength L, of C, is given by

$$L = \int_a^b \sqrt{\left(\frac{dx}{dt}\right)^2 + \left(\frac{dy}{dt}\right)^2 + \left(\frac{dz}{dt}\right)^2}\, dt$$

(a) Let C be the circular helix given by

$$C : \langle x(t), y(t) \rangle = \langle 3\sin(t) + 7,\ 3\cos(t) - 4,\ 2t - 9 \rangle, t \in [0, 2\pi]$$

Find the arclength of C.

(b) Let C be the general circular helix given by

$$C : \langle x(t), y(t) \rangle = \langle R\sin(t) + H, \ R\cos(t) + K, \ at + b \rangle, t \in [0, 2\pi]$$

Find the arclength of C.

10. (Use problem 9) Let $T : \mathbb{R}^3 \to \mathbb{R}^3$ be the linear map represented by the matrix

$$A = \begin{bmatrix} \cos(\theta) & -\sin(\theta) & 0 \\ \sin(\theta) & \cos(\theta) & 0 \\ 0 & 0 & -1 \end{bmatrix}$$

(a) Show that the curve $T(C)$ and C (for the parametric curve C in problem 9 part (a)) have the same length.

(b) Show that the curve $T(C)$ and C (for the parametric curve C in problem 9 part(b)) have the same length.

11. Find the surface area formula $4\pi^2 Rr$ for the *torus* (donut shaped) simple closed parametric surface T given by

$$T : \langle x(u, v), y(u, v), z(u, v) \rangle = \langle R + r\cos(u)) \cos(v), R + r\cos(u)) \sin(v), r\sin(u) \rangle,$$

for $u, v \in [0, 2\pi]$. Can you find a simple formula for its volume in terms of R and r?

Maple Problems

1. Given the map defined by the matrix $A = \begin{bmatrix} 1 & 1 & -2 \\ -2 & 0 & 1 \\ 3 & 3 & 1 \end{bmatrix}$, for each of the following triangles, compute the lengths of all three sides, the area, and then perform the same computations with the transformed triangle.

(a) $T = \{\langle 1, -1, 1 \rangle, \langle 4, -1, -2 \rangle, \langle 1, 1, 0 \rangle\}$

(b) $T = \{\langle 1, 0, 1 \rangle, \langle 4, 2, -3 \rangle, \langle 1, 6, -2 \rangle\}$

(c) $T = \{\langle -2, 2, 5 \rangle, \langle -1, 3, 5 \rangle, \langle 3, 0, -1 \rangle\}$

2. Graph both the original and transformed triangles from problem 1.

3. Perform the Gram-Schmidt orthonormalization process on the vectors formed by the rows of the following matrix:

$$A = \begin{bmatrix} 1 & 1 & 1 \\ 1 & 2 & -1 \\ 2 & 3 & 2 \end{bmatrix}$$

then use the new vectors to construct a new matrix B.

4. Use the matrix B found in problem 3 to transform the following triangles:

(a) $T = \{\langle 1, -1, 1 \rangle, \langle 4, -1, -2 \rangle, \langle 1, 1, 0 \rangle\}$

(b) $T = \{\langle 1, 0, 1 \rangle, \langle 4, 2, -3 \rangle, \langle 1, 6, -2 \rangle\}$

(c) $T = \{\langle -2, 2, 5 \rangle, \langle -1, 3, 5 \rangle, \langle 3, 0, -1 \rangle\}$

Compute the lengths of all three sides, the area, and then perform the same computations with the transformed triangle.

5. Let $T : \mathbb{R}^3 \to \mathbb{R}^3$ be the linear map represented by the matrix

$$A = \begin{bmatrix} -1 & 3 & 5 \\ 4 & 1 & -2 \\ 9 & -10 & 7 \end{bmatrix}$$

For the parametric curve C in *Homework* problem 9 part (a), find the length of $T(C)$ and compare it to that of C, do they differ by a factor approximately $|\det(A)|^{\frac{1}{3}}$? Plot both the curves C and $T(C)$.

6. For the linear map T in *Maple* problem 5, find the surface area and volume of the solid for the torus simple closed parametric surface S and $T(S)$ for S given by

$$S : \langle x(u, v), y(u, v), z(u, v) \rangle = \langle (9 + 4\cos(u))\cos(v), 9 + 4\cos(u))\sin(v), 4\sin(u) \rangle$$

for $u, v \in [0, 2\pi]$. Are their surface areas and volumes related through the value of $\det(A)$? Plot both S and $T(S)$.

10.4   Rotations, Reflections, and Rescalings in Three Dimensions

In this section, we look at those linear maps $T : \mathbb{R}^3 \to R^3$ that are rotations about lines through the origin, reflections about a coordinate axis or the origin, and general rescalings. Let us begin by discussing rotations in $\mathbb{R}^3$, in particular we already know how to rotate objects by an angle θ in the xy-plane. In three dimensions, a rotation in the xy-plane means a rotation about the z-axis. Therefore, we need to construct a 3×3 matrix that keeps all z values the same, and rotates the (x, y) coordinates by an angle of θ. Your first guess should be to put the 2×2 matrix in the upper righthand corner of our hopeful matrix. The question remains, What do we do about the last row and column? Well, we wish for z to stay fixed, therefore we should use the third row and column of the identity matrix. So our matrix looks as follows:

$$A_{\theta_z} = \begin{bmatrix} \cos(\theta) & -\sin(\theta) & 0 \\ \sin(\theta) & \cos(\theta) & 0 \\ 0 & 0 & 1 \end{bmatrix} \tag{10.26}$$

which we can verify works if we perform the multiplication $A_{\theta_z}\vec{v}$, where $\vec{v} \in \mathbb{R}^3$:

$$\begin{bmatrix} \cos(\theta) & -\sin(\theta) & 0 \\ \sin(\theta) & \cos(\theta) & 0 \\ 0 & 0 & 1 \end{bmatrix} \begin{bmatrix} x \\ y \\ z \end{bmatrix} = \begin{bmatrix} x\cos(\theta) - y\sin(\theta) \\ x\sin(\theta) + y\cos(\theta) \\ z \end{bmatrix}$$

The matrix A_{θ_z} can be easily modified to get the two other axis rotation matrices:

$$A_{\theta_x} = \begin{bmatrix} 1 & 0 & 0 \\ 0 & \cos(\theta) & -\sin(\theta) \\ 0 & \sin(\theta) & \cos(\theta) \end{bmatrix}, \quad A_{\theta_y} = \begin{bmatrix} \cos(\theta) & 0 & -\sin(\theta) \\ 0 & 1 & 0 \\ \sin(\theta) & 0 & \cos(\theta) \end{bmatrix} \tag{10.27}$$

Now we want to find the matrix $A_{\vec{v}}$ that allows us to rotate about any line through the origin in three-space that is parallel to the unit vector $\vec{v}$, where no component of $\vec{v}$ is zero. If any component of $\vec{v}$ is zero, then we would simply be rotating about one of the axes. If you analyze what we did in Section 7.1, then you find that the rotation matrix $A_{\vec{v}}$ has columns gotten by revolving respectively each column of I_3 from the 3×3 identity matrix.

> with(linalg): with(plots): with(plottools):
> v:= [alpha,beta,delta]:
> Q:= [1,0,0]:
> C:= evalm(dotprod(v,Q)*v);

$$C := \begin{bmatrix} \alpha^2 & \alpha\beta & \alpha\delta \end{bmatrix}$$

> R:= sqrt(1-alpha^2);

$$R := \sqrt{1 - \alpha^2}$$

> U:= (Q-C)/R;

$$U := \frac{[1, 0, 0] - C}{\sqrt{1 - \alpha^2}}$$

> V:= crossprod(C,U) / norm(crossprod(C,U),frobenius);

$$V := \frac{\left[\begin{array}{ccc} 0 & \alpha\delta\sqrt{1-\alpha^2} + \frac{\alpha^3}{\sqrt{1-\alpha^2}} & -\frac{\alpha^3\beta}{\sqrt{1-\alpha^2}} - \alpha\beta\sqrt{1-\alpha^2} \end{array}\right]}{\sqrt{\left|\alpha\delta\sqrt{1-\alpha^2} + \frac{\alpha^3}{\sqrt{1-\alpha^2}}\right|^2 + \left|\frac{\alpha^3\beta}{\sqrt{1-\alpha^2}} + \alpha\beta\sqrt{1-\alpha^2}\right|^2}}$$

> N:= simplify(evalm(C + R*cos(theta)*U + R*sin(theta)*V), {alpha^2 + beta^2 + delta^2=1});

$$N := \Big[\cos(\theta) + (1 - \cos(\theta))\alpha^2,$$

$$\frac{\left(|\alpha|\sqrt{\frac{|\delta|^2+|\beta|^2}{|-1+\alpha^2|}} - \cos(\theta)|\alpha|\sqrt{\frac{|\delta|^2+|\beta|^2}{|-1+\alpha^2|}}\right)\alpha\beta + \alpha\sin(\theta)\,\delta}{|\alpha|\sqrt{\frac{|\delta|^2+|\beta|^2}{|-1+\alpha^2|}}},$$

$$\frac{-\alpha\sin(\theta)\beta + \left(|\alpha|\sqrt{\frac{|\delta|^2+|\beta|^2}{|-1+\alpha^2|}} - \cos(\theta)|\alpha|\sqrt{\frac{|\delta|^2+|\beta|^2}{|-1+\alpha^2|}}\right)\alpha\delta}{|\alpha|\sqrt{\frac{|\delta|^2+|\beta|^2}{|-1+\alpha^2|}}}\Big]$$

> row1:= simplify(%,{abs(-1 + alpha^2) = abs(delta)^2 + abs(beta)^2});

$$row1 := \Big[\alpha^2 + \cos(\theta) - \cos(\theta)\alpha^2, \frac{\alpha\beta\,|\alpha| - \alpha\cos(\theta)\beta\,|\alpha| + \alpha\sin(\theta)\delta}{|\alpha|},$$

$$\frac{\alpha\delta\,|\alpha| - \alpha\cos(\theta)\delta\,|\alpha| - \alpha\sin(\theta)\beta}{|\alpha|}\Big]$$

This is the first column of the rotation matrix $A_{\vec{v}}$. Now let us find its second and third columns.

> Q:= [0,1,0]:
> C:= evalm(dotprod(v,Q)*v);

$$C := \left[\begin{array}{ccc} \alpha\beta & \beta^2 & \beta\delta \end{array}\right]$$

> R:= sqrt(1-beta^2);

$$R := \sqrt{1 - \beta^2}$$

> U:= (Q-C)/R;

$$U := \frac{[0, 1, 0] - C}{\sqrt{1 - \beta^2}}$$

> V:= crossprod(C,U) / norm(crossprod(C,U),frobenius);

$$V := \frac{\left[\begin{array}{ccc} -\frac{\beta^3\delta}{\sqrt{1-\beta^2}} - \beta\delta\sqrt{1-\beta^2} & 0 & \alpha\beta\sqrt{1-\beta^2} + \frac{\beta^3\alpha}{\sqrt{1-\beta^2}} \end{array}\right]}{\sqrt{\left|\frac{\beta^3\delta}{\sqrt{1-\beta^2}} + \beta\delta\sqrt{1-\beta^2}\right|^2 + \left|\alpha\beta\sqrt{1-\beta^2} + \frac{\beta^3\alpha}{\sqrt{1-\beta^2}}\right|^2}}$$

> N:= simplify(evalm(C+R*cos(theta)*U + R*sin(theta)*V), {alpha^2 + beta^2 + delta^2 = 1});

$$N := \left[\frac{\left(|\beta|\sqrt{\frac{|\delta|^2+|\alpha|^2}{|-\alpha^2-\delta^2|}} - \cos(\theta)\,|\beta|\sqrt{\frac{|\delta|^2+|\alpha|^2}{|-\alpha^2-\delta^2|}}\right)\alpha\beta - \beta\sin(\theta)\delta}{|\beta|\sqrt{\frac{|\delta|^2+|\alpha|^2}{|-\alpha^2-\delta^2|}}},\right.$$

$$1 + (-1+\cos(\theta))\alpha^2 + (-1+\cos(\theta))\delta^2,$$

$$\left.\frac{\alpha\sin(\theta)\beta + \left(|\beta|\sqrt{\frac{|\delta|^2+|\alpha|^2}{|-\alpha^2-\delta^2|}} - \cos(\theta)\,|\beta|\sqrt{\frac{|\delta|^2+|\alpha|^2}{|-\alpha^2-\delta^2|}}\right)\beta\delta}{|\beta|\sqrt{\frac{|\delta|^2+|\alpha|^2}{|-\alpha^2-\delta^2|}}}\right]$$

> row2:= simplify(%, {abs(1-beta^2)= abs(delta)^2 + abs(alpha)^2 });

$$row2 := \left[\frac{\beta\alpha\,|\beta| - \beta\alpha\cos(\theta)\,|\beta| - \beta\sin(\theta)\delta}{|\beta|},\right.$$

$$\left.\beta^2 + \cos(\theta) - \cos(\theta)\beta^2,\ \frac{\alpha\sin(\theta)\beta + \beta\delta\,|\beta| - \beta\delta\cos(\theta)\,|\beta|}{|\beta|}\right]$$

> Q:= [0,0,1]:
> C:= evalm(dotprod(v,Q)*v);

$$C := \left[\begin{array}{ccc} \alpha\delta & \beta\delta & \delta^2 \end{array}\right]$$

> R:= sqrt(1-delta^2);

$$R := \sqrt{1-\delta^2}$$

> U:= (Q-C)/R;

$$U := \frac{[0,0,1]-C}{\sqrt{1-\delta^2}}$$

> V:= crossprod(C,U)/norm(crossprod(C,U), frobenius);

$$V := \frac{\left[\begin{array}{ccc} \beta\delta\sqrt{1-\delta^2} + \frac{\delta^3\,\beta}{\sqrt{1-\delta^2}} & -\frac{\delta^3\,\alpha}{\sqrt{1-\delta^2}} - \alpha\delta\sqrt{1-\delta^2} & 0 \end{array}\right]}{\sqrt{\left|\beta\delta\sqrt{1-\delta^2} + \frac{\delta^3\,\beta}{\sqrt{1-\delta^2}}\right|^2 + \left|\frac{\delta^3\,\alpha}{\sqrt{1-\delta^2}} + \delta\,\alpha\sqrt{1-\delta^2}\right|^2}}$$

> N:= simplify(evalm(C + R*cos(theta)*U + R*sin(theta)*V), {alpha^2 + beta^2 + gama^2=1});

$$N := \left[\frac{\beta\sin(\theta)\delta + \left(|\delta|\sqrt{\frac{|\beta|^2+|\alpha|^2}{|-1+\delta^2|}} - \cos(\theta)\,|\delta|\sqrt{\frac{|\beta|^2+|\alpha|^2}{|-1+\delta^2|}}\right)\delta\,\alpha}{|\delta|\sqrt{\frac{|\beta|^2+|\alpha|^2}{|-1+\delta^2|}}},\right.$$

$$\left.\frac{\left(|\delta|\sqrt{\frac{|\beta|^2+|\alpha|^2}{|-1+\delta^2|}} - \cos(\theta)\,|\delta|\sqrt{\frac{|\beta|^2+|\alpha|^2}{|-1+\delta^2|}}\right)\beta\delta - \alpha\sin(\theta)\delta}{|\delta|\sqrt{\frac{|\beta|^2+|\alpha|^2}{|-1+\delta^2|}}},\ \delta^2 + \cos(\theta) - \delta^2\cos(\theta)\right]$$

> row3:= simplify(%,{abs(-1 + delta^2) = abs(beta)^2 + abs(alpha)^2});

$$row3 := \left[\frac{\beta \sin(\theta)\delta + \delta\,\alpha\,|\delta| - \delta\,\alpha\cos(\theta)\,|\delta|}{|\delta|}, \right.$$

$$\left. \frac{\delta\,\beta\,|\delta| - \delta\beta\cos(\theta)\,|\delta| - \alpha\sin(\theta)\delta}{|\delta|}, \delta^2 + \cos(\theta) - \delta^2\cos(\theta) \right]$$

Next, we have all three columns of $A_{\vec{v}}$. Notice that no component of the unit vector $\vec{v} = \langle \alpha, \beta, \delta \rangle$ parallel to our axis of rotation can be zero since we must divide by their absolute values in $A_{\vec{v}}$.

> rotmatrix:= transpose(matrix([row1,row2,row3]));

$rotmatrix :=$

$$\left[\begin{array}{ccc} \alpha^2 + \cos(\theta) - \cos(\theta)\,\alpha^2 & \frac{\beta\,\alpha\,|\beta| - \beta\,\cos(\theta)\alpha\,|\beta| - \beta\,\sin(\theta)\delta}{|\beta|} & \frac{\delta\,\alpha\,|\delta| - \delta\,\cos(\theta)\alpha\,|\delta| + \beta\,\sin(\theta)\delta}{|\delta|} \\ \frac{\alpha\,\beta\,|\alpha| - \alpha\,\cos(\theta)\beta\,|\alpha| + \alpha\,\sin(\theta)\delta}{|\alpha|} & \beta^2 + \cos(\theta) - \cos(\theta)\,\beta^2 & \frac{\delta\,\beta\,|\delta| - \delta\,\cos(\theta)\beta\,|\delta| - \alpha\,\sin(\theta)\delta}{|\delta|} \\ \frac{-\alpha\,\sin(\theta)\beta + \alpha\,\delta\,|\alpha| - \alpha\,\delta\,\cos(\theta)|\alpha|}{|\alpha|} & \frac{\beta\,\delta\,|\beta| - \beta\,\cos(\theta)\delta\,|\beta| + \alpha\,\sin(\theta)\beta}{|\beta|} & \delta^2 + \cos(\theta) - \cos(\theta)\,\delta^2 \end{array} \right]$$

> A:= (alpha,beta,delta,theta) -> matrix([[alpha^2 + cos(theta) - cos(theta) * alpha^2, (beta* alpha *abs(beta) - beta* alpha* cos(theta)*abs(beta) - beta* sin(theta)* delta) / abs(beta), (delta* alpha* abs(delta) - delta* cos(theta)* alpha* abs(delta) + beta* sin(theta)* delta) / abs(delta)], [(alpha* beta* abs(alpha) - alpha* beta* cos(theta) *abs(alpha) + alpha* sin(theta)* delta) / abs(alpha), 1 - delta^2 + delta^2* cos(theta) - alpha^2 + cos(theta)* alpha^2, (-alpha* sin(theta)* delta + delta* beta* abs(delta) - delta* beta* cos(theta)* abs(delta)) / abs(delta)], [(-alpha* sin(theta)* beta + alpha* delta* abs(alpha) - alpha* delta* cos(theta)* abs(alpha)) / abs(alpha), (alpha* sin(theta)* beta + beta* delta* abs(beta) - beta* delta* cos(theta)* abs(beta)) / abs(beta), cos(theta) + delta^2 - delta^2* cos(theta)]]):

> evalf(A(2/sqrt(38),-5/sqrt(38),3/sqrt(38),9*Pi/7));

$$\left[\begin{array}{ccc} -0.4525961384 & -0.8077236010 & 0.8904895659 \\ -0.8077236010 & 0.4445955941 & -0.3871916093 \\ -0.3778085756 & -0.3871916092 & -0.2389790592 \end{array} \right]$$

The matrix above is an example with $\vec{v} = \dfrac{1}{\sqrt{38}} \langle 2, -5, 3 \rangle$ and $\theta = \dfrac{9}{7}\pi$. The above calculations may seem lengthy, tedious, and complicated, however, if we spend a few minutes working through the details, perhaps we can demystify this whole process. Remember that I_3 is the 3×3 matrix whose columns are the standard basis vectors $\{\vec{e_1}, \vec{e_2}, \vec{e_3}\}$ of $\mathbb{R}^3$. We will rotate $\vec{e_k}$ about $\vec{v} = \langle \alpha, \beta, \delta \rangle$ by an angle of θ, where $k = 1, 2, 3$ and $\alpha^2 + \beta^2 + \delta^2 = 1$. Once again following the pattern of Section 7.1, we must find the point C (or in vector form $\vec{c}$) on $\vec{v}$ closest to the tip of vector $\vec{e_k}$. In vector form, this is simply $\text{proj}_{\vec{v}}(\vec{e_k})$:

$$\vec{c} = \frac{\vec{v} \cdot \vec{e_k}}{|\vec{v}|}\vec{v} \tag{10.28}$$

Under the assumption that $|\vec{v}| = 1$, we find that $\vec{c} = (\vec{v} \cdot \vec{e_k})\vec{v}$. Now that we have the center of rotation, we need to construct our two coordinate axes to rotate about. These coordinate axes will be the plane perpendicular to $\vec{v}$ centered at the point C. To find the first axis, notice that $\vec{e_k} - \vec{c}$ is perpendicular to $\vec{c}$, with $R = |\vec{e_k} - \vec{c}|$ being the distance from the tip of vector $\vec{e_k}$ to the point C. So we define

$$\vec{u} = \frac{1}{R}(\vec{e_k} - \vec{c}) \tag{10.29}$$

to be the unit vector orthogonal to $\vec{c}$ that starts at the point C and points in the direction of $\vec{e_k}$. A unit vector orthogonal to both $\vec{c}$ and $\vec{u}$ is easy to compute by using the cross product:

$$\vec{w} = \frac{\vec{c} \times \vec{u}}{|\vec{c} \times \vec{u}|} \tag{10.30}$$

The set $\{\vec{u}, \vec{w}\}$ forms an orthonormal basis for the plane perpendicular to $\vec{v}$. So to rotate $\vec{e_k}$ by an angle of θ about $\vec{v}$, which we will denote as $\vec{e_k}(\theta_{\vec{v}})$, we simply perform the following vector operation:

$$\vec{e_k}(\theta_{\vec{v}}) = \vec{c} + R\cos(\theta)\vec{u} + R\sin(\theta)\vec{w} \tag{10.31}$$

As it stands, this formula may look simple, however, $\vec{u}$ and $\vec{w}$ both are more complicated expressions. Let us investigate this further. By looking at the definition of $\vec{u}$, we see that

$$R\cos(\theta)\vec{u} = R\cos(\theta)\frac{\vec{e_k} - \vec{c}}{R} \tag{10.32}$$
$$= \cos(\theta)(\vec{e_k} - \vec{c})$$

The final term requires a little more work, but it does simplify after some algebraic maneuvers:

$$R\sin(\theta)\vec{w} = R\sin(\theta)\frac{\vec{c} \times \dfrac{\vec{e_k} - \vec{c}}{R}}{\left|\vec{c} \times \dfrac{\vec{e_k} - \vec{c}}{R}\right|}$$
$$= R\sin(\theta)\frac{\dfrac{1}{R}(\vec{c} \times (\vec{e_k} - \vec{c}))}{\dfrac{1}{R}|\vec{c} \times (\vec{e_k} - \vec{c})|} \tag{10.33}$$
$$= R\sin(\theta)\frac{(\vec{c} \times (\vec{e_k} - \vec{c}))}{|\vec{c} \times (\vec{e_k} - \vec{c})|}$$

Now, we also have that

$$\begin{aligned}
|\vec{c} \times (\vec{e_k} - \vec{c})| &= |\vec{c}| \, |\vec{e_k} - \vec{c}| \sin(\alpha) \\
&= |\vec{c}| \, R \sin\left(\frac{\pi}{2}\right) \\
&= |\vec{c}| \, R
\end{aligned} \tag{10.34}$$

and

$$\begin{aligned}
\vec{c} \times (\vec{e_k} - \vec{c}) &= \vec{c} \times \vec{e_k} - \vec{c} \times \vec{c} \\
&= \vec{c} \times \vec{e_k}
\end{aligned} \tag{10.35}$$

Putting equations (10.34) and (10.35) into (10.33) gives

$$R \sin(\theta) \vec{w} = \frac{\vec{c} \times \vec{e_k}}{|\vec{c}|} \sin(\theta) \tag{10.36}$$

After all this simplification, we can rewrite (10.31) as

$$\vec{e_k}(\theta_{\vec{v}}) = \vec{c} + (\vec{e_k} - \vec{c}) \cos(\theta) + \frac{\vec{c} \times \vec{e_k}}{|\vec{c}|} \sin(\theta) \tag{10.37}$$

For $k = 1, 2, 3$, we can break the above expression into components:

$$\vec{e_1}(\theta_{\vec{v}}) = \langle \alpha^2(1 - \cos(\theta)) + \cos(\theta), \, \alpha\beta(1 - \cos(\theta)) + \frac{\alpha\delta}{|\alpha|}\sin(\theta), \, \alpha\delta(1 - \cos(\theta)) - \frac{\alpha\beta}{|\alpha|}\sin(\theta) \rangle$$

$$\vec{e_2}(\theta_{\vec{v}}) = \langle \alpha\beta(1 - \cos(\theta)) - \frac{\beta\delta}{|\beta|}\sin(\theta), \, \beta^2(1 - \cos(\theta)) + \cos(\theta), \, \beta\delta(1 - \cos(\theta)) - \frac{\alpha\beta}{|\beta|}\sin(\theta) \rangle$$

$$\vec{e_3}(\theta_{\vec{v}}) = \langle \alpha\delta(1 - \cos(\theta)) + \frac{\beta\delta}{|\delta|}\sin(\theta), \, \beta\delta(1 - \cos(\theta)) - \frac{\alpha\delta}{|\delta|}\sin(\theta), \, \delta^2(1 - \cos(\theta)) + \cos(\theta) \rangle$$

So given a vector $\vec{z} = \langle z_1, z_2, z_3 \rangle$, we can express this as $\vec{z} = z_1\vec{e_1} + z_2\vec{e_2} + z_3\vec{e_3}$. If we wish to rotate $\vec{z}$ about a vector $\vec{v}$ by angle θ, the unknown matrix $A_{\vec{v}}$ applied to $\vec{z}$ satisfies:

$$\begin{aligned}
A_{\vec{v}} \vec{z} &= z_1 A_{\vec{v}} \vec{e_1} + z_2 A_{\vec{v}} \vec{e_2} + z_3 A_{\vec{v}} \vec{e_3} \\
&= z_1 \vec{e_1}(\theta_{\vec{v}}) + z_2 \vec{e_2}(\theta_{\vec{v}}) + z_3 \vec{e_3}(\theta_{\vec{v}}) \\
&= \left[\vec{e_1}(\theta_{\vec{v}}) \mid \vec{e_2}(\theta_{\vec{v}}) \mid \vec{e_3}(\theta_{\vec{v}}) \right] \begin{bmatrix} z_1 \\ z_2 \\ z_3 \end{bmatrix}
\end{aligned}$$

Therefore, now we have that the matrix $A_{\vec{v}}$'s columns are $\vec{e_1}$, $\vec{e_2}$, and $\vec{e_3}$:

$$A_{\vec{v}} = \tag{10.38}$$

$$\begin{bmatrix}
\alpha^2(1 - \cos(\theta)) + \cos(\theta) & \alpha\beta(1 - \cos(\theta)) - \frac{\beta\delta}{|\beta|}\sin(\theta) & \alpha\delta(1 - \cos(\theta)) + \frac{\beta\delta}{|\delta|}\sin(\theta) \\
\alpha\beta(1 - \cos(\theta)) + \frac{\alpha\delta}{|\alpha|}\sin(\theta) & \beta^2(1 - \cos(\theta)) + \cos(\theta) & \beta\delta(1 - \cos(\theta)) - \frac{\alpha\delta}{|\delta|}\sin(\theta) \\
\alpha\delta(1 - \cos(\theta)) - \frac{\alpha\beta}{|\alpha|}\sin(\theta) & \beta\delta(1 - \cos(\theta)) - \frac{\alpha\beta}{|\beta|}\sin(\theta) & \delta^2(1 - \cos(\theta)) + \cos(\theta)
\end{bmatrix}$$

Comparing the matrix $A_{\vec{v}}$ above to the matrix labeled *rotmatrix* in the previous set of *Maple* code reveals that we have arrived at the same rotation matrix as that found by *Maple*.

Example 10.4.1. After all of these calculations, it is time to do an example. Let $\vec{w} = \langle -5, 3, 11 \rangle$ be a vector parallel to our axis of rotation that is a line through the origin. We want to rotate the point $P(4, -7, 6)$ about this line through angles θ, which are multiples of $10°$ until we are back at the point P. The plot of these points is essentially the circle of rotation for P, as seen in Figures 10.10 and 10.11.

> P:= [4,-7,6]: w:= [-5,3,11]:

> length_w:= norm(w,frobenius);

$$length\_w := \sqrt{155}$$

> v:= evalm(w/length_w):

> points:= [seq(evalm(evalf(A(op(convert(v,list)),2*Pi*j/36))&*P),j=0..36)]:

> points[1];

$$\begin{bmatrix} 4. & -7. & 6. \end{bmatrix}$$

> plotpts:= seq(sphere(convert(points[j],list),.3),j=1..37):

> text_P:= textplot3d([op(P),"P"], align={ABOVE, RIGHT}, color=black):

> rotaxis:= line(convert(evalm(-10*v),list), convert(evalm(10*v),list), thickness = 2, color = red):

> Center:= convert(evalm(dotprod(v,P)*v),list);

$$Center := \begin{bmatrix} -\dfrac{25}{31}, & \dfrac{15}{31}, & \dfrac{55}{31} \end{bmatrix}$$

> plot_Center:= sphere(Center,.5):

> text_center:= textplot3d([op(Center),"C"], align={ABOVE,RIGHT}, color = black):

> Arrows:= [seq(PLOT3D(arrow(Center, convert(points[j],list), .5, 1.5, .3, cylindrical_arrow, color = blue)), j = 1..37)]:

> display({rotaxis, text_P, text_center, plot_Center, display([plotpts], in-sequence = true), display(Arrows, insequence = true) }, axes = boxed, scaling = constrained);

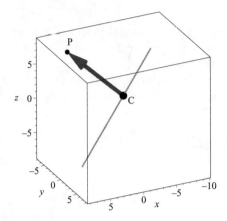

Figure 10.10: First frame of a sphere located at $P(4, -7, 6)$ being rotated about a line in the direction $\vec{w} = \langle -5, 3, 11 \rangle$ in $\mathbb{R}^3$.

An alternative plot of this rotation is below, but it takes much more time and memory to compute since it plots simultaneously 37 spheres at each of the points at 10° of separation in the rotation of the point P.

> display({rotaxis, text_P, text_center, plot_Center, plotpts, display(Arrows, insequence = true)}, axes = boxed, scaling = constrained);

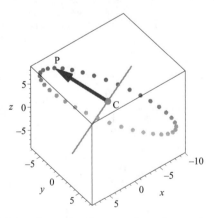

Figure 10.11: All of the spheres in the rotation about the line in the direction $\vec{w} = \langle -5, 3, 11 \rangle$ are depicted here.

Now that we have determined all of the rotational matrices for $\mathbb{R}^3$, we will next focus our attention on rescalings. Similar to the two-dimensional case, a general three-dimensional rescaling is given by matrices of the form

$$A_{(a,b,c)} = \begin{bmatrix} a & 0 & 0 \\ 0 & b & 0 \\ 0 & 0 & c \end{bmatrix}$$

for real entries a, b, and c. The above scaling matrix corresponds to the linear map $T(\langle x, y, z \rangle) = \langle ax, by, cz \rangle$. When a, b, and c are all positive, we simply have a scaling in the same direction as $\langle x, y, z \rangle$. However, if at least one of the components along the diagonal is negative, we have not just a scaling, but a reflection as well. The matrix

$$A_{xy} = \begin{bmatrix} 1 & 0 & 0 \\ 0 & 1 & 0 \\ 0 & 0 & -1 \end{bmatrix}$$

is reflection about the xy-plane, since the x and y coordinates stay fixed, and the z coordinate is replaced by $-z$. The matrix

$$A_{\vec{0}} = \begin{bmatrix} -1 & 0 & 0 \\ 0 & -1 & 0 \\ 0 & 0 & -1 \end{bmatrix}$$

is reflection about the origin. The question of whether all linear maps $T : \mathbb{R}^3 \to \mathbb{R}^3$ can be decomposed into a composite of rotations and general rescalings is obviously a challenging one, and one should consider all of the work done in Section 10.2 to decompose all real 2×2 matrices into a product of rotations, reflections and rescaling matrices.

Homework Problems

1. Prove that the following relations hold, and give a geometric interpretation of this result

$$A_{\theta_x}^{-1} = A_{-\theta_x}, \quad A_{\theta_y}^{-1} = A_{-\theta_y}, \quad A_{\theta_z}^{-1} = A_{-\theta_z}$$

2. Verify that no two rotation matrices commute, (e.g. $A_{\theta_x} A_{\phi_y} \neq A_{\phi_y} A_{\theta_x}$). Remember to use different angle variables for each matrix.

3. As a general rule, scalings and rotations do not commute. However, prove that scalings in just one direction commute with rotations corresponding to the axis of scaling for each of the x, y, and z axes. As an example, show that

$A_{(a,1,1)}A_{\theta_x} = A_{\theta_x}A_{(a,1,1)}$. Can you give a geometric interpretation of this result?

4. For rotations in the plane, it was shown that the rotation matrix A_θ satisfied $A_\theta A_\phi = A_{\theta+\phi}$. Verify that the same property holds for the three rotation matrices A_{θ_x}, A_{θ_y}, and A_{θ_z} for rotations in $\mathbb{R}^3$.

5. Determine the 3×3 matrix that will rotate the point P about the line passing through the origin in the direction of $\vec{v}$ by an angle of θ, then compute the coordinates of the rotated point.

(a) $P(1,1,1)$, $\vec{v} = \langle 3,0,0 \rangle$, $\theta = \frac{\pi}{4}$ (b) $P(1,0,-1)$, $\vec{v} = \langle 0,2,0 \rangle$, $\theta = \frac{\pi}{2}$

(c) $P(1,0,1)$, $\vec{v} = \langle 0,0,1 \rangle$, $\theta = \frac{3}{4}\pi$ (d) $P(0,1,1)$, $\vec{v} = \langle 1,0,0 \rangle$, $\theta = \frac{3}{2}\pi$

6. For each of the following, construct a single matrix that will satisfy each of the following set of criteria, in the sequence specified.

   (a) First rotate about the z-axis by $\theta = \frac{\pi}{3}$, then scale in the x-direction by a factor of 2, and finally, scale in the y-direction by a factor of 3.

   (b) First flip about the xy-plane, next rotate about z-axis by $\theta = \frac{\pi}{4}$, and lastly, scale in z-direction by a factor of 2.

   (c) First scale in y-direction by a factor of 2, then rotate about y-axis by an angle of $\theta = -\frac{\pi}{2}$, then finally, scale in the x-direction by a factor of $\frac{1}{2}$.

7. Describe a procedure for rotating a point $P(x,y,z)$ about the line passing through the origin in the direction of $\vec{v} = \langle \alpha, \beta, 0 \rangle$, where α and β are both non-zero and satisfy $\alpha^2 + \beta^2 = 1$.

8. Decompose an arbitrary 3×3 matrix into the product of rotations, rescalings and reflections.

Maple Problems

1. *Homework* problem 1 illustrated the fact that the inverse rotation matrices can be found by replacing the angle θ with $-\theta$ in the original rotation matrix. We wish to determine if the the same holds true for the rotation matrix $A_{\vec{v}}$.

(a) Replace θ with $-\theta$ in $A_{\vec{v}}$ and perform the matrix multiplications

$$A_{\vec{v}}(\theta)A_{\vec{v}}(-\theta) \text{ and } A_{\vec{v}}(-\theta)A_{\vec{v}}(\theta)$$

(b) Compute $A_{\vec{v}}^{-1}$ using *Maple's* inverse command, simplify and compare the result to $A_{\vec{v}}(-\theta)$.

2. Determine the 3×3 matrix that will rotate the point P about the line passing through the origin in the direction of $\vec{v}$ by an angle of θ.

(a) $P(2,0,3)$, $\vec{v} = \langle 2,-1,2 \rangle$ (b) $P(2,3,-1)$, $\vec{v} = \langle -1,0,0 \rangle$

(c) $P(1,0,1)$, $\vec{v} = \langle 0,0,1 \rangle$ (d) $P(1,-1,0)$, $\vec{v} = \langle 1,2,1 \rangle$

(e) $P(0,1,1)$, $\vec{v} = \langle 1,0,0 \rangle$ (f) $P(1,-1,3)$, $\vec{v} = \langle -1,1,2 \rangle$

3. Plot, and animate, an entire rotation of the point P about the line in the direction $\vec{v}$ for each of parts (a)-(f) of problem 2.

4. Describe and test a procedure for doing rotations about any line in space, not necessarily through the origin.

10.5 Affine Maps

Closely related to linear maps are affine maps. We begin with a definition:

Definition 10.5.1. An *affine map* $S : \mathbb{R}^n \to \mathbb{R}^m$ is a translation of a linear map, that is, $S(\vec{u}) = T(\vec{u}) + \vec{b}$ for all $u \in \mathbb{R}^n$ where $T : \mathbb{R}^n \to \mathbb{R}^m$ is a linear map and $\vec{b} \in \mathbb{R}^m$ is fixed.

If A is the standard $m \times n$ matrix representing the linear map T, then A, along with the fixed vector $\vec{b}$ in $\mathbb{R}^m$, also represents the affine map S, since we have $S(\vec{u}) = A\vec{u} + \vec{b}$ for all $\vec{u} \in \mathbb{R}^n$. Affine maps operate analogously to linear maps, except that they also translate their results by $\vec{b}$. Thus, affine maps will take a subspace $\mathbb{K} \subseteq \mathbb{R}^n$ and return $S(\mathbb{K}) = \mathbb{L} + \vec{b}$ in $\mathbb{R}^m$, where $\mathbb{L} = T(\mathbb{K})$ is the subspace of $\mathbb{R}^m$ corresponding to the image of the subspace $\mathbb{K}$ of $\mathbb{R}^n$ under the map T. Hence, an affine map S translates the image subspace of T by $\vec{b}$. This should remind you of the work we did back in Section 8.1 on subspace translates. The fixed vector $\vec{b}$ of $\mathbb{R}^m$ is $S(\vec{0}_n)$, since

$$S(\vec{0}_n) = T(\vec{0}_n) + \vec{b} = \vec{0}_m + \vec{b} = \vec{b}$$

Note that we used the fact that linear maps take the zero vector of their domain to the zero vector of their range. We can directly relate solving a linear system of equations $A\overrightarrow{x} = -\overrightarrow{b}$, for an $m \times n$ matrix A, to affine maps. This matrix equation can be written as $A\overrightarrow{x} + \overrightarrow{b} = \overrightarrow{0}_m$. The solutions to this matrix equation are the solutions $\overrightarrow{u}$ in $\mathbb{R}^n$ to $S(\overrightarrow{u}) = \overrightarrow{0}_m$, which is to say, solutions to $T(\overrightarrow{u}) + \overrightarrow{b} = \overrightarrow{0}_m$, for the linear map T with standard matrix A representing it. We already know that the solutions $\overrightarrow{x}$ in $\mathbb{R}^n$ to $A\overrightarrow{x} = -\overrightarrow{b}$ are of the form $\overrightarrow{x} = \overrightarrow{x_k} + \overrightarrow{c}$, where $\overrightarrow{x_k} \in \operatorname{Ker}(T)$ corresponds to the the solutions of $A\overrightarrow{x} = \overrightarrow{0}_m$, and $\overrightarrow{c}$ is a fixed vector in $\mathbb{R}^n$, with $A\overrightarrow{c} = -\overrightarrow{b}$.

Example 10.5.1. We next do an example of an affine map. Consider $S : \mathbb{R}^3 \to \mathbb{R}^4$ given by $S(\overrightarrow{u}) = A\overrightarrow{u} + \overrightarrow{b}$, where

$$A = \begin{bmatrix} -3 & 1 & 7 \\ 1 & 9 & 3 \\ -1 & 19 & 13 \\ 4 & 8 & -4 \end{bmatrix}, \quad \overrightarrow{b} = \begin{bmatrix} -8 \\ 5 \\ 2 \\ 13 \end{bmatrix}$$

In the following *Maple* commands, we plot (Figure 10.12) the solutions to $S(\overrightarrow{u}) = \overrightarrow{0}_4$. In other words, we are searching for solutions to $A\overrightarrow{u} = -\overrightarrow{b}$.

```
> with(linalg): with(plots): with(plottools):
> A:= matrix([[-3, 1, 7], [1, 9, 3], [-1, 19, 13], [4, 8, -4]]):
> B:= matrix([[8], [-5], [-2], [-13]]):
> rref(augment(A,B));
```

$$\begin{bmatrix} 1 & 0 & -\frac{15}{7} & -\frac{11}{4} \\ 0 & 1 & \frac{4}{7} & -\frac{1}{4} \\ 0 & 0 & 0 & 0 \\ 0 & 0 & 0 & 0 \end{bmatrix}$$

From the solution given above, we can reconstruct the solution as a line given parametrically as

$$L = \left\{ \langle x, y, z \rangle \ \middle| \ x = \frac{15}{7}t - \frac{11}{4}, y = -\frac{4}{7}t - \frac{1}{4}, z = t \right\} \tag{10.39}$$

for t an arbitrary parameter. We can decompose L into a sum of two components, the first being the line L_0, which passes through the origin, the second being a translation by a particular vector $\overrightarrow{x_p}$. To find the particular vector, notice that all we have to do is set $t = 0$ in the parametric definition of L given above, which yields $\vec{x_p} = \langle -\frac{11}{4}, -\frac{1}{4}, 0 \rangle$. Now that we have $\vec{x_p}$, the line L_0 is simply the remaining portion of the solution given in equation 10.39:

$$L_0 = \left\{ \left\langle \frac{15}{7}t, -\frac{4}{7}t, t \right\rangle \,\middle|\, t \in \mathbb{R} \right\} \tag{10.40}$$

Clearly L_0 is a line though the origin, and is thus a subspace of $\mathbb{R}^3$. As previously discussed, the line L can be realized as a translate of the line L_0 by the particular solution $\vec{x_p}$. Now, let us plot the two lines L and L_0 along with the particular solution vector $\vec{x_p}$.

> plotL:= spacecurve([15/7*t-11/4, -4/7*t-1/4, t], t = -3..3, color=red, thickness=2):

> plotL0:= spacecurve([15/7*t, -4/7*t, t], t = -3..3, color = blue, thickness = 2):

> plotP:= arrow([0,0,0], [-11/4,-1/4,0], .1, .2, .5, cylindrical_arrow, color = black):

> plot_origin:= textplot3d([0,0,0,"O"], align={LEFT, ABOVE}, color = black):

> plot_P:= textplot3d([-11/4,-1/4,0,"P"], align={LEFT, ABOVE}, color = black):

> display({plotL, plotL0, plotP, plot_origin, plot_P}, axes = boxed);

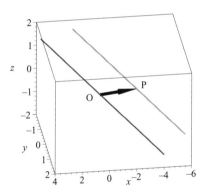

Figure 10.12: The line L can be decomposed into a line through the origin and a particular vector solution.

Recall that a linear map $T : \mathbb{R}^2 \to \mathbb{R}^2$ is uniquely determined by taking a line segment in the domain to another line segment in the range. This is no longer the case for an affine map $S : \mathbb{R}^2 \to \mathbb{R}^2$. As it turns out, an affine map $S : \mathbb{R}^2 \to \mathbb{R}^2$ is uniquely determined by taking a triangle in the domain to another triangle in the range. To see how this is the case, let the triangle in the domain be defined as the interior of the three points that are the terminal points of the following three vectors:

$$T_1 = \{\langle x_1, y_1 \rangle, \langle x_2, y_2 \rangle, \langle x_3, y_3 \rangle\}$$

similarly for the triangle in the range:

$$T_2 = \{\langle z_1, w_1 \rangle, \langle z_2, w_2 \rangle, \langle z_3, w_3 \rangle\}$$

Then,

$$S(\langle x_j, y_j \rangle) = \langle z_j, w_j \rangle, \ 1 \leq j \leq 3$$

If $S(\vec{u}) = A\vec{u} + \vec{b}$, for the 2×2 matrix $A = \begin{bmatrix} a & b \\ c & d \end{bmatrix}$ and $\vec{b} = \begin{bmatrix} \alpha \\ \beta \end{bmatrix}$, then we have the systems of equations

$$\begin{bmatrix} a & b \\ c & d \end{bmatrix} \begin{bmatrix} x_j \\ y_j \end{bmatrix} + \begin{bmatrix} \alpha \\ \beta \end{bmatrix} = \begin{bmatrix} z_j \\ w_j \end{bmatrix}, \quad 1 \leq j \leq 3$$

This gives the new single matrix equation

$$\begin{bmatrix} x_1 & y_1 & 0 & 0 & 1 & 0 \\ 0 & 0 & x_1 & y_1 & 0 & 1 \\ x_2 & y_2 & 0 & 0 & 1 & 0 \\ 0 & 0 & x_2 & y_2 & 0 & 1 \\ x_3 & y_3 & 0 & 0 & 1 & 0 \\ 0 & 0 & x_3 & y_3 & 0 & 1 \end{bmatrix} \begin{bmatrix} a \\ b \\ c \\ d \\ \alpha \\ \beta \end{bmatrix} = \begin{bmatrix} z_1 \\ w_1 \\ z_2 \\ w_2 \\ z_3 \\ w_3 \end{bmatrix} \tag{10.41}$$

This last matrix equation can be solved for our six unknowns $\{a, b, c, d, \alpha, \beta\}$, which determine the affine map uniquely.

Example 10.5.2. As an example, we will now find the affine map S that sends the triangle T_1 to triangle T_2 defined by the following sets of points:

$$T_1 = \{\langle -9, -5 \rangle, \langle 4, 17 \rangle, \langle 1, -6 \rangle\}$$
$$T_2 = \{\langle 3, 11 \rangle, \langle -8, 3 \rangle, \langle -2, -15 \rangle\}$$

> X:= [-9,4,1]: Y:= [-5,17,-6]: Z:= [3,-8,-2]: W:= [11,3,-15]:
> R:= matrix([[Z[1]], [W[1]], [Z[2]], [W[2]], [Z[3]], [W[3]]]);

$$R := \begin{bmatrix} 3 \\ 11 \\ -8 \\ 3 \\ -2 \\ -15 \end{bmatrix}$$

> d:= matrix([[X[1], Y[1], 0, 0, 1, 0], [0, 0, X[1], Y[1], 0, 1], [X[2], Y[2], 0, 0, 1, 0], [0, 0, X[2], Y[2], 0, 1], [X[3], Y[3], 0, 0, 1, 0], [0, 0, X[3], Y[3], 0, 1]]);

$$
\begin{bmatrix}
-9 & -5 & 0 & 0 & 1 & 0 \\
0 & 0 & -9 & -5 & 0 & 1 \\
4 & 17 & 0 & 0 & 1 & 0 \\
0 & 0 & 4 & 17 & 0 & 1 \\
1 & -6 & 0 & 0 & 1 & 0 \\
0 & 0 & 1 & -6 & 0 & 1
\end{bmatrix}
$$

> AB:= evalm(inverse(d)&*R);

$$
AB := \begin{bmatrix}
-\frac{121}{233} \\
-\frac{45}{233} \\
-\frac{580}{233} \\
\frac{258}{233} \\
-\frac{615}{233} \\
-\frac{1367}{233}
\end{bmatrix}
$$

> A:= matrix([[AB[1,1],AB[2,1]], [AB[3,1],AB[4,1]]]);

$$
A := \begin{bmatrix}
-\frac{121}{233} & -\frac{45}{233} \\
-\frac{580}{233} & \frac{258}{233}
\end{bmatrix}
$$

> B:= matrix([[AB[5,1]],[AB[6,1]]]);

$$
B := \begin{bmatrix}
-\frac{615}{233} \\
-\frac{1367}{233}
\end{bmatrix}
$$

> evalm(A&*[X[1],Y[1]]+B);

$$
\begin{bmatrix} 3 \\ 11 \end{bmatrix}
$$

> evalm(A&*[X[2],Y[2]]+B);

$$
\begin{bmatrix} -8 \\ 3 \end{bmatrix}
$$

> evalm(A&*[X[3],Y[3]]+B);

$$
\begin{bmatrix} -2 \\ -15 \end{bmatrix}
$$

The last three *Maple* commands are simply verifications that the vectors $\langle x_k, y_k \rangle$, determining the corners of triangle T_1, were sent to their corresponding counterparts $\langle z_k, w_k \rangle$, of T_2. Hence, our formula works. Now you should see what it takes to determine an affine map $S : \mathbb{R}^3 \to \mathbb{R}^3$ uniquely, and in general when the dimensions are not the same for both domain and range.

$\blacksquare$

We have previously looked at rotations about the origin in $\mathbb{R}^2$ and rotations about a line through the origin in $\mathbb{R}^3$. The next logical question to ask, is what happens if we rotate about a point (a, b) in $\mathbb{R}^2$ which is not the origin, or rotate about a line that does not pass through the origin in $\mathbb{R}^3$? You should have guessed that we get an affine map for both rotations. In Section 8.1, we already expressed this concept in terms of matrix multiplication, so it is fairly straightforward to rewrite the ideas in terms of maps instead.

We will once again define A_θ to be the 2×2 rotation matrix for the fixed angle θ about the origin in $\mathbb{R}^2$, defined in the standard way. If we now wish to rotate a point corresponding to the terminal end of a vector $\vec{u}$ about the point (a, b), we need to use the affine map S. Here, S is defined as follows:

$$
\begin{aligned}
S(\vec{u}) &= A_\theta (\vec{u} - \langle a, b \rangle) + \langle a, b \rangle \\
&= A_\theta \vec{u} + \langle a, b \rangle - A_\theta \langle a, b \rangle \\
&= A_\theta \vec{u} + \vec{b}
\end{aligned}
\tag{10.42}
$$

where

$$
\begin{aligned}
\vec{b} &= \langle a, b \rangle - A_\theta \langle a, b \rangle \\
&= (I_2 - A_\theta) \langle a, b \rangle
\end{aligned}
\tag{10.43}
$$

Notice this is similar to the matrix multiplication version from Section 8.1, where we translated our frame of rotation from the point (a, b) to the origin, and then translated back by adding the point (a, b) to the resulting rotated point. The same can be done for a rotation about a line L not through the origin in $\mathbb{R}^3$. If A_θ is the 3×3 rotation matrix for the fixed angle θ about the line L' parallel to L but through the origin, then the affine map S will rotate about L any point whose represented by the terminal end of a vector $\vec{u} \in \mathbb{R}^3$, and is defined as

$$
\begin{aligned}
S(\vec{u}) &= A_\theta (\vec{u} - \langle a, b, c \rangle) + \langle a, b, c \rangle \\
&= A_\theta \vec{u} + \langle a, b, c \rangle - A_\theta \langle a, b, c \rangle = A_\theta \vec{u} + \vec{b}
\end{aligned}
\tag{10.44}
$$

for

$$
\begin{aligned}
\vec{b} &= \langle a, b, c \rangle - A_\theta \langle a, b, c \rangle \\
&= (I_3 - A_\theta) \langle a, b, c \rangle
\end{aligned}
\tag{10.45}
$$

Here the point (a, b, c) is a fixed point on the line L.

Example 10.5.3. Now, let us do an example of this three-dimensional rotation about a line L that does not pass through the origin. First, we consider the line going through the point $\vec{P}\langle 11, 25, -7\rangle$, which is parallel to the vector $\vec{w} = \langle 5, 2, 9\rangle$. We want to rotate the point $\vec{Q}\langle 4, -7, 13\rangle$ about this line through angles θ that are multiples of $10°$, until we are back at the point $\vec{Q}$. Each rotation through a fixed angle θ is one application of an affine map. The plot of these points, which can be seen in Figures 10.13 and 10.14, is essentially the circle of rotation for $\vec{Q}$. Referring back to Section 4.4, the formula for the rotated point $\vec{Q}_{new}$ will be given by

$$\begin{aligned}
\vec{Q}_{new} &= A_{\vec{v}}(\vec{Q} - \vec{P}) + \vec{P} \\
&= (I_3 - A_{\vec{v}})\,\vec{P} + A_{\vec{v}}\vec{Q}
\end{aligned} \qquad (10.46)$$

In the *Maple* commands below, the lengthy formula defining A below is simply the expression for $A_{\vec{v}}$ given in equation (10.38), and the matrix $\vec{B}$ corresponds to the column matrix $(I_3 - A_{\vec{v}})\,\vec{P}$, giving the formula for $\vec{Q}_{new}$ to now be

$$\vec{Q}_{new} = A_{\vec{v}}\vec{Q} + \vec{B}$$

```
> A:= (alpha,beta,delta,theta) -> matrix([[alpha^2 + cos(theta) - cos(theta) *
alpha^2, (beta* alpha *abs(beta) - beta* alpha* cos(theta)*abs(beta) - beta*
sin(theta)* delta) / abs(beta), (delta* alpha* abs(delta) - delta* cos(theta)*
alpha* abs(delta) + beta* sin(theta)* delta) / abs(delta)], [(alpha* beta*
abs(alpha) - alpha* beta* cos(theta) *abs(alpha) + alpha* sin(theta)* delta) /
abs(alpha), 1 - delta^2 + delta^2* cos(theta) - alpha^2 + cos(theta)* alpha^2,
(-alpha* sin(theta)* delta + delta* beta* abs(delta) - delta* beta* cos(theta)*
abs(delta)) / abs(delta)], [(-alpha* sin(theta)* beta + alpha* delta* abs(alpha)
- alpha* delta* cos(theta)* abs(alpha)) / abs(alpha), (alpha* sin(theta)* beta
+ beta* delta* abs(beta) - beta* delta* cos(theta)* abs(beta)) / abs(beta),
cos(theta) + delta^2 - delta^2* cos(theta) ]]):
> Q:= [4,-7,13]: P:= [11, 25, -7]: w:= [5, 2,9]:
> length_w:= norm(w,frobenius);
```

$$length\_w := \sqrt{110}$$

```
> v:= evalm(w/length_w):
> B:= theta -> evalm((diag(1,1,1)-evalf(A(op(convert(v,list)), theta)))&*P):
> points:= [seq(evalm(evalf(A(op(convert(v,list)), 2*Pi*j / 36)) &*Q + B(
2*Pi*j / 36)), j = 0..36)]:
> points[1];
```

$$\begin{bmatrix} 4. & -7. & 13. \end{bmatrix}$$

> plotpts:= seq(sphere(convert(points[j], list), 1), j=1..37):

> text_Q:= textplot3d([op(Q), "Q"], align={ABOVE,RIGHT}, color=black):

> rotaxis:= line(convert(evalm(-30*v + P), list), convert(evalm(30*v + P), list), thickness = 2, color = red):

> Center:= convert(evalm(dotprod(v,Q)*v+P),list);

$$Center := \left[\frac{365}{22}, \frac{1498}{55}, \frac{337}{110}\right]$$

> plot_Center:= sphere(Center,1):

> text_center:= textplot3d([op(Center), "C"], align={ABOVE, RIGHT}, color = black):

> plot_P:= sphere(P,1):

> text_P:= textplot3d([op(P), "P"],align={BELOW, RIGHT},color=black):

> plot_origin:= sphere([0,0,0],1):

> text_origin:= textplot3d([op([0,0,0]), "O"], align={ABOVE, RIGHT}, color = black):

> Arrows:= [seq(PLOT3D(arrow(Center, convert(points[j], list), 1,5,.3, cylindrical_arrow, color = blue)), j=1..37)]:

> display({rotaxis, text_Q, plot_P, text_P, text_center, plot_Center, plot_origin, text_origin, display([plotpts], insequence = true), display(Arrows, insequence = true)}, axes = boxed, scaling = constrained, view=[-40..60,-20..65,-30..30]);

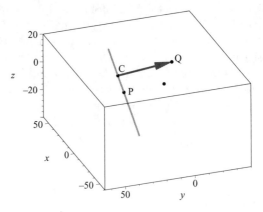

Figure 10.13: The point $\vec{Q}$ (at the tip of the arrow) is rotated about the line passing through the point $\vec{P}$ (the point on the line not at the tail of the vector) in the direction of the vector $\vec{w}$. The origin is also shown for reference and is located directly below the point $\vec{Q}$ in this figure.

> display({rotaxis, text_Q, plot_P, text_P, text_center, plot_Center, plot_origin, text_origin, display([plotpts], insequence = false), display(Arrows[1])}, axes = frame, scaling = constrained, view = [-40..60, -20..65, -30..30], labels = [x, y, z], orientation = [126, 64], style = patchnogrid);

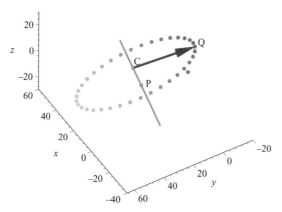

Figure 10.14: All of the rotated points, along with corresponding perpendicular arrows to the line of axis of rotation.

We end this section with a procedure called *RotationsInSpace* which takes a non-zero vector $\vec{v} = \langle \alpha, \beta, \delta \rangle$ parallel to the line L of rotation where the line of rotation goes through the point $\vec{P}$ (which could be the origin) but L is not parallel to any axis or coordinate plane (no component of $\vec{v}$ can be 0), and then rotates a point $\vec{Q}$ through the angle θ about L. The vector $\vec{v}$ and two points $\vec{P}$ and $\vec{Q}$ can be given as lists, although $\vec{v}$ as a vector also works.

> with(linalg):
> RotationsInSpace:= proc(v,P,Q,theta)
    local alpha, beta, delta, A, B, w:
    w:= evalm(v/norm(v, frobenius)):
    alpha:= evalf(w[1]);
    beta:= evalf(w[2]);
    delta:= evalf(w[3]);
    A:= matrix([[alpha^2 + cos(theta) - cos(theta) * alpha^2, (beta* alpha *abs(beta) - beta* alpha* cos(theta)*abs(beta) - beta* sin(theta)* delta) / abs(beta), (delta* alpha* abs(delta) - delta* cos(theta)* alpha* abs(delta) + beta* sin(theta)* delta) / abs(delta)], [(alpha* beta* abs(alpha) - alpha* beta* cos(theta) *abs(alpha) + alpha* sin(theta)* delta) / abs(alpha), 1 - delta^2 + delta^2* cos(theta) - alpha^2 + cos(theta)* alpha^2, (-alpha* sin(theta)* delta + delta* beta* abs(delta) - delta* beta* cos(theta)* abs(delta)) / abs(delta)], [(-alpha* sin(theta)* beta + alpha* delta* abs(alpha) - alpha* delta* cos(theta)

* abs(alpha)) / abs(alpha), (alpha* sin(theta)* beta + beta* delta* abs(beta) -
beta* delta* cos(theta)* abs(beta)) / abs(beta), cos(theta) + delta^2 - delta^2*
cos(theta)]]):
    B:= evalm(P - A&*P):
    evalm(A&*Q + B):
end proc:

> Q1 := RotationsInSpace([5, 2, 9], [11, 25, -7], [4, -7, 13], Pi/3);

$$Q1 := \left[9.340909088 + 15.63678646\sqrt{3}, 9.736363627 - 7.770720100\sqrt{3}, \right.$$
$$\left. - 6.960276898\sqrt{3} + 6.313636364 \right]$$

> v:= [5, 2, 9]:
> w:= evalm(v/norm(v, frobenius));

$$w := \left[\tfrac{1}{22}\sqrt{110} \quad \tfrac{1}{55}\sqrt{110} \quad \tfrac{9}{110}\sqrt{110} \right]$$

> P := [11, 25, -7]: Q := [4, -7, 13]:
> C := evalm(dotprod(w, Q)*w+P);

$$C := \left[\frac{365}{22} \quad \frac{1498}{55} \quad \frac{337}{110} \right]$$

> evalf(norm(Q-C, frobenius)); evalf(norm(Q1-C, frobenius));

$$37.80728645$$
$$37.80728645$$

Homework Problems

1. Find the solutions to the equations $S(\vec{u}) = \vec{0}$, for each of the following affine maps:

(a) $\begin{bmatrix} 1 & -2 \\ 3 & -4 \end{bmatrix}\vec{u} + \begin{bmatrix} -1 \\ 1 \end{bmatrix}$
    (b) $\begin{bmatrix} 4 & 3 \\ -1 & 2 \\ -2 & 5 \end{bmatrix}\vec{u} + \begin{bmatrix} -1 \\ 2 \\ 1 \end{bmatrix}$

(c) $\begin{bmatrix} 0 & 2 \\ 5 & 0 \\ -5 & 2 \end{bmatrix}\vec{u} + \begin{bmatrix} 7 \\ -9 \\ 16 \end{bmatrix}$
    (d) $\begin{bmatrix} 1 & -2 & 2 \\ 0 & 3 & 2 \\ -2 & -4 & -1 \end{bmatrix}\vec{u} + \begin{bmatrix} -1 \\ 1 \\ 4 \end{bmatrix}$

(e) $\begin{bmatrix} 1 & -2 & 2 \\ 0 & 3 & 2 \\ -2 & -4 & -1 \\ 4 & 5 & 6 \end{bmatrix}\vec{u} + \begin{bmatrix} -1 \\ 2 \\ -1 \\ 3 \end{bmatrix}$
    (f) $\begin{bmatrix} 3 & 5 & 1 \\ -2 & 1 & -2 \end{bmatrix}\vec{u} + \begin{bmatrix} -1 \\ 1 \end{bmatrix}$

2. State the dimension for each of the solutions to the affine maps from problem 1.

3. How many points in $\mathbb{R}^n$ does it take to define an affine transformation $S : \mathbb{R}^n \to \mathbb{R}^n$ uniquely?

4. Referring to the matrix equation (10.41), construct the matrix used to determine the number of vectors needed to create a unique affine map from $\mathbb{R}^2$ to $\mathbb{R}^3$ of the form:

$$S(\overrightarrow{u}) = \begin{bmatrix} a & b \\ c & d \\ e & f \end{bmatrix} \overrightarrow{u} + \begin{bmatrix} \alpha \\ \beta \\ \gamma \end{bmatrix}$$

5. Determine the number of vectors needed in $\mathbb{R}^n$ and $\mathbb{R}^m$, with $n, m \geq 1$, which will determine an affine map $S : \mathbb{R}^n \to \mathbb{R}^m$ uniquely. It may help to construct a generic matrix, as in problem 4 and equation (10.41).

6. Express, as an affine map, the rotation of an arbitrary vector $\overrightarrow{u} \in \mathbb{R}^2$ through an angle θ about the point $(1, -1)$.

7. Can a single affine transformation $S : \mathbb{R}^2 \to \mathbb{R}^2$ send the interior and sides of the unit square to the interior and sides of an arbitrary quadrilateral in $\mathbb{R}^2$? If yes, explain why. If no, can it be done for any special type of quadrilateral?

Maple Problems

1. Construct an affine map S that takes the first set of points $\mathbf{S}$ to the second set of points $\mathbf{T}$, with $S(\overrightarrow{s_k}) = \overrightarrow{t_k}$.

(a) $\mathbf{S} = \left\{ \overrightarrow{s_1} = \langle 1, -1, 1 \rangle,\ \overrightarrow{s_2} = \langle 0, 2, -3 \rangle,\ \overrightarrow{s_3} = \langle 2, 4, -3 \rangle,\ \overrightarrow{s_4} = \langle 10, -3, 2 \rangle \right\}$
   $\mathbf{T} = \left\{ \overrightarrow{t_1} = \langle 1, 2 \rangle,\ \overrightarrow{t_2} = \langle 4, 2 \rangle,\ \overrightarrow{t_3} = \langle 4, 5 \rangle,\ \overrightarrow{t_4} = \langle 9, -7 \rangle \right\}$

(b) $\mathbf{S} = \left\{ \overrightarrow{s_1} = \langle 1, 2 \rangle,\ \overrightarrow{s_2} = \langle -2, 3 \rangle,\ \overrightarrow{s_3} = \langle 5, -1 \rangle \right\}$
   $\mathbf{T} = \left\{ \overrightarrow{t_1} = \langle 1, 2, -1 \rangle,\ \overrightarrow{t_2} = \langle 7, 8, -2 \rangle,\ \overrightarrow{t_3} = \langle 3, -5, 6 \rangle \right\}$

(c) $\mathbf{S} = \left\{ \overrightarrow{s_1} = \langle 1, 1, 3 \rangle,\ \overrightarrow{s_2} = \langle -2, 0, 5 \rangle,\ \overrightarrow{s_3} = \langle 4, 3, -3 \rangle,\ \overrightarrow{s_4} = \langle 7, -6, 8 \rangle \right\}$
   $\mathbf{T} = \left\{ \overrightarrow{t_1} = \langle 1, 2, -2, 1 \rangle,\ \overrightarrow{t_2} = \langle 3, 5, 6, 7 \rangle,\ \overrightarrow{t_3} = \langle -4, -6, 9, 0 \rangle,\ \overrightarrow{t_4} = \langle -3, -2, 5, -3 \rangle \right\}$

2. Construct a family of affine maps that takes the first set of points $\mathbf{S}$ to the second set of points $\mathbf{T}$.

(a) $\mathbf{S} = \left\{ \vec{s_1} = \langle 1, -2 \rangle,\ \vec{s_2} = \langle 9, 5 \rangle \right\}$

$\qquad \mathbf{T} = \left\{ \vec{t_1} = \langle 3, -8, -9, 1 \rangle,\ \vec{t_2} = \langle 4, 5, 7, -6 \rangle \right\}$

(b) $\mathbf{S} = \left\{ \vec{s_1} = \langle -4, 5, 0, 3 \rangle,\ \vec{s_2} = \langle 0, 3, -2, 1 \rangle,\ \vec{s_3} = \langle 2, 5, 8, -1 \rangle \right\}$

$\qquad \mathbf{T} = \left\{ \vec{t_1} = \langle 1, 6, 3 \rangle,\ \vec{t_2} = \langle -4, 3, 7 \rangle,\ \vec{t_3} = \langle 2, 3, -2 \rangle \right\}$

3. Rotate the point $(-3, 4)$ about the point $(1, 1)$. Perform a complete rotation, including at least ten frames in your animation.

4. Rotate the point $(-10, -9, 9)$ about the line that passes through the point $(1, 1, 1)$ and points in the direction $\langle 2, -2, 4 \rangle$. Perform a complete rotation, including at least ten frames in your animation.

5. Rotate a sphere of radius 2, with center $(3, 3, 0)$, about the line that passes through the point $(-1, -1, 1)$, and points in the direction perpendicular to the line that passes through the center of the sphere and the point $(-1, -1, 1)$. Perform a complete rotation, including at least ten frames in your animation.

Research Projects

1. An interesting application of affine maps has been used by Michael Barnsley to construct geometric objects called fractals. Research the topic of fractals in Barnsley's book *Fractals Everywhere* and write *Maple* code to generate his Fern fractal example. You might also look at the book *Elementary Linear Algebra with Applications*, 7th edition(or later) by Howard Anton and Chris Rorres for some information on fractals.

Chapter 11

Least-Squares Fits and Pseudoinverses

11.1 Pseudoinverse to a Nonsquare Matrix and Almost Solving an Overdetermined Linear System

Finding the inverse to a matrix A so far has been restricted to A being square. The question is why? If you go back to Section 5.1, you will notice the relationship expressed between a matrix's inverse and determinant. The determinant is a construct that is only realizable when dealing with a matrix of square dimension, and thus, the same must be true of the inverse to a matrix. Now we ask the question: If A is a nonsquare $m \times n$ matrix, is it possible to find some matrix B which, in some way, acts like an inverse matrix to A? It is by no means obvious that such a matrix B should exist, or even how it might behave. Instead, we start with something a bit more obvious. Once again, under the assumption $A \in \mathbb{R}^{m \times n}$, notice that $AA^T \in \mathbb{R}^{m \times m}$ and $A^T A \in \mathbb{R}^{n \times n}$. These two square matrices might have inverses. The question becomes, is one preferred over the other? One of the main reasons for finding an inverse to a matrix is to solve a system of equations. So let us consider a linear problem of the form

$$A\vec{x} = \vec{b} \tag{11.1}$$

where $A \in \mathbb{R}^{m \times n}$, $\vec{x} \in \mathbb{R}^{n \times 1}$, and $\vec{b} \in \mathbb{R}^{m \times 1}$. Here we consider vectors of $\mathbb{R}^m$ and $\mathbb{R}^n$ to be matrices with only one column, as is the standard interpretation. This will make it easier to determine the proper choice for constructing a square system. By examining dimensions of the matrices, notice that we can actually compute $A^T \vec{b}$. This implies we should consider the following new system:

$$A^T A \vec{y} = A^T \vec{b} \tag{11.2}$$

This is now a square system, where we have replaced the unknown solution vector $\vec{x}$, from the original nonsquare system, with $\vec{y}$, for the unknown solution to the new square system. It may seem counterintuitive, but the two systems may not have the same solution sets, which is why we have changed solution variables when going from the nonsquare system to the square system. The main reason why this is so, is that one cannot do any matrix algebra, in terms of matrix multiplication, to arrive back at the original system. However, it should be reasonable to assume that each solution $\vec{x}$ of the original system is a solution to the new system. To see this, consider a specific solution $\vec{x_p}$ to (11.1). If we left multiply both sides by A^T, then the equation still holds, hence $\vec{x_p}$ also satisfies (11.2). Unfortunately, as previously stated, if $\vec{y_p}$ is a solution to (11.2), there is no way, without manual checking, to verify that $\vec{y_p}$ would also be a solution to (11.1), as we cannot compute $\left(A^T\right)^{-1}$ when A is not a square matrix.

Once we have left multiplied both sides of the nonsquare system by A^T to make the system square, we can attempt to find a solution by standard methods. For instance, if $A^T A$ has an inverse, then we arrive at the single solution

$$\vec{y} = (A^T A)^{-1} A^T \vec{b}$$

This solution $\vec{y}$ suggests the following definition:

Definition 11.1.1. The *pseudoinverse* of a matrix $A \in \mathbb{R}^{m \times n}$, denoted by $p(A)$, is defined to be

$$p(A) = (A^T A)^{-1} A^T$$

One should immediately notice that the concept of a pseudoinverse reduces to the standard definition of invertibility if A is square and invertible. To see this, remember that $(AB)^{-1} = B^{-1} A^{-1}$, and hence

$$
\begin{aligned}
p(A) &= (A^T A)^{-1} A^T \\
&= A^{-1} \left(A^T\right)^{-1} A^T \\
&= A^{-1} I \\
&= A^{-1}
\end{aligned}
$$

As a result, the idea of a pseudoinverse generalizes the process of solving square systems of equations to solving nonsquare systems of the form (11.1). In the case of a square system where A is invertible, the pseudoinverse approach and the matrix inverse approach would yield the exact same solution set.

Now that we have defined a pseudoinverse, and we have a process to compute it, we must ask ourselves, can this process be applied to find the pseudoinverses of any matrix? If we consider a matrix $A \in \mathbb{R}^{m \times n}$ with $m < n$, then with a little work, one can show that $\det\left(A^T A\right) = 0$, and hence $p(A)$ does not exist. Therefore, pseudoinverses can only exist if $m \geq n$, as we shall see by way of the following examples. We will first attempt to take the pseudoinverse of arbitrary matrices of dimension 1×3, 2×3, 3×4, and 2×4. To determine if the pseudoinverse can be taken, we will have *Maple* compute the determinant of the matrix $A^T A$. Note the end result in each case.

```
> with(linalg):
> A:= matrix([[a, b, c]]):
> det(evalm(transpose(A)&*A));
```
$$0$$

```
> A:= matrix([[a, b, c], [d, e, f]]):
> det(evalm(transpose(A)&*A));
```
$$0$$

```
> A:= matrix([[a, b, c, d], [e, f, g, h], [i, j, k, l]]):
> det(evalm(transpose(A)&*A));
```
$$0$$

```
> A:= matrix([[a, b, c, d], [e, f, g, h]]):
> det(evalm(transpose(A)&*A));
```
$$0$$

Clearly, the determinant for each of the four matrices given above was 0, and in each case, $m < n$. You should ask yourself why this will always be true when $m < n$.

Example 11.1.1. Next, we put our pseudoinverse to work for us on problems where we can actually apply it. Consider the following problem, expressed in matrix form as follows:

$$\begin{bmatrix} -5 & 2 & -9 \\ 7 & 1 & 3 \\ 8 & -4 & 0 \\ -6 & 10 & -3 \end{bmatrix} \vec{x} = \begin{bmatrix} -11 \\ 6 \\ 1 \\ 5 \end{bmatrix}$$

This problem is in the form of equation (11.1), and since A is not square, we cannot simply multiply both sides by A^{-1}. Obviously, we can still solve this system of equations by augmenting the two matrices and row reducing. However, it should be of interest to try this new method we have discussed as well.

```
> A:= matrix([[-5, 2, -9], [7, 1, 3], [8, -4, 0], [-6, 10, -3]]):
> pseudo_A:= evalm(inverse(transpose(A)&*A)&*transpose(A));
```

$$
pseudo\_A := \begin{bmatrix} \dfrac{1609}{78105} & \dfrac{6404}{78105} & \dfrac{2116}{26035} & \dfrac{1577}{78105} \\[2ex] -\dfrac{143}{15621} & \dfrac{1091}{15621} & \dfrac{32}{5207} & \dfrac{1520}{15621} \\[2ex] -\dfrac{26372}{234315} & -\dfrac{1762}{234315} & -\dfrac{5168}{78105} & -\dfrac{751}{234315} \end{bmatrix}
$$

```
> B:= matrix([[-11], [6], [1], [5]]):
> Y:= evalm(pseudo_A&*B);
```

$$
Y := \begin{bmatrix} \dfrac{34958}{78105} \\[2ex] \dfrac{15815}{15621} \\[2ex] \dfrac{260261}{234315} \end{bmatrix}
$$

```
> rref(augment(A,B));
```

$$
\begin{bmatrix} 1 & 0 & 0 & 0 \\ 0 & 1 & 0 & 0 \\ 0 & 0 & 1 & 0 \\ 0 & 0 & 0 & 1 \end{bmatrix}
$$

This row reduced matrix tells us that the system $A\vec{x} = \vec{b}$ has no solution, since its last row is the equation $0 = 1$. If you recall, a system $A\vec{x} = \vec{b}$, where A is $m \times n$ with $m > n$, is said to be overdetermined, because it has more equations than variables. Such systems invariably have no solutions as we have just seen. Unfortunately, overdetermined linear systems occur very often in real life, as data can be sampled frequently, but rarely will always lie on the graph of a function for which we wish to model the data. Thus, we need an approximate solution to such a system if no actual solution exists.

```
> evalf(evalm(A&*Y));
```

$$
\begin{bmatrix} -10.20962806 \\ 7.477651879 \\ -0.4690608796 \\ 4.106536073 \end{bmatrix}
$$

```
> evalf(evalm(A&*Y-B));
```

$$
\begin{bmatrix} 0.7903719352 \\ 1.477651879 \\ -1.469060880 \\ -0.8934639268 \end{bmatrix}
$$

> norm(evalf(evalm(A&*Y-B)),frobenius);

$$2.400949922$$

Notice that if $\vec{y}$ is the unique solution to the new system (11.2), with $\vec{y} = p(A)\,\vec{b}$, then $A\vec{y} \sim \vec{b}$. Therefore, $\vec{y}$ is approximately a solution to the original system $A\vec{x} = \vec{b}$, even though the system $A\vec{x} = \vec{b}$ has no actual solutions. As seen below, if we take the four 3×3 square subsystems of $A\vec{x} = \vec{b}$, (i.e., remove one of the equations or datapoints), then our solution $\vec{y}$ is close to the unique solutions of all four subsystems. To determine how close, we simply evaluate $\left| A\vec{y} - \vec{b} \right|$, as was done in the very last command in the above section of *Maple* code.

> soln1:= delcols(rref(delrows(augment(A, B), 1..1)), 1..3);

$$soln1 := \begin{bmatrix} \frac{55}{92} \\ \frac{87}{92} \\ \frac{20}{69} \end{bmatrix}$$

> evalf(evalm(Y-soln1));

$$\begin{bmatrix} -0.1502491072 \\ 0.06676700540 \\ 0.8208762081 \end{bmatrix}$$

> soln2:= delcols(rref(delrows(augment(A, B), 2..2)), 1..3);

$$soln2 := \begin{bmatrix} \frac{33}{43} \\ \frac{221}{172} \\ \frac{93}{86} \end{bmatrix}$$

> evalf(evalm(Y-soln2));

$$\begin{bmatrix} -0.3198648808 \\ -0.2724645416 \\ 0.02933593170 \end{bmatrix}$$

> soln3:= delcols(rref(delrows(augment(A, B), 3..3)), 1..3);

$$soln3 := \begin{bmatrix} \frac{22}{171} \\ \frac{169}{171} \\ \frac{37}{27} \end{bmatrix}$$

> evalf(evalm(Y-soln3));

$$\begin{bmatrix} 0.3189220089 \\ 0.02411508574 \\ -0.2596390898 \end{bmatrix}$$

> soln4:= delcols(rref(delrows(augment(A, B), 4..4)), 1..3);

$$soln4 := \begin{bmatrix} \frac{33}{104} \\ \frac{5}{13} \\ \frac{353}{312} \end{bmatrix}$$

> evalf(evalm(Y-soln4));

$$\begin{bmatrix} .1302692874 \\ .6278037947 \\ -.02067897587 \end{bmatrix}$$

So the solution $\overrightarrow{y}$ to the square system is close to the four solutions to each 3×3 subsystem of the original nonsquare system. Now let's plot the four planes that are the four equations of the original system with the solution $\overrightarrow{y}$. We will let the variables in the system be x, y, and z, so that each of the four equations of this system is a plane in three-space. We will then plot these four planes with the solution $\overrightarrow{y}$ to the new system $A^T A \overrightarrow{y} = A^T \overrightarrow{b}$. This is depicted in Figure 11.1.

> with(plots): with(plottools):
> X:= [x,y,z]:
> planes:= evalm(A&*X - B);

$$planes := \begin{bmatrix} -5x + 2y - 9z + 11 \\ 7x + y + 3z - 6 \\ 8x - 4y - 1 \\ -6x + 10y - 3z - 5 \end{bmatrix}$$

> plane1:= plot3d(solve(planes[1,1]=0,z), x=-5..5, y=-5..5, color = red):
> plane2:= plot3d(solve(planes[2,1]=0,z), x=-5..5, y=-5..5, color = tan):
> plane3:= plot3d([t,2*t-1/4,s], t=-5..5, s=-5..5, color = blue):
> plane4:= plot3d(solve(planes[4,1] = 0, z), x = -5..5, y = -5..5, color = yellow):
> plot_Y:= sphere([Y[1,1],Y[2,1],Y[3,1]], 3/5, color = black):

> display({plane1,plane2,plane3,plane4,plot_Y}, axes=boxed, style = patchno-grid, view=[-4..4,-4..4,-4..4], orientation=[-60,54]);

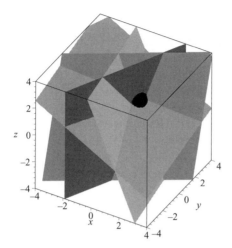

Figure 11.1: The systems of planes and the solution $\overrightarrow{y}$ (sphere) that best approximates the intersection of these planes.

The sphere of radius $\frac{3}{5}$ in this plot has a center at the solution $\overrightarrow{y}$. Notice that $\overrightarrow{y}$ is the point in three-space corresponding to the location of closest intersection of all four planes. In Section 11.3, we will explore in greater detail the reasons why the pseudoinverse solution appears to be the best approximate solution to overdetermined systems.

Pay special attention to how the third plane was defined above. The equation in question is given by $8x - 4y - 1 = 0$. Notice that the other three planes are solved for in terms of z, yet this plane has no z variable in it. Hence, the plane $8x - 4y - 1 = 0$ corresponds to a plane parallel to the line $8x - 4y = 1$ in the xy-plane, independent of z, and sticks out of the xy-plane in a perpendicular fashion. To graph this, we could have solved for another variable instead, or we can graph it parametrically. Solving for y in terms of x gives $y = 2x - \frac{1}{4}$, and hence, if we let $x = t$ the line $y = 2x - \frac{1}{4}$ is given parametrically by

$$(x(t), y(t)) = \left(t, 2t - \frac{1}{4} \right), \ t \in \mathbb{R}$$

Since this is a plane, we need our third coordinate z to be represented somehow. In this case, it does not depend on x or y, and from previous work, we know that surfaces in $\mathbb{R}^3$ must be defined parametrically as a function of two variables. We will call this second variable s, So our final parametrization of the plane

$8x - 4y - 1 = 0$ is given by

$$(x(s,t), y(s,t), z(s,t)) = \left(t, 2t - \frac{1}{4}, s \right), \quad s, t \in \mathbb{R}$$

This parametrization can be found in the definition of *plane3* above.

We end this section with an example that shows how the pseudoinverse method can be used to find solutions to overdetermined systems. If you recall, overdetermined systems rarely have solutions. For instance, three lines in the xy-plane most likely will not intersect one another at the same point. However, if they did, would the pseudoinverse method give us this answer?

Example 11.1.2. Consider the following system of equations:

$$\begin{aligned}
2x - y &= -7 \\
-5x - 2y &= 13 \\
3x + 3y &= -6
\end{aligned} \tag{11.3}$$

which in matrix form $A\overrightarrow{x} = \overrightarrow{b}$ is given by

$$\begin{bmatrix} 2 & -1 \\ -5 & -2 \\ 3 & 3 \end{bmatrix} \begin{bmatrix} x \\ y \end{bmatrix} = \begin{bmatrix} -7 \\ 13 \\ -6 \end{bmatrix}$$

The solution to system (11.3) is $\{x = -3, y = 1\}$. Now we shall have *Maple* solve this overdetermined system, yielding the correct solution.

```
> eqn1:= 2*x-y=-7: eqn2:=-5*x-2*y=13: eqn3:=3*x+3*y=-6:
> solve({eqn1, eqn2, eqn3},{x, y});
```

$$\{x = -3, y = 1\}$$

```
> A:= matrix(3,2,[2,-1,-5,-2,3,3]): B:= matrix(3,1,[-7,13,-6]):
> pseudo_A:= evalm(inverse(transpose(A)&*A)&*transpose(A));
```

$$pseudo\_A := \begin{bmatrix} \frac{5}{27} & -\frac{4}{27} & -\frac{1}{27} \\ -\frac{8}{27} & \frac{1}{27} & \frac{7}{27} \end{bmatrix}$$

```
> soln:= evalm(pseudo_A&*B);
```

$$soln := \begin{bmatrix} -3 \\ 1 \end{bmatrix}$$

From the above output, we see that *Maple* arrived at the correct solution via the pseudoinverse method. Now, let us plot this system of lines and the solution so that we get a geometrical interpretation of the above result. See Figure

11.2 below.

> lineplot:= implicitplot([eqn1, eqn2, eqn3], x=-5..0, y=-2..3, color=[red, black, blue], thickness = 2, labels=[x,y]):

> solnplot:= pointplot([soln[1,1],soln[2,1]], symbol=circle, symbolsize=25):

> display([lineplot, solnplot], scaling = constrained);

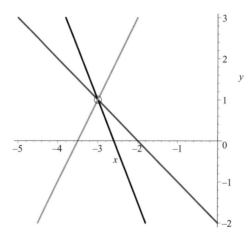

Figure 11.2: Overdetermined system (11.3) consists of three lines in the xy-plane that have a simultaneous solution $(-3, 1)$, found by the pseudoinverse method.

Homework Problems

1. If $A \in \mathbb{R}^{m \times n}$, what are the dimensions of $p(A)$?

2. Compute the pseudoinverse of each of the following matrices. If no pseudoinverse exists, then state so.

(a) $\begin{bmatrix} 3 \\ -2 \end{bmatrix}$
 (b) $\begin{bmatrix} 4 & 3 \end{bmatrix}$
 (c) $\begin{bmatrix} 1 & 4 & 5 \\ 3 & 2 & -2 \end{bmatrix}$

(d) $\begin{bmatrix} 1 & -1 \\ 0 & 2 \\ -1 & 1 \end{bmatrix}$
 (e) $\begin{bmatrix} 2 & 4 \\ -3 & -6 \\ 3 & -2 \end{bmatrix}$
 (f) $\begin{bmatrix} 4 & -2 & 3 & 1 \\ -2 & 1 & 7 & -2 \end{bmatrix}$

(g) $\begin{bmatrix} -1 & 0 & 2 \\ 5 & 2 & 0 \\ 0 & 3 & 4 \\ 2 & 0 & 1 \end{bmatrix}$ (h) $\begin{bmatrix} 1 & -1 & -2 \\ 0 & 2 & 3 \\ -1 & 1 & 5 \end{bmatrix}$ (i) $\begin{bmatrix} -2+i & 1-2i \\ i & -2-2i \\ 3i & 0 \end{bmatrix}$

3. For each of the following nonsquare matrices, perform a row reduction on the augmented matrix $(A\,|\,b)$ and determine if the system has a solution.

(a) $\begin{bmatrix} -2 \\ 4 \end{bmatrix} \vec{x} = \begin{bmatrix} 8 \\ 8 \end{bmatrix}$

(b) $\begin{bmatrix} -2 \\ 4 \end{bmatrix} \vec{x} = \begin{bmatrix} -4 \\ 8 \end{bmatrix}$

(c) $\begin{bmatrix} 3 & -2 \\ 0 & 1 \\ -6 & 3 \end{bmatrix} \vec{x} = \begin{bmatrix} 1 \\ -3 \\ -5 \end{bmatrix}$

(d) $\begin{bmatrix} 3 & -2 \\ 0 & 1 \\ -6 & 3 \end{bmatrix} \vec{x} = \begin{bmatrix} 1 \\ 3 \\ -5 \end{bmatrix}$

(e) $\begin{bmatrix} 3-i & 4-2i \\ 4-2i & 7+6i \\ -5+3i & -10-14i \end{bmatrix} \vec{x} = \begin{bmatrix} 7 \\ 3 \\ 5 \end{bmatrix}$

(f) $\begin{bmatrix} 3-i & 4-2i \\ 4-2i & 7+6i \\ -5+3i & -10-14i \end{bmatrix} \vec{x} = \begin{bmatrix} 7 \\ 3 \\ 1 \end{bmatrix}$

4. Approximately solve each of the nonsquare systems from problem 3 using the pseudoinverse method.

5. For each of the solutions $\vec{y}$ found in problem 4, compute $A\vec{y}$ and compare it to $\vec{b}$ by evaluating $\left| A\vec{y} - \vec{b} \right|$.

6. Let $A \in \mathbb{R}^{1 \times n}$ for $n > 1$. For $n = 2, 3$, find $A^T A$ and $\det\left(A^T A\right)$.

7. Give an argument, based on bases and dimension, as to why no pseudoinverse can exist for $A \in \mathbb{R}^{m \times n}$ with $m < n$.

8. Verify the following properties of the pseudoinverse. Here, you may assume that A is an arbitrary $m \times n$ matrix, and $c \in \mathbb{R}$.

   (a) $Ap(A)A = A$ (b) $p(p(A)) = A$ (c) $p(cA) = \dfrac{1}{c}p(A)$

9. Let $A \in \mathbb{R}^{m \times 1}$ for $m \geq 1$. Find the formula for the pseudoinverse $p(A)$.

10. Is $p(AB) = p(B)p(A)$? If yes, then explain why. If no, then give an example of its failure.

11. Is $p(A^T) = p(A)^T$? If yes, then explain why. If no, then give an example of its failure.

Maple Problems

1. Compute the pseudoinverse of each of the following matrices. If no pseudoinverse exists, then state so.

(a) $\begin{bmatrix} -2 & 8 & 4 \\ 2 & 0 & 1 \\ -4 & 5 & 2 \\ 3 & 1 & -4 \end{bmatrix}$

(b) $\begin{bmatrix} -2 & 8 & 4 \\ 2 & 0 & 1 \\ -4 & 5 & 2 \\ 3 & 1 & -4 \\ 2 & -2 & 2 \end{bmatrix}$

(c) $\begin{bmatrix} 4 & -8 & 3 & 2 \\ 0 & -5 & 6 & 2 \\ 5 & 7 & -2 & 1 \end{bmatrix}$

(d) $\begin{bmatrix} 1 & -2 & 1 & 5 \\ 0 & 2 & -4 & 3 \\ 7 & -6 & 5 & 9 \\ -10 & 4 & -5 & 3 \\ -5 & 2 & 4 & -9 \\ 0 & 2 & 12 & -4 \end{bmatrix}$

(e) $\begin{bmatrix} 2 & 1 & 2 & -8 & 2 \\ -3 & 4 & -6 & 9 & -5 \\ 5 & 4 & 7 & 4 & -1 \\ 8 & -2 & 10 & 0 & 1 \end{bmatrix}$

(f) $\begin{bmatrix} 1-i & -1 & -2+i \\ 2+i & 2-5i & 3 \\ -1 & 1 & 1+5i \\ -i & 0 & 3+2i \end{bmatrix}$

2. For each of the following nonsquare systems, perform a row reduction on the augmented matrix $(A|b)$ and determine if the system has a solution.

(a) $\begin{bmatrix} 2 & -1 & 3 \\ 6 & 4 & 0 \\ -2 & 1 & 1 \\ 6 & 4 & 4 \end{bmatrix} \vec{x} = \begin{bmatrix} 4 \\ -3 \\ 2 \\ 3 \end{bmatrix}$

(b) $\begin{bmatrix} 2 & -1 & 3 \\ 6 & 4 & 0 \\ -2 & 1 & 1 \\ 6 & 4 & 4 \end{bmatrix} \vec{x} = \begin{bmatrix} 1 \\ -3 \\ 2 \\ 2 \end{bmatrix}$

(c) $\begin{bmatrix} 3 & -2 & 1 & 0 \\ 3 & 2 & -1 & 5 \\ 7 & 2 & -4 & 6 \\ 4 & -5 & 6 & 2 \\ 3 & 0 & 11 & -10 \end{bmatrix} \vec{x} = \begin{bmatrix} 2 \\ -3 \\ 4 \\ 2 \\ -2 \end{bmatrix}$

(d) $\begin{bmatrix} 3 & -2 & 1 & 0 \\ 3 & 5 & -20 & 14 \\ 7 & 2 & -4 & 6 \\ 4 & -5 & 6 & 2 \\ 3 & 0 & 11 & -10 \end{bmatrix} \vec{x} = \begin{bmatrix} 2 \\ 6 \\ 4 \\ 2 \\ -2 \end{bmatrix}$

3. Approximately solve each of the nonsquare systems from problem 2 using the pseudoinverse method.

4. For each of the solutions $\vec{y}$ found in problem 3, compute $A\vec{y}$ and compare it to $\vec{b}$ by evaluating $\left| A\vec{y} - \vec{b} \right|$.

5. Create an overdetermined linear system of planes that intersect in a common line. Find the pseudoinverse approximate solution for this system and determine its distance from this line of intersection.

6. Create an overdetermined linear system whose solution set S has dimension greater than one, hence an infinite number of solutions to the system exist. Find the pseudoinverse approximate solution for this system and determine its distance from this solution S.

7. Let $A \in \mathbb{R}^{2 \times n}$. For $n = 3, 4$, find $A^T A$ and $\det\left(A^T A\right)$.

11.2   Fits and Pseudoinverses

In Section 11.1,we found a way to obtain an approximate solution to a system of equations when there were more equations than unknowns. It should be clear that in this situation, the chances of finding an exact solution are small. Normally, one needs the same number of linearly independent equations as unknowns to find a unique solution. If a row reduction is performed on the augmented matrix corresponding to one of these overdetermined systems, one usually arrives at an equation of the form $0 = 1$. Any *rref* matrix with a row corresponding to the equation $0 = 1$ tells us that no solution exists to the system. This is an important piece of information, but it tells us nothing about how close we can come to finding an approximate solution to the system of equations. The pseudoinverse method allows us to find an approximate solution when no exact solution exists, and find exact solutions in the rare instances when exact solutions exist to overdetermined systems.

Consider a set S, consisting of at least three points in the xy-plane:

$$S = \{(x_1, y_1), (x_2, y_2), \ldots, (x_n, y_n)\}$$

If we plot these points, and it appears that they lie fairly close to being on the graph of a line $L : y = ax + b$, for some unknowns a and b, would it not be nice to be able to determine the equation of this line? This line L is called a *line of fit* to the data. What properties should this line of fit satisfy? One would hope that if we plug each point of the set S into the equation defined by L, then $y_j = ax_j + b$ would be approximately true. Notice that with n points, we end up with n equations with only two unknowns, given by

$$ax_1 + b = y_1$$
$$ax_2 + b = y_2$$
$$\vdots$$
$$ax_n + b = y_n$$

(11.4)

We already know that such a system most likely has no solution, but we can use the pseudoinverse to get an approximate solution $\vec{y}$. We must first set up this overdetermined system. To do this, notice that in matrix form, each equation from (11.4) can be expressed in terms of matrix multiplication as

$$\begin{bmatrix} x_j & 1 \end{bmatrix} \begin{bmatrix} a \\ b \end{bmatrix} = y_j$$

Using this for all n points, we can now express the complete system of equations as the matrix equation given by

$$\begin{bmatrix} x_1 & 1 \\ x_2 & 1 \\ \vdots & \vdots \\ x_n & 1 \end{bmatrix} \begin{bmatrix} a \\ b \end{bmatrix} = \begin{bmatrix} y_1 \\ y_2 \\ \vdots \\ y_n \end{bmatrix}$$

(11.5)

The above equation is of the form $A\vec{x} = \vec{b}$, which we can convert to the square form given in (11.2). An approximate solution can now be found, and is given by $\vec{y} = p(A)\vec{b}$, assuming that the pseudoinverse of A exists. This solution, $\vec{y}$, produces a line L of fit to this set. Notice that in equation (11.5), $A \in \mathbb{R}^{n \times 2}$, $\vec{x} \in \mathbb{R}^2$ and $\vec{b} \in \mathbb{R}^n$.

Example 11.2.1. As an example, consider the following situation. A bookstore has the following demand data for each week of the first quarter that are the months January, February, and March:

week	books sold	average price ($)
1	415	13.25
2	372	13.85
3	391	13.60
4	428	13.15
5	403	13.65
6	350	14.05
7	362	13.90

continued on next page

week	books sold	average price ($)
8	410	13.20
9	385	13.60
10	434	13.00
11	465	12.75
12	380	13.65

This data can be interpreted as points of the form (x, p), where x is the number of books sold that week and p is their average selling price, so our data set will have twelve points:

$$S = \{(415, \$13.25), (372, \$13.85), (391, \$13.60), (428, \$13.15)$$
$$(403, \$13.65), (350, \$14.05), (362, \$13.90), (410, \$13.20)$$
$$(385, \$13.60), (434, \$13.00), (465, \$12.75), (380, \$13.65)\}$$

Now that we have our data, we can plot it, along with the line of 'best' fit, which can be found by the pseudoinverse method. We will have *Maple* perform all the necessary computations for us.

> with(linalg): with(plots): with(plottools):

> S:= [[415, 13.25], [372, 13.85], [391, 13.60], [428, 13.15], [403, 13.65], [350, 14.05], [362, 13.90], [410, 13.20], [385, 13.60], [434, 13.00], [465, 12.75], [380, 13.65]]:

> plot_S:= pointplot(S, symbol = cross, symbolsize = 20, color = blue):

> A:= matrix([seq([S[j][1],1], j=1..12)]);

$$
A := \begin{bmatrix}
415 & 1 \\
372 & 1 \\
391 & 1 \\
428 & 1 \\
403 & 1 \\
350 & 1 \\
362 & 1 \\
410 & 1 \\
385 & 1 \\
434 & 1 \\
465 & 1 \\
380 & 1
\end{bmatrix}
$$

> B:= delcols(matrix(S), 1..1):

> pseudoinv:= x -> evalm(inverse(transpose(x)&*x)&*transpose(x)):

> Y:= evalf(evalm(pseudoinv(A)&*B));

$$Y := \left[\begin{array}{c} -0.01172548909 \\ 18.15614336 \end{array} \right]$$

So our fit line to this data using the pseudoinverse is

$$p(x) = -.01172548909x + 18.15614336 \qquad (11.6)$$

where p is used instead of y, since this is really the average price formula per week for p in terms of x, which is how many books are sold that week. Now we want to plot it with the data.

> plot_L:= plot(Y[1,1]*x+Y[2,1], x = 300..500, color = red, thickness = 2):

> display({plot_L, plot_S});

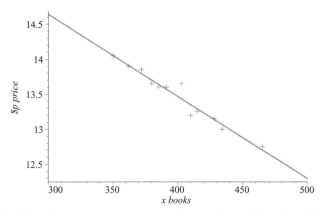

Figure 11.3: The data set representing the number of books sold versus the average selling price is depicted by the crosses, and the line of best fit, L, computed by the pseudoinverse method which approximates the data.

Upon inspection of Figure 11.3, we see that the line determined by the pseudoinverse seems an excellent fit to the data, but is it the best fit to the data? What should best fit even mean? Once again, we will address these questions in Section 8.3, which talks about least-squares fits.

■

Now what do you do if the data set does not seem to be fit by a line, but a curve of some type, such as a polynomial, or a sum of sines and cosines? Now let us look at an example of each of these kinds of fits.

Example 11.2.2. Let us take an approximate data set S of datapoints (t, y), where t is the time in seconds after the launch of a rocket, and y is its height in feet above ground level at time t. So

$$S = \{(0, 925), (2, 1005), (4, 1135), (6, 1215), (8, 1345),$$
$$(10, 1265), (12, 1130), (14, 975), (16, 795)\}$$

First, we want to plot this data, and recognize what type of curve would best fit it. The data is plotted in Figure 11.4.

> S:= [[0, 925], [2, 1005], [4, 1135], [6, 1215], [8, 1345], [10, 1265], [12, 1130], [14, 975], [16, 795]]:

> plot_S:= pointplot(S, symbol = cross, symbolsize = 20, color = blue):

> display(plot_S);

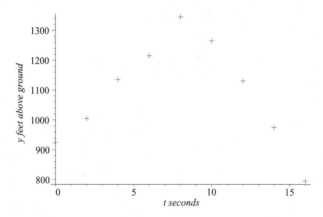

Figure 11.4: The data measured during a rockets flight, a parabola would fit this data more accurately than a line.

This looks roughly like the plot of a standard parabola opening downward, which has the form $y = at^2 + bt + c$. Let us use the pseudoinverse to find a parabola that fits this data and plot the data set with this parabola. To do this, we must first understand how to modify the linear-fit problem from the previous example. The standard form of a parabola has three unknown constants a, b, and c. Our hope is that each point (t_k, y_k) satisfies $y_k = at_k^2 + bt_k + c$. As with the linear case, this is most likely not possible. The best we can hope for is an approximate solution. So, similar to the line of fit problem, we can express

$y_k = at_k^2 + bt_k + c$ in matrix form as

$$\begin{bmatrix} t_k^2 & t_k & 1 \end{bmatrix} \begin{bmatrix} a \\ b \\ c \end{bmatrix} = y_k$$

We can now construct a matrix system of the form (11.1) given by

$$\begin{bmatrix} t_1^2 & t_1 & 1 \\ t_2^2 & t_2 & 1 \\ \vdots & \vdots & \vdots \\ t_9^2 & t_9 & 1 \end{bmatrix} \begin{bmatrix} a \\ b \\ c \end{bmatrix} = \begin{bmatrix} y_1 \\ y_2 \\ \vdots \\ y_9 \end{bmatrix}$$

Compare the above matrix equation to that of (11.5) and see if you can generalize this process to a function with k unknown constants and n known points. Back to our specific example, we have nine points and three unknown constants in our parabola, so $A \in \mathbb{R}^{9 \times 3}$, $\overrightarrow{x} \in \mathbb{R}^3$ and $\overrightarrow{b} \in \mathbb{R}^9$. Our parabola of fit should be given by the solution $\overrightarrow{y} = p(A)\overrightarrow{b}$, which we shall find in the commands below. Pay special attention to how we define the matrix A.

> A:= matrix([seq([S[j][1]^2, S[j][1], 1], j=1..9)]);

$$A := \begin{bmatrix} 0 & 0 & 1 \\ 4 & 2 & 1 \\ 16 & 4 & 1 \\ 36 & 6 & 1 \\ 64 & 8 & 1 \\ 100 & 10 & 1 \\ 144 & 12 & 1 \\ 196 & 14 & 1 \\ 256 & 16 & 1 \end{bmatrix}$$

> B:= delcols(matrix(S),1..1);

$$B := \begin{bmatrix} 925 \\ 1005 \\ 1135 \\ 1215 \\ 1345 \\ 1265 \\ 1130 \\ 975 \\ 795 \end{bmatrix}$$

> Y:= evalf(evalm(pseudoinv(A)&*B));

$$Y := \begin{bmatrix} -6.807359307 \\ 104.1677489 \\ 871.6363636 \end{bmatrix}$$

So now we have that the parabola of fit to this data set has equation

$$y = -6.807359307t^2 + 104.1677489t + 871.6363636$$

Let us plot this parabola, along with the data set.

> plot_parabola:= plot(Y[1,1]*t^2 + Y[2,1]*t + Y[3,1], t=0..22, color = red):
> display({plot_parabola, plot_S});

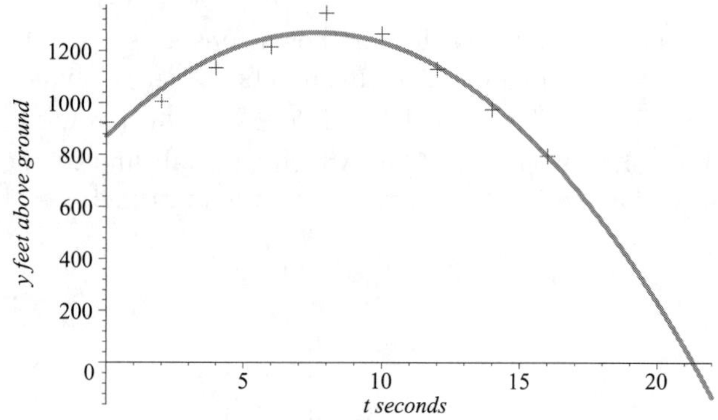

Figure 11.5: Rocket trajectory data set and the parabola approximation.

From Figure 11.5, we see that the fit of this parabola is fairly good and it allows us to say that the rocket will hit a target on ground level at roughly 21 seconds after launch.

Example 11.2.3. Now let us take a data set of stock market data for the price p of a particular stock at half-hour intervals for the first 6 hours after the market has opened, that is, from 9:00 A.M. to 3:00 P.M. The data is given as points (t, v), where v is the stock value at time t hours from the opening of the market.

$$S = \{(0, \$17.83), (.5, \$17.71), (1, \$17.56), (1.5, \$17.60), (2, \$17.65),$$
$$(2.5, \$17.55), (3, \$17.52), (3.5, \$17.58), (4, \$17.60), (4.5, \$17.47),$$
$$(5, \$17.44), (5.5, \$17.61), (6, \$17.69)\}$$

First, we plot the data to determine what type of function, or set of functions, would best approximate it. See Figure 11.6 below.

> S:= [[0, 17.83], [.5, 17.71], [1, 17.56], [1.5, 17.60], [2, 17.65], [2.5, 17.55], [3, 17.52], [3.5, 17.58], [4, 17.60], [4.5, 17.47], [5, 17.44], [5.5, 17.61], [6, 17.69]]:
> plot_S:= pointplot(S, symbol = cross, symbolsize = 20, color = blue):
> display(plot_S);

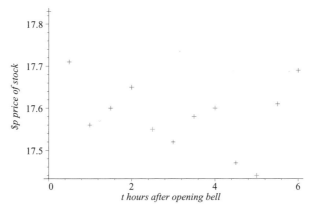

Figure 11.6: Plot of stock market data.

This pattern is similar to a wave pattern that suggests the use of trig functions, such as sine and cosine. Let us try approximating it on the interval $[0, 7]$ since this covers the hours of operation of the stockmarket. Now both $\sin\left(\frac{2\pi t}{7}\right)$ and $\cos\left(\frac{2\pi t}{7}\right)$ have period 7 instead of 2π. So we shall use these functions with positive integer multiples of these angles, as well as the constant function 1 to build our approximating function. Thus, we shall approximate this data set by a function of the form

$$v(t) = a_0 + a_1 \cos\left(\frac{2\pi t}{7}\right) + a_2 \cos\left(\frac{4\pi t}{7}\right) + a_3 \cos\left(\frac{6\pi t}{7}\right)$$
$$+ a_4 \sin\left(\frac{2\pi t}{7}\right) + a_5 \sin\left(\frac{4\pi t}{7}\right) + a_6 \sin\left(\frac{6\pi t}{7}\right)$$

Note that we could have chosen a different function, perhaps with more sine and cosine functions, but the above function is a good first guess. Upon inspection of Figure 11.7, we see that our guess is sufficient. Back to the problem at hand, our overdetermined linear system can be expressed in matrix form (11.1), with

$A =$

$$\begin{bmatrix} 1 & \cos\left(\frac{2\pi t_1}{7}\right) & \cos\left(\frac{4\pi t_1}{7}\right) & \cos\left(\frac{6\pi t_1}{7}\right) & \sin\left(\frac{2\pi t_1}{7}\right) & \sin\left(\frac{4\pi t_1}{7}\right) & \sin\left(\frac{6\pi t_1}{7}\right) \\ 1 & \cos\left(\frac{2\pi t_2}{7}\right) & \cos\left(\frac{4\pi t_2}{7}\right) & \cos\left(\frac{6\pi t_2}{7}\right) & \sin\left(\frac{2\pi t_2}{7}\right) & \sin\left(\frac{4\pi t_2}{7}\right) & \sin\left(\frac{6\pi t_2}{7}\right) \\ \vdots & & & \vdots & & & \vdots \\ 1 & \cos\left(\frac{2\pi t_{13}}{7}\right) & \cos\left(\frac{4\pi t_{13}}{7}\right) & \cos\left(\frac{6\pi t_{13}}{7}\right) & \sin\left(\frac{2\pi t_{13}}{7}\right) & \sin\left(\frac{4\pi t_{13}}{7}\right) & \sin\left(\frac{6\pi t_{13}}{7}\right) \end{bmatrix}$$

> cost:= t -> cos(2*Pi*t/7):

> sint:= t -> sin(2*Pi*t/7):

> A := evalf(matrix([seq([1, cost(S[j][1]), cost(2*S[j][1]), cost(3*S[j][1]), sint(S[j][1]), sint(2*S[j][1]), sint(3*S[j][1])], j=1..13)]));

$A :=$

$$\begin{bmatrix} 1. & 1. & 1. & 1. & 0. & 0. & 0. \\ 1. & .9009688678 & .6234898018 & .2225209335 & .4338837393 & .7818314825 & .9749279123 \\ 1. & .6234898018 & -.2225209335 & -.9009688678 & .7818314825 & .9749279123 & .4338837393 \\ 1. & .2225209335 & -.9009688679 & -.6234898008 & .9749279123 & .4338837392 & -.7818314833 \\ 1. & -.2225209335 & -.9009688678 & .6234898018 & .9749279123 & -.4338837393 & -.7818314825 \\ 1. & -.6234898022 & -.2225209321 & .9009688673 & .7818314822 & -.9749279126 & .4338837403 \\ 1. & -.9009688678 & .6234898018 & -.2225209335 & .4338837393 & -.7818314825 & .9749279123 \\ 1. & -1 & 1 & -1 & 0 & 0 & 0 \\ 1. & -.9009688678 & .6234898018 & -.2225209335 & -.4338837393 & .7818314825 & -.9749279123 \\ 1. & -.6234898008 & -.2225209337 & .9009688662 & -.7818314833 & .9749279122 & -.4338837426 \\ 1. & -.2225209335 & -.9009688678 & .6234898018 & -.9749279123 & .4338837393 & .7818314825 \\ 1. & .2225209333 & -.9009688672 & -.6234898043 & -.9749279123 & -.4338837407 & .7818314805 \\ 1. & .6234898018 & -.2225209335 & -.9009688678 & -.7818314825 & -.9749279123 & -.4338837393 \end{bmatrix}$$

> B:= delcols(matrix(S),1..1):

> Y:= evalf(evalm(pseudoinv(A)&*B));

$$Y := \begin{bmatrix} 17.62004426 \\ 0.119301307 \\ 0.0737797027 \\ 0.010065133 \\ 0.007356146 \\ -0.051888611 \\ -0.058949624 \end{bmatrix}$$

> plot_stock:= plot(Y[1,1]+Y[2,1]*cost(t) + Y[3,1]*cost(2*t) + Y[4,1]*cost(3*t)) + Y[5,1]*sint(t) + Y[6,1]*sint(2*t) + Y[7,1]*sint(3*t), t=0..7, color = red):

> display({plot_S, plot_stock});

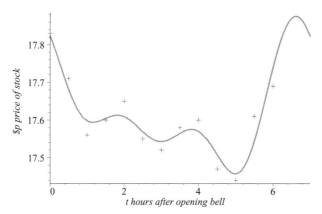

Figure 11.7: Plot of stock market data for the price of a certain stock over the course of the trading day and the trigonometric fit function.

We seem to have a fairly decent approximation using these trig functions, and the approximation can be improved by adding more trig functions to those already used. The question becomes, what types of functions can we use in our approximation? Similar to the case of vectors in $\mathbb{R}^n$ or $\mathbb{C}^n$, we would like the set of functions to be linearly independent. But what does it mean for a set of functions to be linearly independent?

Definition 11.2.1. Given a set of n functions $\{f_1(x), f_2(x), \ldots, f_n(x)\}$ on an interval I, where each of the f_k's is at least $n - 1$ times differentiable, the *Wronskian* W is a function of $x \in I$ defined by

$$W(f_1, f_2, \ldots, f_n)(x) = \det\left(\begin{bmatrix} f_1(x) & f_2(x) & \cdots & f_n(x) \\ f_1'(x) & f_2'(x) & \cdots & f_n'(x) \\ \vdots & \vdots & \ddots & \vdots \\ f_1^{(n-1)}(x) & f_2^{(n-1)}(x) & \cdots & f_n^{(n-1)}(x) \end{bmatrix}\right)$$

Simply stated, the Wronskian of the set of functions $\{f_1(x), f_2(x), \ldots, f_n(x)\}$ is the determinant of the matrix whose kth column consists of the first $n - 1$ derivatives, sequentially, of $f_k(x)$.

Definition 11.2.2. A set of functions $\{f_1(x), f_2(x), \ldots, f_n(x)\}$ on an interval $I = [a, b]$ is said to be a *dependent set of functions* if for all x in the interval I there is at least one function $f_j(x)$ such that $f_j(x)$ is a linear combination of the remaining functions of the set, that is, there are real scalars $a_1, a_2, \ldots, a_{j-1}, a_{j+1}, \ldots, a_n$, such that for all $x \in I$,

$$f_j(x) = a_1 f_1(x) + a_2 f_2(x) + \cdots + a_n f_n(x)$$

Definition 11.2.3. A set of functions $\{f_1(x), f_2(x), \ldots, f_n(x)\}$ on an interval $I = [a, b]$ is said to be an *independent set of functions* if they are not a dependent set. More precisely, this set of functions is an independent set on the interval I if

$$a_1 f_1(x) + a_2 f_2(x) + \cdots + a_n f_n(x) = 0$$

for all $x \in I$ forces all n scalars $a_1, a_2, \ldots, a_n$ to be 0.

Example 11.2.4. So the set of functions $\left\{1, x, x^2, x^3, 8x^2 - 5x + 2\right\}$ is a dependent set of functions on any interval I since the fifth function is clearly a linear combination of the rest of the functions in this set.

The hard part is to determine when a set of functions is independent, happily this can be done through the use of the Wronskian. We can combine the concepts of linearly independent and dependent sets of functions and the Wronskian in the following theorem.

Theorem 11.2.1. *A set of functions* $\{f_1(x), f_2(x), \ldots, f_n(x)\}$, *on an interval* $I = [a, b]$, *where each of the* f_k*'s is at least* $n - 1$ *times differentiable, is linearly independent if there exists* $\hat{x} \in I$ *such that* $W(f_1, f_2, \ldots, f_n)(\hat{x}) \neq 0$.

Proof. We will prove the contrapositive statement that if the set $\{f_1(x), f_2(x), \ldots, f_n(x)\}$ is linearly dependent on I, then $W(f_1, f_2, \ldots, f_n)(x) = 0$ for all $x \in I$. By definition, if the functions $\{f_1(x), f_2(x), \ldots, f_n(x)\}$ are linearly dependent on the interval I, then

$$a_1 f_1(x) + a_2 f_2(x) + \cdots + a_n f_n(x) = 0 \tag{11.7}$$

for constants $a_1, a_2, \ldots, a_n$, of which at least one is nonzero. The kth derivative of equation (11.7) satisfies

$$a_1 f_1^{(k)}(x) + a_2 f_2^{(k)}(x) + \cdots + a_n f_n^{(k)}(x) = 0 \tag{11.8}$$

for $1 \leq k \leq n - 1$. In matrix form, these n equations become:

$$\begin{bmatrix} f_1(x) & f_2(x) & \cdots & f_n(x) \\ f_1'(x) & f_2'(x) & \cdots & f_n'(x) \\ \vdots & \vdots & \ddots & \vdots \\ f_1^{(n-1)}(x) & f_2^{(n-1)}(x) & \cdots & f_n^{(n-1)}(x) \end{bmatrix} \begin{bmatrix} a_1 \\ a_2 \\ \vdots \\ a_n \end{bmatrix} = \begin{bmatrix} 0 \\ 0 \\ \vdots \\ 0 \end{bmatrix}$$

This equation has a nontrivial solution only when the determinant of the matrix on the LHS of the above equation is zero. But this is simply

$$W(f_1, f_2, \ldots, f_n)(x) = 0$$

$\square$

Example 11.2.5. We will compute the Wronskian of the set of functions :

$$\left\{ f_1(x) = 1, f_2(x) = x, f_3 = x^2 \right\}, \quad x \in [0, 20]$$

which was used in Example 11.2.2 to model the trajectory of a rocket.

$$W\left(f_1, f_2, f_3\right)(x) = \det \left(\begin{bmatrix} 1 & x & x^2 \\ 0 & 1 & 2x \\ 0 & 0 & 2 \end{bmatrix} \right) = 2$$

Since the Wronskian is 2 on the entire interval $[0, 20]$, we can conclude that the set $\left\{ f_1(x) = 1, f_2(x) = x, f_3 = x^2 \right\}$ is linearly independent on the interval $[0, 20]$.

Example 11.2.6. It would be nice to see that the Wronskian is zero when we have a linearly dependent set of functions. So we will take the functions from Example 11.2.5, and add to it the function $f_4(x) = x - 3x^2$, which can be written as $f_4(x) = f_2(x) - 3f_3(x)$.

$$W\left(f_1, f_2, f_3, f_4\right)(x) = \det \left(\begin{bmatrix} 1 & x & x^2 & x - 3x^2 \\ 0 & 1 & 2x & 1 - 6x \\ 0 & 0 & 2 & -6 \\ 0 & 0 & 0 & 0 \end{bmatrix} \right) = 0.$$

Notice that Wronskian is zero, independent of the interval we choose since the last row of the matrix given in the formula consists of all zeros. Of course, the main reason that the Wronskian of this set of functions was zero is the fact that four derivatives were taken of polynomials all of degree two or less, which automatically tells us that $f_k^{(4)}(x) = 0$ for $1 \le k \le 4$.

Example 11.2.7. The *Wronskian* command can be found in the *VectorCalculus* package, and we use it on the next two sets of functions:

$$\mathbf{F}_1 = \{\cos(t), \sin(t), \cos(2t), \sin(2t)\}$$
$$\mathbf{F}_2 = \{\cos(t), \sin(t), \cos(2t), \sin(2t), 2\sin(t) - \cos(2t)\}$$

> with(VectorCalculus):
> W1:= simplify(det(Wronskian([cos(t), sin(t), cos(2*t), sin(2*t)], t)));

$$W1 := 18$$

> W2:=det(Wronskian([cos(t), sin(t), cos(2*t), sin(2*t), 2*sin(t)-cos(2*t)], t));

$$W2 := 0$$

No matter what interval we choose, notice that $W1(t) \neq 0$, while $W2(t) = 0$. Therefore, the set $\mathbf{F}_1$ is linearly independent on every interval I, while the set $\mathbf{F}_2$ is linearly dependent on every interval I.

■

Note that there are striking similarities between the work done in this section, and that of Section 4.2. However, there is one important difference. In the aforementioned section, it was required to choose n functions to fit a data set of n points. As long as the resulting square matrix was invertible, we were guaranteed that the linear combination of the n functions would pass through all n points. The pseudoinverse method is a generalization of this process, which does reduce to the square system setting found in Section 4.2.

Homework Problems

1. Determine what types of functions would best fit the following sets of data. You may chose from linear functions, quadratic functions, polynomials of degree greater than 2, a linear combination of sines and cosines, exponential functions.

(a)

x	y
-3	5.1
-1	4.5
0	1.2
1	0.1
2	1.3
4	8.2

(b)

x	y
-2.50	0.81
-0.75	2.41
-0.25	1.49
0.25	.51
1.50	-.88
2.25	.31

(c)

x	y
-2.2	8.51
-1.9	7.46
-1.5	6.68
-0.8	5.12
-0.4	3.26
1.2	0.32

(d)

x	y
-2.5	1.36
-2.0	1.62
-1.0	2.58
-0.5	3.72
0.0	5.57
0.5	8.53
1.0	11.53
1.5	9.74
2.0	-33.1

(e)

x	y
-0.80	0.04
-0.50	4.40
-0.25	5.31
0.00	4.62
0.25	2.37
0.50	-1.60
0.67	-4.01
1.00	-3.42
1.50	4.56

(f)

x	y
-5	4.25
-4	4.50
-3	4.25
-2	3.50
-1	2.25
0	0.50
1	-1.75
2	-4.50
3	-7.75

2. Set up, but do not solve, the nonsquare system of equations for a plane of fit of the form $z = ax + by + c$ corresponding to the following set of points in $\mathbb{R}^3$:

$$S = \{(2, 3, -2), (-1, 4, -9), (0, 2, -1), (3, 4, -2), (-1, -1, 5)\}$$

3. Any vertical line excluding the y-axis can be expressed as $ax = 1$. Using this equation, find the vertical line of fit to the pair of points $\{(-1, 4), (2, 3)\}$. What would you expect the result to be? Does the actual answer agree with your guess?

4. Repeat problem 3, this time use an arbitrary pair of points $\{(x_1, y_1), (x_2, y_2)\}$. Attempt to interpret the answer.

5. (a) Find the Wronskian for the set of functions $\mathbf{S} = \{1, e^x, e^{-x}, e^{2x}\}$.

   (b) Is $\mathbf{S}$ a dependent or independent set of functions for $x \in [-1, 1]$?

6. (a) Find the Wronskian for the sets of functions

$$\mathbf{S} = \{1, e^x, \sinh(x)\}$$
$$\mathbf{T} = \{e^x, e^{-x}, \cosh(x)\}$$
$$\mathbf{R} = \{e^x, \cosh(x), \sinh(x)\}$$

   (b) Are $\mathbf{S}$, $\mathbf{T}$, and $\mathbf{R}$ dependent or independent sets of functions for $x \in [0, 1]$?

Maple Problems

1. Fit each of the following sets of data to a function of fit expressed as a linear combination of the functions $\{1, x, x^2, x^3, x^4\}$.

(a)

x	y
-2.2	34.8
-1.9	22.8
-1.5	12.4
-0.8	4.12
-0.4	2.23
0.1	0.83
0.8	0.62
1.2	2.12

(b)

x	y
-2.0	7.43
-1.5	8.12
-1.0	5.22
-0.5	2.02
0.0	1.01
0.5	2.19
1.5	8.31
2.5	-0.32

(c)

x	y
-1.00	0.80
-0.50	1.48
-0.25	1.76
0.25	2.46
0.75	2.82
1.00	3.22
1.50	4.75
2.00	5.45

2. Plot the data sets and corresponding functions found by the pseudoinverse method for each part of problem 1.

3. For each of the functions found in problem 1, compute the error in the approximation by evaluating $\left| A\vec{y} - \vec{b} \right|$.

4. Fit each of the following sets of data to a function of fit expressed as a linear combination of the functions $\{1, e^{-x}, e^x, e^{-2x}, e^{2x}\}$.

(a)

x	y
-1.5	1.26
-0.5	1.59
-0.25	1.24
0.25	0.72
0.75	0.21
1	0.14
1.5	0.77
2	4.86

(b)

x	y
-2	-2.535
-1.5	1.463
-1	2.245
0	1.845
1	1.341
2	1.131
4	1.018
6	1.002

(c)

x	y
-5	1.963
-4	1.927
-3	1.884
-2	1.473
-1	0.787
0	-1.132
1	-4.102
2	8.342

5. Plot the data sets and corresponding functions found by the pseudoinverse method for each part of problem 4.

6. For each of the functions found in problem 4, compute the error in the approximation by evaluating $\left| A\vec{y} - \vec{b} \right|$.

7. Fit each of the following sets of data to a function of fit expressed as a linear combination of the functions $\{1, \sin(x/2), \cos(x/2), \sin(x/3), \cos(x/3)\}$.

(a)

x	y
-10	1.91
-8	3.64
-6	2.07
-4	-1.69
-1	-4.58
1	-1.93
3	2.26
4	3.83
6	4.46
9	-0.52

(b)

x	y
-30	4.83
-20	-2.89
-15	2.09
-10	2.17
0	-2.00
5	0.09
10	5.86
15	-1.72
20	-0.66
30	2.25

(c)

x	y
-20	0.24
-15	0.97
-10	1.72
0	0.65
5	0.88
10	1.61
15	0.97
20	0.20
25	1.52
30	1.31

8. Plot the data sets and corresponding functions found by the pseudoinverse method for each part of problem 7.

9. For each of the functions found in problem 7, compute the error in the approximation by evaluating $\left| A\vec{y} - \vec{b} \right|$.

10. Fit each of the following sets of data to a function of the form $z = ax+by+c$.

(a)

x	y	z
0	0	2
-1	2	-8
-3	4	-22
3	3	-2
2	-1	13

(b)

x	y	z
0	0	-2.4
1	1	-1.2
-1	-1	-3.3
2	-2	-2.5
-2	2	-2.8

(c)

x	y	z
0	0	2
1	1	-5
-1	-2	9
1	-2	10
3	-2	11

11. Plot the data sets and corresponding functions found by the pseudoinverse method for each part of problem 10.

12. Reuse problem 9, only refer back to problem 10 instead of 7.

13. (a) Find the Wronskian for the sets of functions

$$\mathbf{S} = \left\{ 1, e^x, e^{-x}, e^{2x}, e^{-2x}, e^{3x}, e^{-3x} \right\}$$
$$\mathbf{T} = \left\{ 1, x, x^2, x^3, x^4, x^5, x^6 \right\}$$

(b) Are $\mathbf{S}$ and $\mathbf{T}$ dependent or independent sets of functions on any interval $[a, b]$?

(c) Can you generalize the results of parts (a) and (b)? Explain.

14. (a) For a set of n functions $\{f_1(x, y), f_2(x, y), \ldots, f_n(x, y)\}$, come up with a version of the Wronskian to test if these functions form a dependent or independent set on the rectangle $[a, b] \times [c, d] \subseteq \mathbb{R}^2$.

(b) Use your test in part (a) to decide if the set $\left\{ 1, x, y, xy, xy^2, x^2y \right\}$ is a dependent or independent set on the unit square $[0, 1] \times [0, 1]$.

(c) Use your test in part (a) to decide if the set

$$\{1, \cos(x), \cos(y), \cos(xy), \sin(x), \sin(y), \sin(xy)\}$$

is a dependent or independent set on the square $[0, 2\pi] \times [0, 2\pi]$.

Research Projects

1. Explain the mathematics involved in medical imaging, such as the MRI (magnetic resonance imaging), and how the image is produced.

2. Explain a use for the pseudoinverse and solving overdetermined systems of linear equations in your particular field of study.

11.3   Least-Squares Fits and Pseudoinverses

We have spent a sufficient amount of time utilizing the pseudoinverse to fit curves to data sets. The next step is to determine if this method results in the best possible approximation. To do this, we need to determine a measure of how good a fit is, and then how to get the best one. The simplest measure of how well a function fits a fixed data set is called the *squared deviation D*. To compute the squared deviation for a given function and data set, one simply computes the square of the difference between the function values and the exact values from the data set. In our situation, we are concerned with a linear combination of functions of the form

$$y = a_1 f_1(x) + a_2 f_2(x) + \cdots + a_n f_n(x)$$

and corresponding data set

$$S = \{(x_1, y_1), (x_2, y_2), \ldots, (x_m, y_m)\}$$

We will consider only the case of $m > n$, the case in which the pseudoinverse method is used to find a curve of fit. From the above function and corresponding data set, we arrive at the system of equations:

$$y_k = a_1 f_1(x_k) + a_2 f_2(x_k) + \cdots + a_n f_n(x_k), \quad k = 1, 2, \ldots, m \qquad (11.9)$$

which, in matrix form $F\vec{a} = \vec{y}$, is given by

$$
\begin{bmatrix}
f_1(x_1) & f_2(x_1) & \cdots & f_n(x_1) \\
f_1(x_2) & f_2(x_2) & \cdots & f_n(x_2) \\
\vdots & & \ddots & \vdots \\
f_1(x_m) & f_2(x_m) & \cdots & f_n(x_m)
\end{bmatrix}
\begin{bmatrix}
a_1 \\
\vdots \\
a_n
\end{bmatrix}
=
\begin{bmatrix}
y_1 \\
\vdots \\
y_m
\end{bmatrix}
\qquad (11.10)
$$

To solve system (11.10), we simply apply the pseudoinverse method to get $\vec{a} = p(F)\vec{y}$. The squared deviation is given by

$$D(\vec{a}) = |\vec{y} - F\vec{a}|^2 \qquad (11.11)$$

Now notice that D is actually a function of the vector $\vec{a}$, whose components are the constants $a_1, a_2, \ldots, a_n$ in the linear combination defined in equation (11.9). If we change just one of the a_k's, then $D(\vec{a})$ also changes. Hence, D is a function of n variables. To find the value of $\vec{a}$ that minimizes D, we must take the *gradient* of D, set it equal to zero, and then solve for the unknown coefficients. If we remember our multivariable calculus, the gradient is given by

$$\nabla D(\vec{a}) = \left(\frac{\partial D(\vec{a})}{\partial a_1}, \frac{\partial D(\vec{a})}{\partial a_2}, \ldots, \frac{\partial D(\vec{a})}{\partial a_n} \right)$$

In order to actually perform any computations with $\nabla D(\vec{a})$, we must first rewrite (11.11) in a way that allows us to take a partial derivative with respect to each of the a_k's. The following expression obviously looks more complicated, but is just a quadratic function in terms of the a_k's, and can easily have the gradient operator applied to it.

$$D(a_1, a_2, \ldots, a_n) = |\vec{y} - F\vec{a}|^2 \tag{11.12}$$

$$= \sum_{k=1}^{m} \Big[y_k - \big(a_1 f_1(x_k) + a_2 f_2(x_k) + \cdots + a_n f_n(x_k)\big)\Big]^2$$

$$= \sum_{k=1}^{m} \Big[y_k - \sum_{r=1}^{n} a_r f_r(x_k)\Big]^2.$$

Now we can take the partial derivative of the above expression with respect to an arbitrary coefficient a_j as follows:

$$\frac{\partial D}{\partial a_j} = \frac{\partial}{\partial a_j} \sum_{k=1}^{m} \Big[y_k - \sum_{r=1}^{n} a_r f_r(x_k)\Big]^2$$

$$= \sum_{k=1}^{m} \frac{\partial}{\partial a_j} \Big[y_k - \sum_{r=1}^{n} a_r f_r(x_k)\Big]^2$$

$$= \sum_{k=1}^{m} \Big\{-2 f_j(x_k)\Big[y_k - \sum_{r=1}^{n} a_r f_r(x_k)\Big]\Big\}.$$

Setting $\nabla D = 0$ is equivalent to setting each component to zero. We must now think once again of our main goal. We would like to show that the pseudoinverse solution corresponds to the minimum of the squared deviation. In order to do this, we are going to rewrite $\dfrac{\partial D}{\partial a_j} = 0$ as follows:

$$\sum_{k=1}^{m} \sum_{r=1}^{n} a_r f_j(x_k) f_r(x_k) = \sum_{k=1}^{m} f_j(x_k) y_k$$

To arrive at the above expression, we simply divided both sides of $\dfrac{\partial D}{\partial a_j} = 0$ by -2 and moved the negative portion of the sum to the other side of the equal sign. We now have n of these equations, and thus end up with the very complicated looking system of equations:

$$\sum_{k=1}^{m}\sum_{r=1}^{n} a_r f_1(x_k) f_r(x_k) = \sum_{k=1}^{m} f_1(x_k) y_k$$

$$\sum_{k=1}^{m}\sum_{r=1}^{n} a_r f_2(x_k) f_r(x_k) = \sum_{k=1}^{m} f_2(x_k) y_k \qquad (11.13)$$

$$\vdots \qquad\qquad \vdots$$

$$\sum_{k=1}^{m}\sum_{r=1}^{n} a_r f_n(x_k) f_r(x_k) = \sum_{k=1}^{m} f_n(x_k) y_k.$$

Now remember, the above system of equations is in terms of the unknown constants $a_1, a_2, \ldots, a_n$, and a solution to this system corresponds to a minimum of the square deviation. Our goal now is to rewrite the above system as a giant equation. Let us first focus on the RHS of (11.13), which can be rewritten as:

$$
\begin{bmatrix}
\sum_{k=1}^{m} f_1(x_k) y_k \\
\sum_{k=1}^{m} f_2(x_k) y_k \\
\vdots \\
\sum_{k=1}^{m} f_n(x_k) y_k
\end{bmatrix}
=
\begin{bmatrix}
f_1(x_1) & f_1(x_2) & \cdots & f_1(x_m) \\
f_2(x_1) & f_2(x_2) & \cdots & f_2(x_m) \\
\vdots & & \ddots & \vdots \\
f_n(x_1) & f_n(x_2) & \cdots & f_n(x_m)
\end{bmatrix}
\begin{bmatrix}
y_1 \\
y_2 \\
\vdots \\
y_m
\end{bmatrix}.
$$

We have just shown that the RHS of (11.13), in matrix form, can be represented by $F^T \vec{y}$, what an observation! At this point, we should pause and think back to our pseudoinverse approach to solving nonsquare systems. The first step in this process was to make the system square. In our current case, the nonsquare equation $F \vec{a} = \vec{y}$ is made square by multiplying both sides, on the left, by F^T. Thus our square system is $F^T F \vec{a} = F^T \vec{y}$. It is now our hope that the following matrix equation, corresponding to the LHS of (11.13), is true:

$$
\begin{bmatrix}
\sum_{k=1}^{m}\sum_{r=1}^{n} a_r f_1(x_k) f_r(x_k) \\
\sum_{k=1}^{m}\sum_{r=1}^{n} a_r f_2(x_k) f_r(x_k) \\
\vdots \\
\sum_{k=1}^{m}\sum_{r=1}^{n} a_r f_n(x_k) f_r(x_k)
\end{bmatrix}
=
\begin{bmatrix}
f_1(x_1) & f_1(x_2) & \cdots & f_1(x_m) \\
f_2(x_1) & f_2(x_2) & \cdots & f_2(x_m) \\
\vdots & & \ddots & \vdots \\
f_n(x_1) & f_n(x_2) & \cdots & f_n(x_m)
\end{bmatrix}
$$

$$
\times
\begin{bmatrix}
f_1(x_1) & f_2(x_1) & \cdots & f_n(x_1) \\
f_1(x_2) & f_2(x_2) & \cdots & f_n(x_2) \\
\vdots & & \ddots & \vdots \\
f_1(x_m) & f_2(x_m) & \cdots & f_n(x_m)
\end{bmatrix}
\begin{bmatrix}
a_1 \\
a_2 \\
\vdots \\
a_n
\end{bmatrix}
$$

It should only take a few minutes to realize that the above equation is indeed true. Now we need to ask, What does this mean? Once again, system (11.13) corresponds to the equations that must be satisfied if $\vec{a}$ is a minimum for D. Now we have shown that this system can be expressed in matrix form as $F^T F \vec{a} = F \vec{y}$. The solution to this matrix equation is $\vec{a} = p(F) \vec{y}$, which is the pseudoinverse solution to the nonsquare system $F \vec{a} = \vec{y}$. Therefore, the pseudoinverse method yields a solution corresponding to a minumum of the least-square deviation, which is why this approach is sometimes referred to as the method of least-squares. Note that this approach works as it does due to the fact that the gradient method results in a linear system of equations with respect to the unknown a_k's. The reason that the gradient method gives us this linear system is due to the way in which equation (11.9) is written. If a different combination was chosen such that the a_k's did not appear linearly, system (11.13) will not be the end result of the gradient method, thus negating the relationship between the pseudoinverse method and the least-square deviation.

The most common application of the method of least-squared deviation is the linear case. If our function is given by $y = a_1 x + a_0$, then notice that if we have more than two points in our data set, a perfect fit will be highly unlikely. Following the work done in the arbitrary case, we will consider the case of m points in our data set: $\{(x_1, y_1), (x_2, y_2), \ldots, (x_m, y_m)\}$. The mean-square deviation D will be a function of the variables a_0 and a_1 in this instance, with

$$
D(a_0, a_1) = |\vec{y} - F\vec{a}|^2
$$

$$
= \sum_{k=1}^{m} (y_k - (a_1 x_k + a_0))^2
$$

where $\vec{y}$ is the column of y-coordinates, F is the $m \times 2$ matrix whose rows are $[x_j, 1]$, for $1 \leq k \leq m$ and $\vec{a} = \begin{bmatrix} a_0 \\ a_1 \end{bmatrix}$. Then, $D(a_0, a_1) = 0$ if and only if $y_k = a_1 x_k + a_0$ for each $1 \leq k \leq m$, which is equivalent to stating that the overdetermined system of linear equations $y_k = a_1 x_k + a_0$ is satisfied for all k. As previously pointed out, this is highly unlikely. The best choice of the values of a_0 and a_1 would be those that give the least-squared deviation $D(a_0, a_1)$. Following the more general procedure, we have to solve $\nabla D = 0$, which is equivalent to simultaneously satisfying the two equations $\dfrac{\partial D}{\partial a_0} = 0$, and $\dfrac{\partial D}{\partial a_1} = 0$, where

$$
\begin{aligned}
\frac{\partial D}{\partial a_0} &= \sum_{k=1}^{m} -2\left(y_k - (a_1 x_k + a_0)\right) \\
\frac{\partial D}{\partial a_1} &= \sum_{k=1}^{m} -2 x_k \left(y_k - (a_1 x_k + a_0)\right)
\end{aligned}
\tag{11.14}
$$

Setting both of these partial derivatives to zero and performing some algebraic manipulation, we arrive at the following system of equations:

$$
\begin{aligned}
a_0 \sum_{k=1}^{m} 1 + a_1 \sum_{k=1}^{m} x_k &= \sum_{k=1}^{m} y_k \\
a_0 \sum_{k=1}^{m} x_k + a_1 \sum_{k=1}^{m} x_k^2 &= \sum_{k=1}^{m} x_k y_k
\end{aligned}
\tag{11.15}
$$

You may have seen the above system of equations before, as they are commonly used in statistics when dealing with the topic of *linear regression*.

Example 11.3.1. Now we will have *Maple* compute the solution to the above system, along with the psuedoinverse solution, to Example 11.2.1 from Section 11.2. In the bookstore example, the number of points was $m = 12$.

> with(linalg):

> S:= [[415, 13.25], [372, 13.85], [391, 13.60], [428, 13.15], [403, 13.65], [350, 14.05], [362, 13.90], [410, 13.20], [385, 13.60], [434, 13.00], [465, 12.75], [380, 13.65]]:

> sumx:= add(S[j][1], j=1 .. 12);

$$sumx := 4795$$

> sumy:= add(S[j][2], j=1 .. 12);

$$sumy := 161.65$$

> sumxy:= add(S[j][1]*S[j][2], j=1 .. 12);

$$sumxy := 64452.75$$

> sumxsqrt:= add(S[j][1]^2, j=1 .. 12);

$$sumxsqrt := 1927933$$

> Coeffs:= matrix([[sumxsqrt, sumx], [sumx, 12]]);

$$Coeffs := \begin{bmatrix} 1927933 & 4795 \\ 4795 & 12 \end{bmatrix}$$

> RHS:= matrix([[sumxy], [sumy]]);

$$RHS := \begin{bmatrix} 64452.75 \\ 161.65 \end{bmatrix}$$

> evalm(inverse(Coeffs)&*RHS);

$$\begin{bmatrix} -0.011725489 \\ 18.156143 \end{bmatrix}$$

This says that the least-squares solution for the line of best fit to this bookstore's data is

$$p(x) = -.011725489x + 18.156143$$

Comparing this to the solution given in equation (11.6), we see that the least-squares solution is the same as the pseudoinverse approximate solution to the overdetermined system.

■

If the above approach, and subsequent linear example, is too complicated or confusing in showing that the pseudoinverse solution corresponds to the solution that minimizes the least-squared deviation, then we will try one last approach. Sometimes a geometric representation of a problem can yield a better understanding of how it can be solved. To make this approach as simple as possible, we start with the matrix $A\vec{x} = \vec{b}$. Here, $A \in \mathbb{R}^{m \times n}$, $\vec{x} \in \mathbb{R}^n$ and $\vec{b} \in \mathbb{R}^m$. To make the matrix equation correspond to an overdetermined system, we once again make the assumption that $m > n$. Let $T : \mathbb{R}^n \to \mathbb{R}^m$ be the linear map with standard matrix A representing it, where the rank of the matrix A is n. The rank of the matrix A is n if and only if the n columns of A form an independent set in $\mathbb{R}^m$. These n columns of the matrix A form a basis $\{\vec{w_1}, \vec{w_2}, \ldots, \vec{w_n}\}$ for a subspace $\mathbb{S}$ of $\mathbb{R}^m$. This subspace $\mathbb{S}$ is the image of T, which means

$$\mathbb{S} = \text{Im}(T) = \{A\vec{x} \mid \vec{x} \in \mathbb{R}^n\}$$

If $\overrightarrow{b} \notin \mathbb{S}$, then no solution to the original matrix equation exists. Now the squared variation $D(\overrightarrow{x})$ is the distance squared from $\overrightarrow{b}$ to an element $A\overrightarrow{x}$ of the subspace $\mathbb{S} = \text{Im}(T)$. So $D(\overrightarrow{x}) = \left| \overrightarrow{b} - A\overrightarrow{x} \right|^2$ for all $\overrightarrow{x} \in \mathbb{R}^n$. The question then becomes, what vector in $\mathbb{S}$ lies closest to $\overrightarrow{b}$? The smallest distance between $\overrightarrow{b}$ and $\mathbb{S}$ occurs when $A\overrightarrow{x} = \text{proj}_{\mathbb{S}}(\overrightarrow{b})$, since $\overrightarrow{b} - \text{proj}_{\mathbb{S}}(\overrightarrow{b})$ is orthogonal to all of $\mathbb{S}$. In Section 8.4, we constructed a formula for the projection of a vector onto an entire subspace:

$$\text{proj}_{\mathbb{S}}(\overrightarrow{b}) = A\left[(A^T A)^{-1} A^T\right] \overrightarrow{b} \tag{11.16}$$

What this means, is that since a solution to $A\overrightarrow{x} = \overrightarrow{b}$ can only be satisfied if $\overrightarrow{b} \in \mathbb{S}$, the next best thing, is to solve the equation

$$A\overrightarrow{x} = \text{proj}_{\mathbb{S}}(\overrightarrow{b}) \tag{11.17}$$

instead. Clearly the above equation has a solution, and this solution $\overrightarrow{x}$ gives the least-squared deviation to the original problem. So substituting (11.16) into (11.17) gives the following equation:

$$A\overrightarrow{x} = A\left[(A^T A)^{-1} A^T\right] \overrightarrow{b}$$

Clearly, the pseudoinverse solution $\overrightarrow{x} = p(A)\overrightarrow{b}$ satisfies this equation, since $p(A) = (A^T A)^{-1} A^T$. Thus, the least-squares solution $\overrightarrow{x}$ is the same as the pseudoinverse solution $p(A)\overrightarrow{b}$.

Homework Problems

1. Let $\{(x_1, y_1), (x_2, y_2), \ldots, (x_m, y_m)\}$ be a data set of m points. Let $\overline{x}$ and $\overline{y}$ be the averages, respectively, of the x-coordinates and y-coordinates from this data set, and

$$\overrightarrow{x} = \langle x_1, x_2, \ldots, x_m \rangle, \quad \overrightarrow{y} = \langle y_1, y_2, \ldots, y_m \rangle$$

be the two vectors formed from these coordinates.

(a) Show that the equation of the least-squared line of best fit to this data set is $y = a_0 + a_1 x$, where

$$a_0 = \frac{\overline{y}\,|\overrightarrow{x}|^2 - \overline{x}\,\overrightarrow{x} \cdot \overrightarrow{y}}{|\overrightarrow{x}|^2 - m\overline{x}^2}, \quad a_1 = \frac{\overrightarrow{x} \cdot \overrightarrow{y} - m\overline{x}\,\overline{y}}{|\overrightarrow{x}|^2 - m\overline{x}^2}$$

Hint: Cramer's rule.

(b) Show that if $\bar{x} = 0$, then

$$a_0 = \bar{y}, \quad a_1 = \frac{\vec{x} \cdot \vec{y}}{|\vec{x}|^2}$$

(c) Using part (a), show that the least-squared error in this fit is

$$D = \sum_{k=1}^{m} (y_k - a_0 - a_1 x_k)^2 = |\vec{y}|^2 + a_1^2 |\vec{x}|^2 + 2a_0 m(a_1 \bar{x} - \bar{y}) - 2a_1 \vec{x} \cdot \vec{y} + ma_0^2$$

Does this simplify any farther if you replace a_0 and a_1 by their formulas from part (a)?

(d) Use the formulas above to find the equation of the least-squared line of best fit and the least-squared error in this fit for the data set

$$\{(-3, 11), (2, 4), (5, -1), (9, -7)\}$$

Plot this least-squared line of best fit with the data set.

2. Let $\{(x_1, y_1, z_1), (x_2, y_2, z_2), \ldots, (x_m, y_m, z_m)\}$ be a data set of m points. Let $\bar{x}$, $\bar{y}$, and $\bar{z}$ be the averages, respectively, of their x, y, and z-coordinates from this data set. Let

$$\vec{x} = \langle x_1, x_2, \ldots, x_m \rangle, \quad \vec{y} = \langle y_1, y_2, \ldots, y_m \rangle, \quad \vec{z} = \langle z_1, z_2, \ldots, z_m \rangle$$

be the three vectors formed from these coordinates.

(a) Find the equation $z = a_0 + a_1 x + a_2 y$ of the least-squared plane of best fit to this data set and formulas similar to those in problem 1 for a_0, a_1, and a_2. *Hint: Cramer's rule.*

(b) Find a formula for the least-squared error D in this fit similar to that of problem 1.

(c) Use the formulas of parts (a) and (b) to find the equation of the least-squared plane of best fit and the least-squared error in this fit for the data set

$$\{(-3, 11, 6), (2, 4, -1), (5, -1, -2), (9, -7, 8), (-1, -11, -4)\}$$

3. Thus far, we have not fit datapoints to a circle of radius r and center (a, b), given by the expression $(x - a)^2 + (y - b)^2 = r^2$. The complication in trying to solve this problem is that the unknowns a and b are not linear: $x^2 - 2ax + a^2 + y^2 - 2yb + b^2 = r^2$. However, if we rewrite this expression as

$$2ax + 2yb + c = x^2 + y^2$$

where $c = r^2 - a^2 - b^2$, then one can solve the system for a, b, and c. Explain how this new linear system of equations can be used to solve for the third unknown r from the original equation for the circle.

4. Determine if a similar method to that of problem 3 can be applied to construct a system of linear equations for an ellipse of the form

$$\frac{(x-a)^2}{A^2} + \frac{(y-b)^2}{B^2} = 1$$

for unknown constants a, b, A, and B.

Maple Problems

1. Use *Homework* problem 3 to find the circle of best fit to the given datapoints
(a)   (b)   (c)

x	y		x	y		x	y
1.0	0.75		-4.7	-1.54		-3.9	2.72
1.4	0.92		-4.4	-3.11		-3.4	0.22
2.0	1.03		-4.1	-0.68		-2.7	4.51
2.6	0.93		-3.9	-3.38		-1.9	-0.88
0.2	-1.82		-3.5	-3.42		-1.1	-1.01
0.8	-2.61		-2.9	-0.24		-0.3	4.93
1.7	-2.93		-2.5	-3.61		0.1	-0.76
3.2	-2.67		-2.1	-3.43		0.7	-0.45
2.3	-2.89		-1.8	-0.74		1.1	4.13
3.6	0.23		-1.7	-3.12		1.8	0.97

2. The *Maple* commands to carry out various kinds of least-squared fits are in the *Maple* package *stats*, and it is given by "fit[leastsquare[[x, y], y = a x + b, {a, b}]]([xData, yData])" in order to do least-squared lines $y = ax + b$ of best fit to a data set where all of the x-coordinates are put in the list *xData* while all of the y-coordinates are put in the list *yData* in the same order as their corresponding x-coordinates.

   (a) Use this *Maple* command to find the least-squared line of best fit to the data in *Homework* problem 1. Plot this data with its line of best fit. What is the least-squared error D in this fit?

   (b) Use this *Maple* command to find the least-squared line of best fit to the data set

$$\left\{(4217, 13.72), (3825, 14.03), (4106, 13.89), (4391, 13.44), (3937, 13.95), (4569, 13.15)\right\}.$$

Plot this data with its line of best fit. What is the least-squared error D in this fit?

(c) If the data in part (b) is six consecutive months worth of demand data for a bookstore, where x is the number of books sold per month and y is their average sale price, then how many books will the store sell in a month where its average sale price for the month has been set at $13.50?

3. The *Maple* commands to carry out various kinds of least-squared fits are in the *Maple* package *stats*. For example, in order to do least-squared planes $z = ax + by + c$ of best fit to a data set, we use the command "fit[leastsquare[[x, y, z], z = a x + b y + c, {a, b, c}]]([xData, yData, zData])", where all of the x-coordinates are put in the list *xData*, all of the y-coordinates are put in the list *yData*, and all of the z-coordinates are put in the list *zData*, in the same order as the points in the data set.

(a) Find the equation of the least-squared plane of best fit and the least-squared error in this fit for the data set from *Homework* problem 2 part (c).

(b) Plot this least-squared plane of best fit with this data set.

4. (This problem is related to the *Prime Number Theorem*.) Let x be a positive real variable and the function $\pi(x)$ be the number of primes $\leq x$.

(a) Find the values of $\pi(x)$ and the datapoints $\left(x, \dfrac{x}{\pi(x)}\right)$, for $x = 10^k$ where $k = 6, 7, \ldots, 21$ using the *Maple* package *numtheory* with the function pi(x) for $\pi(x)$.

(b) Plot this data with the equation of the least-squared line of best fit to this data.

(c) Find the least-squared error D in this fit.

(d) Instead of a least-squared line of best fit, use instead the least-squared logarithm of best fit $y = a + b\ln(x)$ to this data. Plot this least-squared logarithm of best fit with the data, and find its least-squared error D.

5. Let $f(x) = \ln\left(\displaystyle\sum_{k=1}^{50} k^x\right)$, where x is a real variable. The reason for using the logarithm is because the values of these sums become very large as x increases and the logarithm will return much smaller more reasonable values to approximate by a best fit.

(a) Plot the function $f(x)$ for $x \in [-5, 10]$. Is there a part(s) of this graph which seems linear?

(b) Construct the data set of the points on the graph of this function for $x \in [0, 10]$ at half units of x. Now find the least-squared line of best fit to this data set as well as its least-squared error D.

(c) Plot this line of best fit with the original function $f(x)$ for $x \in [0, 10]$.

6. Let $f(x) = \sum_{k=1}^{50} k^x$, where x is a real variable.

(a) Plot the function $f(x)$ for $x \in [-10, 0]$. Is there a part(s) of this graph that seems like $\frac{1}{x^2}$?

(b) Construct the data set of the points on the graph of this function for $x \in [-10, -2]$ at half units of x. Now find the least-squared linear combination

$$y = a + \frac{b}{x^2} + \frac{c}{x^4} + \frac{d}{x^6} + \frac{e}{x^8} + \frac{g}{x^{10}}$$

of best fit to this data set as well as its least-squared error D.

(c) Plot this least-squared linear combination of best fit with the original function $f(x)$ for $x \in [-10, -2]$.

7. Find some reasonably accurate world population data from the web by specific years.

(a) Now find a least-squared linear combination of best fit to this data. You might want to plot this data first to have some idea of what sort of functions to use in your linear combination.

(b) Plot this data with your least-squared linear combination of best fit and give its least-squared error D.

Chapter 12

Eigenvalues and Eigenvectors

12.1 What Are Eigenvalues and Eigenvectors, and Why Do We Need Them?

The answer to both of the questions posed in the title above lies in the answer to the following simple question: What is the nicest general type of square matrix for doing arithmetic with matrices, in particular, multiplication and finding inverses? The nicest general type of square matrix is the diagonal matrix, such as the identity matrices. The reason is that when you multiply two diagonal matrices of the same size you merely need to multiply corresponding entries (the order of multiplication will not matter for diagonal matrices) and when you invert a diagonal matrix you only need take the reciprocals of the diagonal entries assuming all of them are nonzero. As well, the determinant of a diagonal matrix is the product of its diagonal entries.

Let us do a few examples to see that this is all true. In the following example, and all that follow it in this chapter, we use *Diag* instead of *D* to denote a diagonal matrix *D* since *Maple* uses the letter *D* as a reserved symbol for the differential operator $\dfrac{d}{dx}$, and so we can not use *D* to name or define anything in *Maple*.

Example 12.1.1. In this first example, we will compute with *Maple* the following products, powers, inverses, and roots of diagonal matrices to see that the result is also a diagonal matrix. This example will show you the extreme simplicity of doing matrix arithmetic with diagonal matrices.

> with(linalg):

> A:= matrix(3,3,[5,0,0,0,-9,0,0,0,7]);

$$A := \begin{bmatrix} 5 & 0 & 0 \\ 0 & -9 & 0 \\ 0 & 0 & 7 \end{bmatrix}$$

> B:= matrix(3,3,[Pi,0,0,0,4/11,0,0,0,-3]);

$$B := \begin{bmatrix} \pi & 0 & 0 \\ 0 & \frac{4}{11} & 0 \\ 0 & 0 & -3 \end{bmatrix}$$

> evalm(A&*B);

$$\begin{bmatrix} 5\pi & 0 & 0 \\ 0 & -\frac{36}{11} & 0 \\ 0 & 0 & -21 \end{bmatrix}$$

> evalm(B&*A);

$$\begin{bmatrix} 5\pi & 0 & 0 \\ 0 & -\frac{36}{11} & 0 \\ 0 & 0 & -21 \end{bmatrix}$$

> evalm(A^(-1));

$$\begin{bmatrix} \frac{1}{5} & 0 & 0 \\ 0 & -\frac{1}{9} & 0 \\ 0 & 0 & \frac{1}{7} \end{bmatrix}$$

> evalm(B^(-1));

$$\begin{bmatrix} \frac{1}{\pi} & 0 & 0 \\ 0 & \frac{11}{4} & 0 \\ 0 & 0 & -\frac{1}{3} \end{bmatrix}$$

> evalm(A^5);

$$\begin{bmatrix} 3125 & 0 & 0 \\ 0 & -59049 & 0 \\ 0 & 0 & 16807 \end{bmatrix}$$

> matrix(3,3,[5^5,0,0,0,(-9)^5,0,0,0,7^5]);

$$\begin{bmatrix} 3125 & 0 & 0 \\ 0 & -59049 & 0 \\ 0 & 0 & 16807 \end{bmatrix}$$

> Diag:= matrix(3,3,[a,0,0,0,b,0,0,0,c]);

$$Diag := \begin{bmatrix} a & 0 & 0 \\ 0 & b & 0 \\ 0 & 0 & c \end{bmatrix}$$

> evalm(Diag^10);

$$
\begin{bmatrix}
a^{10} & 0 & 0 \\
0 & b^{10} & 0 \\
0 & 0 & c^{10}
\end{bmatrix}
$$

> G:= matrix(3,3,[10,0,0,0,100,0,0,0,1000]);

$$
G := \begin{bmatrix}
10 & 0 & 0 \\
0 & 100 & 0 \\
0 & 0 & 1000
\end{bmatrix}
$$

> fourthrootG:= matrix(3,3, [10^.25,0,0,0,100^.25,0,0,0,1000^.25]);

$$
fourthrootG := \begin{bmatrix}
1.778279410 & 0 & 0 \\
0 & 3.162277660 & 0 \\
0 & 0 & 5.623413252
\end{bmatrix}
$$

> evalm(fourthrootG^4);

$$
\begin{bmatrix}
9.999999999 & 0 & 0 \\
0 & 99.99999998 & 0 \\
0 & 0 & 999.9999999
\end{bmatrix}
$$

From Example 12.1.1, we see that all the statements made, about how nice diagonal matrices are, were correct. If only all square matrices were diagonal, then all of our work with matrices would be so much easier! Specifically, if we needed to take the power of a square matrix or even the exponential of a square matrix we could just move the operation in question to the diagonal entries to get our result. We refer back to the *Homework Problems* of Section 5.1 for more information on diagonal matrix properties.

Example 12.1.2. Next, we look at the exponential e^A, starting with a 3×3 diagonal matrix

$$
A = \begin{bmatrix}
a & 0 & 0 \\
0 & b & 0 \\
0 & 0 & c
\end{bmatrix}
\tag{12.1}
$$

First, we need to define what we mean by e^A. The exponential function, in terms of the scalar variable x, can be defined by the Maclaurin series

$$
e^x = \sum_{k=0}^{\infty} \frac{1}{k!} x^k
\tag{12.2}
$$

which converges for all $x \in \mathbb{C}$. There should be no reason why the variable x cannot be replaced by a matrix A. In particular, if we use the definition of A in equation (12.1) with equation (12.2), we get

$$e^A = \sum_{k=0}^{\infty} \frac{1}{k!} A^k \tag{12.3}$$

However, from Example 12.1.1, we know that

$$A^k = \begin{bmatrix} a^k & 0 & 0 \\ 0 & b^k & 0 \\ 0 & 0 & c^k \end{bmatrix}$$

so that (12.3) can now be expressed as follows:

$$
\begin{aligned}
e^A &= \sum_{k=0}^{\infty} \frac{1}{k!} A^k \\[2mm]
&= \sum_{k=0}^{\infty} \frac{1}{k!} \begin{bmatrix} a^k & 0 & 0 \\ 0 & b^k & 0 \\ 0 & 0 & c^k \end{bmatrix} \\[2mm]
&= \sum_{k=0}^{\infty} \begin{bmatrix} \frac{1}{k!} a^k & 0 & 0 \\ 0 & \frac{1}{k!} b^k & 0 \\ 0 & 0 & \frac{1}{k!} c^k \end{bmatrix} \\[2mm]
&= \begin{bmatrix} \sum_{k=0}^{\infty} \frac{1}{k!} a^k & 0 & 0 \\ 0 & \sum_{k=0}^{\infty} \frac{1}{k!} b^k & 0 \\ 0 & 0 & \sum_{k=0}^{\infty} \frac{1}{k!} c^k \end{bmatrix} \\[2mm]
&= \begin{bmatrix} e^a & 0 & 0 \\ 0 & e^b & 0 \\ 0 & 0 & e^c \end{bmatrix}
\end{aligned}
$$

So the exponential operation applied to a diagonal matrix can be moved to its diagonal entries, that is,

$$A = \begin{bmatrix} a & 0 & 0 \\ 0 & b & 0 \\ 0 & 0 & c \end{bmatrix} \rightarrow e^A = \begin{bmatrix} e^a & 0 & 0 \\ 0 & e^b & 0 \\ 0 & 0 & e^c \end{bmatrix} \tag{12.4}$$

Clearly, the prior argument shows that this is independent of the square dimension of A, and can be extended to diagonal matrices of $\mathbb{C}^{n \times n}$. Recall that the exponential e^{a+bi} of a complex number $a + bi$ is defined as

$$e^{a+bi} = e^a \left(\cos(b) + i \sin(b) \right) \tag{12.5}$$

$\blacksquare$

Our next question is: How can we do this for a general square matrix that is not diagonal? From equation (12.3), we see that it is necessary to first find an easy way to compute the powers of A by relating A to some diagonal matrix D. The beginning of our answer lies in the following interesting fact about taking the power of a certain type of product of square matrices. Namely, if D is any square matrix and Q is any invertible square matrix of the same size as D, then for any positive integer n we have

$$\left(QDQ^{-1} \right)^n = QD^nQ^{-1} \tag{12.6}$$

A theorem and proof of this fact follows next. Note that since matrix multiplication is not typically commutative that QDQ^{-1} is not generally D. Two matrices D and QDQ^{-1} are called *similar matrices* since they both have the same determinant and other properties in common, think of this as analogous to similar triangles.

Theorem 12.1.1. *For any square matrix D, and any invertible square matrix Q of the same size as D, we have for any positive integer n that*

$$\left(QDQ^{-1} \right)^n = QD^nQ^{-1}$$

Proof. In order to see that this formula is true, all we need to do is expand out completely $\left(QDQ^{-1} \right)^n$. For simplicity, let us take $n = 3$ so that you get the idea without the mess. One can then apply induction to show it is true for arbitrary n.

$$\begin{aligned} \left(QDQ^{-1} \right)^3 &= \left(QDQ^{-1} \right) \left(QDQ^{-1} \right) \left(QDQ^{-1} \right) \\ &= QD \left(Q^{-1}Q \right) D \left(Q^{-1}Q \right) DQ^{-1} \\ &= QD \left(I \right) D \left(I \right) DQ^{-1} \\ &= QDDDQ^{-1} \\ &= QD^3Q^{-1} \end{aligned}$$

If $n > 3$, then writing out this product will give $n - 1$ products $Q^{-1}Q$, which all disappear to leave D^n between Q and Q^{-1}. $\qquad \square$

This tells us that if we have a square matrix A and we can find two square matrices D and Q of the same size as A with Q invertible and D diagonal where $A = QDQ^{-1}$, then $A^n = QD^nQ^{-1}$, where we know how to compute D^n, since D is diagonal. So now we know how to compute A^n if only we can find matrices D and Q with D diagonal and Q invertible.

How do we find two square matrices D and Q of the same size as A with Q invertible and D diagonal where $A = QDQ^{-1}$? If both Q and D exist for a given square matrix A so that $A = QDQ^{-1}$, then we say that A is diagonalizable. Not every square matrix A is diagonalizable, but almost all are.

But to begin, this equation $A = QDQ^{-1}$ can be expressed as $AQ = QD$ instead, simply by right multiplying both sides of the equation by Q. This tells us what QD looks like for D a diagonal matrix. All we need is an example to see what happens for both of these products, this we will do in Example 12.1.3.

Example 12.1.3. Consider the following matrices:

$$D = \begin{bmatrix} a & 0 & 0 \\ 0 & b & 0 \\ 0 & 0 & c \end{bmatrix}, \ Q = \begin{bmatrix} d & e & f \\ g & h & i \\ j & k & l \end{bmatrix}, \ A = \begin{bmatrix} 1 & 2 & 3 \\ 4 & 5 & 6 \\ 7 & 8 & 9 \end{bmatrix}$$

> Diag_D:= matrix(3,3,[a,0,0,0,b,0,0,0,c]):
> Q:= matrix(3,3,[d,e,f,g,h,i,j,k,l]):
> evalm(Q&*Diag_D);

$$\begin{bmatrix} da & eb & fc \\ ga & hb & ic \\ ja & kb & lc \end{bmatrix}$$

> A:= matrix(3,3,[1,2,3,4,5,6,7,8,9]);

$$A := \begin{bmatrix} 1 & 2 & 3 \\ 4 & 5 & 6 \\ 7 & 8 & 9 \end{bmatrix}$$

> evalm(A&*Q);

$$\begin{bmatrix} d + 2g + 3j & e + 2h + 3k & f + 2i + 3l \\ 4d + 5g + 6j & 4e + 5h + 6k & 4f + 5i + 6l \\ 7d + 8g + 9j & 7e + 8h + 9k & 7f + 8i + 9l \end{bmatrix}$$

> Q_1stcolum:= matrix(3,1,[d,g,j]);

$$Q\_1stcolum := \begin{bmatrix} d \\ g \\ j \end{bmatrix}$$

> evalm(A&*Q_1stcolum);

$$\begin{bmatrix} d + 2g + 3j \\ 4d + 5g + 6j \\ 7d + 8g + 9j \end{bmatrix}$$

> Q_2ndcolum:= matrix(3,1,[e,h,k]);

$$Q\_2ndcolum := \begin{bmatrix} e \\ h \\ k \end{bmatrix}$$

> evalm(A&*Q_2ndcolum);

$$\begin{bmatrix} e + 2h + 3k \\ 4e + 5h + 6k \\ 7e + 8h + 9k \end{bmatrix}$$

> Q_3rdcolum:= matrix(3,1,[f,i,l]);

$$Q\_3rdcolum := \begin{bmatrix} f \\ i \\ l \end{bmatrix}$$

> evalm(A&*Q_3rdcolum);

$$\begin{bmatrix} f + 2i + 3l \\ 4f + 5i + 6l \\ 7f + 8i + 9l \end{bmatrix}$$

So QD, for D a diagonal matrix, is the matrix Q, where each column of Q is multiplied by the corresponding entry of D.

As well, the first column of the product AQ is A times the first column of Q by the way matrix multiplication works and what we see above. Similarly, the second (or third) column of the product AQ is A times the second (or third) column of Q.

In conclusion, we have now learned that $AQ = QD$ is really a system of three equations concerning the columns of Q. It really says that A times the kth column of Q equals the kth entry on the diagonal of D times the kth column of Q for $k = 1, 2, 3$. In other words, we have the three matrix equations

$$A \begin{bmatrix} d \\ g \\ j \end{bmatrix} = a \begin{bmatrix} d \\ g \\ j \end{bmatrix}, \quad A \begin{bmatrix} e \\ h \\ k \end{bmatrix} = b \begin{bmatrix} e \\ h \\ k \end{bmatrix}, \quad A \begin{bmatrix} f \\ i \\ l \end{bmatrix} = c \begin{bmatrix} f \\ i \\ l \end{bmatrix}$$

which we can conveniently express as

$$(A - aI_3) \begin{bmatrix} d \\ g \\ j \end{bmatrix} = \begin{bmatrix} 0 \\ 0 \\ 0 \end{bmatrix}, \quad (A - bI_3) \begin{bmatrix} e \\ h \\ k \end{bmatrix} = \begin{bmatrix} 0 \\ 0 \\ 0 \end{bmatrix}, \quad (A - cI_3) \begin{bmatrix} f \\ i \\ l \end{bmatrix} = \begin{bmatrix} 0 \\ 0 \\ 0 \end{bmatrix}$$

where I_3 is the 3×3 identity matrix. So to find the two matrices D and Q from the matrix A so that $AQ = QD$, we must find the numbers on the diagonal of

D and the columns of Q. Note that these three equations are really the same type of equation and of the form A times the first (second or third) column of Q equals the first (second or third) entry of D times the first (second or third) column of Q.

Note that Q must have an inverse and so no column of Q can be a column of all zeros since then the determinant of Q would be zero and Q would not have an inverse.

■

From Example 12.1.3, we see that a square $n \times n$ matrix A can be written as $A = QDQ^{-1}$ or equivalently $AQ = QD$ for a diagonal matrix D and an invertible matrix Q exactly when $AX_k = D_{k,k}X_k$, where $D_{k,k}$ is the kth entry on the diagonal of D and X_k is the kth column of Q. If we switch notation and let $\lambda_k = D_{k,k}$, then our equation is $AX_k = \lambda_k X_k$ or simply $AX = \lambda X$ if we drop the subscript, which is also the matrix equation $(A - I\lambda)X = 0$ where λ is a complex scalar variable while X is a nonzero column matrix variable in $\mathbb{C}^n$.

Now let λ be one of the diagonal entries of D and X be a column of Q. Then we want to solve the equation $AX = \lambda X$ for both X and λ, where λ goes on the diagonal of D in the same location that X goes into Q as a column. It might seem that we cannot solve for these two unknowns when we have only a single equation, but amazingly we can do it.

By the way, a value for λ is called an eigenvalue of A where the column X, which goes with this λ, is called an eigenvector for the eigenvalue λ. The German word *eigen* means characteristic.

In order to solve the equation $AX = \lambda X$, we need to rewrite it as $(A - \lambda I)X = 0$, where I is the identity matrix of the same size as A, D, and Q, and 0 is the zero column of the same size as X. Remember that X cannot be the zero column since it must go into Q, which has to have an inverse. For the matrix equation $(A - \lambda I)X = 0$ to have a nonzero solution X (think of λ as being fixed for now), the matrix $A - \lambda I$ cannot have an inverse, or equivalently $\det(A - \lambda I) = 0$. If $\det(A - \lambda I) \neq 0$, then $A - \lambda I$ has an inverse and the equation $(A - \lambda I)X = 0$ has as its only solution $X = (A - \lambda I)^{-1}0 = 0$, which is not possible for any column of Q.

Hence, we solve the equation $AX = \lambda X$ for X and λ by first solving the equation $\det(A - \lambda I) = 0$ for λ followed by solving the equation $(A - \lambda I)X = 0$ for a nonzero column X. The λ's will make up the diagonal entries of D while their corresponding X's go in as the columns of Q in the same location as where the λ's are placed. It is usual that for each λ we get 'essentially' a single solution for X.

Example 12.1.4. We will attempt to highlight some of the above concepts with a simple 2×2 matrix. Let $A = \begin{bmatrix} 12 & -2 \\ -7 & -1 \end{bmatrix}$, and we see if A is diagonalizable by looking for the diagonal matrix D and invertible matrix Q, where $A = QDQ^{-1}$. We must first find A's eigenvalues by solving $\det(A - \lambda I) = 0$ for λ. This equation will be a second-order polynomial equation in the variable λ, where the polynomial $P(\lambda) = \det(A - \lambda I)$ is called the *characteristic polynomial* of A, or sometimes the *eigenpolynomial* of A, the eigenvalues of A are the roots (or zeroes) of $P(\lambda)$. Then we find λ's nonzero eigenvector X by solving the matrix equation $(A - \lambda I)X = 0$.

> A:= matrix(2,2,[12,-2,-7,-1]);

$$A := \begin{bmatrix} 12 & -2 \\ -7 & -1 \end{bmatrix}$$

> AminusλI:= λ -> evalm(A - λ*matrix(2,2,[1,0,0,1])):
> eigenpoly:= det(AminusλI(λ));

$$-26 - 11\lambda + \lambda^2$$

> solve(eigenpoly=0,λ);

$$13, -2$$

> λ1:= 13: λ2:= -2:
> X:= matrix(2,1,[x, y]);

$$X := \begin{bmatrix} x \\ y \end{bmatrix}$$

> AminusλI(λ1);

$$\begin{bmatrix} -1 & -2 \\ -7 & -14 \end{bmatrix}$$

> evalm(AminusλI(λ1)&*X);

$$\begin{bmatrix} -x - 2y \\ -7x - 14y \end{bmatrix}$$

> solve({-x-2*y=0, -7*x-14*y=0}, {x, y});

$$\{x = -2y, y = y\}$$

> evalm(AminusλI(λ2)&*X);

$$\begin{bmatrix} 14x - 2y \\ -7x + y \end{bmatrix}$$

> solve({14*x-2*y=0, -7*x+y=0}, {x,y});

$$\{x = x, y = 7x\}$$

> Diag:= matrix(2,2,[λ1,0,0,λ2]);

$$Diag := \begin{bmatrix} 13 & 0 \\ 0 & -2 \end{bmatrix}$$

> Q:= matrix(2,2,[-2,1,1,7]);

$$Q := \begin{bmatrix} -2 & 1 \\ 1 & 7 \end{bmatrix}$$

> evalm(Q&*Diag&*Q^(-1));

$$\begin{bmatrix} 12 & -2 \\ -7 & -1 \end{bmatrix}$$

The eigenvalues of A have been found to be $\lambda_1 = 13$ and $\lambda_2 = -2$, which are the roots of the characteristic polynomial $P(\lambda)$ of A:

$$P(\lambda) = \det(A - \lambda I_2)$$
$$= \det\left(\begin{bmatrix} 12 - \lambda & -2 \\ -7 & -1 - \lambda \end{bmatrix}\right)$$
$$= \lambda^2 - 11\lambda - 26$$

The eigenvector X_1, for eigenvalue $\lambda_1 = 13$, is any nonzero solution to the equation $(A - \lambda_1)X = 0$, which is the same as the system

$$\begin{bmatrix} -1 & -2 \\ -7 & -14 \end{bmatrix} \begin{bmatrix} x \\ y \end{bmatrix} = \begin{bmatrix} 0 \\ 0 \end{bmatrix},$$

or

$$\{-x - 2y = 0, -7x - 14y = 0\}$$

The above system has as its solution

$$\begin{bmatrix} x \\ y \end{bmatrix} = y \begin{bmatrix} -2 \\ 1 \end{bmatrix}$$

which allows us to take $X_1 = \begin{bmatrix} -2 \\ 1 \end{bmatrix}$. The eigenvector X_2, corresponding to $\lambda_2 = -2$, is any nonzero solution to the equation $(A - \lambda_2)X = 0$, which is the same as the system

$$\begin{bmatrix} 14 & -2 \\ -7 & 1 \end{bmatrix} \begin{bmatrix} x \\ y \end{bmatrix} = \begin{bmatrix} 0 \\ 0 \end{bmatrix}$$

or

$$\{14x - 2y = 0, -7x + y = 0\}$$

This system has the following as it solution:

$$\begin{bmatrix} x \\ y \end{bmatrix} = x \begin{bmatrix} 1 \\ 7 \end{bmatrix}$$

which allows us to take $X_2 = \begin{bmatrix} 1 \\ 7 \end{bmatrix}$. Hence, we can take for the diagonal matrix D,

$$D = \begin{bmatrix} 13 & 0 \\ 0 & -2 \end{bmatrix}$$

and correspondingly Q,

$$Q = \begin{bmatrix} -2 & 1 \\ 1 & 7 \end{bmatrix}$$

The *Maple* code has checked that A is diagonalizable, that is, $A = QDQ^{-1}$.

Note that there are an infinite number of solutions for the eigenvectors X for each eigenvalue λ, we merely need to take one that is not zero as a column of Q. So our matrix A is diagonalizable. Let us now verify that $A^3 = QD^3Q^{-1}$ as we expect. Let us also see if $e^A = Qe^DQ^{-1}$, which we expect (and which will be proven in Section 12.2), since the exponentiation moves to D:

> evalm(A^3);

$$\begin{bmatrix} 2050 & -294 \\ -1029 & 139 \end{bmatrix}$$

> evalm(Q&*Diag^3&*Q^(-1));

$$\begin{bmatrix} 2050 & -294 \\ -1029 & 139 \end{bmatrix}$$

> evalf(evalm(sum(A^N/N!, N=0..25)));

$$\begin{bmatrix} 4.125202806 \ 10^5 & -58931.44933 \\ -2.062600727 \ 10^5 & 29465.86000 \end{bmatrix}$$

> eDiag:= matrix(2,2,[exp(λ1),0,0,exp(λ2)]);

$$eDiag := \begin{bmatrix} e^{13} & 0 \\ 0 & e^{-2} \end{bmatrix}$$

> evalf(evalm(Q&*eDiag&*Q^(-1)));

$$\begin{bmatrix} 4.129191749 \ 10^5 & -58988.43421 \\ -2.064595197 \ 10^5 & 29494.35244 \end{bmatrix}$$

In Section 12.2, we will discuss diagonalizability in more detail, as well as provide an example of a matrix A that is not diagonalizable. See *Homework problem 4* if you cannot wait until the next section.

$\blacksquare$

Homework Problems

1. Find the characteristic polynomial $P(\lambda)$, the eigenvalues, and their corresponding independent eigenvectors, for the matrix $A = \begin{bmatrix} 5 & -5 \\ 4 & -7 \end{bmatrix}$. If A is diagonalizable, then find the diagonal matrix D and the invertible matrix Q so that $A = QDQ^{-1}$. Also find e^A.

2. Find the characteristic polynomial $P(\lambda)$, the eigenvalues and their corresponding eigenvectors for the matrix $A = \begin{bmatrix} 1 & 3 & 5 \\ 0 & 4 & 2 \\ 0 & 0 & 7 \end{bmatrix}$. If A is diagonalizable, then find the diagonal matrix D and the invertible matrix Q so that $A = QDQ^{-1}$. Also find e^A.

3. Consider the following complex valued matrix:

$$A = \frac{1}{5} \begin{bmatrix} -1 + 3i & 6 + 2i \\ 14 - 12i & 21 + 7i \end{bmatrix}$$

(a) Compute the eigenvalues of A.

(b) Find the two complex matrices D and Q so that $A = QDQ^{-1}$, and so A is diagonalizable.

(c) Check that $A^2 = QD^2Q^{-1}$

(d) Check that A and D have the same determinant.

(e) Check that $A^{-1} = QD^{-1}Q^{-1}$.

4. Show that both of the following matrices are not diagonalizable.

(a) $\begin{bmatrix} 5 & 2 \\ 0 & 5 \end{bmatrix}$ (b) $\begin{bmatrix} 5 & 0 & 2 \\ 0 & 5 & 0 \\ 0 & 0 & 5 \end{bmatrix}$

5. Let $A = \begin{bmatrix} 1 & 3 & 5 \\ 0 & 4 & 2 \\ 0 & 0 & 7 \end{bmatrix}$. Find A^2, A^3, and A^4. Also, find A^{-2}, A^{-3}, and A^{-4}.

Is there anything you want to conjecture about the form of A^n for any integer n?

6. Let $A = \begin{bmatrix} 5 & 2 \\ 0 & 5 \end{bmatrix}$. Find a formula for A^n for any integer n. Now use it to find a formula for e^A.

7. Extend your last formula to a general exponential formula for e^A, where $A = \begin{bmatrix} \lambda & K \\ 0 & \lambda \end{bmatrix}$, with λ and K any real numbers with $K \neq 0$.

8. Find a formula for A^n for each of the following matrices, then use this to compute e^A.

$$\text{(a)} \begin{bmatrix} 5 & 2 & 0 \\ 0 & 5 & 0 \\ 0 & 0 & 5 \end{bmatrix} \quad \text{(b)} \begin{bmatrix} 5 & 0 & 2 \\ 0 & 5 & 0 \\ 0 & 0 & 5 \end{bmatrix} \quad \text{(c)} \begin{bmatrix} 5 & 0 & 0 \\ 0 & 5 & 2 \\ 0 & 0 & 5 \end{bmatrix}$$

9. Using your answers to problem 8, find a formula for e^A for each of the following matrices. Here, λ and K are any real numbers with $K \neq 0$.

$$\text{(a)} \begin{bmatrix} \lambda & K & 0 \\ 0 & \lambda & 0 \\ 0 & 0 & \lambda \end{bmatrix} \quad \text{(b)} \begin{bmatrix} \lambda & 0 & K \\ 0 & \lambda & 0 \\ 0 & 0 & \lambda \end{bmatrix} \quad \text{(c)} \begin{bmatrix} \lambda & 0 & 0 \\ 0 & \lambda & K \\ 0 & 0 & \lambda \end{bmatrix}$$

10. If A is a square upper or lower triangular matrix of any size $n \times n$, then is e^A also of the same type, upper or lower triangular of size $n \times n$? If you answer yes, then prove it is true. If you answer no, then provide an example to show this is not true.

11. Can you find a formula for e^A when A is either upper or lower triangular of any size $n \times n$ or just the special types mentioned in previous homework problems?

12. Let $A = \begin{bmatrix} a & b \\ c & d \end{bmatrix}$, for real constants a, b, c, and d. Show that A's characteristic polynomial is

$$P(\lambda) = \lambda^2 - \text{trace}(A)\lambda + \det(A)$$

where $\text{trace}(A) = \sum_{j=1}^{2} A_{j,j}$. In general, the *trace* of a square matrix A is simply the sum of A's diagonal entries.

13. What does A's characteristic polynomial look like if A is a general real

3×3 matrix $A = \begin{bmatrix} a & b & c \\ d & e & f \\ g & h & i \end{bmatrix}$ instead of the 2×2 one from problem 12?

14. Let $A = \begin{bmatrix} a & d & e \\ 0 & b & f \\ 0 & 0 & c \end{bmatrix}$ be a real upper triangular matrix. What is A's

characteristic polynomial $P(\lambda)$ and what are its eigenvalues? Repeat this if A is lower triangular or diagonal.

15. Let A be a real $n \times n$ matrix that is diagonalizable. Prove that A is invertible if and only if A does not have 0 as an eigenvalue. If A is diagonalizable with no zero eigenvalue, then give a formula for A^{-1} in terms of D and Q where $A = QDQ^{-1}$.

16. Let A be a real $n \times n$ matrix that is invertible. How are the eigenvalues and eigenvectors of A related to the eigenvalues and eigenvectors of A^{-1}? (See problem 15.)

17. Let A be a real $n \times n$ matrix that is diagonalizable with all real eigenvalues. Is it true that you can find a real invertible matrix Q so that $A = QDQ^{-1}$? Explain the reasoning behind your answer.

18. Let A be a real square matrix. If λ is an eigenvalue of A with an eigenvector $\overrightarrow{x}$, that is, $A\overrightarrow{x} = \lambda \overrightarrow{x}$, then for any positive integer n, give an eigenvalue and a corresponding eigenvector for the matrix A^n. What happens if n is a negative integer?

19. Consider the following real valued matrix:

$$A = \frac{1}{186} \begin{bmatrix} -700 & -184 & 200 & -720 \\ -50 & 970 & 373 & 573 \\ 60 & 324 & -336 & 354 \\ -220 & -196 & 116 & 66 \end{bmatrix}$$

(a) Compute the eigenvalues of A.

(b) Find the two matrices D and Q so that $A = QDQ^{-1}$, and A is diagonalizable.

(c) Check that $A^2 = QD^2Q^{-1}$

(d) Check that A and D have the same determinant.

(e) Check that $A^{-1} = QD^{-1}Q^{-1}$.

Maple Problems

1. Verify your answers to *Homework* problems 1, 2, and 3.

2. Consider the following real valued matrix:

$$A = \frac{1}{384} \begin{bmatrix} -1511 & -20 & 322 & -699 & 921 \\ 1185 & 972 & 402 & -915 & -3423 \\ 494 & 296 & -772 & -42 & -1170 \\ -594 & -216 & 252 & 534 & 558 \\ -759 & -84 & -30 & -75 & 1545 \end{bmatrix}$$

(a) Compute the eigenvalues of A.

(b) Find the two matrices D and Q so that $A = QDQ^{-1}$, and A is diagonalizable.

(c) Check that $A^2 = QD^2Q^{-1}$

(d) Check that A and D have the same determinant.

(e) Check that $A^{-1} = QD^{-1}Q^{-1}$.

3. Use *Maple* to redo *Homework* problem 19.

Research Projects

1. Find an algorithm (from a numerical analysis or numerical linear algebra book) that allows you to find an approximation to an eigenvector X for a given approximate eigenvalue λ of a square matrix A. Now write some *Maple* code that will allow you to test this algorithm on a specific square matrix A.

2. If you are in a major other than mathematics (engineering, biology, chemistry, physics, etc.), then search out an application of eigenvectors and eigenvalues that is directly applicable to your major.

12.2 Summary of Definitions and Methods for Computing Eigenvalues and Eigenvectors as well as the Exponential of a Matrix

Now we have the ability to define eigenvalues and eigenvectors for a square $n \times n$ matrix A whose entries are real or complex numbers. We will not deal with complex square matrices $A \in \mathbb{C}^{n \times n}$ much in this chapter, since in many applications the matrix A is completely real.

Definition 12.2.1. Let A be an $n \times n$ matrix with complex entries. Then a complex number λ is called an *eigenvalue* for the matrix A if there exists a nonzero complex column vector $\vec{x} \in \mathbb{C}^n$ satisfying $A\vec{x} = \lambda\vec{x}$. The vector $\vec{x}$ is called an *eigenvector* of the matrix A for the eigenvalue λ.

The method for computing the eigenvalues and eigenvectors of the matrix A are given after the next definition for the characteristic polynomial $P(\lambda)$ for A, where λ is the variable in this polynomial.

Definition 12.2.2. The *characteristic polynomial* $P(\lambda)$, for $A \in \mathbb{C}^{n \times n}$, is defined to be

$$P(\lambda) = \det(A - \lambda I_n)$$

where I_n is the $n \times n$ identity matrix and λ is the polynomial's variable.

Example 12.2.1. As an example, if $A = \begin{bmatrix} a & b & c \\ d & e & f \\ g & h & i \end{bmatrix}$, then

$$P(\lambda) = \det(A - \lambda I_3) = \det\left(\begin{bmatrix} a - \lambda & b & c \\ d & e - \lambda & f \\ g & h & i - \lambda \end{bmatrix} \right)$$

> with(linalg): with(LinearAlgebra):
> A:=matrix(3,3,[a,b,c,d,e,f,g,h,i]):
> collect(det(A-lambda*IdentityMatrix(3)), lambda);

$$-\lambda^3 + (a + e + i)\lambda^2 + (db - ai + gc - ei - ae + fh)\lambda$$
$$+ aei - gce - dbi - afh + gbf + dch$$

The characteristic polynomial $P(\lambda)$ is of degree n if $A \in \mathbb{C}^{n \times n}$ and so it has n complex roots if we count each root as many times as its multiplicity (or power) in the complete factoring of $P(\lambda)$. The n complex roots of $P(\lambda)$ are the eigenvalues of A. If the matrix A is beyond size 2×2, then its characteristic polynomial $P(\lambda)$ is of degree greater than 2 and so it would be difficult if not impossible to find A's eigenvalues by hand. In general, the Newton-Raphson algorithm can find the roots of any degree polynomial quite efficiently as approximations accurate to any desired number of digits.

For a given eigenvalue λ of A, you compute λ's eigenvectors $\overrightarrow{x}$ as the nonzero solutions to the matrix equation $(A - \lambda I_n)\overrightarrow{x} = \overrightarrow{0}$. Remember that for an eigenvalue λ of A, $\det(A - \lambda I_n) = 0$ and the matrix has no inverse. If you use any approximation of the eigenvalue λ of A, then $\det(A - \lambda I_n) \neq 0$ and the matrix $A - \lambda I_n$ has an inverse that means the matrix equation $(A - \lambda I_n)\overrightarrow{x} = \overrightarrow{0}$ only has the trivial solution $\overrightarrow{x} = \overrightarrow{0}$. Thus, no approximation of an eigenvalue λ of A can be used to find λ's eigenvectors $\overrightarrow{x}$ by solving the matrix equation $(A - \lambda I_n)\overrightarrow{x} = \overrightarrow{0}$ for $\overrightarrow{x}$. There are methods for approximating eigenvectors $\overrightarrow{x}$ for an approximate eigenvalue λ of A, but we will not discuss them here and so we must rely on *Maple* to compute the eigenvectors for us when all we have is an approximation of an eigenvalue λ. See *Research Project* 3 of Section 12.4 if you are interested in one possible method for finding eigenvectors.

For us, the purpose of the eigenvalues and eigenvectors for a square matrix A is to be able to diagonalize it. As such, let us now formally define what we mean by A being diagonalizable.

Definition 12.2.3. Let A be an $n \times n$ square matrix with complex entries. Then A is said to be *diagonalizable* if there exist an $n \times n$ diagonal matrix D with all complex entries and an $n \times n$ matrix Q with all complex entries that are invertible where $A = QDQ^{-1}$.

The n entries on the diagonal of the matrix D are the eigenvalues of A, where an eigenvalue appears on this diagonal as many times as it is a root of A's characteristic polynomial $P(\lambda)$.

The columns of the matrix Q are independent corresponding eigenvectors for the eigenvalues in D. This means that the kth column of the matrix Q is an eigenvector for the eigenvalue at the kth location on the diagonal of D. This matrix Q exists exactly when we can find a basis of the vector space $\mathbb{C}^n$ consisting completely of eigenvectors for A.

As it turns out, almost all square $n \times n$ matrices A are diagonalizable because they normally have n distinct eigenvalues (then each root of A's characterisitic polynomial $P(\lambda)$ has multiplicity one) and each eigenvalue has at least one eigenvector associated with it. The union of independent sets of eigenvectors from different eigenvalues always forms an independent set of column vectors in $\mathbb{C}^n$.

Diagonalizability for square matrices A is a special case of when two square matrices are said to be similar. In fact, A is diagonalizable exactly when it is similar to a diagonal matrix D.

Definition 12.2.4. Two square matrices A and B of the same size are said to be *similar* if there exists an invertible matrix Q of the same size as A and B with $A = QBQ^{-1}$.

Similar matrices have many common properties and so people have classified similar matrices by type or canonical form in order to know when two square matrices can be similar to each other or not. Canonical forms are for an advanced course in linear algebra and so it is left for your future education and interest.

Now we give the definition of e^A for any square matrix A using the Maclaurin series for the function e^x. This definition, although identical to (12.3) in its form, is now applicable to all square matrices.

Definition 12.2.5. Let A be any square matrix. Then its *exponential* e^A is defined by

$$e^A = \sum_{k=0}^{\infty} \frac{1}{k!} A^k$$

As we have seen, when A is diagonalizable with $A = QDQ^{-1}$, then $e^A = Qe^DQ^{-1}$, which makes it much easier to compute e^A than using the infinite series. If A is not diagonalizable, then e^A must be approximately computed using the infinite series by taking a partial sum, or if A is of some very special type of triangular matrix, then we might be able to find a simple formula for e^A from a formula for the powers of A.

Homework Problems

1. If two $n \times n$ matrices A and B are similar, then are their exponentials e^A and e^B also similar matrices? If yes, then explain why. If no, then give an example to verify it.

2. Let A and B be two similar $n \times n$ matrices. Show that A and B have the same eigenvalues. How are their respective eigenvectors related for the same eigenvalue?

3. Let A and B be two similar $n \times n$ matrices where A is diagonalizable. Show that B is also diagonalizable. (First, explain why $(EF)^{-1} = F^{-1}E^{-1}$ for two invertible matrices E and F of the same size.)

4. Let A be a real $n \times n$ matrix and c be a real scalar. How are the eigenvalues and eigenvectors of A related to those of cA? Explain your answer.

5. Let A and B be two $n \times n$ real diagonal matrices. Show that $e^A e^B = e^{A+B}$.

6. Let A and B be two $n \times n$ real matrices with $A = QDQ^{-1}$ and $B = QFQ^{-1}$, for two diagonal matrices D, F, and the same Q. First, show that A and B commute, that is, $AB = BA$. Second, show that $e^A e^B = e^{A+B}$.

7. Give two square matrices A and B of the same size satisfying $e^A e^B \neq e^{A+B}$.

8. Let A be any real diagonalizable matrix and k be any positive integer. Show that $\left(e^A\right)^k = e^{kA}$. Is this also true if $k = 0$ or k is negative?

9. Let A be any real square matrix. Show that we have $e^{KAK^{-1}} = Ke^A K^{-1}$, for any real invertible matrix K the same size as A.

10. Is there a way to define the logarithm, $\ln(A)$, of a general real square matrix A, and if this fails can you define $\ln(A)$ for some specific type of real square matrix A? If so, give your definition and check that it works with a specific example for A, that is, check if $\ln\left(e^A\right) = e^{\ln(A)} = A$. If you do not believe you can define $\ln(A)$ for any type of real square matrix A, then please explain your reasoning.

11. Let A be any square $n \times n$ matrix. Define the trace of A as

$$\text{trace}(A) = \sum_{j=1}^{n} A_{j,j}$$

Explain why for any two square $n \times n$ matrices A and B, $\text{trace}(AB) = \text{trace}(BA)$. *Hint: Use* Maple *to help you understand why.*

12. Show that if A and B are two similar matrices, then $\text{trace}(A) = \text{trace}(B)$ and $\det(A) = \det(B)$. *Hint: See* Homework *problem 11.*

13. Give an example of two same size square matrices A and B that are not similar, where $\det(A) = \det(B)$ and $\text{trace}(A) = \text{trace}(B)$. Hence, the converse of problem 12 is false. *Hint: See problem 11.*

14. Are any two square diagonal matrices A and B of the same size always similar? If yes, explain why. If no, explain why not and give an example.

15. Show that

$$A = \begin{bmatrix} \lambda & K & 0 \\ 0 & \lambda & 0 \\ 0 & 0 & 1 \end{bmatrix}$$

for $K \neq 0$ is not diagonalizable for any value of λ. Can you give two other versions of this matrix A that are also not diagonalizable?

Research Projects

1. Investigate the topic of *Markov chains* from statistics and see how eigenvalues and eigenvectors are helpful in their study.

12.3 Applications of the Diagonalizability of Square Matrices

It is clear from the last two sections that the ability to diagonalize a square matrix A enables us to easily take powers of the matrix A by $A^n = QD^nQ^{-1}$ if $A = QDQ^{-1}$. Simply put, this tells us that similar matrices have similar powers for the same similarity matrix Q. As a consequence, this also allows us to find the exponential of the square matrix A by $e^A = Qe^DQ^{-1}$. This exponential formula will be very useful to us in solving square systems of first-order linear differential equations in Section 12.4. The power formula $A^n = QD^nQ^{-1}$ is also useful in statistics when studying linear Markov processes, which we leave for you to discover in a statistics course. Also, eigenvalues and eigenvectors are helpful in physics, specifically in Quantum Mechanics.

Before we start in with examples, let us prove that if $A = QDQ^{-1}$, then $e^A = Qe^DQ^{-1}$. As before, this simply says that similar matrices have similar exponentials for the same similarity matrix Q.

Theorem 12.3.1. *If A is a $n \times n$ matrix such that $A = QDQ^{-1}$, then $e^A = Qe^DQ^{-1}$.*

Proof. As in Section 12.2, we start with the definition of the matrix exponential in Definition (12.2.5). Notice that in the following string of equalities, Theorem 12.1.1 is used.

$$e^A = \sum_{k=0}^{\infty} \frac{1}{k!} A^k$$

$$= \sum_{k=0}^{\infty} \frac{1}{k!} \left(QDQ^{-1}\right)^k$$

$$= \sum_{k=0}^{\infty} \frac{1}{k!} QD^kQ^{-1}$$

$$= Q \left(\sum_{k=0}^{\infty} \frac{1}{k!} D^k \right) Q^{-1}$$

$$= Qe^DQ^{-1}$$

where we know from Section 12.2 that if D is diagonal, then e^D is also diagonal, and gotten by taking the exponential of the diagonal entries of D. $\square$

Something else of interest is that you can find the formulas for $\sin(A)$ and $\cos(A)$ for any square diagonalizable matrix A using the Maclaurin series for $\sin(x)$ and $\cos(x)$. This is left as an exercise.

Now let us do an example of diagonalizability where the eigenvalues and eigenvectors are complex, not real, as we saw in Section 12.2. It turns out to be the case that if A is a real symmetric matrix, then its eigenvalues are all real, which allows us to take only real eigenvectors for these eigenvalues, but this is a rare case. Even when A is a real matrix, its eigenvalues and eigenvectors are almost always all complex occurring in complex conjugate pairs as we shall see in the example to follow.

Example 12.3.1. We will investigate whether or not the matrix

$$A = \begin{bmatrix} 3 & -5 & 1 \\ 0 & 2 & -2 \\ -4 & 7 & 0 \end{bmatrix}$$

is diagonalizable, and if so, what the matrices, D and Q, that diagonalize it are. If D and Q exist, we will use them to find e^A and A^4.

> with(linalg): with(LinearAlgebra):

> A:= matrix(3,3,[3,-5,1,0,2,-2,-4,7,0]);

$$A := \begin{bmatrix} 3 & -5 & 1 \\ 0 & 2 & -2 \\ -4 & 7 & 0 \end{bmatrix}$$

> evalm(A - lambda*IdentityMatrix(3));

$$\begin{bmatrix} 3-\lambda & -5 & 1 \\ 0 & 2-\lambda & -2 \\ -4 & 7 & -\lambda \end{bmatrix}$$

> AminusλI:= lambda -> evalm(A - lambda*IdentityMatrix(3)):

> eigenpoly:= det(AminusλI(lambda));

$$eigenpoly := -24\lambda + 5\lambda^2 + 10 - \lambda^3$$

> λlist_exact:= solve(eigenpoly=0, lambda);

$$\lambda list\_exact := -\frac{1}{3}\left(280 + 3\sqrt{20247}\right)^{1/3} + \frac{47}{3}\frac{1}{\left(280 + 3\sqrt{20247}\right)^{1/3}} + \frac{5}{3},$$

$$\frac{1}{6}\left(280 + 3\sqrt{20247}\right)^{1/3} - \frac{47}{6}\frac{1}{\left(280 + 3\sqrt{20247}\right)^{1/3}} + \frac{5}{3}$$

$$+\frac{1}{2}I\sqrt{3}\left(-\frac{1}{3}\left(280 + 3\sqrt{20247}\right)^{1/3} - \frac{47}{3}\frac{1}{\left(280 + 3\sqrt{20247}\right)^{1/3}}\right),$$

$$\frac{1}{6}\left(280 + 3\sqrt{20247}\right)^{1/3} - \frac{47}{6}\frac{1}{\left(280 + 3\sqrt{20247}\right)^{1/3}} + \frac{5}{3}$$

$$-\frac{1}{2}I\sqrt{3}\left(-\frac{1}{3}\left(280 + 3\sqrt{20247}\right)^{1/3} - \frac{47}{3}\frac{1}{\left(280 + 3\sqrt{20247}\right)^{1/3}}\right)$$

> λlist_approx:= [fsolve(eigenpoly=0, lambda, complex)];

$$\lambda list\_approx := [0.4560429226, 2.271978539 - 4.0946152511I, 2.271978539 + 4.0946152511I]$$

> X:= matrix(3,1,[x, y, z]);

$$X := \begin{bmatrix} x \\ y \\ z \end{bmatrix}$$

> AminusλI(λlist_approx[1]);

$$\begin{bmatrix} 2.543957077 & -5. & 1. \\ 0. & 1.543957077 & -2. \\ -4 & 7. & -0.4560429226 \end{bmatrix}$$

> det(AminusλI(λlist_approx[1]));

$$-1.\ 10^{-8}$$

> evalm(AminusλI(λlist_approx[1])&*X);

$$\begin{bmatrix} 2.543957077x - 5.y + 1.z \\ 1.543957077y - 2.z \\ -4.x + 7.y - 0.4560429226z \end{bmatrix}$$

> solve({2.543957077*x - 5.*y + 1.*z = 0, 1.543957077*y - 2.*z = 0, -4.*x + 7.*y - .4560429226*z = 0}, {x, y, z});

$$\{x = 0., y = 0., z = 0.\}$$

Notice above that if we solve for the eigenvector, denoted X, for the approximate real eigenvalue $\lambda = 0.4560429226$, then we get the zero column that we cannot use in Q. The problem is that when you use an approximation to an eigenvalue instead of its exact value as we just did, then the matrix $A - \lambda I$ has a nonzero determinant that says the system of linear equations $(A - \lambda I)X = 0$ only has the solution $X = 0$, which cannot be used as an eigenvector.

One remedy is to not approximate the eigenvalues λ, but use their exact values. The exact values of a matrix's eigenvalues can generally only be found when the size of the matrix is not more than 4. This would be very cumbersome to do by hand, but with *Maple* it is almost a pleasure to do.

> AminusλI(λlist_exact[1]);

$$\left[\left[\frac{4}{3} + \frac{1}{3}\sqrt[3]{280 + 3\sqrt{20247}} - \frac{47}{3}\frac{1}{\sqrt[3]{280+3\sqrt{20247}}}, -5, 1\right]\right.$$

$$\left[0, \frac{1}{3} + \frac{1}{3}\sqrt[3]{280 + 3\sqrt{20247}} - \frac{47}{3}\frac{1}{\sqrt[3]{280+3\sqrt{20247}}}, -2\right]$$

$$\left.\left[-4, 7, \frac{1}{3}\sqrt[3]{280 + 3\sqrt{20247}} - \frac{47}{3}\frac{1}{\sqrt[3]{280+3\sqrt{20247}}} - \frac{5}{3}\right]\right]$$

> det(AminusλI(λlist_exact[1]));

$$0$$

> evalm(AminusλI(λlist_exact[1])&*X);

$$\left[\begin{array}{c}\left(4/3 + 1/3\sqrt[3]{280 + 3\sqrt{20247}} - \frac{47}{3}\frac{1}{\sqrt[3]{280+3\sqrt{20247}}}\right)x - 5y + z \\[2ex] \left(1/3 + 1/3\sqrt[3]{280 + 3\sqrt{20247}} - \frac{47}{3}\frac{1}{\sqrt[3]{280+3\sqrt{20247}}}\right)y - 2z \\[2ex] -4x + 7y + \left(1/3\sqrt[3]{280 + 3\sqrt{20247}} - \frac{47}{3}\frac{1}{\sqrt[3]{280+3\sqrt{20247}}} - 5/3\right)z\end{array}\right]$$

> solve({%[1,1]=0, %[2,1]=0, %[3,1]=0},{x,y,z});

$$\left\{x = -\frac{1}{2}\frac{\left(-29\left(280 + 3\sqrt{20247}\right)^{1/3} + \left(280 + 3\sqrt{20247}\right)^{2/3} - 47\right)y}{4\left(280 + 3\sqrt{20247}\right)^{1/3} + \left(280 + 3\sqrt{20247}\right)^{2/3} - 47},\right.$$

$$\left. y = y, z = \frac{1}{6}\frac{\left(\left(280 + 3\sqrt{20247}\right)^{1/3} + \left(280 + 3\sqrt{20247}\right)^{2/3} - 47\right)y}{\left(280 + 3\sqrt{20247}\right)^{1/3}}\right\}$$

> evalf(%);

$$\{x = 1.661986164y, y = y, z = 0.7719785388y\}$$

This tells us that we can use as eigenvector X, for the real eigenvalue $\lambda = 0.4560429226$, the column $X = \begin{bmatrix} 1.661986164 \\ 1 \\ .771978538 \end{bmatrix}$. Next we need to find the remaining eigenvectors X for the other two complex eigenvalues.

> AminusλI(λlist_exact[2]);

$$\left[\left[\frac{4}{3} - \frac{1}{6}\left(280 + 3\sqrt{20247}\right)^{1/3} + \frac{47}{6}\frac{1}{\left(280 + 3\sqrt{20247}\right)^{1/3}} \right.\right.$$

$$\left. - \frac{1}{2}I\sqrt{3}\left(-\frac{1}{3}\left(280 + 3\sqrt{20247}\right)^{1/3} - \frac{47}{3}\frac{1}{\left(280 + 3\sqrt{20247}\right)^{1/3}}\right), -5, 1\right]$$

$$\left[0, \frac{1}{3} - \frac{1}{6}\left(280 + 3\sqrt{20247}\right)^{1/3} + \frac{47}{6}\frac{1}{\left(280 + 3\sqrt{20247}\right)^{1/3}}\right.$$

$$\left. - \frac{1}{2}I\sqrt{3}\left(-\frac{1}{3}\left(280 + 3\sqrt{20247}\right)^{1/3} - \frac{47}{3}\frac{1}{\left(280 + 3\sqrt{20247}\right)^{1/3}}\right), -2\right]$$

$$\left[-4, 7, -\frac{1}{6}\left(280 + 3\sqrt{20247}\right)^{1/3} + \frac{47}{6}\frac{1}{\left(280 + 3\sqrt{20247}\right)^{1/3}} - \frac{5}{3}\right.$$

$$\left.\left. - \frac{1}{2}I\sqrt{3}\left(-\frac{1}{3}\left(280 + 3\sqrt{20247}\right)^{1/3} - \frac{47}{3}\frac{1}{\left(280 + 3\sqrt{20247}\right)^{1/3}}\right)\right]\right]$$

> det(AminusλI(λlist_exact[2]));

$$0$$

> evalm(AminusλI(λlist_exact[2])&*X);

$$\left[\left[\left(\frac{4}{3} - \frac{1}{6}\left(280 + 3\sqrt{20247}\right)^{1/3} + \frac{47}{6}\frac{1}{\left(280 + 3\sqrt{20247}\right)^{1/3}} \right.\right.\right.$$

$$\left.\left. - \frac{1}{2}I\sqrt{3}\left(-\frac{1}{3}\left(280 + 3\sqrt{20247}\right)^{1/3} - \frac{47}{3}\frac{1}{\left(280 + 3\sqrt{20247}\right)^{1/3}}\right)\right) x - 5y + z\right]$$

$$\left[\left[\left(\frac{1}{3} - \frac{1}{6}\left(280 + 3\sqrt{20247}\right)^{1/3} + \frac{47}{6}\frac{1}{\left(280 + 3\sqrt{20247}\right)^{1/3}} \right.\right.\right.$$

$$\left.\left. - \frac{1}{2}I\sqrt{3}\left(-\frac{1}{3}\left(280 + 3\sqrt{20247}\right)^{1/3} - \frac{47}{3}\frac{1}{\left(280 + 3\sqrt{20247}\right)^{1/3}}\right)\right) y - 2z\right]$$

$$\left[-4x + 7y + \left(-\frac{1}{6}\left(280 + 3\sqrt{20247}\right)^{1/3} + \frac{47}{6}\frac{1}{\left(280 + 3\sqrt{20247}\right)^{1/3}} - \frac{5}{3} \right.\right.$$

$$\left.\left. -\frac{1}{2}I\sqrt{3}\left(-\frac{1}{3}\left(280 + 3\sqrt{20247}\right)^{1/3} - \frac{47}{3}\frac{1}{\left(280 + 3\sqrt{20247}\right)^{1/3}} \right) \right) z \right]\right]$$

> solve({%[1,1]=0, %[2,1]=0, %[3,1]=0},{x,y,z});

$$\left\{ x = -\frac{1}{2}\frac{\left(-58\left(280 + 3\sqrt{20247}\right)^{1/3} - \left(280 + 3\sqrt{20247}\right)^{2/3} + 47 + I\sqrt{3}\left(280 + 3\sqrt{20247}\right)^{2/3} + 47\,I\sqrt{3}\right)y}{8\left(280 + 3\sqrt{20247}\right)^{1/3} + 47 + I\sqrt{3}\left(280 + 3\sqrt{20247}\right)^{2/3} + 47\,I\sqrt{3} - \left(280 + 3\sqrt{20247}\right)^{2/3}}, \right.$$

$$\left. y = y, z = \frac{1}{12}\frac{\left(2\left(280 + 3\sqrt{20247}\right)^{1/3} - \left(280 + 3\sqrt{20247}\right)^{2/3} + 47 + I\sqrt{3}\left(280 + 3\sqrt{20247}\right)^{2/3} + 47\,I\sqrt{3}\right)y}{\left(280 + 3\sqrt{20247}\right)^{1/3}} \right\}$$

> evalf(%)

$$\{x = (-0.2684930819 - 1.302065680I)y, y = y, z = (-0.1359892693 + 2.047307627I)y\}$$

This tells us that we can use as eigenvector X, for the complex eigenvalue $\lambda = 2.271978539 - 4.094615251i$, the column $X = \begin{bmatrix} -.2684930816 - 1.302065680i \\ 1 \\ -.1359892693 + 2.047307627i \end{bmatrix}$.

> AminusλI(λlist_exact[3]);

$$\left[\left[\frac{4}{3} - \frac{1}{6}\left(280 + 3\sqrt{20247}\right)^{1/3} + \frac{47}{6}\frac{1}{\left(280 + 3\sqrt{20247}\right)^{1/3}} \right.\right.$$

$$\left.\left. +\frac{1}{2}I\sqrt{3}\left(-\frac{1}{3}\left(280 + 3\sqrt{20247}\right)^{1/3} - \frac{47}{3}\frac{1}{\left(280 + 3\sqrt{20247}\right)^{1/3}} \right), -5, 1 \right]\right.$$

$$\left[0, \frac{1}{3} - \frac{1}{6}\left(280 + 3\sqrt{20247}\right)^{1/3} + \frac{47}{6}\frac{1}{\left(280 + 3\sqrt{20247}\right)^{1/3}} \right.$$

$$\left.\left. +\frac{1}{2}I\sqrt{3}\left(-\frac{1}{3}\left(280 + 3\sqrt{20247}\right)^{1/3} - \frac{47}{3}\frac{1}{\left(280 + 3\sqrt{20247}\right)^{1/3}} \right), -2 \right]\right.$$

$$\left[-4, 7, -\frac{1}{6}\left(280 + 3\sqrt{20247}\right)^{1/3} + \frac{47}{6}\frac{1}{\left(280 + 3\sqrt{20247}\right)^{1/3}} - \frac{5}{3} \right.$$

$$\left.\left.\left. +\frac{1}{2}I\sqrt{3}\left(-\frac{1}{3}\left(280 + 3\sqrt{20247}\right)^{1/3} - \frac{47}{3}\frac{1}{\left(280 + 3\sqrt{20247}\right)^{1/3}} \right) \right]\right]\right]$$

> det(AminusλI(λlist_exact[3]));

0

> evalm(AminusλI(λlist_exact[3])&*X);

$$
\left[\left[\left(\frac{4}{3}-\frac{1}{6}\left(280+3\sqrt{20247}\right)^{1/3}+\frac{47}{6}\frac{1}{\left(280+3\sqrt{20247}\right)^{1/3}}\right.\right.\right.
$$

$$
\left.\left.+\frac{1}{2}I\sqrt{3}\left(-\frac{1}{3}\left(280+3\sqrt{20247}\right)^{1/3}-\frac{47}{3}\frac{1}{\left(280+3\sqrt{20247}\right)^{1/3}}\right)\right)x-5\,y+z\right]
$$

$$
\left[\left(\frac{1}{3}-\frac{1}{6}\left(280+3\sqrt{20247}\right)^{1/3}+\frac{47}{6}\frac{1}{\left(280+3\sqrt{20247}\right)^{1/3}}\right.\right.
$$

$$
\left.\left.+\frac{1}{2}I\sqrt{3}\left(-\frac{1}{3}\left(280+3\sqrt{20247}\right)^{1/3}-\frac{47}{3}\frac{1}{\left(280+3\sqrt{20247}\right)^{1/3}}\right)\right)y-2\,z\right]
$$

$$
\left[-4\,x+7\,y+\left(-\frac{1}{6}\left(280+3\sqrt{20247}\right)^{1/3}+\frac{47}{6}\frac{1}{\left(280+3\sqrt{20247}\right)^{1/3}}-\frac{5}{3}\right.\right.
$$

$$
\left.\left.\left.+\frac{1}{2}I\sqrt{3}\left(-\frac{1}{3}\left(280+3\sqrt{20247}\right)^{1/3}-\frac{47}{3}\frac{1}{\left(280+3\sqrt{20247}\right)^{1/3}}\right)\right)z\right]\right]
$$

> solve({%[1,1]=0, %[2,1]=0, %[3,1]=0},{x,y,z});

$$
\left\{x=-\frac{1}{2}\frac{\left(58\left(280+3\sqrt{20247}\right)^{1/3}+\left(280+3\sqrt{20247}\right)^{2/3}-47+I\sqrt{3}\left(280+3\sqrt{20247}\right)^{2/3}+47\,I\sqrt{3}\right)y}{-8\left(280+3\sqrt{20247}\right)^{1/3}+47+I\sqrt{3}\left(280+3\sqrt{20247}\right)^{2/3}-47\,I\sqrt{3}+\left(280+3\sqrt{20247}\right)^{2/3}},\right.
$$

$$
z=-\frac{1}{12}\frac{\left(-2\left(280+3\sqrt{20247}\right)^{1/3}+\left(280+3\sqrt{20247}\right)^{2/3}-47+I\sqrt{3}\left(280+3\sqrt{20247}\right)^{2/3}+47\,I\sqrt{3}\right)y}{\left(280+3\sqrt{20247}\right)^{1/3}}
$$

$$
\left.y=y\right\}
$$

> evalf(%)

$\{x=(-0.2684930819+1.302065680I)y,\ y=y,\ z=(-0.1359892693-2.047307627I)y\}$

This tells us that we can use as eigenvector X, for the complex eigenvalue

$$\lambda=2.271978539+4.094615251i,\ \text{vector}\ X=\begin{bmatrix}-.2684930816+1.302065680i\\1\\-.1359892693-2.047307627i\end{bmatrix}.$$

Note that these are the complex conjugates of the previous values for λ and $\bar{X}$.
Next, we can put the eigenvalues and eigenvectors together to form D and Q.

> Diag:= DiagonalMatrix(λlist_approx);

$$Diag:=\begin{bmatrix}0.4560429226 & 0 & 0\\0 & 2.271978539-4.094615251I & 0\\0 & 0 & 2.271978539+4.094615251I\end{bmatrix}$$

> Q:= transpose(matrix(3,3,[1.661986164, 1, 0.771978538, -0.2684930816
- 1.302065680*I, 1, -0.1359892693 + 2.047307627*I, -0.2684930816
+ 1.302065680*I, 1, -0.1359892693 - 2.047307627*I]));

$$Q := \begin{bmatrix} 1.661986164 & -0.2684930816 - 1.302065680I & -0.2684930816 + 1.302065680I \\ 1 & 1 & 1 \\ 0.771978538 & -0.1359892693 + 2.047307627I & -0.1359892693 - 2.047307627I \end{bmatrix}$$

> evalm(Q&*Diag&*Q^(-1));

$$\begin{bmatrix} 2.999999997 + 0.I & -4.999999996 + 0.I & 0.9999999977 + 0.I \\ -1.34\ 10^{-9} + 0.I & 2.000000000 + 0.I & -1.999999999 + 0.I \\ -4.000000001 + 0.I & 7.000000005 + 0.I & 2.27\ 10^{-9} + 0.I \end{bmatrix}$$

> print(A);

$$\begin{bmatrix} 3 & -5 & 1 \\ 0 & 2 & -2 \\ -4 & 7 & 0 \end{bmatrix}$$

Now we will compute both e^A and A^4, since we now know that A is diagonalizable.

> evalm(A^4);

$$\begin{bmatrix} -275 & 532 & -97 \\ 8 & -180 & 216 \\ 428 & -736 & 32 \end{bmatrix}$$

> evalm(Q&*Diag^4&*Q^(-1));

$$\begin{bmatrix} -274.9999998 + 0.I & 531.9999997 + 0.I & -96.99999976 + 0.I \\ 8.000000184 + 0.I & -180.0000001 + 0.I & 215.9999999 + 0.I \\ 428.0000001 + 0.I & -736.0000006 + 0.I & 31.99999966 + 0.I \end{bmatrix}$$

> eDiag:= matrix(3,3,[exp(λlist_approx[1]), 0, 0, 0, exp(λlist_approx[2]), 0, 0,
0, exp(λlist_approx[3])]);

$$eDiag := \begin{bmatrix} 1.577818067 & 0 & 0 \\ 0 & -5.617623744 + 7.905983803I & 0 \\ 0 & 0 & -5.617623744 - 7.905983803I \end{bmatrix}$$

> evalm(Q&*eDiag&*Q^(-1));

$$\begin{bmatrix} -4.578515012 + 0.I & 10.52197660 + 0.I & -0.3759640848 + 0.I \\ 1.471005308 + 0.I & -4.570302257 + 0.I & 4.797192762 + 0.I \\ 8.858882876 + 0.I & -13.11266142 + 0.I & -0.5086121531 + 0.I \end{bmatrix}$$

> evalf(evalm(sum(A^N/N!, N=0..25)));

$$\begin{bmatrix} -4.578515015 & 10.52197660 & -0.3759640867 \\ 1.471005306 & -4.570302255 & 4.797192763 \\ 8.858882874 & -13.11266141 & -0.5086121444 \end{bmatrix}$$

Now you have seen an example of both the power and the pitfalls of using eigenvalues and eigenvectors. Clearly, if the size of the matrix A gets any bigger it becomes hopeless to do anything by hand without driving yourself nuts with silly errors. Happily, *Maple* has built in *eigenvalues* and *eigenvectors* commands that now we will use. These commands use numerical approximation methods that we leave for a numerical analysis course to explain to you.

> eigenvalues(A);

$$-\frac{1}{3}\left(280 + 3\sqrt{20247}\right)^{1/3} + \frac{47}{3}\frac{1}{\left(280 + 3\sqrt{20247}\right)^{1/3}} + \frac{5}{3},$$

$$\frac{1}{6}\left(280 + 3\sqrt{20247}\right)^{1/3} - \frac{47}{6}\frac{1}{\left(280 + 3\sqrt{20247}\right)^{1/3}} + \frac{5}{3}$$

$$+\frac{1}{2}I\sqrt{3}\left(-\frac{1}{3}\left(280 + 3\sqrt{20247}\right)^{1/3} - \frac{47}{3}\frac{1}{\left(280 + 3\sqrt{20247}\right)^{1/3}}\right)$$

$$\frac{1}{6}\left(280 + 3\sqrt{20247}\right)^{1/3} - \frac{47}{6}\frac{1}{\left(280 + 3\sqrt{20247}\right)^{1/3}} + \frac{5}{3}$$

$$-\frac{1}{2}I\sqrt{3}\left(-\frac{1}{3}\left(280 + 3\sqrt{20247}\right)^{1/3} - \frac{47}{3}\frac{1}{\left(280 + 3\sqrt{20247}\right)^{1/3}}\right)$$

> eigvals:= [evalf(%)];

$eigvals := [0.456042924, 2.271978539 - 4.094615252I, 2.271978539 + 4.094615252I]$

> eigvects:= [evalf(eigenvectors(A))];

$eigvects := \big[\,[0.456042924, 1.0, \{\,[1.661986163, 1.0, 0.7719785391]\,\}], [2.271978539$
$\quad - 4.094615252\,I, 1.0, \{\,[-0.2684930831 - 1.302065682\,I, 1.0, -0.1359892692$
$\quad + 2.047307626\,I]\,\}\,], [2.271978539 + 4.094615252\,I, 1.0, \{\,[-0.2684930831$
$\quad + 1.302065682\,I, 1.0, -0.1359892692 - 2.047307626\,I]\,\}\,]\,\big]$

Notice that the eigenvalues of A are given also in the result of the *eigenvectors* command since you need to know which eigenvalue has what eigenvectors. So in general you only need to use the eigenvectors command in order to get both A's eigenvalues and their associated independent eigenvectors.

> vals:= [eigvects[1][1], eigvects[2][1], eigvects[3][1]];

$vals := [0.456042924, 2.271978539 - 4.0946152521, 2.271978539 + 4.0946152521]$

> Diag2:= DiagonalMatrix(vals);

$$Diag2 := \begin{bmatrix} 0.456042924 & 0 & 0 \\ 0 & 2.271978539 - 4.094615252I & 0 \\ 0 & 0 & 2.271978539 + 4.094615252I \end{bmatrix}$$

> Qlist:= [op(convert(op(eigvects[1][3]), list)), op(convert(op(eigvects [2][3]), list)), op(convert(op(eigvects[3][3]), list))];

$$Qlist := [1.661986163, 1., 0.7719785391, -0.2684930831 - 1.3020656821I, 1.,$$
$$-0.1359892692 + 2.0473076261I, -0.2684930831 + 1.3020656821I,$$
$$1., -0.1359892692 - 2.0473076261I]$$

> Q2:= transpose(matrix(3,3,Qlist));

$$Q2 := \begin{bmatrix} 1.661986163 & -0.2684930831 - 1.3020656821I & -0.2684930831 + 1.3020656821I \\ 1. & 1. & 1. \\ 0.7719785391 & -0.1359892692 + 2.0473076261I & -0.1359892692 - 2.0473076261I \end{bmatrix}$$

The last matrix $Q2$ above is the matrix Q gotten from the results of the eigenvectors command while $Diag2$ is the diagonal eigenvalue matrix D gotten from the same command.

∎

The following example will give a square matrix A that is not diagonalizable. Section 12.5 will explain in better detail how the rare case of nondiagonalizability can occur.

Example 12.3.2. Let $A = \begin{bmatrix} -10 & 11 & -6 \\ -15 & 16 & -10 \\ -3 & 3 & -2 \end{bmatrix}$. Let us see if A is diagonalizable or not. In the following example, we also use the *LinearAlgebra* package in *Maple* since it has an identity matrix command *IdentityMatrix*. The *LinearAlgebra* package of *Maple* also has its own *Eigenvalues* and *Eigenvectors* commands that you should investigate since they give somewhat better outputs than the corresponding commands, *eigenvalues* and *eigenvectors*, of the *linalg* package.

> A:= matrix(3,3,[-10,11,-6,-15,16,-10,-3,3,-2]);

$$A := \begin{bmatrix} -10 & 11 & -6 \\ -15 & 16 & -10 \\ -3 & 3 & -2 \end{bmatrix}$$

> AminusλI:= evalm(A - λ*IdentityMatrix(3));

$$Aminus\lambda I := \begin{bmatrix} -10 - \lambda & 11 & -6 \\ -15 & 16 - \lambda & -10 \\ -3 & 3 & -2 - \lambda \end{bmatrix}$$

> eigenpoly:= det(AminusλI);

$$eigenpoly := 2 - 5\lambda + 4\lambda^2 - \lambda^3$$

> factor(%);
$$-(\lambda - 2)(\lambda - 1)^2$$
> λlist:= solve(eigenpoly=0,λ);
$$\lambda list := 2, 1, 1$$
> eigenvectors(A);
$$\left[1, 2, \{[1, 1, 0]\}\right], \left[2, 1, \left\{\left[\frac{26}{3}, 10, 1\right]\right\}\right]$$

The reason that A is not diagonalizable is that A has only two eigenvalues, $\lambda = 1$ and $\lambda = 2$, with $\lambda = 1$ occurring twice as a root of the characteristic polynomial, while $\lambda = 2$ occurs only once. You might think that this is no problem, that we can use the eigenvalue $\lambda = 1$ twice in the diagonal matrix D. From the result of the *eigenvectors* command we see that there is essentially only one independent eigenvector for the eigenvalue $\lambda = 1$, namely $\vec{v} = \begin{bmatrix} 1 \\ 1 \\ 0 \end{bmatrix}$, since all others are multiples of this single eigenvector. The reason for this is that the set of all eigenvectors for the eigenvalue $\lambda = 1$, with the zero vector thrown in, is a subspace of $\mathbb{C}^3$ which has dimension one. The eigenvector $\vec{v}$ is a basis for this subspace of $\mathbb{C}^3$ and we need this subspace to have dimension two if we are to use $\lambda = 1$ as an eigenvalue twice in D. We also need two independent eigenvectors for the eigenvalue $\lambda = 1$ if we are to use it twice on the diagonal of D and these eigenvectors as corresponding columns of Q. The eigenvectors that form the columns of Q must form an independent set since a square matrix Q is invertible if and only if its columns (or rows) form an independent set and a square matrix.

The last example of this section will look at diagonalizing a 4×4 matrix A, all of whose entries are complex numbers. We have rarely, if at all, looked at problems in this text involving complex numbers, but you should still know that this type of problem can occur and works very much like the real case. Note that in A below, we have placed a decimal point in the real part of $A_{1,1}$ so that *Maple* will automatically approximate A's eigenvalues since it believes that $A_{1,1}$ is now an approximate value. You should remove this decimal point from $A_{1,1}$ in order to see what the *eigenvalues* command produces instead.

Example 12.3.3. Consider $A = \begin{bmatrix} 2-i & 5 & i & 0 \\ 3+i & -i & i\pi & -2 \\ 1 & 2 & 3 & 4 \\ -2i & 0 & -1 & i \end{bmatrix}$. Let us see if A is diagonalizable or not, and find A^5 and e^A if it is. In this example, we use the *DiagonalMatrix* command from the *LinearAlgebra* package to create our diagonal matrix of eigenvalues.

> A:= matrix(4,4,[2.-I,5,I,0,3+I,-I,Pi*I,-2,1,2,3,4,-2*I,0,-1,I]);

$$
A := \begin{bmatrix}
2. - 1.I & 5 & I & 0 \\
3 + I & -I & I\pi & -2 \\
1 & 2 & 3 & 4 \\
-2I & 0 & -1 & I
\end{bmatrix}
$$

> eigenvalues(A);

$$
6.210840714 + 1.007820110I, -3.073876918 - 1.082974056I,
$$
$$
1.212861620 + 2.303514220I, .6501745852 - 3.228360273I
$$

> AminusλI:= evalm(A - λ*IdentityMatrix(4));

$$
AminusλI := \begin{bmatrix}
2. - 1.I - \lambda & 5 & I & 0 \\
3 + I & -I - \lambda & I\pi & -2 \\
1 & 2 & 3 - \lambda & 4 \\
-2I & 0 & -1 & I - \lambda
\end{bmatrix}
$$

> eigenpoly:= sort(simplify(det(AminusλI)));

$$
eigenpoly := \lambda^4 - 5.0\,\lambda^3 + 1.\,I\lambda^3 - 15.28318531\,I\lambda^2 - 4.\,\lambda^2
$$
$$
+ 17.\,\lambda + 9.858407346\,I\lambda - 171.8052988 - 37.28318531\,I
$$

> factor(%);

$$
(\lambda + 3.073876916 + 1.082974054\,I)\,(\lambda - 0.6501745822 + 3.228360273\,I)
$$
$$
(\lambda - 1.212861618 - 2.303514217\,I)\,(\lambda - 6.210840715 - 1.007820110\,I)
$$

> λlist:= solve(eigenpoly=0,λ);

$$
\lambda list := 6.210840715 + 1.007820110I, 1.212861618 + 2.303514217I,
$$
$$
- 3.073876916 - 1.082974054I, 0.6501745824 - 3.228360273I
$$

> eigvects:= eigenvectors(A);

$$
eigvects := [6.210840714 + 1.007820110\,I, 1, \{[0.01788303683
$$
$$
+ 0.6732215534\,I, -0.1792482580 + 0.5269681537\,I, 0.2358938819
$$
$$
+ 0.3801638650\,I, 0.1787235928 - 0.06719341181\,I]\}]
$$
$$
[-3.073876918 - 1.082974056\,I, 1, \{[-0.05872748151 - 0.6427523253\,I,
$$
$$
0.05554552910 + 0.6851502587\,I, -0.1596322452 + 0.03308339733\,I,
$$

$0.2382628740 - 0.1889037739\,I]\}]$

$[1.212861620 + 2.303514220\,I, 1, \{[0.2367700961 + 0.1140482068\,I,$
$\quad - 0.00733911250 + 0.2468358222\,I, -0.5417774520 + 0.5264351389\,I,$
$\quad - 0.1166278534 - 0.6991310888\,I]\}]$

$[0.6501745852 - 3.228360273\,I, 1, \{[-0.5563321961 + 0.6738575000\,I,$
$\quad 0.3136887417 + 0.1394588156\,I, -0.3671755020 - 0.6841049146\,I,$
$\quad - 0.3541960221 + 0.4600316962\,I]\}]$

> vals:= [eigvects[1][1], eigvects[2][1], eigvects[3][1], eigvects[4][1]];

$$vals := [-3.073876918 - 1.082974056I, 1.212861620 + 2.303514220I,$$
$$6.210840714 + 1.007820110I, 0.6501745852 - 3.228360273I]$$

> Diag:= DiagonalMatrix(vals);

$$Diag := \big[\,[-3.073876918 - 1.082974056I, 0, 0, 0], \;[0, 1.212861620$$
$$+ 2.303514220I, 0, 0], \;[0, 0, 6.210840714 + 1.007820110I, 0], \;[0, 0, 0,$$
$$0.6501745852 - 3.228360273I]\,\big]$$

> Qlist:= [op(convert(op(eigvects[1][3]),list)), op(convert(op(eigvects[2][3]), list)), op(convert(op(eigvects[3][3]), list)), op(convert(op(eigvects[4][3]), list))];

$$Qlist := [-0.05872748151 - 0.6427523253\,I, 0.05554552910 + 0.6851502587\,I,$$
$$\quad - 0.1596322452 + 0.03308339733\,I, 0.2382628740 - 0.1889037739\,I,$$
$$\quad 0.2367700961 + 0.1140482068\,I, -0.00733911250 + 0.2468358222\,I,$$
$$\quad - 0.5417774520 + 0.5264351389\,I, -0.1166278534 - 0.6991310888\,I,$$
$$\quad 0.01788303683 + 0.6732215534\,I, -0.1792482580 + 0.5269681537\,I,$$
$$\quad 0.2358938819 + 0.3801638650\,I, 0.1787235928 - 0.06719341181\,I,$$
$$\quad - 0.5563321961 + 0.6738575000\,I, 0.3136887417 + 0.1394588156\,I,$$
$$\quad - 0.3671755020 - 0.6841049146\,I, -0.3541960221 + 0.4600316962\,I]$$

> Q:= transpose(matrix(4,4,Qlist));

$$Q := \big[\,[-0.05872748151 - 0.6427523253\,I, 0.2367700961 + 0.1140482068\,I,$$
$$\quad 0.01788303683 + 0.6732215534\,I, -0.5563321961 + 0.6738575000\,I]$$
$$\quad [0.05554552910 + 0.6851502587\,I, -0.00733911250 + 0.2468358222\,I,$$
$$\quad - 0.1792482580 + 0.5269681537\,I, 0.3136887417 + 0.1394588156\,I]$$
$$\quad [-0.1596322452 + 0.03308339733\,I, -0.5417774520 + 0.5264351389\,I,$$
$$\quad 0.2358938819 + 0.3801638650\,I, -0.3671755020 - 0.6841049146\,I]$$
$$\quad [0.2382628740 - 0.1889037739\,I, -0.1166278534 - 0.6991310888\,I$$

$$0.1787235928 - 0.06719341181\,I, -0.3541960221 + 0.4600316962\,I]\,]$$

> evalm(Q&*Diag&*Q^(-1));

$$\big[\,\big[2.000000002 - 1.000000001I, 4.999999999 + 3.4\ 10^{-9}I, 0. + 0.9999999955I,$$
$$4.1\ 10^{-9} - 6.\ 10^{-9}I\big], \big[3.000000003 + 1.000000001I, -4.6\ 10^{-9} - 1.000000003I, -9.0\ 10^{-10}$$
$$+ 3.141592656I, -2.000000007 - 1.4\ 10^{-9}I\big], \big[0.9999999985 - 7.\ 10^{-10}I, 2.000000003$$
$$+ 3.\ 10^{-9}I, 3.000000004 + 1.0\ 10^{-9}I, 4.000000009 - 3.7\ 10^{-9}I\big], \big[2.\ 10^{-10} - 2.000000002I,$$
$$- 3.0\ 10^{-9} + 3.1\ 10^{-9}I, -0.9999999973 - 1.9\ 10^{-9}I, -4.\ 10^{-10} + 1.000000006I\big]\,\big]$$

> print(A);

$$\begin{bmatrix} 2. - 1.I & 5 & I & 0 \\ 3 + I & -i & I\pi & -2 \\ 1 & 2 & 3 & 4 \\ -2i & 0 & -1 & I \end{bmatrix}$$

> evalm(Q&*Diag^5&*Q^(-1));

$$\big[\,\big[3572.038382 + 1915.141471\,I, 3201.218440 + 2107.698960\,I$$
$$- 838.5827964 + 3720.122762\,I, -1462.377066 + 1829.185730\,I\big]$$
$$\big[2081.931295 + 2707.502284\,I, 1978.967418 + 2078.072280\,I,$$
$$- 1854.982423 + 2737.086169\,I, 1755.362051 + 852.6912926\,I\big]$$
$$\big[2631.680817 + 93.91142393\,I, 2410.193544 + 114.2476896\,I,$$
$$995.8861365 + 2358.089425\,I, -181.4889212 + 1713.575131\,I\big]$$
$$\big[316.4247475 - 1145.433600\,I, 218.6725269 - 957.7610407\,I,$$
$$1069.301361 - 176.0857684\,i, 741.7875670 + 86.7699225\,I\big]\,\big]$$

> evalf(evalm(A^5));

$$\big[\,\big[3572.038383 + 1915.141472\,I, 3201.218448 + 2107.698962\,I$$
$$- 838.5827962 + 3720.122765\,I, -1462.377070 + 1829.185731\,I\big]$$
$$\big[2081.931296 + 2707.502286\,I, 1978.967423 + 2078.072285\,I,$$
$$- 1854.982422 + 2737.086170\,I, 1755.362053 + 852.6912925\,I\big]$$
$$\big[2631.680819 + 93.91142568\,I, 2410.193549 + 114.247683\,I,$$
$$995.8861362 + 2358.089427\,I, -181.4889225 + 1713.575131\,I\big]$$
$$\big[316.4247479 - 1145.433599\,I, 218.6725275 - 957.7610416\,I,$$
$$1069.301359 - 176.0857683\,I, 741.7875657 + 86.76992327\,I\big]\,\big]$$

So we see that $A^5 = QD^5Q^{-1}$.

> eDiag:= matrix(4,4,[exp(vals[1]), 0, 0, 0, 0,exp(vals[2]), 0, 0, 0, 0,exp(vals[3]),
0, 0, 0, 0, exp(vals[4])]);

$$eDiag := \big[\, [0.02167355838 - 0.04084771931I, 0, 0, 0],\ [0, -2.249548797$$
$$+ 2.499987362I, 0, 0],\ [0, 0, 265.8492828 + 421.2452319I, 0],\ [0, 0, 0,$$
$$-1.908667859 + .1660274281I]\,\big]$$

> evalm(Q&*eDiag&*Q^(-1));

$$\big[\, [150.3611282 + 137.1169073\,I, 136.1063734 + 124.1448840\,I$$
$$- 77.84613452 + 181.9403021\,I, -95.10771796 + 76.34298896\,I]$$
$$[77.28135855 + 149.6810527\,I, 69.99510691 + 136.1425217\,I,$$
$$- 112.0864000 + 118.1406732\,I, 93.98746668 + 32.60744508\,I]$$
$$[132.2915019 + 27.39271192\,I, 120.4627634 + 26.07491129\,I,$$
$$14.99972386 + 131.3762826\,I, -26.46719880 + 76.13049351\,I]$$
$$[23.18786384 - 53.29126219\,I, 20.01174729 - 48.85032507\,I,$$
$$54.79416792 + 4.096073126\,I, 26.35678080 + 19.23468731\,I]\,\big]$$

> evalf(evalm(sum(A^N/N!, N=0..25)));

$$\big[\, [150.3611286 + 137.1169081\,I, 136.1063743 + 124.1448848\,I$$
$$- 77.84613453 + 181.9403021\,I, -95.10771796 + 76.34298896\,I]$$
$$[77.28135854 + 149.6810527\,I, 69.99510689 + 136.1425218\,I,$$
$$- 112.0864000 + 118.1406733\,I, -93.98746667 + 32.60744507\,I]$$
$$[132.2915019 + 27.39271191\,I, 120.4627634 + 26.07491133\,I,$$
$$14.99972387 + 131.3762826\,I, -26.46719880 + 76.13049351\,I]$$
$$[23.18786384 - 53.29126218\,I, 20.01174731 - 48.85032507\,I,$$
$$54.79416792 + 4.096073126\,I, 26.35678081 + 19.23468731\,I]\,\big]$$

Homework Problems

1. Find formulas for $\sin(A)$ and $\cos(A)$ for a square real matrix A using the Maclaurin series for $\sin(x)$ and $\cos(x)$. If the matrix A is diagonalizable, then find a formula for $\sin(A)$ and $\cos(A)$ using this diagonalizability. Check this formula on the diagonalizable real symmetric matrix $A = \begin{bmatrix} 5 & 2 \\ 2 & -3 \end{bmatrix}$ to see if

it works. Is the identity

$$\sin^2(A) + \cos^2(A) = I_2$$

for the identity matrix I_2, still true for this matrix A?

2. Let $A = \dfrac{1}{5} \begin{bmatrix} 7 & 26 & 34 \\ 4 & 27 & 8 \\ 3 & -16 & -4 \end{bmatrix}$. See if A is diagonalizable, and if so find the

matrices D and Q that diagonalize it. If D and Q exist, then use them to find e^A, and A^7.

3. If you did problem 1, then use your results of problem 1 on the matrix A of problem 2.

4. Let A be a square nondiagonalizable matrix. Explain why KAK^{-1} is also nondiagonalizable for any invertible matrix K of the same size as A.

5. Let A be a square diagonalizable matrix. Explain why KAK^{-1} is also diagonalizable for any invertible matrix K of the same size as A.

6. Let $A = \begin{bmatrix} 6 & 0 & 0 \\ 1 & 6 & 0 \\ -4 & 7 & 6 \end{bmatrix}$. Show that A is not diagonalizable.

7. Let $A = \begin{bmatrix} 9 & 0 & 0 & 0 \\ 5 & 9 & 0 & 0 \\ 0 & 0 & 11 & 0 \\ 0 & 0 & 2 & 11 \end{bmatrix}$. Show that A is not diagonalizable.

8. Show that the upper and lower triangular square matrices with all identical diagonal entries are all nondiagonalizable unless they are actually diagonal matrices.

9. Let A and B be two real 2×2 diagonalizable matrices with $A = QDQ^{-1}$ and $B = KEK^{-1}$. Show that the block 4×4 matrix created from A and B with A and B on its diagonal, that is, the 4×4 matrix $\begin{bmatrix} A & 0 \\ 0 & B \end{bmatrix}$, where 0 is the zero 2×2 matrix, is also diagonalizable.

10. If you did problem 9, show that if at least one of the matrices A and B is not diagonalizable, then the block matrix $\begin{bmatrix} A & 0 \\ 0 & B \end{bmatrix}$ is also not diagonalizable.

Maple Problems

1. Let $A = \begin{bmatrix} 2+i & -4-i & 1+2i & 5-i \\ 3-i & 1-i & -i & -2 \\ i & 7 & -3 & i \\ -2i & 8+i & -1 & 3+i \end{bmatrix}$. See if A is diagonalizable or

not, and find A^3 and e^A if it is diagonalizable. Of course, check that both A^3 and e^A are correct.

2. If you did *Homework* problem 1, then use your results from this problem on the matrix A of *Maple* problem 1.

3. Show that the following matrix is not diagonalizable:

$$G = \begin{bmatrix} 4-11i & 4-2i & 6-3i \\ -4+2i & 14-16i & 12-6i \\ 2-i & -4+2i & -9i \end{bmatrix}$$

4. Let A and B be defined as follows:

$$A = \begin{bmatrix} 2-5i & 8-2i & 30-15i \\ -11-10i & 10-4i & -21-12i \\ 20+15i & -16+6i & -12+9i \end{bmatrix}, B = \begin{bmatrix} i & 1-2i & 1+i \\ 2+2i & 1-3i & 2-i \\ 1-i & 2+i & -5i \end{bmatrix}$$

(a) Show that the 6×6 block matrix $G = \begin{bmatrix} A & 0 \\ 0 & B \end{bmatrix}$ is diagonalizable using the diagonalizability of both A and B, and using this diagonalizability, find e^G and G^3.

(b) Can you also find $\sqrt{G}$? Do so if it is possible, and then check your result. If not, then explain why not.

5. Let A and B be the two matrices from the previous problem. Now let $G = \begin{bmatrix} A & I_3 \\ 0 & B \end{bmatrix}$. Is G still diagonalizable? If yes, then diagonalize G. If not, then explain why not.

6. Define G as follows:

$$G = \begin{bmatrix} 1 & -1 & 2 & 1 & 0 & 3 \\ -1 & -2 & 0 & 1 & 0 & 1 \\ 2 & 1 & -1 & 1 & 0 & 1 \\ 1 & -1 & 2 & 1 & 1 & 1 \\ 0 & 1 & -1 & 1 & 0 & 2 \\ 3 & 0 & 1 & -1 & 2 & 0 \end{bmatrix}$$

Use the diagonalizability of G to find both e^G and $\sqrt[3]{G}$, check your results.

Research Projects

1. For a real $n \times n$ matrix A, define the $n \times n$ matrix $e^{\sin(A)}$ and see if diagonalizability can help you find $e^{\sin(A)}$ as it did for $\sin(A)$ and e^A. Is the same thing possible for $e^{\cos(A)}$? Can you find a way to decide if

$$\frac{d}{dt} e^{\sin(A\,t)} = A \cos(A\,t) e^{\sin(A\,t)}$$

for a real variable t?

12.4   Solving a Square First-Order Linear System of Differential Equations

Now we will apply the diagonalizability of a square matrix A to solving a square *first-order linear system of homogeneous differential equations*. The form of a square 3×3 first-order linear system of homogeneous differential equations is

$$\frac{dX}{dt} = AX(t) \tag{12.7}$$

where

$$X(t) = \begin{bmatrix} x(t) \\ y(t) \\ z(t) \end{bmatrix}, \quad \frac{dX}{dt} = \begin{bmatrix} \frac{dx}{dt} \\ \frac{dy}{dt} \\ \frac{dz}{dt} \end{bmatrix}$$

and A is a square 3×3 matrix of real constants. Hence, the matrix equation can be expressed as

$$\frac{dx}{dt} = A_{1,1}x(t) + A_{1,2}y(t) + A_{1,3}z(t)$$

$$\frac{dy}{dt} = A_{2,1}x(t) + A_{2,2}y(t) + A_{2,3}z(t)$$

$$\frac{dz}{dt} = A_{3,1}x(t) + A_{3,2}y(t) + A_{3,3}z(t)$$

This is called a linear system since the RHS of these equations involve linear combinations of the unknown functions $x(t)$, $y(t)$, and $z(t)$. You will see later why it is called homogeneous.

In general, the derivative of a matrix of functions is the matrix of the corresponding derivatives of the original matrix's entries. The usual derivative rules also apply to matrix functions, such as the product rule.

First let us do an example so that you can see what we have in mind where all our eigenvalues and eigenvectors are real.

Example 12.4.1. Let us solve the 3×3 square first-order linear system of homogeneous differential equations given by the three equations

$$\frac{dx}{dt} = 5x(t) - 9y(t) + z(t)$$

$$\frac{dy}{dt} = -9x(t) + 2y(t) + \pi z(t) \qquad (12.8)$$

$$\frac{dz}{dt} = x(t) + \pi y(t) - 4z(t)$$

where $x(t)$, $y(t)$, and $z(t)$ are three unknown functions of t with the initial conditions:

$$x(0) = 7, \; y(0) = 15, \; z(0) = 21 \qquad (12.9)$$

This system of three differential equations can be written as a single matrix differential equation of the form (12.7), with

$$A = \begin{bmatrix} 5 & -9 & 1 \\ -9 & 2 & \pi \\ 1 & \pi & -4 \end{bmatrix}$$

The initial conditions become the single initial condition

$$X(0) = \begin{bmatrix} x(0) \\ y(0) \\ z(0) \end{bmatrix} = \begin{bmatrix} 7 \\ 15 \\ 21 \end{bmatrix}$$

The derivative matrix $\dfrac{dX}{dt}$ of the matrix $X(t)$ is the matrix of the corresponding derivatives of the entry functions of $X(t)$, and this is the definition for the derivative matrix.

In order to solve this matrix differential equation for the column $X(t)$, we first look at the simple analogous situation that we already know how to solve, namely,

$$\frac{dw}{dt} = \alpha \, w(t) \qquad (12.10)$$

where $w(t)$ is a normal function of t and α is a real constant. This simple differential equation has as its solution

$$w(t) = C \, e^{\alpha t} = e^{\alpha t} \, C$$

where C is an arbitrary constant. This solution $w(t)$ can be found by knowing the derivative rules and realizing that $w(t)$ must be an exponential function, or by using the *method of separation of variables* if you have had previous experience solving differential equations.

In case you do not recall the method of separation of variables, let us use it now to solve this differential equation. When considering equation (12.10), if we treat $\dfrac{dw}{dt}$ as a fraction (it is not a fraction), then we can multiply the equation by dt to get

$$dw = \alpha\, w\, dt$$

notice we have expressed $w(t)$ as w in the above expression. Next, we separate the variables w and t to different sides of the equation getting

$$\frac{dw}{w} = \alpha\, dt$$

Then we integrate both sides of this equation, with respect to the appropriate variable on each side:

$$\int \frac{1}{w}\, dw = \int \alpha\, dt$$

Both of these are simple integrals to compute, which gives

$$\ln(|w|) = \alpha\, t + D$$

for an arbitrary constant D. This equation can be solved for $w(t)$ to get

$$|w(t)| = C\, e^{\alpha t}$$

where $C = e^D$ is a positive arbitrary constant. If we allow C to be negative and 0, then our solution is

$$w(t) = C\, e^{\alpha t}, \tag{12.11}$$

where C is an arbitrary constant. You can easily check that this solution works by plugging it back into the original differential equation (12.10).

Now if the differential equation (12.10) has the initial condition $w(0) = w_0 \in \mathbb{R}$, then the constant C is no longer arbitrary, but a specific value determined by the initial condition. So we can determine C's value by plugging the initial condition information into our solution (12.11). This gives

$$w(0) = w_0 = C e^{\alpha 0} = C$$

which tells us that in the case of an initial condition $w(0) = w_0$, the solution to our differential equation is

$$w(t) = w_0\, e^{\alpha t} \tag{12.12}$$

This tells us that we should expect the matrix differential equation (12.7) with initial condition $X(0) = X_0 = \begin{bmatrix} x(0) \\ y(0) \\ z(0) \end{bmatrix}$ to have solution

$$X(t) = e^{At}X_0 \qquad (12.13)$$

where X_0 is the constant 3×1 column of initial conditions. The column X_0 must be placed on the RHS in the product since the other order of multiplication is impossible because e^{At} is a 3×3 matrix of functions of t since A is 3×3.

If the matrix A is diagonalizable, then we know that $A = QDQ^{-1}$ and so $At = Q(Dt)Q^{-1}$, since t is a real scalar variable. We must also have

$$e^{At} = Qe^{Dt}Q^{-1}$$

where Dt corresponds to scalar multiplication of D by the scalar variable t. In other words, if $D = \begin{bmatrix} \lambda_1 & 0 & 0 \\ 0 & \lambda_2 & 0 \\ 0 & 0 & \lambda_3 \end{bmatrix}$, then $e^{Dt} = \begin{bmatrix} e^{\lambda_1 t} & 0 & 0 \\ 0 & e^{\lambda_2 t} & 0 \\ 0 & 0 & e^{\lambda_3 t} \end{bmatrix}$, since

$Dt = \begin{bmatrix} \lambda_1 t & 0 & 0 \\ 0 & \lambda_2 t & 0 \\ 0 & 0 & \lambda_3 t \end{bmatrix}$, where the λ_1, λ_2, and λ_3 are the eigenvalues of A.

Now let us have *Maple* compute all of this for us by diagonalizing A. Since A is a real *symmetric matrix*, all of its eigenvalues and eigenvectors will be real.

```
> with(linalg): with(LinearAlgebra):
> A:= matrix(3,3,[5,-9,1,-9,2,Pi,1,Pi,-4]);
```

$$A := \begin{bmatrix} 5 & -9 & 1 \\ -9 & 2 & \pi \\ 1 & \pi & -4 \end{bmatrix}$$

```
> eigvects:= evalf(eigenvectors(A));
```

$eigvects := [12.72322141 - 1.\ 10^{-9}I, 1., \{[1, -0.8696437714$
$\qquad + 1.6628582711\ 10^{-10}I, -0.1035725380 + 2.131809784\ 10^{-11}I]\}],$
$\qquad [-7.991173640 - 1.632050808\ 10^{-9}I, 1., \{[1, 1.301772246$
$\qquad + 2.904280729\ 10^{-10}I, -1.275223427 + 2.797044013\ 10^{-10}I]\}],$
$\qquad [-1.732047776 + 1.832050808\ 10^{-9}I, 1., \{[1, 0.9419793895$
$\qquad - 5.430170884\ 10^{-10}I, 1.745766726 - 2.212317177\ 10^{-9}I]\}]$

```
> vals:= map(Re, [eigvects[1][1], eigvects[2][1], eigvects[3][1]]);
```

$$vals := [12.72322141, -7.991173640, -1.732047776]$$

> Diag:= DiagonalMatrix(vals);

$$Diag := \begin{bmatrix} 12.72322141 & 0 & 0 \\ 0 & -7.991173640 & 0 \\ 0 & 0 & -1.732047776 \end{bmatrix}$$

> Qlist:= map(Re, [op(convert(op(eigvects[1][3]), list)), op(convert(op(eigvects [2][3]), list)), op(convert(op(eigvects [3][3]), list))]);

$$Qlist := [1., -0.8696437714, -0.1035725380, 1., 1.301772246,$$
$$- 1.275223427, 1., 0.9419793895, 1.745766726]$$

> Q:= transpose(matrix(3,3,Qlist));

$$Q := \begin{bmatrix} 1. & 1. & 1. \\ -0.8696437714 & 1.301772246 & 0.9419793895 \\ -0.1035725380 & -1.275223427 & 1.745766726 \end{bmatrix}$$

> evalm(Q&*Diag&*Q^(-1));

$$\begin{bmatrix} 5.000000003 & -9.000000003 & 1.000000001 \\ -8.999999994 & 1.999999989 & 3.141592651 \\ 0.9999999992 & 3.141592654 & -3.999999999 \end{bmatrix}$$

> print(A);

$$\begin{bmatrix} 5 & -9 & 1 \\ -9 & 2 & \pi \\ 1 & \pi & -4 \end{bmatrix}$$

We see from the last two computations that A is diagonalizable for the above Q and D (or $Diag$). Now we can solve our first-order linear system of differential equations.

> eDiagt:=matrix(3,3,[exp(vals[1]*t),0,0,0,exp(vals[2]*t),0,0,0,exp(vals[3]*t)]);

$$eDiagt := \begin{bmatrix} e^{12.72322141\ t} & 0 & 0 \\ 0 & e^{-7.991173640\ t} & 0 \\ 0 & 0 & e^{-1.732047776\ t} \end{bmatrix}$$

> eAt:= evalm(Q&*eDiagt&*Q^(-1));

$eAt :=$

$\big[\ [0.5659285350e^{12.72322141\,t} + 0.2314383135e^{-7.991173640\,t} + 0.2026331516e^{-1.732047776\,t},$

$\quad -0.4921562260e^{12.72322141\,t} + 0.3012799734e^{-7.991173640\,t} + 0.1908762527e^{-1.732047776\,t},$

$\quad -0.05861465461e^{12.72322141\,t} - 0.2951355591e^{-7.991173640\,t} + 0.3537502137e^{-1.732047776\,t}]$

$\quad [-0.4921562255e^{12.72322141\,t} + 0.3012799732e^{-7.991173640\,t} + 0.1908762524e^{-1.732047776\,t},$

$\quad 0.4280005965e^{12.72322141\,t} + 0.3921979076e^{-7.991173640\,t} + 0.1798014960e^{-1.732047776\,t},$

$\quad 0.05097386929e^{12.72322141\,t} - 0.3841992796e^{-7.991173640\,t} + 0.3332254103e^{-1.732047776\,t}]$

$\quad [0.05861465470e^{12.72322141\,t} - 0.2951355593e^{-7.991173640\,t} + 0.3537502136e^{-1.732047776\,t},$

$\quad 0.05097386942e^{12.72322141\,t} - 0.3841992802e^{-7.991173640\,t} + 0.3332254107e^{-1.732047776\,t},$

$\quad 0.006070868542e^{12.72322141\,t} + 0.3763637791e^{-7.991173640\,t} + 0.6175653524e^{-1.732047776\,t}]\ \big]$

> X0:= matrix(3,1,[7,15,21]);

$$X0 := \begin{bmatrix} 7 \\ 15 \\ 21 \end{bmatrix}$$

> X:= evalm(eAt&*X0);

$X :=$

$$\begin{bmatrix} -4.651751392\,e^{12.72322141\,t} - 0.058578946\,e^{-7.991173640\,t} + 11.71033034\,e^{-1.732047776\,t} \\ 4.045366625\,e^{12.72322141\,t} - 0.076256446\,e^{-7.991173640\,t} + 11.03088982\,e^{-1.732047776\,t} \\ 0.4817936978\,e^{12.72322141\,t} + 0.074701243\,e^{-7.991173640\,t} + 20.44350506\,e^{-1.732047776\,t} \end{bmatrix}$$

Note that the solution column $X(t)$ above is written in a particularly nice way, namely, if the three eigenvalues of A are called λ_1, λ_2, and λ_3, then

$$X(t) = e^{\lambda_1 t}Z_1 + e^{\lambda_2 t}Z_2 + e^{\lambda_3 t}Z_3 \tag{12.14}$$

for constant column vectors Z_1, Z_2, and Z_3. If we compute the derivative $\dfrac{dX}{dt}$ for this linear combination, we get

$$\frac{dX}{dt} = \lambda_1 e^{\lambda_1 t}Z_1 + \lambda_2 e^{\lambda_2 t}Z_2 + \lambda_3 e^{\lambda_3 t}Z_3$$

since the Z's are constant columns.

Now switching to compute the other side of the differential equation we have

$$AX(t) = e^{\lambda_1 t}AZ_1 + e^{\lambda_2 t}AZ_2 + e^{\lambda_3 t}AZ_3$$

If we compare these two equations we see that the solution given by (12.14) is going to be a solution to our differential equation (12.7) if we have $AZ_k = \lambda_k Z_k$, for $k = 1, 2, 3$. In particular, it is now easy to see that for eigenvectors $X_1, X_2,$

and X_3 (forming the columns of Q) for the corresponding eigenvalues λ_1, λ_2, and λ_3 of A, each of the functions $e^{\lambda_k t} X_k$, $k = 1, 2, 3$ are solutions to our system of differential equations.

In a homogeneous linear system of differential equations, the linearity of the system will tell us that any linear combination of solutions is also a solution, that is, if $U(t)$ and $V(t)$ are two solutions and a, b are two real arbitrary constants, then $X(t) = aU(t) + bV(t)$ is also a solution since

$$
\begin{aligned}
\frac{dX}{dt} &= a\frac{dU}{dt} + b\frac{dV}{dt} \\
&= aAU(t) + bAV(t) \\
&= A(aU(t) + bV(t)) \\
&= AX(t)
\end{aligned}
$$

So from what we have seen above we now expect that the solutions $X(t)$ can be written as

$$X(t) = a_1 e^{\lambda_1 t} X_1 + a_2 e^{\lambda_2 t} X_2 + a_3 e^{\lambda_3 t} X_3 \tag{12.15}$$

for real arbitrary constants a_1, a_2, and a_3, where the eigenvectors X_1, X_2, and X_3 are for the corresponding eigenvalues λ_1, λ_2, and λ_3 of A, when A is diagonalizable.

Now the initial condition $X(0) = X_0$ will determine uniquely the three values of the arbitrary constants a_1, a_2, and a_3 since

$$
\begin{aligned}
X_0 = X(0) &= a_1 e^{\lambda_1 \cdot 0} X_1 + a_2 e^{\lambda_2 \cdot 0} X_2 + a_3 e^{\lambda_3 \cdot 0} X_3 \\
&= a_1 X_1 + a_2 X_2 + a_3 X_3
\end{aligned}
$$

So you can solve the linear system:

$$a_1 X_1 + a_2 X_2 + a_3 X_3 = X_0$$

for a unique solution for three unknown coefficients a_1, a_2, and a_3. This linear system can also be written as the matrix equation

$$
Q \begin{bmatrix} a_1 \\ a_2 \\ a_3 \end{bmatrix} = X_0
$$

giving

$$
\begin{bmatrix} a_1 \\ a_2 \\ a_3 \end{bmatrix} = Q^{-1} X_0 \tag{12.16}
$$

for the columns of Q the eigenvectors X_1, X_2, and X_3 in this order. We will verify all of this for our current example shortly.

So, now we have another way to solve the $n \times n$ homogeneous first-order linear system of differential equations of the form (12.7) with initial condition $X(0) = X_0$ for A diagonalizable. The solution $n \times 1$ column $X(t)$ is given by

$$X(t) = \sum_{k=1}^{n} a_k e^{\lambda_k t} X_k$$

where A's n eigenvalues are the λ_k's, their corresponding eigenvectors are the X_k's, and the coefficients a_k's are gotten by solving the linear system

$$\sum_{k=1}^{n} a_k X_k = X_0$$

where, similar to equation 12.16, the column matrix of coefficients is given by

$$\begin{bmatrix} a_1 \\ a_2 \\ \vdots \\ a_n \end{bmatrix} = Q^{-1} X_0 \tag{12.17}$$

with the columns of the matrix Q being the eigenvectors $X_1, X_2, \ldots, X_n$, in this order. You can check that this solution works by plugging it into the differential equation (12.7) and also checking that $X(0) = X_0$. One thing to remember in solving differential equations with initial conditions is that their solution is unique, and so any method you use for getting the answer will give the same result as any other method although the answers may appear to be different or rearranged.

```
> x:= unapply(X[1,1], t);
```

$$x := t \rightarrow -4.651751392 e^{12.72322141\,t} - 0.058578946 e^{-17.991173640\,t} + 11.71033034 e^{-1.732047776\,t}$$

```
> x(0);
```

$$7.000000002$$

```
> y:= unapply(X[2,1], t);
```

$$y := t \rightarrow 4.045366625 e^{12.72322141\,t} - 0.076256446 e^{-7.991173640\,t} + 11.03088982 e^{-1.732047776\,t}$$

```
> y(0);
```

$$15.00000000$$

```
> z:= unapply(X[3,1], t);
```

$$z := t \rightarrow 0.4817936978 e^{12.72322141\,t} + 0.074701243 e^{-7.991173640\,t} + 20.44350506 e^{-1.732047776\,t}$$

```
> z(0);
```

$$21.00000000$$

> dXdt:= map(diff, X, t);

$dX dt :=$

$$\begin{bmatrix} -59.18526290e^{12.72322141\,t} + 0.4681145291e^{-7.991173640\,t} - 20.28285162e^{-1.732047776\,t} \\ 51.47009525e^{12.72322141\,t} + 0.6093785012e^{-7.991173640\,t} - 19.10602818e^{-1.732047776\,t} \\ 6.129967891e^{12.72322141\,t} - 0.5969506039e^{-7.991173640\,t} - 35.40912747e^{-1.732047776\,t} \end{bmatrix}$$

> evalf(evalm(A&*X));

$$\begin{bmatrix} -59.18526288e^{12.72322141\,t} + 0.468114527e^{-7.991173640\,t} - 20.28285162e^{-1.732047776\,t} \\ 51.47009532e^{12.72322141\,t} + 0.6093784983e^{-7.991173640\,t} - 19.10602814e^{-1.732047776\,t} \\ 6.129967887e^{12.72322141\,t} - 0.5969506086e^{-7.991173640\,t} - 35.40912747e^{-1.732047776\,t} \end{bmatrix}$$

The last two outputs show that our solution X does satisfy the matrix differential equation (12.7), as expected. Now let us recompute the solution $X(t)$ using (12.15), where $a_1 X_1 + a_2 X_2 + a_3 X_3 = X_0$, or $\begin{bmatrix} a_1 \\ a_2 \\ a_3 \end{bmatrix} = Q^{-1} X_0$, and A's eigenvalues are λ_1, λ_2 and λ_3 with corresponding eigenvectors X_1, X_2, and X_3 forming the columns of Q. We will compute the a_k's directly as well as using the formula for them of $Q^{-1} X_0$.

> X1:= col(Q,1);

$$X1 := \begin{bmatrix} 1. & -0.8696437714 & -0.1035725380 \end{bmatrix}$$

> X2:= col(Q,2);

$$X2 := \begin{bmatrix} 1. & 1.301772246 & -1.275223427 \end{bmatrix}$$

> X3:= col(Q,3);

$$X3 := \begin{bmatrix} 1. & 0.9419793895 & 1.745766726 \end{bmatrix}$$

> leftsidesystem:= convert(evalm(a*X1+b*X2+c*X3), list);

$leftsidesystem := \big[1. a + 1. b + 1. c, -0.8696437714 a + 1.301772246 b$
$+ 0.9419793895 c, -0.1035725380 a - 1.275223427 b + 1.745766726 c\big]$

> coeffs_a:= solve({leftsidesystem[1] = X0[1,1], leftsidesystem[2] = X0[2,1], leftsidesystem[3] = X0[3,1]}, [a, b, c])[1];

$coeffs\_a := [a = -4.651751392, b = -0.05857894720, c = 11.71033034]$

> rhs(coeffs_a[2]);

$$-0.05857894720$$

> evalm(Q^(-1)&*X0);

$$\begin{bmatrix} -4.651751392 \\ -0.058578946 \\ 11.71033034 \end{bmatrix}$$

> new_X:= evalm(add(rhs(coeffs_a[K])*exp(vals[K]*t)*col(Q,K), K=1..3));

$new\_X :=$

$[-4.651751392e^{12.72322141\ t} - 0.05857894720e^{-7.991173640\ t} + 11.71033034e^{-1.732047776\ t},$

$4.045366624e^{12.72322141\ t} - 0.07625644766e^{-7.991173640\ t} + 11.03088982e^{-1.732047776\ t},$

$0.4817936978e^{12.72322141\ t} + 0.07470124580e^{-7.991173640\ t} + 20.44350506e^{-1.732047776\ t}]$

> print(X);

$[-4.651751392e^{12.72322141\ t} - 0.05857894720e^{-7.991173640\ t} + 11.71033034e^{-1.732047776\ t},$

$4.045366624e^{12.72322141\ t} - 0.07625644766e^{-7.991173640\ t} + 11.03088982e^{-1.732047776\ t},$

$0.4817936978e^{12.72322141\ t} + 0.07470124580e^{-7.991173640\ t} + 20.44350506e^{-1.732047776\ t}]$

Our three solution functions $x(t)$, $y(t)$, and $z(t)$ to the linear system of differential equations are $X[1,1]$, $X[2,1]$, and $X[3,1]$, respectively, from above. It is also checked that $X(t)$ is the solution to this system of differential equations and satisfies the initial conditions.

Maple can also solve systems of differential equations using its *dsolve* command. Let us use it now to check again that our result above is correct. The option "method = laplace" should always be used with *dsolve* if you want to solve linear systems of differential equations without getting unnecessary complex numbers when you know your results must be completely real. You should redo this computation without "method = laplace" in order to see what you would get instead. The option 'method = laplace' uses the *Laplace transform*.

Note that we must now unassign the names x, y, and z since we used them above as function names for our solutions.

> x:= 'x': y:= 'y': z:= 'z':
> evalf(dsolve({diff(x(t), t) = 5.*x(t)-9.*y(t)+z(t), diff(y(t), t) = -9.*x(t) + 2.*y(t)+Pi*z(t), diff(z(t), t) = x(t)+Pi*y(t)-4.*z(t), x(0) = 7., y(0) = 15., z(0) = 21.}, {x(t), y(t), z(t)}, method = laplace));

$\{x(t) = -0.05857895046e^{-7.991173640\ t} + 11.71033034e^{-1.732047775\ t} - 4.651751394e^{12.72322141\ t},$

$y(t) = -0.07625645190e^{-7.991173640\ t} + 11.03088982e^{-1.732047775\ t} + 4.045366631e^{12.72322141\ t},$

$z(t) = 0.07470124992e^{-7.991173640\ t} + 20.44350505e^{-1.732047775\ t} + 0.4817936984e^{12.72322141\ t}\}$

As the final part of this example, let us plot this solution $X(t)$ as a curve in space for $t \in [-0.25, 0.25]$. This plot will be a parametric plot that gives a space curve. Also, the initial condition point at X_0 is plotted as a box. The *point* command is in the *plottools* package, while the *spacecurve* command is in the *plots* package. See Figure 12.1 below.

> with(plots): with(plottools):

> plot_X := spacecurve([X[1,1], X[2,1], X[3,1]], t = -0.25..0.25, axes = boxed, thickness = 2, color = red):

> point_X0:= point([X0[1,1],X0[2,1],X0[3,1]], symbol = box, symbolsize = 25, color = black):

> display3d({plot_X, point_X0});

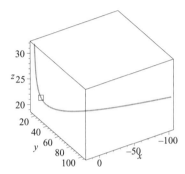

Figure 12.1: Solution to the ODE system (12.8), with initial conditions given in equation (12.9).

The last example took a great deal of time and effort since we needed to use it to illustrate how several methods might be applied to solve the same problem. Happily, in Example 12.4.1, the eigenvalues and eigenvectors were all real since the coefficient matrix A was a real symmetric matrix. The next example will not be as nice as the last one. We will choose a system whose corresponding real-valued matrix A has complex eigenvalues and eigenvectors. However, the system of homogeneous differential equations will still be completely real and we still seek a completely real set of solution functions.

Example 12.4.2. Now let us solve the 3×3 square first-order linear system of homogeneous differential equations given by the three equations

$$\frac{dx}{dt} = -2x(t) + 9y(t) - z(t)$$
$$\frac{dy}{dt} = -4x(t) + y(t) - \pi z(t) \qquad (12.18)$$
$$\frac{dz}{dt} = 5x(t) - 3y(t) + 4z(t)$$

where $x(t)$, $y(t)$, and $z(t)$ are three unknown functions of t with the initial conditions

$$x(0) = -6,\ y(0) = 10,\ z(0) = 7 \tag{12.19}$$

This system of three differential equations can be written as the single matrix differential equation (12.7), where

$$X(t) = \begin{bmatrix} x(t) \\ y(t) \\ z(t) \end{bmatrix},\ \frac{dX}{dt} = \begin{bmatrix} \frac{dx}{dt} \\ \frac{dy}{dt} \\ \frac{dz}{dt} \end{bmatrix},\ A = \begin{bmatrix} -2 & 9 & -1 \\ -4 & 1 & -\pi \\ 5 & -3 & 4 \end{bmatrix}$$

The three scalar initial conditions become the single column vector initial condition

$$X(0) = \begin{bmatrix} x(0) \\ y(0) \\ z(0) \end{bmatrix} = \begin{bmatrix} -6 \\ 10 \\ 7 \end{bmatrix}$$

Now let us have *Maple* compute the eigenvalues and eigenvectors of A. Since A is not a real symmetric matrix, just a real 3×3 matrix, some of its eigenvalues and eigenvectors will be real and the rest must be in complex conjugate pairs since the degree of A's characteristic polynomial is three, which is odd.

> A:= matrix(3,3,[-2,9,-1,-4,1,-Pi,5,-3,4]):
> eigvects:= evalf(eigenvectors(A));

   $eigvects := \big[\, 0.260559645,\, 1.0,\, \{[-0.8021012527,\, -0.09035530313,\, 1.]\}\,\big],\, [1.369720178$
      $- 4.794298811\, I,\, 1.,\, \{\, [\, - 0.9421769983 - 0.8482872741 I,\, -0.6935350565$
      $+ 0.1842874795 I,\, 1.]\,\}\,\big],\, [1.369720178 + 4.794298811 I,\, 1.,\, \{\, [\, - 0.9421769983$
      $+ 0.8482872741 I,\, -0.6935350565 - 0.1842874795 I,\, 1.]\,\}\,\big]$

> vals:= [eigvects[1][1], eigvects[2][1], eigvects[3][1]];

   $vals := [0.260559645,\, 1.369720178 - 4.794298811I,\, 1.369720178 + 4.794298811I]$

> Diag:= DiagonalMatrix(vals);

$$Diag := \begin{bmatrix} 0.260559645 & 0 & 0 \\ 0 & 1.369720178 - 4.794298811I & 0 \\ 0 & 0 & 1.369720178 + 4.794298811I \end{bmatrix}$$

> Qlist:= [op(convert(op(eigvects[1][3]), list)), op(convert(op(eigvects[2][3]), list)), op(convert(op(eigvects[3][3]), list))];

   $Qlist := \big[-0.8021012527,\, -0.09035530313,\, 1.,$
      $- 0.9421769983 - 0.8482872741I,\, -0.6935350565 + 0.1842874795I,\, 1.,$
      $- 0.9421769983 + 0.8482872741I,\, -0.6935350565 - 0.1842874795I,\, 1.\big]$

> Q:= transpose(matrix(3,3,Qlist));

$$Q := \begin{bmatrix} -0.8021012527 & -0.9421769983 - 0.8482872741I & -0.9421769983 + 0.8482872741I \\ -0.09035530313 & -0.6935350565 + 0.1842874795I & -0.6935350565 - 0.1842874795I \\ 1. & 1. & 1. \end{bmatrix}$$

> evalm(Q&*Diag&*Q^(-1));

$$\begin{bmatrix} -2.000000000 + 0.I & 9.000000012 + 0.I & -0.9999999937 + 0.I \\ -4.000000000 + 0.I & 1.000000001 + 0.I & -3.141592652 + 0.I \\ 5.000000001 + 0.I & -3.000000002 + 0.I & 3.999999999 + 0.I \end{bmatrix}$$

> print(A);

$$\begin{bmatrix} -2 & 9 & -1 \\ -4 & 1 & -\pi \\ 5 & -3 & 4 \end{bmatrix}$$

We see from the last two computations that A is diagonalizable for the above defined Q and D (or $Diag$). Now we can solve our first-order linear system of differential equations.

> eDiagt:=matrix(3,3,[exp(vals[1]*t),0,0,0,exp(vals[2]*t),0,0,0,exp(vals[3]*t)]);

$eDiagt :=$

$$\begin{bmatrix} e^{0.260559645t} & 0 & 0 \\ 0 & e^{(1.369720178-4.794298811I)t} & 0 \\ 0 & 0 & e^{(1.369720178+4.794298811I)t} \end{bmatrix}$$

> eAt:= evalm(Q&*eDiagt&*Q^(-1));

$eAt := [[(- 0.2750170080 + 0. I) e^{0.260559645\,t} + (0.6375085039$

$- 0.3832425068I) e^{(1.369720178-4.794298811I)\,t} + (0.6375085039$

$+ 0.3832425068I) e^{(1.369720178+4.794298811I)\,t}, (- 1.265921205 - 0. I) e^{0.260559645\,t}$

$+ (0.6329606023 + 0.7921794691I) e^{(1.369720178-4.794298811I)\,t} + (0.6329606023$

$- 0.7921794691I) e^{(1.369720178+4.794298811I)\,t}, (- 1.137075434 + 0. I) e^{0.260559645\,t}$

$+ (0.5685377167 - 0.2358216786\,I) e^{(1.369720178-4.794298811\,I)t} + (0.5685377167$

$+ 0.2358216786\,I) e^{(1.369720178+4.794298811\,I)t}], [(- 0.03098018491$

$+ 0.\,I) e^{0.260559645\,t} + (0.01549009246 - 0.4207457813\,I) e^{(1.369720178-4.794298811\,I)t}$

$+ (0.01549009246 + 0.4207457813\,I) e^{(1.369720178+4.794298811\,I)t}, (- 0.1426038094$

$- 0.\,I) e^{0.260559645\,t} + (0.5713019046 - 0.05505400424\,I) e^{(1.369720178-4.794298811\,I)t}$

$+ (0.5713019046 + 0.05505400424\,I) e^{(1.369720178+4.794298811\,I)t}, (- 0.1280895587$

$+ 0.\,I) e^{0.260559645\,t} + (0.06404477938 - 0.3424551395\,I) e^{(1.369720178-4.794298811\,I)t}$

$+ (0.06404477938 + 0.3424551395\,I) e^{(1.369720178+4.794298811\,I)t}], [[(0.3428706876$

$$- 0.I) e^{0.260559645\,t} - (0.1714353438 - 0.5611142367\,I) e^{(1.369720178 - 4.794298811\,I)t}$$

$$- (0.1714353438 + 0.5611142367\,I) e^{(1.369720178 + 4.794298811\,I)t},\ (1.578256112$$

$$+ 0.I) e^{0.260559645\,t} - (0.7891280561 + 0.1303069187\,I) e^{(1.369720178 - 4.794298811\,I)t}$$

$$- (0.7891280561 - 0.1303069187\,I) e^{(1.369720178 + 4.794298811\,I)t},\ (1.417620818$$

$$- 0.I) e^{0.260559645\,t} - (0.2088104088 - 0.4382965110\,I) e^{(1.369720178 - 4.794298811\,I)t}$$

$$- (0.2088104088 + 0.4382965110\,I) e^{(1.369720178 + 4.794298811\,I)t}]\,]$$

> X0:= matrix(3,1,[-6,10,7]);

$$X0 := \begin{bmatrix} -6 \\ 10 \\ 7 \end{bmatrix}$$

> X:= evalm(eAt&*X0);

$$X := [[(-18.96863804 + 0.I)\,e^{0.260559645\,t}$$

$$+ (6.484319017 + 8.570497980\,I)\,e^{(1.369720178 - 4.794298811\,I)t}$$

$$+ (6.484319017 - 8.570497980\,I)\,e^{(1.369720178 + 4.794298811\,I)t}]$$

$$[(-2.136783895 + 0.I)\,e^{0.260559645\,t}$$

$$+ (6.068391947 - 0.423251330\,I)\,e^{(1.369720178 - 4.794298811\,I)t}$$

$$+ (6.068391947 + 0.423251330\,I)\,e^{(1.369720178 + 4.794298811\,I)t}]$$

$$[(23.64868272 + 0.I)\,e^{0.260559645\,t}$$

$$+ (-8.324341360 + 1.601679030\,I)\,e^{(1.369720178 - 4.794298811\,I)t}$$

$$+ (-8.324341360 + 1.601679030\,I)\,e^{(1.369720178 + 4.794298811\,I)t}]]$$

Our solution functions $x(t)$, $y(t)$, and $z(t)$ must be all real functions since A is real and the initial conditions are all real. In order to get real functions from the solution X above, we apply the real part command, Re, to the components of X. We must also tell *Maple* to assume that t is a real variable, which it is. In order to avoid thereafter *Maple* displaying t as $t \sim$ in order to indicate t has an assumption placed upon it, we use "interface(showassumed = 0)" to tell *Maple* to not use $t \sim$ in place of t. We now recall *Euler's Formula*:

$$e^{a+ib} = e^a\,(\cos(a) + i\sin(b))$$

where a and b are both real constants or real functions of t. Also, the real part of the product of two complex numbers $a + ib$ and $c + id$ is $ac - bd$ while its imaginary part is $ad + bc$. The same two formulas work if one of the two factors is not a complex number but a complex column vector.

> assume(t, real); interface(showassumed = 0):

> x:= unapply(Re(X[1,1]), t);

$$x := t \to -18.96863804e^{0.260559645\,t} + 12.96863803e^{1.369720178\,t}\cos\left(4.794298811\,t\right)$$
$$+ 17.14099596e^{1.369720178\,t}\sin\left(4.794298811\,t\right)$$

> x(0);
$$-6.00000001$$

> y:= unapply(Re(X[2,1]), t);

$$y := t \to -2.136783895e^{0.260559645\,t} + 12.13678389e^{1.369720178\,t}\cos\left(4.794298811\,t\right)$$
$$- 0.846502660e^{1.369720178\,t}\sin\left(4.794298811\,t\right)$$

> y(0);
$$9.999999995$$

> z:= unapply(Re(X[3,1]), t);

$$z := t \to 23.64868272e^{0.260559645\,t} - 16.64868272e^{1.369720178\,t}\cos\left(4.794298811\,t\right)$$
$$- 3.203358060e^{1.369720178\,t}\sin\left(4.794298811\,t\right)$$

> z(0);
$$7.00000000$$

As the next part of this example, let us plot this real solution $X(t)$ as a curve in space for $t \in [-1, 1]$. This plot, depicted in Figure 12.2, will be a parametric plot that gives a space curve. Also, the initial condition point at X_0 is plotted as a box.

> plot_X:= spacecurve([x(t),y(t),z(t)], t = -1..1, axes = boxed, thickness = 2, color = red):

> point_X0:= point([X0[1,1], X0[2,1], X0[3,1]], symbol = box, symbolsize = 25, color = black):

> display3d({plot_X, point_X0});

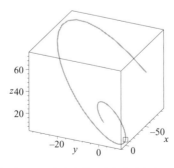

Figure 12.2: Solution to (12.18) with initial conditions (12.19).

Now we redefine our solution column X by the column of these three functions x, y, and z of t.

> X:= matrix(3,1,[x(t), y(t), z(t)]);

$$X := \big[\,[-18.96863804e^{0.260559645\,t} + 12.96863803e^{1.369720178\,t} \cos{(4.794298811\,t)}$$
$$+ 17.14099596e^{1.369720178\,t} \sin{(4.794298811\,t)}],$$
$$[-2.136783895e^{0.260559645\,t} + 12.13678389e^{1.369720178\,t} \cos{(4.794298811\,t)}$$
$$- 0.846502660e^{1.369720178\,t} \sin{(4.794298811\,t)}\,],$$
$$[23.64868272e^{0.260559645\,t} - 16.64868272e^{1.369720178\,t} \cos{(4.794298811\,t)}$$
$$- 3.203358060e^{1.369720178\,t} \sin{(4.794298811\,t)}\,]\,\big]$$

> dXdt:= map(diff, X, t);

$$dXdt := \big[\,[-4.942461594e^{0.260559645\,t} + 99.94246174e^{1.369720178\,t} \cos{(4.794298811\,t)}$$
$$- 38.69715785e^{1.369720178\,t} \sin{(4.794298811\,t)}],$$
$$[-0.5567596531e^{0.260559645\,t} + 12.56561109e^{1.369720178\,t} \cos{(4.794298811\,t)}$$
$$- 59.34684034e^{1.369720178\,t} \sin{(4.794298811\,t)}],$$
$$[6.161892374e^{0.260559645\,t} - 38.16189240e^{1.369720178\,t} \cos{(4.794298811\,t)}$$
$$+ 75.43105560e^{1.369720178\,t} \sin{(4.794298811\,t)}\,]\,\big]$$

> evalf(evalm(A&*X));

$$\big[\,[-4.94246170e^{0.260559645\,t} + 99.94246166e^{1.369720178\,t} \cos{(4.794298811\,t)}$$
$$- 38.69715780e^{1.369720178\,t} \sin{(4.794298811\,t)}], [-0.55675965e^{0.260559645\,t}$$
$$+ 12.56561110e^{1.369720178\,t} \cos{(4.794298811\,t)}$$
$$- 59.34684035e^{1.369720178\,t} \sin{(4.794298811\,t)}],$$
$$[6.161892365e^{0.260559645\,t} - 38.16189240e^{1.369720178\,t} \cos{(4.794298811\,t)}$$
$$+ 75.43105554e^{1.369720178\,t} \sin{(4.794298811\,t)}\,]\,\big]$$

Our three solution functions $x(t)$, $y(t)$, and $z(t)$ to the linear system of differential equations are $X[1, 1]$, $X[2, 1]$, and $X[3, 1]$, respectively, from above. It is also checked that $X(t)$ is the solution to this system of differential equations and satisfies the initial conditions.

Maple can now solve this system using *dsolve* if we take the real part of the three solutions. Compare the result to X computed previously.

> x:= 'x': y:= 'y': z:= 'z':

> map(Re, evalf(dsolve({diff(x(t),t) = -2.*x(t)+9.*y(t)-z(t), diff(y(t), t) = -4.*x(t)+y(t)-Pi*z(t), diff(z(t),t) = 5*x(t)-3*y(t)+4.*z(t), x(0) = -6., y(0) =

10., z(0) = 7.},{x(t), y(t), z(t)}, method = laplace)));

$\{\Re\left(x\left(t\right)\right) = -18.96863801e^{0.2605596442\,t} + 12.96863801e^{1.369720178\,t}\cos\left(4.794298809\,t\right)$
$\qquad + 17.14099594e^{1.369720178\,t}\sin\left(4.794298809\,t\right),$
$\Re\left(y\left(t\right)\right) = -2.136783878e^{0.2605596442\,t} + 12.13678388e^{1.369720178\,t}\cos\left(4.794298809\,t\right)$
$\qquad - 0.8465026612e^{1.369720178\,t}\sin\left(4.794298809\,t\right),$
$\Re\left(z\left(t\right)\right) = 23.64868269e^{0.2605596442\,t} - 16.64868269e^{1.369720178\,t}\cos\left(4.794298809\,t\right)$
$\qquad - 3.203358060e^{1.369720178\,t}\sin\left(4.794298809\,t\right)\}$

> print(X);

$\Big[\,\big[-18.96863804e^{0.260559645\,t} + 12.96863803e^{1.369720178\,t}\cos\left(4.794298811\,t\right)$
$+ 17.14099596e^{1.369720178\,t}\sin\left(4.794298811\,t\right)\big],$
$\big[-2.136783895e^{0.260559645\,t} + 12.13678389e^{1.369720178\,t}\cos\left(4.794298811\,t\right)$
$- 0.846502660e^{1.369720178\,t}\sin\left(4.794298811\,t\right)\big],$
$\big[23.64868272e^{0.260559645\,t} - 16.64868272e^{1.369720178\,t}\cos\left(4.794298811\,t\right)$
$- 3.203358060e^{1.369720178\,t}\sin\left(4.794298811\,t\right)\big]\,\Big]$

Now let us recompute the solution $X(t)$ using

$$X(t) = \Re\left(a_1 e^{\lambda_1 t}X_1 + a_2 e^{\lambda_2 t}X_2 + a_3 e^{\lambda_3 t}X_3\right)$$

where

$$a_1 X_1 + a_2 X_2 + a_3 X_3 = X_0,$$

which results in the following matrix equation to find the a_k's:

$$\begin{bmatrix} a_1 \\ a_2 \\ a_3 \end{bmatrix} = \Re\left(Q^{-1}X_0\right)$$

> X1:= col(Q,1);

$$X1 := \begin{bmatrix} -0.8021012527 & -0.09035530313 & 1. \end{bmatrix}$$

> X2:= col(Q,2);

$$X2 := \begin{bmatrix} -0.9421769983 - 0.8482872741I & -0.6935350565 + 0.1842874795I & 1. \end{bmatrix}$$

> X3:= col(Q,3);

$$X3 := \begin{bmatrix} -0.9421769983 + 0.8482872741I & -0.6935350565 - 0.1842874795I & 1. \end{bmatrix}$$

> leftsidesystem:= convert(evalm(a*X1+b*X2+c*X3), list);

$leftsidesystem := \big[-0.8021012527\,a + (-.9421769983 - .8482872741I)\,b$
$+ (-0.9421769983 + 0.8482872741I)\,c, \;-0.09035530313\,a + (-0.6935350565$
$+ 0.1842874795I)\,b + (-0.6935350565 - 0.1842874795I)\,c, 1.\,a + 1.\,b + 1.\,c\big]$

> coeffs_a:= solve({leftsidesystem[1] = X0[1,1], leftsidesystem[2] = X0[2,1], leftsidesystem[3] = X0[3,1]}, [a, b, c]);

$$coeffs\_a := \left[\,\left[\,a = 23.64868272, b = -8.324341362 - 1.601679030I,\right.\right.$$
$$\left.\left. c = -8.324341362 + 1.601679030I\,\right]\,\right]$$

> evalm(Re(Q^(-1)&*X0));

$$\begin{bmatrix} 23.64868272 \\ -8.324341360 \\ -8.324341360 \end{bmatrix}$$

> new_X:= map(Re, evalm(add(rhs(coeffs_a[1,K])*exp(vals[K]*t)*col(Q, K), K=1..3)));

$new\_X := [\, -18.96863803e^{0.260559645\,t} + 12.96863804e^{1.369720178\,t}\cos(4.794298811\,t)$
$+ 17.14099597e^{1.369720178\,t}\sin(4.794298811\,t), -2.136783896e^{0.260559645\,t}$
$+ 12.13678390e^{1.369720178\,t}\cos(4.794298811\,t) - 0.8465026630e^{1.369720178\,t}\sin(4.794298811\,t),$
$23.64868272e^{0.260559645\,t} - 16.64868272e^{1.369720178\,t}\cos(4.794298811\,t)$
$- 3.203358060e^{1.369720178\,t}\sin(4.794298811\,t)]$

> print(X);

$$\left[\,\left[-18.96863804e^{0.260559645\,t} + 12.96863803e^{1.369720178\,t}\cos\left(4.794298811\,t\right)\right.\right.$$
$$\left.+ 17.14099596e^{1.369720178\,t}\sin\left(4.794298811\,t\right)\right],$$
$$\left[-2.136783895e^{0.260559645\,t} + 12.13678389e^{1.369720178\,t}\cos\left(4.794298811\,t\right)\right.$$
$$\left.- 0.846502660e^{1.369720178\,t}\sin\left(4.794298811\,t\right)\right],$$
$$\left[23.64868272e^{0.260559645\,t} - 16.64868272e^{1.369720178\,t}\cos\left(4.794298811\,t\right)\right.$$
$$\left.\left.- 3.203358060e^{1.369720178\,t}\sin\left(4.794298811\,t\right)\right]\,\right]$$

In our next example of solving first-order linear systems of differential equations, we do a *nonhomogeneous* example. In the nonhomogeneous case, the system of linear differential equations is of the form

$$\frac{dX}{dt} = AX(t) + F(t) \tag{12.20}$$

where $F(t)$ is a column of real functions of t. In a homogeneous system, $F(t)$ does not appear or you can say $F(t)$ is the constant zero column.

In solving a homogeneous system of linear differential equations, a linear combination $Y(t) = aX_1(t) + bX_2(t)$ of two solutions $X_1(t)$ and $X_2(t)$ to the system is also a solution, while in a nonhomogeneous system this is not the case. This is the same when solving homogeneous $AX = 0$ versus nonhomogeneous $AX = B$ linear systems of equations, for $B \neq 0$.

Example 12.4.3. Let us solve the 3×3 square first-order linear system of nonhomogeneous differential equations given by the three equations

$$\frac{dx}{dt} = 5x(t) - 9y(t) + z(t) + \sin(t)$$

$$\frac{dy}{dt} = -9x(t) + 2y(t) + \pi z(t) - e^t \quad (12.21)$$

$$\frac{dz}{dt} = x(t) + \pi y(t) - 4z(t) + \cos(3t)$$

where $x(t)$, $y(t)$, and $z(t)$ are three unknown functions of t with the initial conditions

$$x(0) = -2, \; y(0) = 4, \; z(0) = 10 \quad (12.22)$$

This system of three differential equations can be written as a single matrix differential equation of the form (12.20), where

$$X(t) = \begin{bmatrix} x(t) \\ y(t) \\ z(t) \end{bmatrix}, \; A = \begin{bmatrix} 5 & -9 & 1 \\ -9 & 2 & \pi \\ 1 & \pi & -4 \end{bmatrix}, \; F(t) = \begin{bmatrix} \sin(t) \\ -e^t \\ \cos(3t) \end{bmatrix}, \; X(0) = \begin{bmatrix} -2 \\ 4 \\ 10 \end{bmatrix}$$

In order to solve this matrix differential equation (12.20) for the column $X(t)$, we first look at the simple analogous situation that we already know how to solve, namely,

$$\frac{dw}{dt} = \alpha \, w(t) + f(t) \quad (12.23)$$

where $w(t)$ and $f(t)$ are normal functions of t and $\alpha \in \mathbb{R}$. This first-order linear differential equation (12.23) has as its solution

$$w(t) = e^{\alpha t} \int e^{-\alpha t} f(t) \, dt + e^{\alpha t} C \quad (12.24)$$

where C is an arbitrary constant. You can easily check that this works by plugging it back into the differential equation for $w(t)$.

This solution was arrived at using the *method of integrating factors*. In case you have not seen this method or do not recall it now, let us go though the method. If you rewrite the differential equation (12.23) as

$$\frac{dw}{dt} - \alpha w(t) = f(t) \quad (12.25)$$

then you can see that the LHS of our new equation looks very much like the result of the product derivative rule, but not quite. If you multiply equation (12.25) by $e^{-\alpha t}$, then we have

$$e^{-\alpha t} \frac{dw}{dt} - \alpha e^{-\alpha t} w(t) = e^{-\alpha t} f(t)$$

or

$$\frac{d}{dt}\left(e^{-\alpha t}w(t)\right) = e^{-\alpha t}f(t),$$ (12.26)

by the product derivative rule. To get rid of the $\dfrac{d}{dt}$ on the LHS of equation (12.26), we must integrate both sides with respect to t. This yields

$$e^{-\alpha t}w(t) = \int e^{-\alpha t}f(t)\,dt + C$$

for C an arbitrary constant. Upon multiplying both sides of the above expression by $e^{\alpha t}$, we arrive at the solution given in equation (12.24).

Now if the differential equation (12.23) has the initial condition $w(0) = w_0$, then the constant C is no longer arbitrary, but a specific value determined by the initial condition. So we can determine C's value by plugging the initial condition information into our solution (12.24). This gives

$$w(0) = w_0 = e^{\alpha \cdot 0}\left(\int e^{-\alpha t}f(t)\,dt\right)\Bigg|_{t=0} + e^{\alpha \cdot 0}C$$

$$= \left(\int e^{-\alpha t}f(t)\,dt\right)\Bigg|_{t=0} + C$$

Solving for the unknown constant C gives

$$C = w_0 - \left(\int e^{-\alpha t}f(t)\,dt\right)\Bigg|_{t=0}$$

So in the case of an initial condition $w(0) = w_0$, the solution to our differential equation (12.23) is

$$w(t) = e^{\alpha t}\int e^{-\alpha t}f(t)\,dt + e^{\alpha t}\left[w_0 - \left(\int e^{-\alpha t}f(t)\,dt\right)\Bigg|_{t=0}\right]$$ (12.27)

This tells us that we should expect the matrix differential equation (12.20) with initial condition $X(0) = \begin{bmatrix} x(0) \\ y(0) \\ z(0) \end{bmatrix}$ to have the solution

$$X(t) = e^{At}(Z(t) + P_0)$$ (12.28)

where

$$Z(t) = \int e^{-At}F(t)\,dt, \quad P_0 = X_0 - Z(0)$$ (12.29)

This solution is correct regardless of A being diagonalizable or not.

When A is diagonalizable with $A = Qe^{D}Q^{-1}$, then $e^{-At} = Qe^{-Dt}Q^{-1}$, which allows our solution X(t) to be simplified to

$$X(t) = Qe^{Dt}(W(t) + R_0)$$ (12.30)

where

$$W(t) = \int e^{-Dt} Q^{-1} F(t) \ dt, \quad R_0 = Q^{-1} X_0 - W(0) \tag{12.31}$$

Let us now have *Maple* compute all of this for us by diagonalizing A again (Note that this is the same A as in Example 12.4.1), and then integrating for us. The integral of a matrix, $M(t)$, of functions of t is the corresponding matrix of the integrals of $M(t)$'s entries.

Since A is a real symmetric matrix, all of its eigenvalues and eigenvectors will be real. The eigenvalues of $-A$ are the negatives of the eigenvalues of A for the same eigenvectors of A, that is, if $A = QDQ^{-1}$, then $-A = Q(-D)Q^{-1}$.

> F:= matrix(3,1,[sin(t), -exp(t), cos(3*t)]);

$$F := \begin{bmatrix} \sin(t) \\ -e^t \\ \cos(3t) \end{bmatrix}$$

> A:= matrix(3,3,[5,-9,1,-9,2,Pi,1,Pi,-4]);

$$A := \begin{bmatrix} 5 & -9 & 1 \\ -9 & 2 & \pi \\ 1 & \pi & -4 \end{bmatrix}$$

> eigvects:= evalf(eigenvectors(A)):
> vals:= map(Re, [eigvects[1][1], eigvects[2][1], eigvects[3][1]]):
> Qlist:= map(Re, [op(convert(op(eigvects[1][3]), list)), op(convert(op(eigvects[2][3]), list)), op(convert(op(eigvects[3][3]), list))]):
> Q:= transpose(matrix(3,3,Qlist)):
> Diag:= DiagonalMatrix(vals):
> evalm(Q&*Diag&*Q^(-1));

$$\begin{bmatrix} 5.000000003 & -9.000000003 & 1.000000001 \\ -8.999999994 & 1.999999989 & 3.141592651 \\ 0.9999999992 & 3.141592654 & -3.999999999 \end{bmatrix}$$

> print(A);

$$\begin{bmatrix} 5 & -9 & 1 \\ -9 & 2 & \pi \\ 1 & \pi & -4 \end{bmatrix}$$

We see from the last two computations that A is diagonalizable for the above defined Q and D (or *Diag*). Now we can solve our first-order linear system of differential equations.

> eDiagt:= matrix(3, 3, [exp(vals[1]*t), 0, 0, 0, exp(vals[2]*t), 0, 0, 0, exp(
vals[3]*t)]):

> eAt:= evalm(Q&*eDiagt&*Q^(-1)):

> cnegDiagt:= matrix(3,3, [exp(-vals[1]*t), 0, 0, 0, exp(-vals[2]*t), 0, 0, 0,
exp(-vals[3]*t)]);

$$enegDiagt := \begin{bmatrix} e^{-12.72322141\ t} & 0 & 0 \\ 0 & e^{7.991173640\ t} & 0 \\ 0 & 0 & e^{1.732047776\ t} \end{bmatrix}$$

> enegAt:= evalm(Q&*enegDiagt&*Q^(-1));

$enegAt :=$

$[\,[0.5659285350e^{-12.72322141\ t} + 0.2314383135e^{7.991173640\ t} + 0.2026331516e^{1.732047776\ t},$

$-\,0.4921562260e^{-12.72322141\ t} + 0.3012799734e^{7.991173640\ t} + 0.1908762527e^{1.732047776\ t},$

$-\,0.05861465461e^{-12.72322141\ t} - 0.2951355591e^{7.991173640\ t} + 0.3537502137e^{1.732047776\ t}]$

$[-0.4921562255e^{-12.72322141\ t} + 0.3012799732e^{7.991173640\ t} + 0.1908762524e^{1.732047776\ t},$

$0.4280005965e^{-12.72322141\ t} + 0.3921979076e^{7.991173640\ t} + 0.1798014960e^{1.732047776\ t},$

$0.05097386929e^{-12.72322141\ t} - 0.3841992796e^{7.991173640\ t} + 0.3332254103e^{1.732047776\ t}]$

$[0.05861465470e^{-12.72322141\ t} - 0.2951355593e^{7.991173640\ t} + 0.3537502136e^{1.732047776\ t},$

$0.05097386942e^{-12.72322141\ t} - 0.3841992802e^{7.991173640\ t} + 0.3332254107e^{1.732047776\ t},$

$0.006070868542e^{-12.72322141\ t} + 0.3763637791e^{7.991173640\ t} + 0.6175653524e^{1.732047776\ t}]\,]$

> X0:= matrix(3,1,[-2,4,10]);

$$X0 := \begin{bmatrix} -2 \\ 4 \\ 10 \end{bmatrix}$$

> Z:= map(int, evalm(enegAt&*F), t);

$Z := [\,[-0.003474504381 \cos(t)\, e^{-12.72322141\ t} - 0.04420688853 \sin(t)\, e^{-12.72322141\ t}$

$-\,0.003568337884 \cos(t)\, e^{7.991173640\ t} + 0.02851520764 \sin(t)\, e^{7.991173640\ t}$

$-\,0.05065842090 \cos(t)\, e^{1.732047776\ t} + 0.08774280525 \sin(t)\, e^{1.732047776\ t}$

$-\,0.04198131288 e^{-11.72322141\ t} - 0.03350841453 e^{8.991173640\ t}$

$-\,0.06986563499 e^{2.732047776\ t} + 0.004364265239 \cos(3.0\,t)\, e^{-12.72322141\ t}$

$-\,0.001029047228 \sin(3.0\,t)\, e^{-12.72322141\ t} - 0.03237052604 \cos(3.0\,t)\, e^{7.991173640\ t}$

$-\,0.01215235490 \sin(3.0\,t)\, e^{7.991173640\ t} + 0.05105940059 \cos(3.0\,t)\, e^{1.732047776\ t}$

$+\,0.08843763082 \sin(3.0\,t)\, e^{1.732047776\ t}]\,, [0.003021581094 \cos(t)\, e^{-12.72322141\ t}$

$+\,0.03844424526 \sin(t)\, e^{-12.72322141\ t} - 0.004645163222 \cos(t)\, e^{7.991173640\ t}$

$+\,0.03712030590 \sin(t)\, e^{7.991173640\ t} - 0.04771918838 \cos(t)\, e^{1.732047776\ t}$

$+\, 0.08265191411\sin\left(t\right)e^{1.732047776\,t} + 0.03650878726e^{-11.72322141\,t}$

$-\, 0.04362032403e^{8.991173640\,t} - 0.06581198820e^{2.732047776\,t}$

$-\, 0.003795356081\cos\left(3.0t\right)e^{-12.72322141t} + .0008949045118\sin\left(3.0t\right)e^{-12.72322141t}$

$-\, 0.04213905238\cos\left(3.0\,t\right)e^{7.991173640\,t} - 0.01581959833\sin\left(3.0\,t\right)e^{7.991173640\,t}$

$+\, 0.04809690299\cos\left(3.0\,t\right)e^{1.732047776\,t} + 0.08330642548\sin\left(3.0\,t\right)e^{1.732047776\,t}\Big],$

$\Big[0.0003598632371\cos\left(t\right)e^{-12.72322141t} + 0.004578619642\sin\left(t\right)e^{-12.72322141t}$

$+\, 0.004550428066\cos\left(t\right)e^{7.991173640\,t} - 0.03636326081\sin\left(t\right)e^{7.991173640\,t}$

$-\, 0.08843778559\cos\left(t\right)e^{1.732047776\,t} + 0.1531784698\sin\left(t\right)e^{1.732047776\,t}$

$+\, 0.004348111124e^{-11.72322141\,t} + 0.04273071521e^{8.991173640\,t}$

$-\, 0.1219691008e^{2.732047776\,t} - 0.0004520180273\cos\left(3.0\,t\right)e^{-12.72322141\,t}$

$+\, 0.0001065810331\sin\left(3.0\,t\right)e^{-12.72322141\,t} + 0.04127965315\cos\left(3.0\,t\right)e^{7.991173640\,t}$

$+\, 0.01549696766\sin\left(3.0\,t\right)e^{7.991173640\,t} + 0.08913780260\cos\left(3.0\,t\right)e^{1.732047776\,t}$

$+\, 0.1543914732\sin\left(3.0\,t\right)e^{1.732047776\,t}\Big]\Big]$

> P0:= evalm(X0 - evalf(map(eval, Z, t = 0)));

$$P0 := \begin{bmatrix} -1.819996514 \\ 4.120103801 \\ -9.971547669 \end{bmatrix}$$

> X:= simplify(evalm(eAt&*(Z + P0)));

$X := \big[\,\big[0.07205112435\sin\left(t\right) - 0.1453553623\,e^{t} - 3.645536968\,e^{12.72322141\,t}$

$-\, 2.139665046\,e^{-7.991173640\,t} + 3.965205500\,e^{-1.732047776\,t} - 0.05770126316\cos\left(t\right)$

$+\, 0.02305313978\cos\left(3.0\,t\right) + 0.07525622868\sin\left(3.0\,t\right)$

$+\, 2.683336537\ 10^{-11}\sin\left(t\right)e^{-6.259125864\,t} + 1.599477400\ 10^{-16}\cos\left(t\right)e^{20.71439505\,t}$

$-\, 1.912027166\ 10^{-12}\sin\left(t\right)e^{20.71439505\,t} - 7.635442550\ 10^{-12}\cos\left(t\right)e^{14.45526919\,t}$

$+\, 1.085374411\ 10^{-11}\sin\left(t\right)e^{14.45526919\,t} - 1.321608833\ 10^{-12}\sin\left(3.0\,t\right)e^{20.71439505\,t}$

$+\, 6.699338768\ 10^{-12}\sin\left(3.0\,t\right)e^{14.45526919\,t} - 5.097729913\ 10^{-12}\sin\left(t\right)e^{-20.71439505\,t}$

$-\, 7.794567873\ 10^{-12}\cos\left(t\right)e^{-6.259125864\,t} - 4.247282042\ 10^{-12}\cos\left(3.0\,t\right)e^{20.71439505\,t}$

$+\, 3.936959924\ 10^{-12}\cos\left(3.0\,t\right)e^{14.45526919\,t} - 2.417829065\ 10^{-13}\cos\left(t\right)e^{-20.71439505\,t}$

$+\, 6.640907445\ 10^{-13}\cos\left(3.0\,t\right)e^{-20.71439505\,t} + 2.215894065\ 10^{-12}\cos\left(3.0\,t\right)e^{6.259125864\,t}$

$-\, 2.429500981\ 10^{-13}\sin\left(3.0\,t\right)e^{-20.71439505\,t} + 7.023651771\ 10^{-12}\cos\left(3.0\,t\right)e^{-6.259125864\,t}$

$+\, 1.642523818\ 10^{-11}\sin\left(3.0\,t\right)e^{-6.259125864\,t} + 5.087480625\ 10^{-13}\cos\left(t\right)e^{-14.45526919\,t}$

$+\, 4.653263549\ 10^{-12}\sin\left(t\right)e^{-14.45526919\,t} + 3.701642904\ 10^{-13}\cos\left(t\right)e^{6.259125864\,t}$

$-\, 4.750359430\ 10^{-13}\sin\left(t\right)e^{6.259125864\,t} - 3.601678703\ 10^{-13}\cos\left(3.0\,t\right)e^{-14.45526919\,t}$

$+\, 1.731494053\ 10^{-14}\sin\left(3.0\,t\right)e^{-14.45526919\,t} - 2.182631762\ 10^{-12}\,e^{15.45526919\,t}$

$+\, 3.352045248\ 10^{-12}\,e^{7.259125864\,t} - 4.514638424\ 10^{-12}\,e^{-19.71439505\,t}$

$-\, 3.045704596\ 10^{-11}\,e^{-5.259125864\,t} + 4.936575193\ 10^{-12}\,e^{-13.45526919\,t}$

$-\, 6.906126388\ 10^{-12}\,e^{21.71439505\,t} + 2.287081110\ 10^{-13}\sin\left(3.0\,t\right)e^{6.259125864\,t}\big],$

$\big[0.1582164652\sin\left(t\right) - 0.07292352496\,e^{t} + 3.170318517\,e^{12.72322141\,t}$

$- 2.785356572 \, e^{-7.991173640 \, t} + 3.735141855 \, e^{-1.732047776 \, t}$

$- 0.04934277051 \, \cos{(t)} + 0.00216249452 \, \cos{(3.0\,t)}$

$+ 0.06838173165 \, \sin{(3.0\,t)} + 3.989937946 \, 10^{-11} \, \sin{(t)} \, e^{-6.259125864 \, t}$

$- 1.079335879 \, 10^{-13} \, \cos{(t)} \, e^{20.71439505 \, t} + 2.524187005 \, 10^{-12} \, \sin{(t)} \, e^{20.71439505 \, t}$

$+ 5.836738749 \, 10^{-12} \, \cos{(t)} \, e^{14.45526919 \, t} - 8.047404818 \, 10^{-12} \, \sin{(t)} \, e^{14.45526919 \, t}$

$+ 7.822232664 \, 10^{-13} \, \sin{(3.0\,t)} \, e^{20.71439505 \, t} - 4.423533062 \, 10^{-12} \, \sin{(3.0\,t)} \, e^{14.45526919 \, t}$

$- 9.359379323 \, 10^{-12} \, \sin{(t)} \, e^{-20.71439505 \, t} - 1.301529260 \, 10^{-11} \, \cos{(t)} \, e^{-6.259125864 \, t}$

$+ 2.715754114 \, 10^{-12} \, \cos{(3.0\,t)} \, e^{20.71439505 \, t} - 2.614017356 \, 10^{-12} \, \cos{(3.0\,t)} \, e^{14.45526919 \, t}$

$- 5.287876680 \, 10^{-13} \, \cos{(t)} \, e^{-20.71439505 \, t} + 1.133348612 \, 10^{-12} \, \cos{(3.0\,t)} \, e^{-20.71439505 \, t}$

$+ 1.272264469 \, 10^{-12} \, \cos{(3.0\,t)} \, e^{6.259125864 \, t} - 3.796585347 \, 10^{-13} \, \sin{(3.0\,t)} \, e^{-20.71439505 \, t}$

$+ 1.203444095 \, 10^{-11} \, \cos{(3.0\,t)} \, e^{-6.259125864 \, t} + 2.638971295 \, 10^{-11} \, \sin{(3.0\,t)} \, e^{-6.259125864 \, t}$

$+ 6.283885417 \, 10^{-13} \, \cos{(t)} \, e^{-14.45526919 \, t} + 6.281053101 \, 10^{-12} \, \sin{(t)} \, e^{-14.45526919 \, t}$

$+ 2.588394462 \, 10^{-13} \, \cos{(t)} \, e^{6.259125864 \, t} + 2.705143930 \, 10^{-13} \, \sin{(t)} \, e^{6.259125864 \, t}$

$- 5.266259536 \, 10^{-13} \, \cos{(3.0\,t)} \, e^{-14.45526919 \, t} + 6.048668653 \, 10^{-14} \, \sin{(3.0\,t)} \, e^{-14.45526919 \, t}$

$+ 7.901345130 \, 10^{-13} \, e^{15.45526919 \, t} + 2.313844411 \, 10^{-12} \, e^{7.259125864 \, t}$

$- 8.463223910 \, 10^{-12} \, e^{-19.71439505 \, t} - 4.360428891 \, 10^{-11} \, e^{-5.259125864 \, t}$

$+ 6.452384225 \, 10^{-12} \, e^{-13.45526919 \, t} + 4.993627521 \, 10^{-12} \, e^{21.71439505 \, t}$

$- 9.054754200 \, 10^{-14} \, \sin{(3.0\,t)} \, e^{6.259125864 \, t}], \, [0.1213938286 \, \sin{(t)}$

$- 0.07489027448 \, e^{t} + 0.3775775161 \, e^{12.72322141 \, t} + 2.728550992 \, e^{-7.991173640 \, t}$

$+ 6.922323823 \, e^{-1.732047776 \, t} - 0.08352749428 \, \cos{(t)} + 0.1299654377 \, \cos{(3.0\,t)}$

$+ 0.1699950219 \, \sin{(3.0\,t)} - 3.954072877 \, 10^{-11} \, \sin{(t)} \, e^{-6.259125864 \, t}$

$+ 9.287483332 \, 10^{-15} \, \cos{(t)} \, e^{20.71439505 \, t} + 1.236832310 \, 10^{-13} \, \sin{(t)} \, e^{20.71439505 \, t}$

$+ 9.277899806 \, 10^{-13} \, \cos{(t)} \, e^{14.45526919 \, t} - 1.361384627 \, 10^{-12} \, \sin{(t)} \, e^{14.45526919 \, t}$

$+ 1.685683131 \, 10^{-13} \, \sin{(3.0\,t)} \, e^{20.71439505 \, t} - 9.329809580 \, 10^{-13} \, \sin{(3.0\,t)} \, e^{14.45526919 \, t}$

$+ 6.014652059 \, 10^{-12} \, \sin{(t)} \, e^{-20.71439505 \, t} + 1.301259028 \, 10^{-11} \, \cos{(t)} \, e^{-6.259125864 \, t}$

$+ 5.243043757 \, 10^{-13} \, \cos{(3.0\,t)} \, e^{20.71439505 \, t} - 5.458128980 \, 10^{-13} \, \cos{(3.0\,t)} \, e^{14.45526919 \, t}$

$+ 2.701220791 \, 10^{-13} \, \cos{(t)} \, e^{-20.71439505 \, t} - 7.988752459 \, 10^{-13} \, \cos{(3.0\,t)} \, e^{-20.71439505 \, t}$

$+ 7.700845450 \, 10^{-12} \, \cos{(3.0\,t)} \, e^{6.259125864 \, t} + 2.985004022 \, 10^{-13} \, \sin{(3.0\,t)} \, e^{-20.71439505 \, t}$

$- 1.205382212 \, 10^{-11} \, \cos{(3.0\,t)} \, e^{-6.259125864 \, t} - 2.631018700 \, 10^{-11} \, \sin{(3.0\,t)} \, e^{-6.259125864 \, t}$

$+ 9.140857782 \, 10^{-13} \, \cos{(t)} \, e^{-14.45526919 \, t} + 8.453430115 \, 10^{-12} \, \sin{(t)} \, e^{-14.45526919 \, t}$

$+ 1.068683161 \, 10^{-12} \, \cos{(t)} \, e^{6.259125864 \, t} - 4.205274410 \, 10^{-12} \, \sin{(t)} \, e^{6.259125864 \, t}$

$- 6.613397368 \, 10^{-13} \, \cos{(3.0\,t)} \, e^{-14.45526919 \, t} + 3.790765990 \, 10^{-14} \, \sin{(3.0\,t)} \, e^{-14.45526919 \, t}$

$+ 4.149600754 \, 10^{-13} \, e^{15.45526919 \, t} + 9.819017275 \, 10^{-12} \, e^{7.259125864 \, t}$

$+ 5.295552240 \, 10^{-12} \, e^{-19.71439505 \, t} + 4.307736027 \, 10^{-11} \, e^{-5.259125864 \, t}$

$+ 8.931416612 \, 10^{-12} \, e^{-13.45526919 \, t} + 8.026545522 \, 10^{-13} \, e^{21.71439505 \, t}$

$+ 1.838012613 \, 10^{-12} \, \sin{(3.0\,t)} \, e^{6.259125864 \, t}] \,]$

The above solution X should strike you as rather odd looking. By this we mean that there are a great many terms in X with very small coefficient values involving negative powers of 10. The question should be, are these terms actually necessary or can we do away with them and still have the solution to our system of differential equations? We should suspect that they are all superfluous. Are these superfluous terms due to our particular solution method, due to rounding error, or due to something entirely different?

Let us take $X1$ to be the above solution X with these superfluous terms removed, and then we can check to see if $X1$ really is our solution by seeing directly if it solves our nonhomogeneous system of linear differential equations and its initial conditions.

```
> x1:=unapply(0.07205112437*sin(t) - 0.1453553624*exp(t)
- 3.645536968*exp(12.72322141*t) - 2.139665046*exp(-7.991173640*t)
+ 3.965205500*exp(-1.732047776*t) - 0.05770126316*cos(t)
+ 0.02305313978*cos(3.*t) + 0.07525622868*sin(3.*t), t);
```

$$x1 := t \to 0.07205112437\sin(t) - 0.1453553624\,e^t - 3.645536968\,e^{12.72322141\,t}$$
$$- 2.139665046\,e^{-7.991173640\,t} + 3.965205500\,e^{-1.732047776\,t} - 0.05770126316\cos(t)$$
$$+ 0.02305313978\cos(3.\,t) + 0.07525622868\sin(3.\,t)$$

```
> x1(0);
```
$$-1.999999999$$

```
> y1:= unapply(0.1582164653*sin(t)-0.07292352494*exp(t)
+3.170318517*exp(12.72322141*t)-2.785356572*exp(-7.991173640*t)
+3.735141855*exp(-1.732047776*t)-0.04934277051*cos(t)
+0.00216249452*cos(3.*t)+0.06838173165*sin(3.*t), t);
```

$$y1 := t \to 0.1582164653\sin(t) - 0.07292352494\,e^t + 3.170318517\,e^{12.72322141\,t}$$
$$- 2.785356572\,e^{-7.991173640\,t} + 3.735141855\,e^{-1.732047776\,t} - 0.04934277051\cos(t)$$
$$+ 0.00216249452\cos(3.\,t) + 0.06838173165\sin(3.\,t)$$

```
> y1(0);
```
$$3.999999999$$

```
> z1:= unapply(0.1213938286*sin(t)-0.07489027446*exp(t)
+0.3775775161*exp(12.72322141*t)+2.728550992*exp(-7.991173640*t)
+6.922323823*exp(-1.732047776*t)-0.08352749427*cos(t)
+0.1299654377*cos(3.*t)+0.1699950219*sin(3.*t), t);
```

$$z1 := t \to 0.1213938286\sin(t) - 0.07489027446\,e^t + 0.3775775161\,e^{12.72322141\,t}$$
$$+ 2.728550992\,e^{-7.991173640\,t} + 6.922323823\,e^{-1.732047776\,t} - 0.08352749427\cos(t)$$
$$+ 0.1299654377\cos(3.\,t) + 0.1699950219\sin(3.\,t)$$

```
> z1(0);
```
$$10.00000000$$

```
> X1:= matrix(3,1,[x1(t), y1(t), z1(t)]);
```

$$X1 := \big[\,[0.07205112437\sin(t) - 0.1453553624\,e^t - 3.645536968\,e^{12.72322141\,t}$$
$$- 2.139665046\,e^{-7.991173640\,t} + 3.965205500\,e^{-1.732047776\,t} - 0.05770126316\cos(t)$$

$+ 0.02305313978 \cos{(3.0\,t)} + 0.07525622868 \sin{(3.0\,t)}\big], \big[0.1582164653 \sin{(t)}$

$- 0.07292352494\, e^t + 3.170318517\, e^{12.72322141\, t} - 2.785356572\, e^{-7.991173640\, t}$

$+ 3.735141855\, e^{-1.732047776\, t} - 0.04934277051 \cos{(t)} + 0.00216249452 \cos{(3.0\,t)}$

$+ 0.06838173165 \sin{(3.0\,t)}\big], \big[0.1213938286 \sin{(t)} - 0.07489027446\, e^t$

$+ 0.3775775161\, e^{12.72322141\, t} + 2.728550992\, e^{-7.991173640\, t} + 6.922323823\, e^{-1.732047776\, t}$

$- 0.08352749427 \cos{(t)} + 0.1299654377 \cos{(3.0\,t)} + 0.1699950219 \sin{(3.0\,t)}\big]\,\big]$

> dX1dt:= map(diff, X1, t);

$dX1dt := \big[\, \big[0.07205112437 \cos{(t)} - 0.1453553624\, e^t$

$- 46.38297400\, e^{12.72322141\, t} + 17.09843491\, e^{-7.991173640\, t} - 6.867925368\, e^{-1.732047776\, t}$

$+ 0.05770126316 \sin{(t)} - 0.06915941934 \sin{(3.0\,t)} + 0.2257686860 \cos{(3.0\,t)}\big],$

$\big[0.1582164653 \cos{(t)} - 0.07292352494\, e^t + 40.33666443\, e^{12.72322141\, t}$

$+ 22.25826802\, e^{-7.991173640\, t} - 6.469444143\, e^{-1.732047776\, t} + 0.04934277051 \sin{(t)}$

$- 0.00648748356 \sin{(3.0\,t)} + 0.2051451950 \cos{(3.0\,t)}\big],$

$\big[0.1213938286 \cos{(t)} - 0.07489027446\, e^t + 4.804002337\, e^{12.72322141\, t}$

$- 21.80432476\, e^{-7.991173640\, t} - 11.98979558\, e^{-1.732047776\, t} + 0.08352749427 \sin{(t)}$

$- 0.3898963131 \sin{(3.0\,t)} + 0.5099850657 \cos{(3.0\,t)}\big]\,\big]$

> AX1F:= simplify(evalm(A&*X1 + F));

$AX1F := \big[\, \big[0.05770126240 \sin{(t)} - 0.1453553620\, e^t - 46.38297397\, e^{12.72322141\, t}$

$+ 17.09843491\, e^{-7.991173640\, t} - 6.867925380\, e^{-1.732047776\, t} + 0.07205112450 \cos{(t)}$

$+ 0.2257686859 \cos{(3.0\,t)} - 0.06915941950 \sin{(3.0\,t)}\big], \big[0.04934277150 \sin{(t)}$

$- 0.07292352410\, e^t + 40.33666449\, e^{12.72322141\, t} + 22.25826802\, e^{-7.991173640\, t}$

$- 6.469444120\, e^{-1.732047776\, t} + 0.1582164650 \cos{(t)} + 0.2051451954 \cos{(3.0\,t)}$

$- 0.006487482800 \sin{(3.0\,t)}\big], \big[0.08352749510 \sin{(t)} - 0.07489027490\, e^t$

$+ 4.804002332\, e^{12.72322141\, t} - 21.80432476\, e^{-7.991173640\, t} - 11.98979558\, e^{-1.732047776\, t}$

$+ 0.1213938285 \cos{(t)} + 0.5099850659 \cos{(3.0\,t)} - 0.3898963131 \sin{(3.0\,t)}\big]\,\big]$

Our three solution functions $x1(t)$, $y1(t)$, and $z1(t)$, to the linear system of differential equations, are $X1[1,1]$, $X1[2,1]$, and $X1[3,1]$, respectively, from above. It is also checked that $X1(t)$ is the solution to this system of differential equations and satisfies the initial conditions. Let us plot, as shown in Figure 12.3, this real solution $X1(t)$ as a curve in space for $t \in [-0.25, 0.25]$.

> plot_X:= spacecurve([x1(t), y1(t), z1(t)], t = -0.25..0.25, axes = boxed, thickness = 2, color = red):

> point_X0:= point([X0[1,1], X0[2,1], X0[3,1]], symbol = box, symbolsize = 25, color = black):

> display3d({plot_X, point_X0});

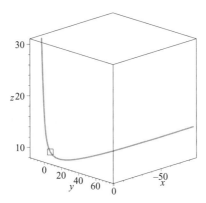

Figure 12.3: Solution to the ODE system (12.21), with initial conditions given in equation (12.22).

   Maple can also solve systems of differential equations using its *dsolve* command. Let us use it now to check again that our result $X1$ above is correct. Again, the option "method = laplace" should always be used with *dsolve* if you want to solve nonhomogeneous linear systems of differential equations without getting unnecessary complex numbers when you know your results should be completely real. You should redo this computation without "method = laplace" in order to see what you would get instead. The Laplace transform is used by *Maple* in "method = laplace". Note that we must now unassign the names x, y, and z since we used them above as function names for our solutions.

   As you will see below, the solution from *Maple*'s *dsolve* command is virtually identical to the solution $X1$ where we have eliminated the superfluous terms from the solution $X(t)$ gotten by directly using our solution formula (12.28) with $Z(t)$ and P_0 defined as in (12.29).

   Now you should use the alternate solution formula (12.30) along with (12.31) when A is diagonalizable. The hope is that this alternate solution formula might eliminate most, if not all, of the superfluous terms in the solution $X(t)$ we received above using the first and more general formula.

> x:= 'x': y:= 'y': z:= 'z':
> evalf(dsolve({diff(x(t),t) = 5.*x(t) - 9.*y(t) + z(t) + sin(t), diff(y(t),t) = -9.*x(t) + 2.*y(t) + Pi*z(t) - exp(t), diff(z(t),t) = x(t) + Pi*y(t) - 4.*z(t) + cos(3*t), x(0) = -2., y(0) = 4.,z(0) = 10.}, {x(t), y(t), z(t)}, method = laplace));

$$\{x\,(t) = -2.139665049\,e^{-7.991173640\,t} + 3.965205503\,e^{-1.732047775\,t}$$
$$- 3.645536970\,e^{12.72322141\,t} - 0.05770126330\,\cos{(t)} + 0.02305313980\,\cos{(3.0\,t)}$$
$$- 0.1453553624\,e^{t} + 0.07205112460\,\sin{(t)} + 0.07525622878\,\sin{(3.0\,t)},$$

$$y(t) = -0.07292352487\, e^{t} - 2.785356578\, e^{-7.991173640\, t}$$
$$+\, 3.735141857\, e^{-1.732047775\, t} + 3.170318522\, e^{12.72322141\, t} - 0.04934277062\, \cos(t)$$
$$+\, 0.002162494481\, \cos(3.0\, t) + 0.1582164654\, \sin(t) + 0.06838173168\, \sin(3.0\, t)\,,$$
$$z(t) = 2.728550994\, e^{-7.991173640\, t} + 6.922323825\, e^{-1.732047775\, t}$$
$$+\, 0.3775775166\, e^{12.72322141\, t} - 0.08352749450\, \cos(t) + 0.1299654376\, \cos(3.0\, t)$$
$$-\, 0.07489027450\, e^{t} + 0.1213938289\, \sin(t) + 0.1699950219\, \sin(3.0\, t)\}$$

> print(X1);

$$\big[\, \big[0.07205112437\, \sin(t) - 0.1453553624\, e^{t} - 3.645536968\, e^{12.72322141\, t}$$
$$-\, 2.139665046\, e^{-7.991173640\, t} + 3.965205500\, e^{-1.732047776\, t} - 0.05770126316\, \cos(t)$$
$$+\, 0.02305313978\, \cos(3.0\, t) + 0.07525622868\, \sin(3.0\, t)\big], \big[0.1582164653\, \sin(t)$$
$$-\, 0.07292352494\, e^{t} + 3.170318517\, e^{12.72322141\, t} - 2.785356572\, e^{-7.991173640\, t}$$
$$+\, 3.735141855\, e^{-1.732047776\, t} - 0.04934277051\, \cos(t) + 0.00216249452\, \cos(3.0\, t)$$
$$+\, 0.06838173165\, \sin(3.0\, t)\big], \big[0.1213938286\, \sin(t) - 0.07489027446\, e^{t}$$
$$+\, 0.3775775161\, e^{12.72322141\, t} + 2.728550992\, e^{-7.991173640\, t} + 6.922323823\, e^{-1.732047776\, t}$$
$$-\, 0.08352749427\, \cos(t) + 0.1299654377\, \cos(3.0\, t) + 0.1699950219\, \sin(3.0\, t)\big]\,\big]$$

■

In the last example, we will look at solving a single kth order *constant coefficient linear nonhomogeneous differential equation* of the form

$$\frac{d^k y}{dt^k} + a_{k-1}\frac{d^{k-1}y}{dt^{k-1}} + \cdots + a_1 \frac{dy}{dt} + a_0 y = f(t) \tag{12.32}$$

for real constants $a_{k-1}, a_{k-2}, \ldots, a_0$ with the initial conditions

$$y(0) = b_0,\ \frac{dy}{dt}(0) = b_1, \ldots,\ \frac{d^{k-1}y}{dt^{k-1}}(0) = b_{k-1} \tag{12.33}$$

for the b's all real constants. This single equation can be equivalently written as a nonhomogeneous first-order linear system of differential equations in the new k unknown functions

$$x_1(t) = y(t),\ x_2(t) = \frac{dy}{dt}, \ldots,\ x_k(t) = \frac{d^{k-1}y}{dt^{k-1}}$$

The nonhomogeneous first-order linear system of differential equations is then

$$\frac{dx_1}{dt} = x_2(t),\ \frac{dx_2}{dt} = x_3(t), \ldots,\ \frac{dx_{k-1}}{dt} = x_k(t),$$
$$\frac{dx_k}{dt} = -\left(a_0 x_1(t) + a_1 x_2(t) + \cdots + a_{k-1}x_k(t)\right) + f(t) \tag{12.34}$$

with the initial conditions

$$x_1(0) = b_0,\ x_2(0) = b_1, \ldots,\ x_k(0) = b_{k-1} \tag{12.35}$$

Example 12.4.4. The nonhomogeneous constant coefficient differential equation we will solve is the third-order equation

$$\frac{d^3y}{dt^3} + 5\frac{d^2y}{dt^2} - 2\frac{dy}{dt} + 3y(t) = \sin(t) \tag{12.36}$$

with the initial conditions

$$y(0) = 4, \quad \frac{dy}{dt}(0) = -9, \quad \frac{d^2y}{dt^2}(0) = 7 \tag{12.37}$$

This third-order equation becomes the first-order linear system (12.20), where

$$X(t) = \begin{bmatrix} x_1(t) \\ x_2(t) \\ x_3(t) \end{bmatrix}, A = \begin{bmatrix} 0 & 1 & 0 \\ 0 & 0 & 1 \\ -3 & 2 & -5 \end{bmatrix}, F(t) = \begin{bmatrix} 0 \\ 0 \\ \sin(t) \end{bmatrix}, X_0 = \begin{bmatrix} 4 \\ -9 \\ 7 \end{bmatrix}$$

for $x_1(t) = y(t)$, $x_2(t) = \dfrac{dy}{dt}$, and $x_3(t) = \dfrac{d^2y}{dt^2}$. The solution formula we shall use here is (12.30) with (12.31) since A is diagonalizable. We will employ this formula since it is hoped that this form of the solution will not give us the superfluous terms that the general formula does with unnecessary very small coefficients.

> F:= matrix(3,1,[0, 0, sin(t)]);

$$F := \begin{bmatrix} 0 \\ 0 \\ \sin(t) \end{bmatrix}$$

> A:= matrix(3,3,[0,1,0,0,0,1,-3,2,-5]);

$$A := \begin{bmatrix} 0 & 1 & 0 \\ 0 & 0 & 1 \\ -3 & 2 & -5 \end{bmatrix}$$

> eigvects:= evalf(eigenvectors(A));

$$eigvects := \big[-5.466280376, 1., \{[1., -5.466280376, 29.88022115]\}\big],$$
$$\big[0.233140187 - 0.7031819975I, 1., \{[1., 0.233140187 - 0.7031819975I,$$
$$- 0.4401105748 - 0.3278799648I]\}\big], \big[0.233140187 + 0.7031819975I, 1.,$$
$$\{[1., 0.233140187 + 0.7031819975I, -0.4401105748 + 0.3278799648I]\}\big]$$

> vals:= [eigvects[1][1], eigvects[2][1], eigvects[3][1]];

$$vals := [-5.466280376, 0.233140187 - 0.7031819975I, 0.233140187 + 0.7031819975I]$$

> Diag:= DiagonalMatrix(vals);

$$Diag := \begin{bmatrix} -5.466280376 & 0 & 0 \\ 0 & 0.233140187 - 0.7031819975I & 0 \\ 0 & 0 & 0.233140187 + 0.7031819975I \end{bmatrix}$$

> Qlist:= [op(convert(op(eigvects[1][3]), list)), op(convert(op(eigvects[2][3]), list)), op(convert(op(eigvects[3][3]), list))];

$$Qlist := \big[1., -5.466280376, 29.88022115, 1., 0.233140187 - 0.7031819975I,$$
$$- 0.4401105748 - 0.3278799648I, 1., 0.233140187 + 0.7031819975I,$$
$$- 0.4401105748 + 0.3278799648\,I\big]$$

> Q:= transpose(matrix(3,3,Qlist));

$$Q := \begin{bmatrix} 1. & 1. & 1. \\ -5.466280376 & 0.233140187 - 0.7031819975I & 0.233140187 + 0.7031819975I \\ 29.88022115 & -0.4401105748 - 0.3278799648I & -0.4401105748 + 0.3278799648I \end{bmatrix}$$

> evalm(Q&*Diag&*Q^(-1));

$$\begin{bmatrix} -4.\,10^{-11} + 0.I & 0.9999999998 + 0.I & -4.\,10^{-11} + 0.I \\ 0. + 0.I & -1.\,10^{-10} + 0.I & 1.000000000 + 0.I \\ -2.999999997 + 0.I & 1.999999991 + 0.I & -5.000000002 + 0.I \end{bmatrix}$$

> print(A);

$$\begin{bmatrix} 0 & 1 & 0 \\ 0 & 0 & 1 \\ -3 & 2 & -5 \end{bmatrix}$$

We see from the last two computations that A is diagonalizable for the above defined Q and D (or Diag). Now we can solve our first-order linear system of differential equations.

> eDiagt:=matrix(3,3,[exp(vals[1]*t),0,0,0,exp(vals[2]*t),0,0,0,exp(vals[3]*t)]);

$$eDiagt := \begin{bmatrix} e^{-5.466280376t} & 0 & 0 \\ 0 & e^{(0.233140187-0.7031819975I)\,t} & 0 \\ 0 & 0 & e^{(0.233140187+0.7031819975I)\,t} \end{bmatrix}$$

> QeDiagt:= evalm(Q&*eDiagt);

$$QeDiagt := \big[\,[1.e^{-5.466280376t}, 1.e^{(0.233140187-0.7031819975I)\,t},$$
$$1.e^{(0.233140187+0.7031819975I)\,t}], [-5.466280376e^{-5.466280376\,t},$$
$$(0.233140187 - 0.7031819975I)\,e^{(0.233140187-0.7031819975I)\,t},$$
$$(0.233140187 + 0.7031819975I)\,e^{(0.233140187+0.7031819975I)\,t}],$$
$$[29.88022115e^{-5.466280376t}, (-0.4401105748 - 0.3278799648I)\,e^{(0.233140187-0.7031819975I)t},$$
$$(-0.4401105748 + 0.3278799648I)\,e^{(0.233140187+0.7031819975I)\,t}]\,\big]$$

> enegDiagt:= matrix(3,3,[exp(-vals[1]*t), 0, 0, 0, exp(-vals[2]*t), 0, 0, 0, exp(-vals[3]*t)]);

$enegDiagt :=$

$$\begin{bmatrix} e^{5.466280376\,t} & 0 & 0 \\ 0 & e^{(-0.233140187+0.7031819975I)\,t} & 0 \\ 0 & 0 & e^{(-0.233140187-0.7031819975I)\,t} \end{bmatrix}$$

> enegDiagtQinv:= evalm(enegDiagt&*Q^(-1));

$enegDiagtQinv := \Big[\big[(0.01664205239 - 0.I)e^{5.466280376t},$

$\quad(-0.01413919456 + 0.I)e^{5.466280376t},\ (0.03032337483 - 0.I)e^{5.466280376t}\big],$

$\quad\big[(0.4916789738 - 0.09833167795I)e^{(-0.233140187+0.7031819975I)\,t},$

$\quad(0.007069597278 + 0.6537530731I)e^{(-0.233140187+0.7031819975I)\,t},$

$\quad(-0.01516168741 + 0.1228882897I)e^{(-0.233140187+0.7031819975I)\,t}\big],$

$\quad\big[(0.4916789738 + 0.09833167795I)e^{(-0.233140187-0.7031819975I)\,t},$

$\quad(0.007069597278 - 0.6537530731I)e^{(-0.233140187-0.7031819975I)\,t},$

$\quad(-0.01516168741 - 0.1228882897I)e^{(-0.233140187-0.7031819975I)\,t}\big]\Big]$

> X0:= matrix(3,1,[4,-9,7]);

$$X0 := \begin{bmatrix} 4 \\ -9 \\ 7 \end{bmatrix}$$

> W:= map(expand, map(int, evalm(enegDiagtQinv&*F), t));

$W := \Big[\big[-0.0009819675411\,e^{5.466280376t}\cos(t)$

$\quad+0.005367709900\,e^{5.466280376t}\sin(t)\big],$

$\quad\big[0.1158755991\,e^{-0.2331401870t}e^{0.7031819975\,I\,t}\cos(t)$

$\quad-0.1516281582\,I\,e^{-0.2331401870\,t}e^{0.7031819975\,I\,t}\cos(t)$

$\quad-0.07960693233\,e^{-0.2331401870t}e^{0.7031819975\,I\,t}\sin(t)$

$\quad-0.1168322524\,I\,e^{-0.2331401870\,t}e^{0.7031819975\,I\,t}\sin(t)\big],$

$\quad\big[0.1158755991\,e^{-0.2331401870t}e^{-0.7031819975\,I\,t}\cos(t)$

$\quad+0.1516281582\,I\,e^{-0.2331401870\,t}e^{-0.7031819975\,I\,t}\cos(t)$

$\quad-0.07960693233\,e^{-0.2331401870t}e^{-0.7031819975\,I\,t}\sin(t)$

$\quad+0.1168322524\,I\,e^{-0.2331401870\,t}e^{-0.7031819975\,I\,t}\sin(t)\big]\Big]$

> R0:= evalm(Q^(-1)&*X0 - evalf(map(eval, W, t = 0)));

$$R0 := \begin{bmatrix} 0.4070665519 - 0.I \\ 1.681082109 - 5.265258184I \\ 1.681082109 + 5.265258184I \end{bmatrix}$$

> X:= simplify(evalm(QeDiagt&*(W + R0)));

$X :=$

$\big[\, [0.2307692307\cos(t) - 0.1538461547\sin(t) + 0.4070665519e^{-5.466280376t}$

$+ 1.681082109e^{(0.2331401870 - 0.7031819975I)t} - 5.265258184I \; e^{(0.2331401870 - 0.7031819975I)t}$

$+ 1.681082109e^{(0.2331401870 + 0.7031819975I)t}] + 5.265258184I \; e^{(0.2331401870 + 0.7031819975I)t}$

$[-0.1538461548\cos(t) - 0.2307692307\sin(t) - 2.225139904e^{-5.466280376t}$

$- 3.310506970e^{(0.2331401870 - 0.7031819975I)t} + 2.409649953I \; e^{(0.2331401870 + 0.7031819975I)t}$

$- 2.409649953I \; e^{(0.2331401870 - 0.7031819975I)t} - 3.310506970e^{(0.2331401870 + 0.7031819975I)t}]$

$[-0.2307692307\cos(t) + 0.1538461547\sin(t) + 12.16323859e^{-5.466280376t}$

$- 2.466234681e^{(0.2331401870 - 0.7031819975I)t} - 1.766102663I \; e^{(0.2331401870 + 0.7031819975I)t}$

$+ 1.766102663I \; e^{(0.2331401870 - 0.7031819975I)t} - 2.466234681e^{(0.2331401870 + 0.7031819975I)t}] \,\big]$

> assume(t, real): interface(showassumed = 0):
> evalf(map(eval, X, t = 0));

$$\begin{bmatrix} 4.000000001 + 0.I \\ -8.999999999 + 0.I \\ 6.999999997 + 0.I \end{bmatrix}$$

Our solution $y(t)$ to the original third-order constant coefficient differential equation is the first entry in X above if we take its real part, we call it $x1$ below. We also check below that it satisfies this differential equation and its initial conditions. As well, we check that it is correct by solving this differential equation with *dsolve*, but since *dsolve* gives us a complex result we must take its real part.

So this form of the solution $X(t)$ will avoid the superfluous terms that the general solution produces, but it can only be used when A is diagonalizable. In Figure 12.4, we plot the solution function $y(t)$ for $t \in [-0.5, 5.0]$

> x1:= Re(X[1,1]);

$x1 := 0.2307692307\cos(t) - 0.1538461547\sin(t) + 0.4070665519\, e^{-5.466280376t}$

$\qquad + 3.362164218\, e^{0.2331401870t}\cos(0.7031819975\, t)$

$\qquad - 10.53051637\, e^{0.2331401870t}\sin(0.7031819975\, t)$

> eval(x1, t=0);

$$4.000000001$$

> eval(diff(x1,t), t=0);

$$-9.000000001$$

> eval(diff(x1,t$2), t=0);

$$6.999999996$$

> diff(x1,t$ 3)+5*diff(x1,t$ 2)-2*diff(x1,t)+3*x1;

$$1.000000002 \sin(t) + 2.\,10^{-9} \cos(t) + 1.\,10^{-8} e^{-5.466280376t}$$
$$+\, 8.\,10^{-8} e^{0.2331401870t} \cos(0.7031819975t)$$
$$+\, 1.\,10^{-8} e^{0.2331401870t} \sin(0.7031819975t)$$

> Re(evalf(dsolve({ diff(y(t), t$ 3) + 5*diff(y(t), t$ 2) - 2*diff(y(t), t) + 3*y(t) = sin(t), y(0)=4., D(y)(0)=-9., (D@@2)(y)(0)=7.}, y(t), method = laplace)));

$$\Re(y(t)) = 0.4070665524 e^{-5.466280376t} + 3.362164216 e^{0.2331401880t} \cos(0.7031819976t)$$
$$-\, 10.53051637 e^{0.2331401880t} \sin(0.7031819976t) + 0.2307692308 \cos(t)$$
$$-\, 0.1538461538 \sin(t)$$

> plot(x1, t=-0.5..5, thickness=2);

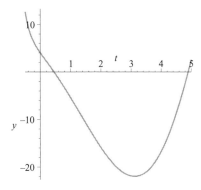

Figure 12.4: Solution to the ODE (12.36), with initial conditions given in equation (12.37).

The last part of this section will summarize what we have discovered about solving homogeneous and nonhomogeneous first-order square linear systems of differential equations. But it is more than a summary, since it will also recap what we have done in a purely matrix approach based on the diagonalizability of the coefficient matrix A.

As well, in this final portion of this section we have formulas that solve the homogeneous and nonhomogeneous problems that do not depend on the diagonalizability of the matrix A. But, without the diagonalizability of the matrix A, we may have great difficulty solving the problem unless A is of some special form, such as upper or lower triangular.

We begin with the homogeneous case (12.7), which we rewrite for completeness, along with the initial condition:

$$\frac{dX}{dt} = AX(t), \quad X(0) = X_0 \tag{12.38}$$

where A is an $n \times n$ square matrix, and $X(t)$ is an $n \times 1$ column matrix, which we can also think of as a vector in $\mathbb{R}^n$ or $\mathbb{C}^n$. As usual, we assume that A is diagonalizable with $A = QDQ^{-1}$, for D the diagonal matrix of A's eigenvalues $\lambda_1, \lambda_2, \ldots, \lambda_n$, and Q an $n \times n$ matrix of A's eigenvector columns $X_1, X_2, \ldots, X_n$ corresponding to these eigenvalues for which Q is invertible, that is, these n eigenvector columns of Q must be an independent set in $\mathbb{C}^n$. This diagonalizability also says that $e^{At} = Qe^{Dt}Q^{-1}$.

We have seen previously that all of the solutions $X(t)$ to this system can be written as

$$X(t) = a_1 e^{\lambda_1 t} X_1 + a_2 e^{\lambda_2 t} X_2 + \cdots + a_n e^{\lambda_n t} X_n \tag{12.39}$$

for arbitrary constants $a_1, a_2, \ldots, a_n$ if we have no initial condition. This is called the *general solution* to our system of differential equations. If we have the initial condition $X(0) = X_0$, then these arbitrary constants have unique values given by

$$\begin{bmatrix} a_1 \\ a_2 \\ \vdots \\ a_n \end{bmatrix} = Q^{-1}X_0$$

$$a_1 X_1 + a_2 X_2 + \cdots + a_n X_n = X_0$$

The general solution given in (12.39) can be written as the alternative matrix equation

$$X(t) = Qe^{Dt} \begin{bmatrix} a_1 \\ a_2 \\ \vdots \\ a_n \end{bmatrix} = e^{At}Q \begin{bmatrix} a_1 \\ a_2 \\ \vdots \\ a_n \end{bmatrix}$$

In order to see that this is true, we could multiply out this product of matrices or simply check that it satisfies our system of differential equations without an initial condition. We leave it to you to do this multiplication as your check.

Let us now perform our check. For

$$X(t) = Qe^{Dt} \begin{bmatrix} a_1 \\ a_2 \\ \vdots \\ a_n \end{bmatrix} \tag{12.40}$$

we have

$$\frac{dX}{dt} = QDe^{Dt} \begin{bmatrix} a_1 \\ a_2 \\ \vdots \\ a_n \end{bmatrix}$$

which follows from the fact that

$$\frac{d}{dt}e^{Dt} = De^{Dt}$$

which is left as an exercise for you to verify. Now

$$AX(t) = AQe^{Dt} \begin{bmatrix} a_1 \\ a_2 \\ \vdots \\ a_n \end{bmatrix} = QDe^{Dt} \begin{bmatrix} a_1 \\ a_2 \\ \vdots \\ a_n \end{bmatrix} = \frac{dX}{dt}$$

since $AQ = QD$ from A's diagonalizability, although the equation $AQ = QD$ is true even if Q is not invertible.

If we have the initial condition $X(0) = X_0$, then the general solution given in (12.39) at $t = 0$ gives

$$X_0 = X(0) = Qe^{D0} \begin{bmatrix} a_1 \\ a_2 \\ \vdots \\ a_n \end{bmatrix} = Q \begin{bmatrix} a_1 \\ a_2 \\ \vdots \\ a_n \end{bmatrix}$$

since the exponential of the zero matrix is the identity matrix. Solving for the unknown column matrix of the a_k's gives

$$\begin{bmatrix} a_1 \\ a_2 \\ \vdots \\ a_n \end{bmatrix} = Q^{-1}X_0.$$

It follows that our unique solution is

$$X(t) = Qe^{Dt}Q^{-1}X_0$$

to our homogeneous system under the initial condition $X(0) = X_0$. Of course, $e^{At} = Qe^{Dt}Q^{-1}$, and so this solution can also be expressed as

$$X(t) = e^{At}X_0$$

which is of the same form as the solution to the simple differential equation given in (12.10). Do not forget to take the real part of this solution $X(t)$ as your real answer to the problem when A is real and the initial conditions are also real.

After working through examples of the homogeneous case, we looked at an example of a nonhomogeneous linear differential equation. This was given by

$$\frac{dX}{dt} = AX(t) + F(t), \quad X(0) = X_0 \tag{12.41}$$

In order to find the general solution to the nonhomogeneous case (i.e., we do not consider the initial condition $X(0) = X_0$), when A is diagonalizable, we use the method called variation of parameters. The method of variation of parameters says that we should seek a general solution $X(t)$ to the nonhomogeneous case of the form given in equation (12.40), and modified as

$$X(t) = Qe^{Dt} \begin{bmatrix} a_1(t) \\ a_2(t) \\ \vdots \\ a_n(t) \end{bmatrix} \tag{12.42}$$

where we have taken the general solution of the homogeneous case and turned its arbitrary constants into unknown functions of t. If we compute the derivative of this potential solution function $X(t)$, we get from the product rule that

$$\frac{dX}{dt} = QDe^{Dt} \begin{bmatrix} a_1(t) \\ a_2(t) \\ \vdots \\ a_n(t) \end{bmatrix} + Qe^{Dt} \begin{bmatrix} \frac{da_1}{dt}(t) \\ \frac{da_2}{dt}(t) \\ \vdots \\ \frac{da_n}{dt}(t) \end{bmatrix}$$

Setting $\dfrac{dX}{dt}$ equal to $AX(t) + F(t)$ to satisfy (12.41) gives

$$QDe^{Dt} \begin{bmatrix} a_1(t) \\ a_2(t) \\ \vdots \\ a_n(t) \end{bmatrix} + Qe^{Dt} \begin{bmatrix} \frac{da_1}{dt}(t) \\ \frac{da_2}{dt}(t) \\ \vdots \\ \frac{da_n}{dt}(t) \end{bmatrix} = AQe^{Dt} \begin{bmatrix} a_1(t) \\ a_2(t) \\ \vdots \\ a_n(t) \end{bmatrix} + F(t)$$

However, since $AQ = QD$, the first term on the left cancels with the first term on the right in the above equation, resulting in the simpler equation

$$
Qe^{Dt}
\begin{bmatrix}
\dfrac{da_1}{dt}(t) \\[6pt]
\dfrac{da_2}{dt}(t) \\[6pt]
\vdots \\[6pt]
\dfrac{da_n}{dt}(t)
\end{bmatrix}
= F(t)
$$

Solving for the column matrix of derivatives of the $a_k(t)$'s gives

$$
\begin{bmatrix}
\dfrac{da_1}{dt}(t) \\[6pt]
\dfrac{da_2}{dt}(t) \\[6pt]
\vdots \\[6pt]
\dfrac{da_n}{dt}(t)
\end{bmatrix}
= e^{-Dt}Q^{-1}F(t)
$$

Clearly, it is time to integrate in order to find the unknown functions $a_1(t)$, $a_2(t)$, $\ldots, a_n(t)$, and when you integrate a matrix of functions you integrate its entry functions. So

$$
\begin{bmatrix}
a_1(t) \\
a_2(t) \\
\vdots \\
a_n(t)
\end{bmatrix}
= \int e^{-Dt}Q^{-1}F(t)\,dt + C
$$

where C is a column of n arbitrary constants that can be used to satisfy an initial condition, when given.

Let $Z(t) = \int e^{-Dt}Q^{-1}F(t)\,dt$. Then the general solution $X(t)$ to our non-homogeneous system of differential equation is

$$
X(t) = Qe^{Dt}(Z(t) + C) \tag{12.43}
$$

for C a column of n arbitrary real constants. This general solution can also be written as

$$
X(t) = e^{At}\left[\int e^{-At}F(t)\,dt + C\right]
$$

since $e^{-At} = Qe^{-Dt}Q^{-1}$.

If our nonhomogeneous system of differential equations has the initial condition $X(0) = X_0$, then C is no longer arbitrary but a particular column of real numbers. In order to find this particular column C, we plug in $t = 0$ into our general solution for $X(t)$. This gives us

$$X_0 = X(0) = Qe^{D0}(Z(0) + C) = Q(Z(0) + C)$$

since the exponential of the zero matrix is the identity of the same size. Upon solving for C we have

$$C = Q^{-1}X_0 - Z(0) \qquad (12.44)$$

So the solution to the nonhomogeneous system of differential equations (12.41), including the initial condition, is now given by

$$X(t) = Qe^{Dt}\left(Z(t) + Q^{-1}X_0 - Z(0)\right) \qquad (12.45)$$

This form of the solution will avoid the creation of superfluous terms and so should be used when A is diagonalizable. Do not forget to take the real part of this solution $X(t)$ as your real answer to the problem when A is real and the initial conditions are also real.

We can write this solution, using the original matrix A, as

$$X(t) = e^{At}\left[\int e^{-At}F(t)\,dt + X_0 - W(0)\right]$$

where $W(t) = \int e^{-At}F(t)\,dt$. This form of the solution should only be used when A is not diagonalizable since it creates superfluous terms in the solution with very small coefficient values, these superfluous terms occur because $QQ^{-1} \approx I_n$ because Q is not exact and we get very small values instead of 0 in this identity matrix.

This concludes our application of eigenvalues and eigenvectors in solving square homogeneous and nonhomogeneous first-order linear systems of differential equations with initial conditions when the coefficient matrix A is diagonalizable. You are asked in the exercises to look at a problem, where A is not diagonalizable but it is upper triangular of a form where e^{At} can be computed. In the research projects, you are asked to look at the situation where A is not a constant matrix, but its entries are functions of t.

In Section 12.5, we will look at a collection of important facts about eigenvalues and their eigenvectors with proofs for most of this information.

Homework Problems

Note: It is permissible to use Maple *to find the matrices D and Q, when the coefficient matrix A is diagonalizable, if the system is larger than* 2×2.

1. Find the real solutions to the homogeneous first-order system of linear differential equations given by

$$\frac{dx}{dt} = 2x(t) + 5y(t) - z(t)$$

$$\frac{dy}{dt} = 5x(t) - 3y(t) + 8z(t)$$

$$\frac{dz}{dt} = -x(t) + 8y(t) - 7z(t)$$

where $x(t)$, $y(t)$, and $z(t)$ are three unknown functions of t with the initial conditions $x(0) = -12$, $y(0) = 4$, $z(0) = 7$.

2. Find the real solutions to the nonhomogeneous first-order system of linear differential equations given by

$$\frac{dx}{dt} = 2x(t) + 5y(t) - z(t) + \cos(t)$$

$$\frac{dy}{dt} = 5x(t) - 3y(t) + 8z(t) - \sin(t)$$

$$\frac{dz}{dt} = -x(t) + 8y(t) - 7z(t) + e^{3t}$$

where $x(t)$, $y(t)$, and $z(t)$ are three unknown functions of t with the initial conditions $x(0) = -12, y(0) = 4, z(0) = 7$.

3. Find the real solutions to the homogeneous first-order system of linear differential equations given by

$$\frac{dx}{dt} = 7x(t) - 5y(t)$$

$$\frac{dy}{dt} = 15x(t) - 3y(t)$$

where $x(t)$ and $y(t)$ are two unknown functions of t with the initial conditions $x(0) = 8, y(0) = -1$.

4. Find the real solutions to the nonhomogeneous first-order system of linear differential equations given by

$$\frac{dx}{dt} = 7x(t) - 5y(t) + t^2$$

$$\frac{dy}{dt} = 15x(t) - 3y(t) - \cos(4t)$$

where $x(t)$ and $y(t)$ are two unknown functions of t with the initial conditions $x(0) = 8, y(0) = -1$.

5. Find by hand, using back substitution, the real solutions to the homogeneous first-order system of linear differential equations given by

$$\frac{dx}{dt} = 5x(t) + 2y(t)$$

$$\frac{dy}{dt} = 5y(t)$$

$$\frac{dz}{dt} = 5z(t)$$

where $x(t)$, $y(t)$, and $z(t)$ are three unknown functions of t with the initial conditions $x(0) = 2, y(0) = -1, z(0) = 3$.

6. The upper triangular coefficient matrix A of the system from problem 5 is not diagonalizable. Can you use the methods of this section to get the solution to this system of differential equations? You will need to find both e^{At} and e^{-At}. If yes, do so and compare what you get here with the results of problem 5 to see if it has worked. If no, explain why not.

7. Find using matrix methods (check with *dsolve*) and plot from $t \in [-1, 1]$ the solution function $y(t)$ to the fourth-order constant coefficient nonhomogeneous linear differential equation

$$\frac{d^4y}{dt^4} - 8\frac{d^3y}{dt^3} + 5\frac{dy}{dt} + 2y(t) = \cos(t)$$

with the initial conditions

$$y(0) = -1, \quad \frac{dy}{dt}(0) = 10, \quad \frac{d^2y}{dt^2}(0) = -7, \quad \frac{d^3y}{dt^3}(0) = 3$$

8. Show that $\frac{d}{dt}e^{At} = Ae^{At}$ is a correct derivative rule for A any real diagonalizable matrix. If the matrix A is not diagonalizable, then check that this derivative rule is still correct using the definition of the exponential of a square matrix as an infinite sum. You may use

$$\frac{d}{dt}\sum_{n=0}^{\infty}\frac{1}{n!}(At)^n = \sum_{n=1}^{\infty}\frac{1}{(n-1)!}A(At)^{n-1}$$

but you should justify why this derivative formula should be correct.

9. For the nonhomogeneous system $\dfrac{dX}{dt} = AX(t) + F(t)$ of first-order linear differential equations, we found that the general solution to this system is

$$X(t) = e^{At}\left(\int e^{-At}F(t)\, dt + C\right)$$

Now verify this directly by computing the derivative of this solution and seeing that it satisfies the equation.

Maple Problems

1. Compare your answers to *Homework* problems 1, 2, and 5 using matrix methods to the answer from *dsolve* using the option "method = laplace". Plot the solution as a parametric spacecurve with the initial condition point.

2. Compare your answer to *Homework* problems 3 and 4 using matrix methods to the answer from *dsolve* using the option "method = laplace". Plot the solution as a parametric curve with the initial condition point.

Research Projects

1. For the homogeneous system

$$\frac{dX}{dt} = A\,X(t)$$

of first-order linear differential equations, we found that the general solution to this system is

$$X(t) = e^{A\,t}\,C$$

where C is an arbitrary constant column. If we change this system to one in which the real square matrix A is no longer constant but a matrix function $A(t)$, then we have the new system

$$\frac{dX}{dt} = A(t)X(t)$$

Prove or disprove that this system has the general solution

$$X(t) = e^{\int A(t)\,dt}\,C$$

If it does work then check it on an example.

2. For the nonhomogeneous system

$$\frac{dX}{dt} = A\,X(t) + F(t)$$

of first-order linear differential equations, we found that the general solution to this system is

$$X(t) = e^{At}\left(\int e^{-At} F(t)\, dt + C \right),$$

where C is an arbitrary constant column. If we change this system to one in which the real square matrix A is no longer constant but a matrix function $A(t)$, then we have the new system

$$\frac{dX}{dt} = A(t)\, X(t) + F(t)$$

Prove or disprove that this system has the general solution

$$X(t) = e^{\int A(t)\, dt}\left(\int e^{-\int A(t)\, dt} F(t)\, dt + C \right)$$

If it does work then check it on an example.

3. (a) Is there a way, using matrix methods, to solve the following system of nonhomogeneous constant coefficient linear differential equations

$$\frac{d^2x}{dt^2} - 8\frac{dx}{dt} + 3\frac{dy}{dt} + 4x(t) - y(t) = \sin(t)$$

$$\frac{d^2y}{dt^2} + 7\frac{dx}{dt} - 2\frac{dy}{dt} + 9x(t) - 3y(t) = \cos(t)$$

with initial conditions

$$x(0) = -6,\ y(0) = 2,\ \frac{dx}{dt}(0) = 1,\ \frac{dy}{dt}(0) = -5$$

If yes, then solve this problem for its two real solutions $x(t)$ and $y(t)$. If no, then explain why not.

(b) Can *dsolve* solve this problem?

(c) If the answer to part (a) was yes, then generalize your method to any size square system.

12.5   Basic Facts About Eigenvalues, Eigenvectors, and Diagonalizability

Now we turn to some purely theoretical material concerning eigenvalues, eigenvectors, and diagonalizability. The few facts that we give without a proof would require a great deal of material to be developed for their proofs, and so we leave their proofs to a future course in advanced linear algebra if you should decide to take one or as a project to satisfy your curiosity.

Let us begin with some of the most basic facts and then move on to the more complicated ones. In all that follows, let A be a square $n \times n$ matrix with all complex entries unless otherwise specified. The typical case in many applied situations is that A has all real entries, but complex eigenvalues and eigenvectors.

Theorem 12.5.1. *Let $A \in \mathbb{C}^{n \times n}$ and $\lambda \in \mathbb{C}$ be an eigenvalue of A. Also, let $\mathbb{E}_\lambda$ be the subset of $\mathbb{C}^n$ consisting of $\overrightarrow{0}$ and all the eigenvectors $\overrightarrow{x}$ of A for the eigenvalue λ, that is,*

$$\mathbb{E}_\lambda = \left\{ \overrightarrow{x} \in \mathbb{C}^n \mid A\overrightarrow{x} = \lambda\overrightarrow{x} \right\}$$

Then $\mathbb{E}_\lambda$ is a subspace of $\mathbb{C}^n$, where $\mathbb{E}_{\lambda_1} \cap \mathbb{E}_{\lambda_2} = \left\{ \overrightarrow{0} \right\}$ when $\lambda_1 \neq \lambda_2$. As well,

$$1 \leq dim\left(\mathbb{E}_\lambda\right) \leq M_\lambda$$

where M_λ is the algebraic multiplicity of λ as a root of A's characteristic polynomial $P(\lambda)$.

Proof. Let us show that $\mathbb{E}_\lambda$ is a subspace of $\mathbb{C}^n$ for any eigenvalue of A. Since $\mathbb{E}_\lambda$ is a nonempty subset of the vector space $\mathbb{C}$ over the field of complex numbers, all we need show is that any linear combination of the elements of $\mathbb{E}_\lambda$ is also in $\mathbb{E}_\lambda$. In other words, we need to verify that for any two vectors $\overrightarrow{x}, \overrightarrow{y} \in \mathbb{E}_\lambda$, and $a, b \in \mathbb{C}^n$, we have that $a\overrightarrow{x} + b\overrightarrow{y} \in \mathbb{E}_\lambda$. It is sufficient to show

$$A(a\overrightarrow{x} + b\overrightarrow{y}) = \lambda(a\overrightarrow{x} + b\overrightarrow{y})$$

Now

$$\begin{aligned}
A(a\overrightarrow{x} + b\overrightarrow{y}) &= a(A\overrightarrow{x}) + b(A\overrightarrow{y}) \\
&= a(\lambda\overrightarrow{x}) + b(\lambda\overrightarrow{y}) \\
&= \lambda(a\overrightarrow{x} + b\overrightarrow{y})
\end{aligned}$$

With the above simple matrix calculation, we have shown that $\mathbb{E}_\lambda$ is a subspace of $\mathbb{C}^n$.

Let $\lambda_1 \neq \lambda_2$ be two different eigenvalues for the matrix A. Assume that $\mathbb{E}_{\lambda_1} \cap \mathbb{E}_{\lambda_2} \neq \left\{ \overrightarrow{0} \right\}$. Then there exists some nonzero eigenvector $\overrightarrow{x} \in \mathbb{E}_{\lambda_1} \cap \mathbb{E}_{\lambda_2}$.

So both $A\vec{x} = \lambda_1 \vec{x}$ and $A\vec{x} = \lambda_2 \vec{x}$. When we combine these two equations, we have $\lambda_1 \vec{x} = \lambda_2 \vec{x}$ or $(\lambda_1 - \lambda_2) \vec{x} = \vec{0}$. Since $\vec{x} \neq \vec{0}$, we must have that $\lambda_1 - \lambda_2 = 0$ or $\lambda_1 = \lambda_2$. This is a contradiction, and so $\mathbb{E}_{\lambda_1} \cap \mathbb{E}_{\lambda_2} = \{\vec{0}\}$.

Next, we prove the second part of the theorem, namely, that $1 \leq \dim (\mathbb{E}_\lambda) \leq M_\lambda$. Clearly, $1 \leq \dim (\mathbb{E}_\lambda)$ since only the subspace consisting of the zero vector has dimension zero. Also $\mathbb{E}_\lambda \neq \{\vec{0}\}$ since $\mathbb{E}_\lambda$ must contain an eigenvector $\vec{x} \neq \vec{0}$ corresponding to the eigenvalue λ. The proof of $\dim (\mathbb{E}_\lambda) \leq M_\lambda$ will not be given in this text. (If you combine the two future Theorems 12.5.3 and 12.5.7, then you should begin to see why this is true. It is left as a project for you to see how these two facts, with other information, can complete this proof.) □

Next, we look at two facts which relate the eigenvalues of A to the determinant of A, and a fact about upper and lower triangular matrices' eigenvalues. In all that follows in this section, we let $A \in \mathbb{C}^{n \times n}$, unless otherwise stated.

Theorem 12.5.2. *The determinant of A is the constant term of A's characteristic polynomial.*

Proof. The characteristic polynomial of A is $P(\lambda) = \det(A - \lambda I_n)$. If you expand out this determinant, you will get

$$P(\lambda) = (-1)^n \lambda^n + a_{n-1}\lambda^{n-1} + a_{n-2}\lambda^{n-2} + \cdots + a_1 \lambda + a_0 \qquad (12.46)$$

where the a_k's are the coefficients of $P(\lambda)$. Then plugging in 0 for λ in this polynomial gives

$$P(0) = \det(A - 0\,I_n) = \det(A)$$

by its definition, and $P(0) = a_0$ in the expansion given in (12.46). Therefore, we have that $a_0 = \det(A)$. □

Theorem 12.5.3. *The determinant of A is the product of A's eigenvalues where each eigenvalue is a factor in this product M_λ times, where M_λ is the algebraic multiplicity of λ as a root of A's characteristic polynomial. As well, the sum of the multiplicities of A's distinct eigenvalues is n. In general, the algebraic multiplicity of a root of a polynomial is 1 and the polynomial has as many distinct roots as its degree.*

Proof. Now, let us factor completely the characteristic polynomial as it is expressed in (12.46). We get

$$P(\lambda) = (-1)^n \prod_{j=1}^{k} (\lambda - \lambda_j)^{M_{\lambda_j}}$$

where $\lambda_1, \lambda_2, \ldots, \lambda_k$ are the k distinct roots of $P(\lambda)$ corresponding to the k distinct eigenvalues of A. Their respective algebraic multiplicities are the positive integers $M_{\lambda_1}, M_{\lambda_2}, \ldots, M_{\lambda_k}$. Then since the characteristic polynomial $P(\lambda)$ has degree n, we must have

$$M_{\lambda_1} + M_{\lambda_2} + \cdots + M_{\lambda_k} = n$$

Now from the Theorem 12.5.2,

$$\det(A) = P(0) = (-1)^n \prod_{j=1}^{k} (-\lambda_j)^{M_{\lambda_j}}$$

$$= (-1)^n (-1)^{M_{\lambda_1} + M_{\lambda_2} + \cdots + M_{\lambda_k}} \prod_{j=1}^{k} \lambda_j^{M_{\lambda_j}}$$

$$= (-1)^{2n} \prod_{j=1}^{k} \lambda_j^{M_{\lambda_j}}$$

$$= \prod_{j=1}^{k} \lambda_j^{M_{\lambda_j}}$$

So the determinant of A is the product of A's eigenvalues, where each eigenvalue is a factor in this product M_λ times, and M_λ is the algebraic multiplicity of λ as a root of A's characteristic polynomial. □

Theorem 12.5.4. *If A is a lower or upper triangular matrix, then A's eigenvalues are its diagonal entries and its characteristic polynomial is*

$$P(\lambda) = \prod_{k=1}^{n} (A_{k,k} - \lambda) \tag{12.47}$$

where A's diagonal entries are $A_{1,1}, A_{2,2}, \ldots, A_{n,n}$.

Proof. In order to compute A's characteristic polynomial

$$P(\lambda) = \det(A - \lambda I_n)$$

when A is upper triangular, expand along the first column of $A - \lambda I_n$ to find this determinant and then continue expanding along first columns at each future determinant computation. This gives exactly (12.47). From this factoring of the characteristic polynomial we have that A's diagonal entries are its eigenvalues. If A is lower triangular, then expand along rows instead of columns to find the characteristic polynomial. □

Now let us turn to facts related to the diagonalizability of A, and when A is not diagonalizable.

Theorem 12.5.5. *Let A be diagonalizable with $A = QDQ^{-1}$. Then A and D have the same characteristic polynomial $P(\lambda)$.*

Proof. Let $A = QDQ^{-1}$, and define $P_A(\lambda) = \det(A - \lambda I_n)$ and $P_D(\lambda) = \det(D - \lambda I_n)$. Then,

$$
\begin{aligned}
P_A(\lambda) &= \det(A - \lambda I_n) \\
&= \det(QDQ^{-1} - \lambda I_n) \\
&= \det(Q(D - \lambda I_n)Q^{-1}) \\
&= \det(Q)\det(D - \lambda I_n)\det(Q^{-1}) \\
&= \det(Q)\det(D - \lambda I_n)\det(Q)^{-1} \\
&= \det(D - \lambda I_n) \\
&= P_D(\lambda)
\end{aligned}
$$

So A and D have the same characteristic polynomial $P(\lambda)$.   □

In particular, we now have $\det(A) = \det(D)$ by Theorem 12.5.3. Also, as a consequence of Theorem 12.5.4, we have that each distinct eigenvalue λ_k of A must appear on the diagonal of D exactly M_{λ_k} times for M_{λ_k} the algebraic multiplicity of λ_k as a root of A's characteristic polynomial $P(\lambda)$.

Before we can discuss the conditions under which the square matrix A is or is not diagonalizable, we need to know something about the independence of a set of eigenvectors of A with one eigenvector chosen for each eigenvalue of A. This will relate to the ability of the matrix Q being invertible or not in our diagonalizability discussion for A.

Theorem 12.5.6. *Let $\lambda_1, \lambda_2, \ldots, \lambda_k \in \mathbb{C}$ be the k distinct roots of the characteristic polynomial $P(\lambda)$ of A, that is, they are the k distinct eigenvalues of A. Now let $\vec{x_1}, \vec{x_2}, \ldots, \vec{x_k}$ be k eigenvectors of A corresponding to the k distinct eigenvalues of A. Then the set $\{\vec{x_1}, \vec{x_2}, \ldots, \vec{x_k}\}$ is an independent subset of $\mathbb{C}^n$.*

Proof. We will show by finite induction on the size s of the subset $\{\vec{x_{j_1}}, \vec{x_{j_2}}, \ldots, \vec{x_{j_s}}\}$ that the full set $\{\vec{x_1}, \vec{x_2}, \ldots, \vec{x_k}\}$ is independent. That is, we will show if $s = 1$, then $\{\vec{x_{j_1}}\}$ is independent, and if for all $s < k$ and $\{\vec{x_{j_1}}, \vec{x_{j_2}}, \ldots, \vec{x_{j_s}}\}$ is independent for all subsets of size s or less, then $\{\vec{x_{j_1}}, \vec{x_{j_2}}, \ldots, \vec{x_{j_s}}, \vec{x_{j_{s+1}}}\}$ is also independent for any subset of size $s + 1$ of $\{\vec{x_1}, \vec{x_2}, \ldots, \vec{x_k}\}$. From this, it would follow that $\{\vec{x_1}, \vec{x_2}, \ldots, \vec{x_k}\}$ must be independent.

When $s = 1$, then $\{\vec{x_{j_1}}\}$ is independent since any nonzero vector automatically forms an independent set. Next we let $s < k$ and assume that the set $\{\vec{x_{j_1}}, \vec{x_{j_2}}, \ldots, \vec{x_{j_s}}\}$ is independent for all subsets with s distinct elements from the set $\{\vec{x_1}, \vec{x_2}, \ldots, \vec{x_k}\}$. Now we need to show that any subset of $s + 1$ distinct elements of $\{\vec{x_1}, \vec{x_2}, \ldots, \vec{x_k}\}$ is also independent. Assume that $\{\vec{x_{j_1}}, \vec{x_{j_2}}, \ldots, \vec{x_{j_s}}, \vec{x_{j_{s+1}}}\}$ is a dependent set for some subset of size $s + 1$. We seek a contradiction that then forces our subset to be independent. In order that we do not end up going blind due to this double subscripting, without loss of generality we will say that our dependent subset is $\{\vec{x_1}, \vec{x_2}, \ldots, \vec{x_s}, \vec{x_{s+1}}\}$.

This dependence implies that there exist complex numbers $b_1, b_2, \ldots, b_s, b_{s+1}$ with at least one of the b_k's nonzero such that

$$b_1 \vec{x_1} + b_2 \vec{x_2} + \cdots + b_s \vec{x_s} + b_{s+1} \vec{x_{s+1}} = 0 \qquad (12.48)$$

Now if any of these b_k's are zero, then we have remaining a linear combination of s or less of our $\vec{x_k}$'s that equals 0, which forces all the rest of the b_k's to be zero by the independence of any subset of the $\vec{x_k}$'s of size s or less. Hence, all the b_k's must be nonzero.

Now multiply equation (12.48) by the matrix A, and use the fact that $A\vec{x_k} = \lambda_k \vec{x_k}$ to get the new equation

$$b_1 \lambda_1 \vec{x_1} + b_2 \lambda_2 \vec{x_2} + \cdots + b_s \lambda_s \vec{x_s} + b_{s+1} \lambda_{s+1} \vec{x_{s+1}} = 0 \qquad (12.49)$$

Since the eigenvalues we are using are all distinct, we cannot have two of them that are both 0. There must be at least one nonzero eigenvalue in the set $\{\lambda_1, \lambda_2, \ldots, \lambda_s, \lambda_{s+1}\}$ since s is at least 1. Let us say that $\lambda_{s+1} \neq 0$ for convenience.

Let us now look closely at the two equations (12.48) and (12.49). We can solve each of these two equations for the last eigenvector, $\vec{x_{s+1}}$, since the coefficients, b_{s+1} and $b_{s+1}\lambda_{s+1}$, are both nonzero as we have already argued. Equation (12.48) yields

$$\vec{x_{s+1}} = -\frac{1}{b_{s+1}} \left(b_1 \vec{x_1} + b_2 \vec{x_2} + \cdots + b_s \vec{x_s} \right)$$

while equation (12.49) gives

$$\vec{x_{s+1}} = -\frac{1}{b_{s+1}\lambda_{s+1}} \left(b_1 \lambda_1 \vec{x_1} + b_2 \lambda_2 \vec{x_2} + \cdots + b_s \lambda_s \vec{x_s} \right)$$

Now the set $\{\vec{x_1}, \vec{x_2}, \ldots, \vec{x_s}\}$ has s elements, and thus is independent. If two linear combinations of an independent set's elements are equal, then all of their coefficients must equal for corresponding vectors in the set. This is our situation, which forces the two coefficients of $\vec{x_1}$ to equal. This results in the equation

$$-\frac{b_1}{b_{s+1}} = -\frac{b_1 \lambda_1}{b_{s+1}\lambda_{s+1}}$$

If we simplify this equation we get that $\lambda_1 = \lambda_{s+1}$. Finally, we have arrived at the desired contradiction since all the eigenvalues of A we are using are distinct, no duplications are allowed. Since this holds for arbitrary $s < k$, we can now conclude that the set $\{\vec{x_1}, \vec{x_2}, \ldots, \vec{x_k}\}$ is an independent subset of $\mathbb{C}^n$. $\square$

Theorem 12.5.7. Let $\lambda_1, \lambda_2, \ldots, \lambda_k \in \mathbb{C}$ be the k distinct eigenvalues of A. Also, let $B_1, B_2, \ldots, B_k$ be bases for the eigenspaces $\mathbb{E}_{\lambda_1}, \mathbb{E}_{\lambda_2}, \ldots, \mathbb{E}_{\lambda_k}$, respectively. Then $B_1 \cup B_2 \cup \cdots \cup B_k$ is an independent set in $\mathbb{C}^n$.

We leave the proof of Theorem 12.5.7 as an exercise. It is similar in spirit to, and should be viewed as an extension of, Theorem 12.5.6.

If we now combine several of these facts, we will see when a square matrix A is or is not diagonalizable.

Theorem 12.5.8. *A square matrix A is diagonalizable if and only if for each eigenvalue λ of A, we have $dim\,(\mathbb{E}_\lambda) = M_\lambda$.*

Proof. First, we prove that if a square matrix A is diagonalizable, then for each eigenvalue λ of A, we must have $\dim\,(\mathbb{E}_\lambda) = M_\lambda$. So let A be diagonalizable with $A = QDQ^{-1}$ for diagonal matrix D with the eigenvalues of A on its diagonal and Q an invertible matrix whose columns are eigenvectors of A for the corresponding eigenvalues of A in D. So by Theorem 12.5.5, each distinct eigenvalue λ_k of A must appear on the diagonal of D exactly M_{λ_k} times. Then corresponding to each distinct eigenvalue λ_k of A we must have M_{λ_k} independent eigenvectors for the eigenvalue λ_k appearing as columns in Q in the same location as λ_k appears in D. The independence is necessary since if any subset of the columns of Q form a dependent set, then Q has no inverse. So $\dim\,(\mathbb{E}_\lambda) \geq M_\lambda$ since the dimension of a vector space is the size of its largest independent subset. By Theorem 12.5.1, $\dim\,(\mathbb{E}_\lambda) \leq M_\lambda$ for each eigenvalue λ of A. Combining these two inequalities, we have $\dim\,(\mathbb{E}_\lambda) = M_\lambda$ for each eigenvalue λ of A.

Next, we prove that if for each eigenvalue λ of A, we have $\dim\,(\mathbb{E}_\lambda) = M_\lambda$, then A is diagonalizable.

In order to prove that A is diagonalizable, we only need to construct the diagonal matrix D of eigenvalues of A and the invertible matrix Q of corresponding eigenvectors of A. The diagonal matrix D is always constructible since we can place each distinct eigenvalue λ of A on D's diagonal M_λ times, giving a total of n eigenvalues since the sum of all the multiplicities of A's distinct eigenvalues is n by Theorem 12.5.3.

As to the construction of Q, since $\dim\,(\mathbb{E}_\lambda) = M_\lambda$ for each distinct eigenvalue of A, we can find a basis $\mathbf{B}_\lambda$ of each subspace $\mathbb{E}_\lambda$ consisting of M_λ elements. Now place in the matrix Q the elements of the basis $\mathbf{B}_\lambda$ in the same locations you have placed their eigenvalue λ in D. This completes Q since Q must have n columns and the sum of all the multiplicities of A's distinct eigenvalues is n. The final piece is that Q must be invertible: This fact follows from Theorem 12.5.7, since a square matrix is invertible if and only if its columns form an independent set. $\qquad\square$

Theorem 12.5.9. *A square matrix A is not diagonalizable if and only if for some eigenvalue λ of A, we have $dim\,(\mathbb{E}_\lambda) < M_\lambda$.*

Proof. This part follows from Theorems 12.5.1 and 12.5.8. Note that a matrix A is not diagonalizable if and only if the matrix Q is not constructible from independent eigenvectors of A to have an inverse Q^{-1}. $\qquad\square$

Theorem 12.5.10. *A square $n \times n$ matrix A is diagonalizable if and only if there is a basis for $\mathbb{C}^n$ consisting entirely of eigenvectors of A.*

Proof. If A is diagonalizable, then $A = QDQ^{-1}$. The Q's columns $\overrightarrow{x_1}, \overrightarrow{x_2}, \ldots, \overrightarrow{x_n}$, are n eigenvectors of A and elements of $\mathbb{C}^n$. Q has an inverse if and only if its columns (and rows) form a basis of $\mathbb{C}^n$. So Q^{-1} existing forces its columns to be a basis of $\mathbb{C}^n$.

Let $\{\overrightarrow{x_1}, \overrightarrow{x_2}, \ldots, \overrightarrow{x_n}\}$ be a basis for $\mathbb{C}^n$ consisting entirely of eigenvectors of A. Then the matrix Q with these n vectors as its columns has an inverse matrix Q^{-1}. If we also let D be the diagonal matrix with the corresponding eigenvalues of A on its diagonal to these $\overrightarrow{x_k}$'s, then A is diagonalizable with $A = QDQ^{-1}$, since $AQ = QD$ is automatically true for any $n \times n$ matrix Q of eigenvectors of A and the corresponding diagonal matrix of their eigenvalues D. $\qquad\square$

The following fact is a culmination of the previous facts related to A's diagonalizability. It gives the typical situation when A is diagonalizable and some examples of general matrices that are not diagonalizable.

Theorem 12.5.11. *If a square $n \times n$ matrix A has n distinct eigenvalues, then A is diagonalizable.*

Proof. If a square $n \times n$ matrix A has n distinct eigenvalues, then A is diagonalizable follows from Theorems 12.5.8 and 12.5.1. In this case all the eigenvalues of A must have algebraic multiplicity 1 since the characteristic polynomial $P(\lambda)$ of A has degree n, so $1 \leq \dim(\mathbb{E}_\lambda) \leq M_\lambda = 1$, for each eigenvalue λ of A, giving $\dim(\mathbb{E}_\lambda) = 1 = M_\lambda$. $\qquad\square$

Nondiagonalizability is fairly rare for square $n \times n$ matrices A since most characteristic polynomials $P(\lambda)$ have n distinct roots, with each root having algebraic multiplicity 1. As a consequence of Theorem 12.5.3, if A is upper or lower triangular with distinct diagonal entries, then A is diagonalizable.

Example 12.5.1. As some examples of nondiagonalizable square matrices, consider the following families of upper triangular matrices

$$A = \begin{bmatrix} a & b \\ 0 & a \end{bmatrix}, B = \begin{bmatrix} a & b & c \\ 0 & a & d \\ 0 & 0 & a \end{bmatrix} \tag{12.50}$$

where $b \neq 0$ in the matrix A, and at least one of b, c, and d is not zero in B. These two types of upper triangular matrices have their only eigenvalue a along their diagonals. We leave it as an exercise to check that the 3×3 matrix B is nondiagonalizable, while we now show that A is nondiagonalizable.

The characteristic polynomial for A is

$$P(\lambda) = \det(A - \lambda I_2) = (a - \lambda)^2 = (\lambda - a)^2$$

Now solving $P(\lambda) = 0$ tells us that a is the only eigenvalue of A and it has algebraic multiplicity $M_a = 2$.

Now let us find a basis of the eigenspace $\mathbb{E}_a$. In order to do this, we must solve the linear system of equations given by

$$(A - a\,I_2)\vec{x} = \vec{0} \tag{12.51}$$

for the eigenvectors $\vec{x}$ for eigenvalue a. Equation (12.51) really is

$$\begin{bmatrix} 0 & b \\ 0 & 0 \end{bmatrix} \vec{x} = \vec{0}$$

which has solutions $\vec{x} = \begin{bmatrix} c \\ 0 \end{bmatrix}$. These solutions $\vec{x}$ form the eigenspace $\mathbb{E}_a$, and give us a basis vector $\langle 1, 0 \rangle$. So $\dim(\mathbb{E}_\lambda) = 1 < M_a = 2$, and by Theorem 12.5.9, A is not diagonalizable.

$\blacksquare$

You should look in an advanced linear algebra text to see the complete description of nondiagonalizable square matrices, as well as a discussion of *canonical forms* of which diagonalizability is only one. Canonical forms for square matrices classify when two square matrices are similar or not: The square matrix A and the diagonal matrix D of A's eigenvalues are similar matrices when the matrix Q exists and is invertible in order for A to be diagonalizable.

Our final facts (hopefully, you have not fallen asleep yet) concerning A go back to its eigenvalues and eigenvectors and then moves on to its diagonalizability. These theorems will only be true when A is a real matrix because the realness of A allows certain nice behavior in the roots of A's characteristic polynomial.

Theorem 12.5.12. *If A is a real matrix, then A's complex eigenvalues occur in complex conjugate pairs λ and $\overline{\lambda}$. As well, if $\vec{x}$ is an eigenvector for λ, then $\overline{\vec{x}}$ is an eigenvector for $\overline{\lambda}$. Moreover, if A is $n \times n$ and n is odd, then A has at least one real eigenvalue λ and its eigenspace $\mathbb{E}_\lambda$ has a real basis since A is real.*

Proof. Let $P(\lambda)$ be A's characteristic polynomial. Then all of the coefficients of the characteristic polynomial $P(\lambda)$ are real since A is real. Now the complex conjugation operation has several useful properties (given without proof, but you should do as exercises), which are that for any two complex numbers c and d, $\overline{c + d} = \overline{c} + \overline{d}$ and $\overline{c \cdot d} = \overline{cd}$. These two properties apply to differences as well as sums, divisions as well as products and to more terms or factors than just two.

Let r be a root of the characteristic polynomial $P(\lambda)$. Then $P(r) = 0$. If we take the conjugate of this equation, then we get $P(\overline{r}) = 0$ by using the conjugation properties above since the polynomial's coefficients are all real.

So any real coefficient polynomial has complex roots that occur in complex conjugate pairs. It follows that $P(\lambda)$ has at least one real root r if its degree is odd since it must have an even number of complex roots while its total number of roots is odd. Then if A is $n \times n$ and n is odd, then A has at least one real eigenvalue since $P(\lambda)$ has degree n.

Let $\overrightarrow{x}$ be an eigenvector for λ. Then $A\overrightarrow{x} = \lambda\overrightarrow{x}$. If we take the complex conjugate of this matrix equation we have

$$\overline{A\overrightarrow{x}} = \overline{A}\,\overline{\overrightarrow{x}} = A\overline{\overrightarrow{x}}$$

for the conjugate of the LHS. While on the RHS, $\overline{\lambda\overrightarrow{x}} = \overline{\lambda}\,\overline{\overrightarrow{x}}$. If we combine these two we get that $\overline{\overrightarrow{x}}$ is an eigenvector for $\overline{\lambda}$. □

Theorem 12.5.13. *If A is a real symmetric matrix, then all of A's eigenvalues will be real and there exists a real basis for each of A's eigenspaces $\mathbb{E}_\lambda$. Also, the eigenspaces of A are perpendicular (or orthogonal) to each other, meaning that if $\overrightarrow{x_1} \in \mathbb{E}_{\lambda_1}$ and $\overrightarrow{x_2} \in \mathbb{E}_{\lambda_2}$ are two real eigenvectors for two distinct eigenvalues λ_1 and λ_2 of A, then the vectors $\overrightarrow{x_1}$ and $\overrightarrow{x_2}$ are perpendicular to each other, hence $\overrightarrow{x_1} \cdot \overrightarrow{x_2} = 0$.*

Proof. Let λ be an eigenvalue of A with eigenvector $\overrightarrow{x}$, that is, $A\overrightarrow{x} = \lambda\overrightarrow{x}$, where $\overrightarrow{x} \in \mathbb{C}^n$ and $\overrightarrow{x} \neq 0$. Now we need the complex dot product on $\mathbb{C}^n$.

Recall that if $a \in \mathbb{C}$, then $|a|^2 = a\,\overline{a}$. As a consequence, for $\overrightarrow{v} \in \mathbb{C}^n$,

$$|\overrightarrow{v}|^2 = \overrightarrow{v} \cdot \overrightarrow{v}$$
$$= \overrightarrow{v}^T \overline{\overrightarrow{v}}$$
$$= \sum_{k=1}^{n} v_k \,\overline{v_k}$$
$$= \sum_{k=1}^{n} |v_k|^2$$

which implies that $|\overrightarrow{v}|^2 \in \mathbb{R}^+$, and $|\overrightarrow{v}|^2 = 0$ if and only if $\overrightarrow{v} = \overrightarrow{0}$. We will also make use of the last two dot product properties listed in the table at the end of Section 6.2. Let $\alpha, \beta \in \mathbb{C}$ and $\overrightarrow{u}, \overrightarrow{v} \in \mathbb{C}^n$, then we have:

$$(\alpha\overrightarrow{u}) \cdot \overrightarrow{v} = \alpha(\overrightarrow{u} \cdot \overrightarrow{v}) = \overrightarrow{u} \cdot (\overline{\alpha}\overrightarrow{v}),$$
$$\overrightarrow{u} \cdot (\beta\overrightarrow{v}) = \overline{\beta}(\overrightarrow{u} \cdot \overrightarrow{v}) = (\overline{\beta}\overrightarrow{u}) \cdot \overrightarrow{v}$$

Now let B be any complex $n \times n$ matrix. Then we take the complex dot product of $B\vec{y}$ and $\vec{z}$ for any $\vec{y}, \vec{z} \in \mathbb{C}^n$ to get

$$
\begin{aligned}
(B\vec{y}) \cdot \vec{z} &= (B\vec{y})^T \overline{\vec{z}} \\
&= (\vec{y}^T B^T) \overline{\vec{z}} \\
&= \vec{y}^T \left(B^T \overline{\vec{z}} \right) \\
&= \vec{y}^T \left(\overline{\overline{B^T} \vec{z}} \right) \\
&= \vec{y} \cdot \left(\overline{B^T} \vec{z} \right)
\end{aligned}
$$

We have used here that the conjugate of a product is the product of the conjugates. This is true for the product of either complex numbers or matrices with complex entries.

Finally, we can prove our theorem. We now look at the complex dot product of $A\vec{x}$ and $\vec{x}$. This first gives us

$$
\begin{aligned}
(A\vec{x}) \cdot \vec{x} &= (\lambda \vec{x}) \cdot \vec{x} \\
&= \lambda (\vec{x} \cdot \vec{x}) \\
&= \lambda |\vec{x}|^2
\end{aligned}
$$

where $|\vec{x}|^2 \neq 0$, since $\vec{x} \neq \vec{0}$. Now from the previous results above, we also have that

$$
\begin{aligned}
(A\vec{x}) \cdot \vec{x} &= \vec{x} \cdot \left(\overline{A^T} \vec{x} \right) \\
&= \vec{x} \cdot (A\vec{x}) \\
&= \vec{x} \cdot (\lambda \vec{x}) \\
&= \overline{\lambda} (\vec{x} \cdot \vec{x}) \\
&= \overline{\lambda} |\vec{x}|^2
\end{aligned}
$$

because A is both real and symmetric. If we put these two pieces of our puzzle together, we get

$$
\lambda |\vec{x}|^2 = \overline{\lambda} |\vec{x}|^2
$$

and so $\lambda = \overline{\lambda}$. This implies that λ is real, so we can now conclude that all of A's eigenvalues are real.

Next, we turn our attention to A's eigenvectors. Remember that for the eigenvalue λ of A, $\vec{x}$ is an eigenvector of A for λ if $\vec{x} \in \mathbb{C}^n$ is a solution to the linear system $(A - \lambda I_n) \vec{x} = \vec{0}$. Since the matrix $A - \lambda I_n$ is now a real matrix, this system of linear equations can be solved to produce a real basis for its solution space, so in the rest of this proof, we can treat $\mathbb{E}_\lambda$ to be a subspace of $\mathbb{R}^n$ instead of $\mathbb{C}^n$.

We want to show that if $\overrightarrow{x_1} \in \mathbb{E}_{\lambda_1}$ and $\overrightarrow{x_2} \in \mathbb{E}_{\lambda_2}$ are two real eigenvectors for two distinct eigenvalues λ_1 and λ_2 of A, then the vectors $\overrightarrow{x_1}$ and $\overrightarrow{x_2}$ are orthogonal to each other. We need to compute the real dot product of $A\left(\overrightarrow{x_1} + \overrightarrow{x_2}\right)$ with itself in two different ways. First,

$$
\begin{aligned}
A\left(\overrightarrow{x_1} + \overrightarrow{x_2}\right) \cdot A\left(\overrightarrow{x_1} + \overrightarrow{x_2}\right) &= \left(\lambda_1 \overrightarrow{x_1} + \lambda_2 \overrightarrow{x_2}\right) \cdot \left(\lambda_1 \overrightarrow{x_1} + \lambda_2 \overrightarrow{x_2}\right) \\
&= \lambda_1^2 \left|\overrightarrow{x_1}\right|^2 + 2\lambda_1 \lambda_2 \overrightarrow{x_1} \cdot \overrightarrow{x_2} + \lambda_2^2 \left|\overrightarrow{x_2}\right|^2
\end{aligned} \tag{12.52}
$$

Second, we have

$$
\begin{aligned}
A\left(\overrightarrow{x_1} + \overrightarrow{x_2}\right) \cdot A\left(\overrightarrow{x_1} + \overrightarrow{x_2}\right) &= \left(\overrightarrow{x_1} + \overrightarrow{x_2}\right) \cdot A^2\left(\overrightarrow{x_1} + \overrightarrow{x_2}\right) \\
&= \left(\overrightarrow{x_1} + \overrightarrow{x_2}\right) \cdot \left(\lambda_1^2 \overrightarrow{x_1} + \lambda_2^2 \overrightarrow{x_2}\right) \\
&= \lambda_1^2 \left|\overrightarrow{x_1}\right|^2 + \left(\lambda_1^2 + \lambda_2^2\right) \overrightarrow{x_1} \cdot \overrightarrow{x_2} + \lambda_2^2 \left|\overrightarrow{x_2}\right|^2
\end{aligned} \tag{12.53}
$$

If we equate these two calculations and cancel equal terms, we get that

$$
2\lambda_1 \lambda_2 \, \overrightarrow{x_1} \cdot \overrightarrow{x_2} = \left(\lambda_1^2 + \lambda_2^2\right) \overrightarrow{x_1} \cdot \overrightarrow{x_2}
$$

Assume that our two eigenvectors are not orthogonal, then $\overrightarrow{x_1} \cdot \overrightarrow{x_2} \neq 0$ and our last equation simplifies to

$$
2\lambda_1 \lambda_2 = \lambda_1^2 + \lambda_2^2
$$

which is the same as

$$
\left(\lambda_1 - \lambda_2\right)^2 = 0
$$

This forces $\lambda_1 = \lambda_2$, which is not possible since we said that these two eigenvalues are distinct. This is a contradiction, and so our two eigenvectors are orthogonal, or $\overrightarrow{x_1} \cdot \overrightarrow{x_2} = 0$. □

As a consequence of the proof of this theorem, we also have the following theorem which we state without further proof.

Theorem 12.5.14. *Let B be any complex $n \times n$ matrix with $\overrightarrow{y}, \overrightarrow{z} \in \mathbb{C}^n$. Then*

$$
\left(B\overrightarrow{y}\right) \cdot \overrightarrow{z} = \overrightarrow{y} \cdot \left(\overline{B^T} \overrightarrow{z}\right)
$$

for the complex dot product. As a consequence, if the matrix B is Hermitian, that is, $B = \overline{B^T}$, then all of B's eigenvalues are real, but B's eigenvectors remain complex.

Our final (we promise) result of this section gives the best possible situation for the real symmetric matrices, they are all diagonalizable. We already know from Theorem 12.5.13 that if they are diagonalizable, then their diagonal matrix D is real and so is their invertible matrix Q of eigenvectors. Theorem 12.5.15 will now also tell us that Q can be made into a very special type of matrix, an *orthogonal matrix* where $Q^{-1} = Q^T$. We will not prove Theorem 12.5.15, although Theorem 12.5.13 and Gram-Schmidt orthonormalization explain why Q can be made orthogonal.

Theorem 12.5.15. *All real symmetric matrices A are diagonalizable with $A = QDQ^{-1}$, where Q and D are real valued and the eigenvector matrix Q can be made an orthogonal matrix, that is, the columns of Q are mutually perpendicular unit vectors.*

A proof of the portion involving D and Q follows directly from Theorem 12.5.13 and the Gram-Schmidt orthonormalization process, but we give no proof here that A must be diagonalizable.

Note that a square matrix Q is orthogonal if and only if $Q^{-1} = Q^T$, or equivalently that $Q^T Q = I_n$, where I_n is the identity matrix of the correct size. If $\det(Q) = 1$ for an orthogonal matrix Q (orthogonal matrices always have a determinant which is ± 1), then Q is called a rotation matrix since multiplication by Q will rotate all vectors about the origin (the zero vector) through some fixed angle θ. In particular, if Q is a 2×2 rotation matrix, then $Q = \begin{bmatrix} \cos(\theta) & -\sin(\theta) \\ \sin(\theta) & \cos(\theta) \end{bmatrix}$ for some fixed angle θ.

So if A is a real symmetric matrix, then A is diagonalizable with $A = QDQ^T$, where D is a real diagonal matrix of the eigenvalues of A and Q is a real rotation matrix of corresponding eigenvectors of A.

Hopefully, these facts about eigenvalues, eigenvectors, and diagonalizability have been helpful in clarifying this material as well as giving you topics for further study. We also hope that the few facts that have been given in this section without proof will encourage the reader to further investigations into linear algebra, and beyond.

Homework Problems

1. Let A be a $n \times n$ matrix that is diagonalizable. Let $P(\lambda)$ be A's characteristic polynomial. Show that $P(A)$ is the $n \times n$ zero matrix where $P(A)$ is the characteristic polynomial with λ replaced by A. Note that it is not valid to replace λ by A in the determinant formula for $P(\lambda)$, why not?

Comment: This is a version of the Cayley-Hamilton theorem *which says that any square matrix A plugged into its characteristic polynomial gives the zero matrix back.*

2. Let A be a $n \times n$ matrix with n distinct eigenvalues $\lambda_1, \lambda_2, \ldots, \lambda_n$. Let $P(\lambda)$ be A's characteristic polynomial. Show that $P(A)$ is the $n \times n$ zero matrix where $P(A)$ is the characteristic polynomial with λ replaced by A.

3. Let A be a diagonalizable matrix with all non-negative real eigenvalues. Define $\sqrt[k]{A}$, for k any positive integer and check with an example that your definition works.

4. Let A be a diagonalizable $n \times n$ matrix where all of A's eigenvalues satisfy $|\lambda| < 1$. Show that

$$\lim_{k \to \infty} A^k = 0_n$$

where 0_n is the zero $n \times n$ matrix.

5. Let A be any diagonalizable $n \times n$ matrix with $A = QDQ^{-1}$ for D the diagonal matrix having the eigenvalues $\lambda_1, \lambda_2, \ldots, \lambda_n$ on its diagonal. Show that e^A must be invertible and that

$$\det\left(e^A\right) = e^{\lambda_1 + \lambda_2 + \cdots + \lambda_n}$$

6. Let λ be any real number and K be any nonzero real number. Then we know that the 2×2 matrix $A = \begin{bmatrix} \lambda & K \\ 0 & \lambda \end{bmatrix}$ is nondiagonalizable. Show that

$$e^A = e^\lambda \begin{bmatrix} 1 & K \\ 0 & 1 \end{bmatrix}$$

7. Let λ be any real number and K be any nonzero real number. Show that the following 3×3 matrices are nondiagonalizable.

(a) $\begin{bmatrix} \lambda & K & 0 \\ 0 & \lambda & 0 \\ 0 & 0 & \lambda \end{bmatrix}$ (b) $\begin{bmatrix} \lambda & 0 & K \\ 0 & \lambda & 0 \\ 0 & 0 & \lambda \end{bmatrix}$ (c) $\begin{bmatrix} \lambda & 0 & 0 \\ 0 & \lambda & K \\ 0 & 0 & \lambda \end{bmatrix}$

8. We know that the 3×3 matrices from problem 7 are all nondiagonalizable. Find the formulas for e^A for each of them.

9. Let λ be any real number and K, L and M be any real numbers with at least one of K, L, and M being nonzero. Show that the 3×3 matrix $A = \begin{bmatrix} \lambda & K & L \\ 0 & \lambda & M \\ 0 & 0 & \lambda \end{bmatrix}$ is nondiagonalizable.

10. For the nondiagonalizable matrix A in problem 9, find a formula for e^A. You may have to look at specializing the situation to two of the three real numbers K, L, and M being equal and the third number being 0 and/or all three of K, L, and M being equal to each other.

11. Use the methods of Section 12.4, along with problem 10 above to solve the following homogeneous first-order system of linear differential equations given

by

$$\frac{dx}{dt} = 5x(t) + 2y(t) + 2z(t)$$

$$\frac{dy}{dt} = 5y(t)$$

$$\frac{dz}{dt} = 5z(t)$$

where $x(t)$, $y(t)$, and $z(t)$ are three unknown functions of t with the initial conditions $x(0) = 2$, $y(0) = -1$, $z(0) = 3$.

12. Let A be a real square matrix. Remember that $\left(Q^{-1}\right)^T = \left(Q^T\right)^{-1}$ and $(BC)^T = C^T B^T$.

(a) Show that A is diagonalizable if and only if A^T is also diagonalizable.

(b) If A is diagonalizable, how are the eigenvalues and eigenvectors of A^T related to those of A?

(c) Is $e^{A^T} = \left(e^A\right)^T$ if A is diagonalizable? If yes, then prove it. If no, then give a counterexample. If not, then how can we alter things appropriate to the complex matrix case?

12. Let Q be a real $n \times n$ orthogonal matrix, that is, $QQ^T = I_n$. Show that the following are all true.

(a) $\det(Q) = \pm 1$

(b) $Q^T Q = I_n$ makes Q^T orthogonal as well, and $Q^T = Q^{-1}$.

(c) If S is also a real $n \times n$ orthogonal matrix, then the product QS is also orthogonal.

(d) Are all of these facts still correct if we delete real from everything?

Maple Problems

1. Find, using *Maple*'s *dsolve* command, the real solutions to the homogeneous first-order system of linear differential equations given in *Homework* problem 11.

2. Use *Maple* wherever possible to verify the results discussed in *Homework Problems*.

Research Projects

1. Find an advanced linear algebra text and go over the proof that the dimension of an eigenspace $\mathbb{E}_\lambda$ is less than or equal to the eigenvalue λ's algebraic multiplicity M_λ.

2. Read the article "A Method for Finding the Eigenvectors of an $n \times n$ Matrix Corresponding to Eigenvalues of Multiplicity 1" by M. Carchidi, in *The American Mathematical Monthly* of October 1986 Volume 93, Number 8, on finding an eigenvector for an eigenvalue of algebraic multiplicity 1. Now write some *Maple* code to carry out this article's algorithm and test it out on an example to see if it works.

3. Investigate the topic of canonical forms that categorizes square matrices by similarity types. Specifically, look into *Jordan and rational canonical forms* in an advanced linear algebra book. See if you can get *Maple* to do some computational work related to these types of canonical forms.

4. Investigate the topic of inner products and inner product spaces. Specifically, look into when a square matrix A is orthogonally diagonalizable like the real symmetric matrices A.

12.6   The Geometry of the Ellipse Using Eigenvalues and Eigenvectors

The purpose of this section is to look more closely at matrix multiplication by a real square matrix A and how we can interpret this multiplication geometrically. This geometric interpretation will involve conic sections, specifically the ellipse. For our analysis, we will also need some information from multivariable calculus involving optimizing functions of several variables subject to constraints and the *method of Lagrange multipliers* that solves such problems. The method of Lagrange multipliers is only needed for Example 12.6.1 since in Example 12.6.2 we use only linear algebra.

Our first example will look at a specific real square matrix A and see how multiplication by it has some interesting geometry related to eigenvalues and eigenvectors.

Example 12.6.1. As usual, we begin with an example to illustrate what we have in mind. We start with looking at multiplication by the matrix $A = \begin{bmatrix} 12 & -2 \\ -7 & -1 \end{bmatrix}$ geometrically by seeing what it does to all the vectors on the unit circle. This particular matrix A is chosen so that it is diagonalizable with real D and Q. All of the vectors on the unit circle can be written as the columns of the form $\begin{bmatrix} \cos(t) \\ \sin(t) \end{bmatrix}$. We are going to multiply the vectors of the unit circle by A, and plot the resulting vector endpoints in the plane in order to see what parametric curve they form, as illustrated in Figure 12.5.

> with(linalg): with(plots):

> A:= matrix(2, 2, [12, -2, -7, -1]):

> UnitCircle:= matrix(2,1,[cos(t),sin(t)]);

$$UnitCircle := \begin{bmatrix} \cos(t) \\ \sin(t) \end{bmatrix}$$

> UnitCirclePlot:= plot([UnitCircle[1,1], UnitCircle[2,1], t = 0..2*Pi], color = blue, thickness = 3):

> ATimesUnitCircle:= unapply([evalm(A&*UnitCircle)[1,1], evalm(A&* UnitCircle)[2,1]], t);

$$ATimesUnitCircle := t \rightarrow [12\cos(t) - 2\sin(t), -7\cos(t) - \sin(t)]$$

> ATimesUnitCircle(t)[1];

$$12\cos(t) - 2\sin(t)$$

> ATimesUnitCircle(0)[1];

$$12$$

> ATimesUnitCirclePlot:= plot([ATimesUnitCircle(t)[1], ATimesUnitCircle (t)[2], t = 0..2*Pi], color = red, thickness = 3):

> display({ATimesUnitCirclePlot, UnitCirclePlot}, scaling = constrained);

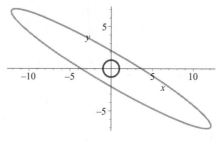

Figure 12.5: Plot of unit circle via the *UnitCircle* vector and its transformation under multiplication by matrix A.

This plot tells us that multiplication by the matrix A has turned the unit circle and its vectors of length one into an ellipse with vectors of varying length. Multiplication by A has also rotated the vectors of the unit circle to vectors on the ellipse, otherwise we should expect the ellipse to have its axes parallel to the coordinate axes. Note that both the ellipse and unit circle have the origin as center.

> evalm(A&*matrix(2,1,[1,0]));

$$\begin{bmatrix} 12 \\ -7 \end{bmatrix}$$

> angle1:= evalf(angle(col(A,1), vector([1,0])));

$$angle1 := 0.5280744484$$

> length1:= norm(col(A,1), frobenius);

$$length1 := \sqrt{193}$$

> evalm(A&*matrix(2,1,[0,1]));

$$\begin{bmatrix} -2 \\ -1 \end{bmatrix}$$

> angle2:= evalf(angle(col(A,2), vector([0,1])));

$$angle2 := 2.034443936$$

> length2:= norm(col(A,2), frobenius);

$$length2 := \sqrt{5}$$

These two angles tell us that multiplication by A is rotating the points of the unit circle but the angle of rotation is not constant since *angle1* and *angle2* are not the same. As well, the effect of A on the lengths of vectors is not constant since *length1* and *length2* are also not the same though both come from A times a unit vector.

Let us create an animation of 100 frames that displays, on the unit circle, a short arrow for the unit vector $\vec{x}$, while on the ellipse, it displays the corresponding vector $A\vec{x}$ as a long arrow. This animation will allow us to visualize how the multiplication by a matrix A works geometrically.

> UnitCircleArrows:= [seq(arrow([cos(2*Pi*(K-1)/100), sin(2*Pi*(K-1)/100)],
shape = arrow, color = blue, thickness = 3), K = 1..101)]:

> ATimesUnitCircleArrows:= [seq(arrow([ATimesUnitCircle(2*Pi*(K-1)/100)
[1], ATimesUnitCircle(2*Pi*(K-1)/100)[2]], shape = arrow, color = red, thickness = 3), K = 1..101)]:

> display([seq(display({ ATimesUnitCirclePlot, UnitCirclePlot, UnitCircleAr-
rows[K], ATimesUnitCircleArrows[K] }, scaling = constrained), K = 1..101)],
insequence − true);

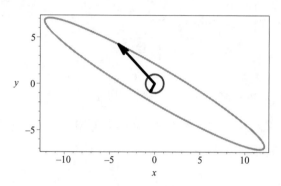

Figure 12.6: A single frame in the animation of transformed unit vectors under
matrix multiplication by A.

The frame of the animation, depicted in Figure 12.6, illustrates the geometric nature of multiplication by A in the sense of what it does to unit vectors.
Now we want to see how eigenvalues and their eigenvectors come into play in
this geometry. What we should be most curious about is the ellipse that is created by A from the unit circle. Let us find this ellipse's rectangular equation
in x and y, as well as plot two eigenvectors for A, one for each real eigenvalue
gotten by multiplying the unit eigenvector by its eigenvalue so that we get each
eigenvector to terminate on the ellipse.

> eigenvectors(A);
$$[-2, 1, \{[1, 7]\}], [13, 1, \{[-2, 1]\}]$$

> AEigvec1:= evalf(evalm(13*matrix(2, 1, [-2, 1])/sqrt(5)));

$$AEigvec1 := \begin{bmatrix} -11.62755348 \\ 5.813776740 \end{bmatrix}$$

> AEigvec2:= evalf(evalm(-2*matrix(2, 1, [1, 7])/sqrt(50)));

$$AEigvec2 := \begin{bmatrix} -0.2828427124 \\ -1.979898987 \end{bmatrix}$$

> Aarrow1:= arrow([AEigvec1[1, 1], AEigvec1[2, 1]], shape = arrow, thickness
= 3):

> Aarrow2:= arrow([AEigvec2[1, 1], AEigvec2[2, 1]], shape = arrow, thickness = 3):

> display({ ATimesUnitCirclePlot, Aarrow1, Aarrow2, UnitCirclePlot}, scaling = constrained, axes=boxed);

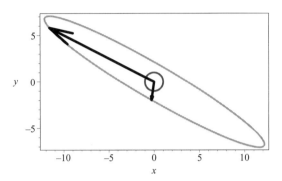

Figure 12.7: Plot of the unit circle, and its image under the map A. The two vectors are in the direction of A's eigenvectors, with each having magnitude the corresponding eigenvalues.

Upon inspection of Figure 12.7, notice that the two eigenvectors for A are nearly the major and minor axes of the transformed ellipse. Why are they so close, but not exactly there? First, the axes of an ellipse are perpendicular and these two eigenvectors are not and so they could not provide the axes. Second, perhaps in the cases when the two eigenvectors are perpendicular they do follow the axes.

Let us continue this problem by finding the rectangular equation of our ellipse and locating its major and minor axes.

> sincos:= solve({x = ATimesUnitCircle(t)[1], y = ATimesUnitCircle(t)[2]}, [sin(t), cos(t)]);

$$sincos := \left[\left[\sin(t) = -\frac{6}{13}y - \frac{7}{26}x, \cos(t) = 1/26\,x - 1/13\,y \right] \right]$$

> g:= unapply(simplify(rhs(sincos[1][1])^2 +rhs(sincos[1][2])^2-1), [x, y]);

$$g := (x, y) \rightarrow \frac{37}{169}y^2 + \frac{41}{169}yx + \frac{25}{338}x^2 - 1$$

> plot_g:= implicitplot(g(x,y)=0, x=-13..13, y=-12..12, grid=[50,50]):

> display(plot_g, thickness = 3, axes=normal);

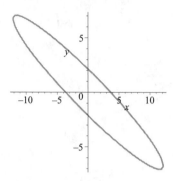

Figure 12.8: Plot of ellipse $g(x, y)$ algebraically solved for above.

The ellipse in Figure 12.8 is clearly our previous ellipse with center at the origin. In order to find the major and minor axes for this ellipse, we need to find the points on the ellipse closest to and farthest from the origin, these points are the endpoints of the major and minor axes.

We need to apply the method of Lagrange multipliers to solve this problem of maximizing and minimizing the *objective function* $f(x, y) = x^2 + y^2$ subject to the *constraint* that the points (x, y) must be on the ellipse, that is, they have to satisfy the equation

$$\frac{25}{338}x^2 + \frac{41}{169}xy + \frac{37}{169}y^2 = 1$$

or the equation $g(x, y) = 0$ for

$$g(x, y) = \frac{25}{338}x^2 + \frac{41}{169}xy + \frac{37}{169}y^2 - 1$$

The objective function $f(x, y)$ is the square of the distance from a point (x, y) to the origin, and maximizing or minimizing it is the same as doing it for the distance itself while avoiding the distraction of square roots.

The method of Lagrange multipliers from calculus says the following. A solution (x, y) to our problem of maximizing (or minimizing) the objective function $f(x, y)$ subject to the constraint $g(x, y) = 0$ must satisfy

$$\left[\frac{\partial g}{\partial x}, \frac{\partial g}{\partial y}\right] = \lambda \left[\frac{\partial f}{\partial x}, \frac{\partial f}{\partial y}\right]$$

for some constant λ called a *Lagrange multiplier*. In our case, the equation involving the multiplier λ becomes

$$\left[\frac{50}{338}x + \frac{41}{169}y, \frac{41}{169}x + \frac{74}{169}y\right] = \lambda\,[2x, 2y]$$

or when you divide by 2 we get

$$\left[\frac{25}{338}x + \frac{20.5}{169}y, \frac{20.5}{169}x + \frac{37}{169}y\right] = \lambda\,[x,\,y]$$

You should not be surprised that we can write this last equation as the matrix equation

$$\begin{bmatrix} \frac{25}{338} & \frac{41}{338} \\ \frac{41}{338} & \frac{37}{169} \end{bmatrix}\begin{bmatrix} x \\ y \end{bmatrix} = \lambda\begin{bmatrix} x \\ y \end{bmatrix} \tag{12.54}$$

This matrix equation says that our solution $X = \begin{bmatrix} x \\ y \end{bmatrix}$ is an eigenvector for

the matrix $B = \begin{bmatrix} \frac{25}{338} & \frac{41}{338} \\ \frac{41}{338} & \frac{37}{169} \end{bmatrix}$ for an eigenvalue λ of B. The two components
x and y of the eigenvector X must also satisfy $g(x, y) = 0$ in order to be a
solution of our optimization problem.

> B:= matrix(2,2,[25/338,41/(2*169),41/(2*169),37/169]);

$$B := \begin{bmatrix} \frac{25}{338} & \frac{41}{338} \\ \frac{41}{338} & \frac{37}{169} \end{bmatrix}$$

> evalf(eigenvectors(B));

$$[0.2877586772, 1., \{[0.5673764124, 1.]\}],$$
$$[0.0051407310, 1., \{[-1.762498364, 1.]\}]$$

Since our eigenvectors X of B will give the major and minor axes of our
ellipse if they are chosen to satisfy the ellipse $g(x, y) = 0$, we need to find a
scalar α so that αX satisfies $g(x, y) = 0$. This works for this problem because
any multiple of an eigenvector is also an eigenvector for the same eigenvalue.

> BEigvec1:= vector([.5673764124, 1]);

$$BEigvec1 := [0.5673764124, 1]$$

> alpha1:= fsolve(g(alpha*BEigvec1[1], alpha*BEigvec1[2]) = 0, alpha, max-
sols = 1);

$$\alpha1 := -1.621376437$$

> BEigvec2:= vector([-1.762498364, 1]);

$$BEigvec2 := [-1.762498364, 1]$$

> alpha2:= fsolve(g(alpha*BEigvec2[1], alpha*BEigvec2[2]) = 0, alpha, max-sols = 1);

$$\alpha2 := -6.882669443$$

> dotprod(BEigvec1, BEigvec2);

$$1.4\,10^{-9}$$

> Barrow1:= arrow([alpha1*BEigvec1[1], alpha1*BEigvec1[2]], shape = arrow, color = black):

> Barrow2:= arrow([alpha2*BEigvec2[1], alpha2*BEigvec2[2]], shape = arrow, color = blue):

> display({ Barrow1, Barrow2, plot_g}, scaling = constrained, thickness = 3);

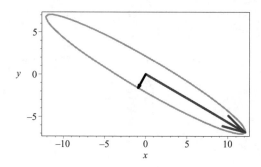

Figure 12.9: Plot of ellipse and Largrange eigenvectors solutions.

So we see from Figure 12.9 that the two eigenvectors of B do determine the major and minor axes of this ellipse. These two eigenvectors are also perpendicular as was checked above when we saw that their dot product is effectively zero. The eigenvector *BEigvec1* is the direction of the minor axis while *BEigvec2* is the direction of the major axis.

Now let us compute the lengths of these two new eigenvectors, $\alpha1$ *BEigvec1* and $\alpha2$ *BEigvec2*, which will allow us to write the equation of this ellipse in the standard form

$$\frac{x^2}{a^2} + \frac{y^2}{b^2} = 1 \tag{12.55}$$

where we let a be the length of the semimajor axis and b is the length of the semiminor axis. There are always two possible ways of writing the equation of a standard form ellipse, either that of equation (12.55), or

$$\frac{x^2}{b^2} + \frac{y^2}{a^2} = 1 \tag{12.56}$$

Notice the only difference is the division by the constants a and b. In this manner, we will always assume that a is the length of the semimajor axis and b is the length of the semiminor axis. It is up to you to pick which one of these you want since you can rotate any ellipse into either form.

Now we will plot the two ellipses to see if we are correct that one is a rotation about the origin (their mutual center) of the other. This is depicted in Figure 12.10.

> a:= norm(alpha2*BEigvec2, frobenius);

$$a := 13.94721716$$

> b:= norm(alpha1*BEigvec1, frobenius);

$$b := 1.864171164$$

> plot_strdform:= implicitplot(x^2/a^2 + y^2/b^2 = 1, x=-15..15, y=-13..13, color = blue, grid = [50,50]):
> display({plot_g, plot_strdform}, thickness = 3, scaling = constrained);

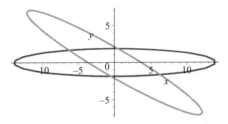

Figure 12.10: Plot of transformed ellipse and the ellipse with same major and minor axes corresponding to the x- and y-axes instead.

As the very last part of this example we compute the rotation angle needed to rotate the original ellipse counterclockwise in order to get the standard form ellipse. The eigenvector $-BEigvec2$ is rotated to the x-axis that allows us to compute this angle, which seems under $45°$. The calculation below tells us that this angle is close to $30°$.

> convert(angle([-BEigvec2[1], -BEigvec2[2]], [1,0]), 'units','radians','degrees');

$$29.56955413$$

■

Next, we will look at the situation of Example 12.6.1 in full generality. What this means is that we will theoretically study the nonstandard form ellipse generated by all the vectors $\langle \cos(t), \sin(t) \rangle$ of the unit circle multiplied by a 2×2 real matrix $A = \begin{bmatrix} \alpha & \beta \\ \gamma & \delta \end{bmatrix}$. Our goal is to know when this result is truly a nonstandard form ellipse or something else, and when it is a nonstandard form ellipse we wish to find two vectors that go from its center at the origin to the ends of its major and minor axes. This will also allow us to find a standard form equation of the ellipse that you get by rotating the original ellipse so that its axes become the coordinate axes, and of course we want to find this rotation angle θ.

If we multiply A times all of the vectors of the unit circle given parametrically as $\langle \cos(t), \sin(t) \rangle$ for $t \in [0, 2\pi]$, and label the result as $X = \begin{bmatrix} x \\ y \end{bmatrix}$, then we have

$$X = \begin{bmatrix} \alpha \cos(t) + \beta \sin(t) \\ \gamma \cos(t) + \delta \sin(t) \end{bmatrix} \tag{12.57}$$

In terms of x- and y-coordinates, this yields

$$x = \alpha \cos(t) + \beta \sin(t), y = \gamma \cos(t) + \delta \sin(t) \tag{12.58}$$

If $\det(A) = 0$, then the two rows of A form a dependent set of vectors and so one of them must be a scalar multiple of the other, say $\langle \alpha, \beta \rangle = \kappa \langle \gamma, \delta \rangle$ for some scalar κ. This tells us that the two equations (12.58) become the single equation $x = \kappa y$ that is the equation of a line through the origin. So when $\det(A) = 0$ we do not get an ellipse, but the degenerate case of a line though the origin. Clearly, this case is not very interesting and so we move on to the case of $\det(A) \neq 0$.

If $\det(A) \neq 0$, then we want to solve the equation (12.57) for $\cos(t)$ and $\sin(t)$ in terms of x and y, as we did in Example 12.6.1. Since A is invertible, we get $\begin{bmatrix} \cos(t) \\ \sin(t) \end{bmatrix} = A^{-1}X$, which can be written out in full as

$$\begin{bmatrix} \cos(t) \\ \sin(t) \end{bmatrix} = \frac{1}{\det(A)} \begin{bmatrix} \delta x - \beta y \\ -\gamma x + \alpha y \end{bmatrix}$$

So

$$\cos(t) = \frac{1}{\det(A)} (\delta x - \beta y), \sin(t) = \frac{1}{\det(A)} (-\gamma x + \alpha y)$$

If we now square these two equations and add them together we have

$$(\det(A))^2 = (\delta x - \beta y)^2 + (-\gamma x + \alpha y)^2$$

Finally, simplifying this equation we have

$$(\delta^2 + \gamma^2) x^2 - 2(\beta \delta + \alpha \gamma)xy + (\alpha^2 + \beta^2) y^2 = (\det(A))^2 \tag{12.59}$$

Back in Section 4.1, we discussed that all conic sections (ellipse, hyperbola and parabola) have equations of the form given by equation (4.1), which we reproduce here:

$$a\,x^2 + b\,xy + c\,y^2 + d\,x + e\,y + f = 0 \qquad (12.60)$$

for real constants a through f where not all of a, b and c can be zero, since if they were you would only have a line, which is a degenerate conic section. This quadratic equation in x and y is an ellipse exactly when its discriminant $b^2 - 4ac$ satisfies $b^2 - 4ac < 0$. It is a hyperbola if $b^2 - 4ac > 0$, and a parabola if $b^2 - 4ac = 0$. The degenerate conic sections of points and lines are also possible. If both the constants d and e are zero, then the center of this conic section is the origin.

Now let us compute the discriminant of our conic section, given in equation (12.59), and see if our equation is an ellipse.

$$
\begin{aligned}
b^2 - 4ac &= 4(\beta\,\delta + \alpha\,\gamma)^2 - 4\left(\delta^2 + \gamma^2\right)\left(\alpha^2 + \beta^2\right) \\
&= 4\left(\beta^2\delta^2 + 2\alpha\beta\gamma\delta + \alpha^2\gamma^2\right) - 4\left(\alpha^2\delta^2 + \alpha^2\gamma^2 + \beta^2\delta^2 + \beta^2\gamma^2\right) \\
&= 4\left(2\alpha\beta\gamma\delta\right) - 4\left(\alpha^2\delta^2 + \beta^2\gamma^2\right) \\
&= -4\det(A)^2 < 0
\end{aligned}
$$

So equation (12.59) is that of an ellipse with center at the origin since this equation has no terms involving a multiple of x or of y. If we divide this equation by $(\det(A))^2$ so that the RHS is 1, the equation now has the form

$$ax^2 + 2bxy + cy^2 = 1 \qquad (12.61)$$

where

$$a = \frac{\delta^2 + \gamma^2}{(\det(A))^2}, \quad b = -\frac{\beta\,\delta + \alpha\,\gamma)}{(\det(A))^2}, \quad c = \frac{\alpha^2 + \beta^2}{(\det(A))^2}$$

We want to rotate this ellipse about the origin so that it is in standard form with its major and minor axes the coordinate axes.

The equation (12.61), corresponding to our ellipse, can be written in the following matrix form:

$$\begin{bmatrix} x & y \end{bmatrix}\begin{bmatrix} a & b \\ b & c \end{bmatrix}\begin{bmatrix} x \\ y \end{bmatrix} = 1 \qquad (12.62)$$

Upon setting $B = \begin{bmatrix} a & b \\ b & c \end{bmatrix}$ and $X = \begin{bmatrix} x \\ y \end{bmatrix}$, we can express equation (12.62) simply as

$$X^T B X = 1 \qquad (12.63)$$

Note that in order to achieve this new form for our equation that we are using, we really did require the $2b$ coefficient in front of the xy term in equation (12.61), instead of just b as the coefficient. This matrix B is a real symmetric matrix and so it has two distinct real eigenvalues λ_1 and λ_2 with corresponding real perpendicular unit eigenvectors $\overrightarrow{x_1}$ and $\overrightarrow{x_2}$; if $\overrightarrow{x_1}$ and $\overrightarrow{x_2}$ are not both unit vectors then divide each by their lengths to make them unit vectors. The matrix B is diagonalizable with $B = QDQ^{-1}$ where $D = \begin{bmatrix} \lambda_1 & 0 \\ 0 & \lambda_2 \end{bmatrix}$ and Q is the matrix whose columns are the eigenvectors $\overrightarrow{x_1}$ and $\overrightarrow{x_2}$. Also, $Q^{-1} = Q^T$ since Q is an orthogonal matrix with perpendicular unit column vectors. This tells us that the equation $B = QDQ^{-1}$ can be written as $BQ = QD$ or $D = Q^T BQ$. Also note that neither of these two eigenvalues can be zero since

$$\det(B) = ac - b^2 = -\frac{1}{4}\left(4b^2 - 4ac\right) > 0$$

where $4b^2 - 4ac < 0$ is the discriminant of the ellipse given in equation (12.61), and $\det(B) = \lambda_1 \lambda_2$.

Note that the matrix Q is orthogonal, therefore there is some angle θ so that

$$Q = \begin{bmatrix} \cos(\theta) & -\sin(\theta) \\ \sin(\theta) & \cos(\theta) \end{bmatrix}$$

and multiplication by Q is a rotation of all vectors about the origin through the angle θ, where the vector $\langle 1, 0 \rangle$ is sent to the vector $\langle \cos(\theta), \sin(\theta) \rangle$ on the unit circle.

Let u and v be two new perpendicular coordinate axes with the same origin as the x and y axes (the center of our ellipse), with the u, v coordinate axes corresponding to the major and minor axes of our ellipse. Our ellipse should be in standard form in the u, v coordinate system:

$$\frac{u^2}{K^2} + \frac{v^2}{L^2} = 1 \tag{12.64}$$

where K and L are the lengths of the semimajor and semiminor axes of our ellipse. The claim is that

$$X = \begin{bmatrix} x \\ y \end{bmatrix} = Q \begin{bmatrix} u \\ v \end{bmatrix}$$

will work to place our ellipse in standard form. Let us replace X by $Q \begin{bmatrix} u \\ v \end{bmatrix}$ in equation (12.63) and see if we get an equation of the standard form given in equation (12.64). For convenience, we will set $U = \begin{bmatrix} u \\ v \end{bmatrix}$, and thus $X = QU$.

Carrying out the desired replacement gives

$$1 = X^T B X$$
$$= (QU)^T B Q U$$
$$= U^T Q^T B Q U$$
$$= U^T \left(Q^T B Q \right) U$$
$$= U^T D U$$
$$= U^T \begin{bmatrix} \lambda_1 & 0 \\ 0 & \lambda_2 \end{bmatrix} U$$
$$= \lambda_1 u^2 + \lambda_2 v^2$$

So $\dfrac{1}{\sqrt{|\lambda_1|}}$ and $\dfrac{1}{\sqrt{|\lambda_2|}}$ are the lengths of the semimajor and semiminor axes of our ellipse depending on which one is the larger. This also tells us that the unit perpendicular eigenvectors $\overrightarrow{x_1}$ and $\overrightarrow{x_2}$, which make up Q, satisfy

$$\overrightarrow{x_1} = Q \overrightarrow{e_1}, \quad \overrightarrow{x_2} = Q \overrightarrow{e_2}$$

for $\overrightarrow{e_1} = \langle 1, 0 \rangle$, and $\overrightarrow{e_2} = \langle 0, 1 \rangle$. So $\dfrac{\overrightarrow{x_1}}{\sqrt{|\lambda_1|}}$ and $\dfrac{\overrightarrow{x_2}}{\sqrt{|\lambda_2|}}$ are vectors that go from the origin to the ends of the semimajor and semiminor axes of our original ellipse.

The rotation angle θ can now be found by the angle between the vectors $\overrightarrow{u_1}$ and $\overrightarrow{x_1}$, since $\overrightarrow{x_1} = Q \overrightarrow{u_1}$. If this angle is not between $0°$ and $90°$, then use $-\overrightarrow{x_1}$ instead of $\overrightarrow{x_1}$. This is necessary since the direction of $\overrightarrow{x_1}$ might need to be reversed in order to get the correct angle of rotation as an angle between $0°$ and $90°$.

In the next example that follows, we will carry out the procedure outlined above to put into standard form the equation of the ellipse $5x^2 - 8xy + 17y^2 = 1$, which has center at the origin. In the exercises you will have to deal with the situation where this ellipse has center off the origin, as well as placing in standard form the other types of conic sections: the hyperbola and parabola.

Example 12.6.2. The following equation

$$5x^2 - 8xy + 17y^2 = 1 \tag{12.65}$$

is an ellipse since its discriminant is $b^2 - 4ac = -276 < 0$ (see Figure 12.11). We wish to place this ellipse in standard form, finding a standard form equation for it, determine the rotation angle it takes to get it into standard form, and of course, plot both ellipses to verify our results.

In order to do this we need to find the eigenvalues and eigenvectors of the matrix $B = \begin{bmatrix} 5 & -4 \\ -4 & 17 \end{bmatrix}$. Recall that the equation of our ellipse is given by $X^T B X = 1$, where $X = \begin{bmatrix} x \\ y \end{bmatrix}$. Also, the lengths of the semimajor and semiminor axes are $\dfrac{1}{\sqrt{|\lambda_1|}}$ and $\dfrac{1}{\sqrt{|\lambda_2|}}$ for the eigenvalues λ_1 and λ_2 of B, where their corresponding eigenvectors $\vec{x_1}$ and $\vec{x_2}$ are the direction vectors for the major and minor axes of our nonstandard form ellipse. These two eigenvectors are not necessarily unit vectors, and so to get vectors which follow the semimajor and semiminor axes we need to take the unit eigenvectors $\dfrac{\vec{x_1}}{|\vec{x_1}|}$ and $\dfrac{\vec{x_2}}{|\vec{x_2}|}$, and then change their lengths to those of the semimajor and semiminor axes. So the vectors $\dfrac{1}{\sqrt{|\lambda_1|}}\dfrac{\vec{x_1}}{|\vec{x_1}|}$ and $\dfrac{1}{\sqrt{|\lambda_2|}}\dfrac{\vec{x_2}}{|\vec{x_2}|}$ go from the center, at the origin, to the ends of the semimajor and semiminor axes of our nonstandard form ellipse.

> plot_orig_ellipse:= implicitplot(5*x^2 - 8*x*y + 17*y^2 = 1, x=-1..1, y=-1..1, color = red, grid=[50,50], thickness = 2):

> display(plot_orig_ellipse, scaling=constrained);

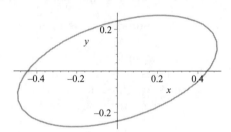

Figure 12.11: Plot of the ellipse given by equation (12.65).

> B:= matrix(2,2,[5,-4,-4,17]);

$$B := \begin{bmatrix} 5 & -4 \\ -4 & 17 \end{bmatrix}$$

> eigeninfo:= evalf(eigenvectors(B));

$$eigeninfo :=[18.21110255, 1., \{[1., -3.302775638]\}],$$
$$[3.788897450, 1., \{[1., 0.302775638]\}]$$

> eigvals:= [eigeninfo[1][1], eigeninfo[2][1]];

$$eigvals := [18.21110255, 3.788897450]$$

> Qlist:= [op(convert(op(eigeninfo[1][3]), list)), op(convert(op(eigeninfo[2][3]), list))];

$$Qlist := [1., -3.302775638, 1., 0.302775638]$$

> Q:= transpose(matrix(2,2, Qlist));

$$Q := \begin{bmatrix} 1. & 1. \\ -3.302775638 & 0.302775638 \end{bmatrix}$$

> dotprod(col(Q,1),col(Q,2));

$$-1. \, 10^{-9}$$

Note that Q is not an orthogonal matrix since its columns are not unit vectors, although they are perpendicular. Q can be made orthogonal if you divide each column of Q by its corresponding length. We do this next, as it will be needed for future calculations.

> L1:= sqrt(dotprod(col(Q,1),col(Q,1)));

$$L1 := 3.450844376$$

> L2:= sqrt(dotprod(col(Q,2),col(Q,2)));

$$L2 := 1.044831607$$

> Q:= matrix(2,2,[Q[1,1]/L1, Q[1,2]/L2, Q[2,1]/L1, Q[2,2]/L2]);

$$Q := \begin{bmatrix} 0.2897841488 & 0.9570920264 \\ -0.9570920268 & 0.2897841489 \end{bmatrix}$$

> evalm(transpose(Q)&*Q);

$$\begin{bmatrix} 1.000000001 & -2. \, 10^{-10} \\ -2. \, 10^{-10} & 1.000000000 \end{bmatrix}$$

Since $Q^T Q \approx I_2$, Q is an orthogonal matrix with unit perpendicular columns. If we had used exact values for the entries of each eigenvector, we would have a matrix that satisfies $Q^T Q = I_2$, but for our example, Q is close enough to orthogonal.

> K:= 1/sqrt(eigvals[1]);

$$K := 0.2343321515$$

> L:= 1/sqrt(eigvals[2]);

$$L := 0.5137402285$$

So the length of the semimajor axis is $L = 0.5137402285$ while the length of the semiminor axis is $K = 0.2343321515$. We can take as the standard form equation, $\dfrac{x^2}{L^2} + \dfrac{y^2}{K^2} = 1$. The standard form ellipse and the original are depicted together in Figure 12.12.

> plot_new_ellipse:= implicitplot(x^2/L^2 + y^2/K^2 = 1, x=-1..1, y=-1..1, color = blue, grid=[50,50]):

> display({plot_orig_ellipse, plot_new_ellipse}, scaling = constrained, thickness = 2);

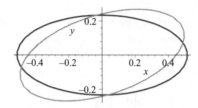

Figure 12.12: Plot of the ellipse given by equation (12.65) and the ellipse with same major and minor axes corresponding to the x- and y-axes instead.

> arrow1:= arrow([K*col(Q,1)[1], K*col(Q,1)[2]], shape = arrow, thickness = 2, color = blue):

> arrow2:= arrow([L*col(Q,2)[1], L*col(Q,2)[2]], shape = arrow, thickness = 2, color = black):

> display({plot_orig_ellipse, arrow1, arrow2}, scaling = constrained, thickness = 2, view = [-0.5..0.5, -0.5..0.5]);

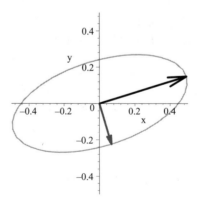

Figure 12.13: Plot of the ellipse given by equation (12.65) along with the transformed major and minor axes.

From what we see in Figure 12.13, the semimajor and semiminor axes of the original ellipse are the vectors

$$L\vec{x_2} = \frac{\vec{x_2}}{\sqrt{|\lambda_2|}}, \ \ K\vec{x_1} = \frac{\vec{x_1}}{\sqrt{|\lambda_1|}}$$

where $\vec{x_1}$ and $\vec{x_2}$ are the eigenvectors of B and the columns of Q. Remember that these formulas require Q to be made orthogonal. If Q is not orthogonal, simply make the columns of Q into unit vectors.

As to the value of the rotation angle θ, it is the angle between the vectors $\vec{x_2}$ and the unit vector along the x-axis, $\langle 1, 0 \rangle$. From the calculation below, $\theta \approx 16.845°$.

> theta:= convert(angle(col(Q,2),[1,0]), 'units', 'radians', 'degrees');

$$\theta := 16.84503378$$

■

What we have done in this section for the ellipse can be extended to the other two types of conic sections, the hyperbola and the parabola, using eigenvalues and eigenvectors to find their standard form equations where their axes are the coordinate axes if their centers are the origin.

As well, this material can be moved into three dimensions for the quadric surfaces of the ellipsoid, hyperboloid, and paraboloid. These quadric surfaces have equations of the general form

$$a\,x^2 + 2b\,xy + 2c\,xz + d\,y^2 + 2e\,yz + f\,z^2 + g\,x + h\,y + i\,z = j \qquad (12.66)$$

for real constants a through j. These surfaces will have centers at the origin as long as all of the coefficients g, h, and i are zero. In the exercises, you will be asked to apply the ideas developed here for the ellipse to other conic sections as well as quadric surfaces.

The fact that we can do all of this for the conic sections and quadric surfaces using eigenvalues and eigenvectors is called the *Principal Axis theorem*. This theorem can also be extended to the higher dimensions beyond three for quadratic polynomial equations in n variables.

Now let us state the *General Principal Axis theorem* for n real variables $x_1, x_2, \ldots, x_n$, which tells us how general quadratic polynomial equations in these n variables can be rewritten without any cross terms, this is sometimes called diagonalizing the equation. A general quadratic polynomial equation in the n variables given has the form

$$\sum_{i,j=1}^{n} g_{i,j} x_i x_j + \sum_{i=1}^{n} h_i x_i + k = 0 \qquad (12.67)$$

for real constants $g_{i,j}$, h_i, and k, where $g_{i,j} = g_{j,i}$ for all $i \neq j$. In other words, the matrix G whose i, j^{th} entries are the $g_{i,j}$ is a symmetric matrix. We need this matrix G to be real symmetric so that it is rotationally diagonalizable. A translation of our n variables can be made to eliminate the terms $\sum_{i=1}^{n} h_i x_i$ from this equation, and for simplicity we assume this has been done for the quadratic polynomial equation we use in the following theorem.

Theorem 12.6.1. *General Principal Axis Theorem Let $x_1, x_2, \ldots, x_n$ be n real variables. We are given the quadratic polynomial equation*

$$\sum_{i,j=1}^{n} g_{i,j} x_i x_j + k = 0 \tag{12.68}$$

for real constants $g_{i,j}$ and k, where the matrix $G = (g_{i,j})$ is real and symmetric. Let $\lambda_1, \lambda_2, \ldots, \lambda_n$ be the n real eigenvalues of G and Q be the real rotation matrix that diagonalizes G, that is, $G = QDQ^T$, where D is the diagonal matrix of the eigenvalues $\lambda_1, \lambda_2, \ldots, \lambda_n$, and Q is the orthogonal matrix with determinant 1 of the corresponding eigenvectors of G. Then in the new n real variables $u_1, u_2, \ldots, u_n$ defined by $U = Q^T X$ (or alternatively $X = QU$), where U and X are the column vectors whose components are the u_k and x_k variables respectively, the original quadratic equation (12.68) becomes the new quadratic equation:

$$\sum_{j=1}^{n} \lambda_j u_j^2 + k = 0 \tag{12.69}$$

Proof. The proof is very much analogous to what we did in the discussion following Example 12.6.1 if you realize that our quadratic equation (12.68) can be written as $X^T G X = -k$, for the real symmetric matrix G and variable column X given above. □

Homework Problems

1. Explain why the *General Principal Axis theorem* is correct.

Maple Problems

1. Let $A = \begin{bmatrix} -5 & 2 \\ 9 & 1 \end{bmatrix}$.

(a) Find and plot the equation of the ellipse gotten by multiplying all of the points of the unit circle by A.

(b) Find a standard form equation for this ellipse, and plot it with the original ellipse.

(c) Compute the rotation angle θ for rotating the original ellipse into your standard form one.

(d) Plot the eigenvectors as arrows that go from the origin to the ends of the major and minor axes for the original ellipse.

2. Let $A = \begin{bmatrix} 3 & -2 \\ 7 & 5 \end{bmatrix}$.

(a) Find and plot the equation of the ellipse gotten by multiplying all of the points of the unit circle by A.

(b) Find a standard form equation for this ellipse, and plot it with the original ellipse.

(c) Compute the rotation angle θ for rotating the original ellipse into your standard form one.

(d) Plot the eigenvectors as arrows that go from the origin to the ends of the major and minor axes for the original ellipse.

3. Let $7x^2 - 10xy + 12y^2 = 1$ be the nonstandard form equation of our original ellipse.

(a) Find a standard form equation for this ellipse, and plot it with the original ellipse.

(b) Compute the rotation angle θ for rotating the original ellipse into your standard form one.

(c) Plot the eigenvectors as arrows that go from the origin to the ends of the major and minor axes for the original ellipse.

4. Let $9x^2 + 8xy + 3y^2 = 1$ be the nonstandard form equation of our original ellipse.

(a) Find a standard form equation for this ellipse, and plot it with the original ellipse.

(b) Compute the rotation angle θ for rotating the original ellipse into your standard form one.

(c) Plot the eigenvectors as arrows that go from the origin to the ends of the major and minor axes for the original ellipse.

5. Let $9x^2 + 4xy + 3y^2 + 58x - 28y + 256 = 0$ be the nonstandard form equation of our original ellipse.

(a) Find values of K and L so that $x = u + K$ and $y = v + L$ is a translation of this ellipse's equation into u, v coordinates in which the equation of this ellipse has the form $9u^2 + 4uv + 3v^2 + M = 0$ for some constant M.

(b) Now find a standard form equation for this ellipse, and plot it with the original ellipse.

(c) Compute the rotation angle θ for rotating (and translating) the original ellipse into your standard form one.

(d) Plot the eigenvectors as arrows that go from its center to the ends of the major and minor axes for the original ellipse.

6. Let $9x^2 + 16xy + 3y^2 = 1$ be the nonstandard form equation of our original hyperbola.

(a) Find a standard form equation for this hyperbola, and plot it with the original hyperbola. The two possible standard form equations of a hyperbola are

$$\frac{x^2}{K^2} - \frac{y^2}{L^2} = 1, \quad \frac{y^2}{K^2} - \frac{x^2}{L^2} = 1$$

(b) Compute the rotation angle θ for rotating the original hyperbola into your standard form one.

(c) Plot the eigenvectors as arrows with the original hyperbola to see that they are parallel to the hyperbola's axes.

Research Projects

1. Find a discriminant test similar to that for conic sections that allows you to decide when a quadratic polynomial equation in the variables x, y, and z is an

ellipsoid, that is, when an equation of the form

$$ax^2 + by^2 + cz^2 + 2dxy + 2exz + 2fyz + gx + hy + iz + k = 0$$

is an ellipsoid. Now use this discriminant test as well as plotting the equation to decide that the equation

$$3x^2 + 4y^2 + 5z^2 + 2xy + 4xz + 6yz = 1$$

is an ellipsoid.

Next, find a standard form equation

$$\frac{x^2}{K^2} + \frac{y^2}{L^2} + \frac{z^2}{M^2} = 1$$

for this ellipsoid using eigenvalues and eigenvectors, plot this standard form ellipsoid with the original ellipsoid. Also, plot this original ellipsoid with three eigenvectors as arrows moving from its center at the origin to the ends of its three axes.

2. (Continuation of project 1.) For the 3×3 matrix,

$$A = \begin{bmatrix} -2 & 9 & -1 \\ -4 & 1 & -\pi \\ 5 & -3 & 4 \end{bmatrix}$$

find the equation of the ellipsoid you get if you multiply all the points of the unit sphere by A. Plot this ellipsoid as a parametric surface in space. Now find its standard form equation.

3. Investigate the topic of *quadratic forms*, which is related to inner products as well as conic sections and quadric surfaces.

12.7 A Maple Eigen–Procedure

We end this chapter with a few *Maple* procedures that you may find useful in your attempts to work, and solve, problems involving eigenvalues and eigenvectors. The following procedure, called *eigen*, is a function of the two variables, C and z, where C is a square matrix defined using either the *Matrix* command from *LinearAlgebra* or the *matrix* command from *linalg*, and z is either 1, 2 or 3. This function *eigen* has the following outputs:

$$eigen(C, 1) = \begin{cases} diagonalizable, & \text{if } C \text{ is diagonalizable} \\ not\ diagonalizable, & \text{if } C \text{ is not diagonalizable} \end{cases}$$

With option $z = 2$, $eigen(C, 2) = D$, where D is the diagonal matrix of C's eigenvalues appearing as many times as their multiplicity as a root of C's characteristic polynomial. Finally, $eigen(C, 3) = Q$, for the matrix Q of C's corresponding eigenvectors to those eigenvalues appearing in D when C is diagonalizable.

The matrix C should be created before the use of this procedure using the command *Matrix* from *LinearAlgebra* or the *matrix* command from *linalg* since the procedure *eigen* will convert whatever you give it to the format of a matrix from *LinearAlgebra*.

```
> restart: with(LinearAlgebra): with(linalg):
> eigen := proc (C, z)
      local B, eigvalsvecs, Diag, Q:
      B := convert(C, Matrix):
      eigvalsvecs := [evalf(Eigenvectors(B))]:
      Diag := DiagonalMatrix(eigvalsvecs[1]):
      Q := eigvalsvecs[2]:
      if Equal(Transpose(B), B) = true and Equal(map(conjugate, B), B) = true
      then
          Diag := map(Re, Diag):
          Q := map(Re, Q):
      end if:
      if Equal(HermitianTranspose(B), B) = true then
          Diag := map(Re, Diag):
      end if:
      if z = 2 then
          Diag;
      elif z = 3 then
          Q;
      elif z = 1 then
          if Determinant(Q) <> 0 then
              diagonalizable
          else
              not diagonalizable
          end if:
      end if:
  end proc:
```

```
> A:= matrix([[5, 2*I, -I], [-2*I, 3, 3+4*I], [I, 3-4*I, 1]]);
```

$$
A := \begin{bmatrix} 5 & 2I & -I \\ -2I & 3 & 3+4I \\ I & 3-4I & 1 \end{bmatrix}
$$

> eigen(A, 1);

$$diagonalizable$$

> eigen(A, 2);

$$\begin{bmatrix} 4.744339266 & 0 & 0 \\ 0 & 7.759700429 & 0 \\ 0 & 0 & -3.504039696 \end{bmatrix}$$

> eigen(A, 3);

$$\begin{bmatrix} 1.799386794 - 0.9571974480I & -0.8757192987 + 0.1357692854I & -0.1559131674 + 0.2436930543I \\ 0.6223589026 + 0.2300162732I & 0.6873412733 + 1.2083614621 & -0.5361877031 - 0.6629458822I \\ 1. & 1. & 1. \end{bmatrix}$$

> evalm(A&*eigen(A, 3));

$$\begin{bmatrix} 8.536901424 - 4.541269435I & -6.795319418 + 1.053528974I & 0.5463259270 - 0.853910134I \\ 2.952681812 + 1.091275232I & 5.333562391 + 9.376522983I & 1.878823000 + 2.322988688I \\ 4.744339249 + 3.10^{-9}I & 7.759700383 - 5.910^{-9}I & -3.504039692 - 1.610^{-9}I \end{bmatrix}$$

> E:= evalm(eigen(A, 3)&*eigen(A, 2)&*inverse(eigen(A,3)));

$$E:=$$

$$\begin{bmatrix} 5.000000001 - 3.212195983\ 10^{-9}I & 3.0\ 10^{-9} + 2.000000000I & -6.\ 10^{-10} - .9999999991I \\ -9.1\ 10^{-9} - 1.999999998I & 2.999999985 + 7.327982314\ 10^{-10}I & 2.999999989 + 3.9999999941 \\ -4.4\ 10^{-9} + .999999999I & 3.000000013 - 4.000000017I & 1.000000013 + 1.985250153\ 10^{-9}I \end{bmatrix}$$

The next procedure will take a complex matrix C and test all of its entries to see if any of their real and/or imaginary parts are so small that they should be eliminated in order to make the matrix more reflective of its true value without irrelevant very small terms. This procedure is called *superfluous*(C, K) with two variables, C, which is a matrix, and tolerance, K, which is a positive integer; where *superfluous* will eliminate from C's entries any real or imaginary part $\leq 10^{-K}$ in absolute value. The matrix C should be created before the use of this procedure using the command *Matrix* from *LinearAlgebra* or the *matrix* command from *linalg* since the procedure *superfluous* will convert whatever you give it to the format of a matrix from *LinearAlgebra*.

```
> superfluous := proc (C, K)
      local B, N, M, R, J, i, j:
      B:= convert(C, Matrix):
      N:= RowDimension(B):
      M:= ColumnDimension(B):
      R:= map(Re, B):
      J:= map(Im, B):
      for i to N do
          for j to M do
```

```
        if abs(R[i, j]) <= 10^(-K) then
            R[i, j] := 0:
        end if:
        if abs(J[i, j]) <= 10^(-K) then
            J[i, j] := 0:
        end if:
      end do:
    end do:
    simplify(evalm(R+I*J));
  end proc:
```

> superfluous(E, 8);

$$
\begin{bmatrix}
5.000000001 & 2.000000001I & -0.9999999991I \\
-1.999999998I & 2.999999985 & 2.999999989 + 3.999999994I \\
0.9999999994I & 3.000000013 - 4.000000017I & 1.000000013
\end{bmatrix}
$$

The following procedure called *deleteterms(E, K)* will eliminate those superfluous terms from an expression E that have, as their leading constant coefficients, the values C that are very small. That is, $|C| \leq 10^{-K}$, for a given positive integer K.

```
> deleteterms:= proc (E, K)
    local N, L, M, j, i:
    N:= nops(E):
    L:= NULL:
    for j to N do
        if abs(op(1, op(j, E))) <= 10^(-K) then
            L:= L, j = 0:
        end if:
    end do:
    M:= subsop(L, E):
    L:= NULL:
    if abs(op(1, M)) <= 10^(-K) then
        L:= L, 1 = 0:
    end if:
    subsop(L, M):
  end proc:
```

> q := -(3.87152*x^3*10^(-8)*(x^2))*cos(4*x) - 3*10^(-11) + (9*x*y*7)*exp(x + y) + 5*x*y*z - 4.15*10^(-9):

> deleteterms(q, 7);

$$
63\,xy\,e^{x+y} + 5\,xyz
$$

Suggested Reaading

Anton H., & Rorres, C. (2005). *Elementary Linear Algebra with Applications.* (9th ed.) NY: John Wiley & Sons, Inc.

Apostol T. (1997). *Linear Algebra: A First Course, with Applications to Differential Equations.* NY: John Wiley & Sons, Inc.

Barnsley M. (1993). *Fractals Everywhere* San Francisco: Morgan Kaufmann Publishers

Bauldry W., Evans B., & Johnson, J. (1995). *Linear Algebra With Maple.* NY: John Wiley & Sons, Inc.

Carchidi, M. (1986) "A Method for Finding the Eigenvectors of an $n \times n$ Matrix Corresponding to Eigenvalues of Multiplicity 1," *The American Mathematical Monthly* 93(8), 647–649

Hardy, K. (2005). *Linear Algebra for Engineers and Scientists Using Matlab.* Boston: Pearson Education, Inc.

Herman, E., & Pepe M. (2005). *Visual Linear Algebra.* Boston: Pearson Education, Inc.

Lawson T. (1996). *Linear Algebra, Mat Labs.* NY: John Wiley & Sons, Inc.

Lax P. (2007). *Linear Algebra and Its Applications.* (2nd ed.) NY: John Wiley & Sons, Inc.

Leon S. (2002). *Linear Algebra with Applications.* (6th ed.) Upper Saddle River, NJ: Prentice-Hall, Inc.

Meade D., May M., Cheung C-K., & Keough G. (2009). *Getting Started with Maple.* (3rd ed.) NY: John Wiley & Sons, Inc.

Olver, P., & Shakiban C. (2006). *Applied Linear Algebra.* Upper Saddle River, NJ: Prentice-Hall, Inc.

Penney R. (2008). *Linear Algebra: Ideas and Applications.* (3rd ed.) NY: John Wiley & Sons, Inc.

Sauer, T. (2006). *Numerical Analysis.* Boston: Pearson Education, Inc.

Sewell G. (2005). *Computational Methods of Linear Algebra.* (2nd ed.) NY: John Wiley & Sons, Inc.

Szabo, F. (2002). *Linear Algebra: An Introduction Using Maple.* Burlington, MA: Harcourt/Academic Press.

Wackerly, D., & Mendenhall III, W., & Scheaffer, R. (1996). *Mathematical Statistics with Applications.* (5th ed.) Belmont, CA: Duxbury Press

Weiner, J., & Wilkens, G. (2005). "Quaternions and Rotations in $\mathbb{E}^4$," *The American Mathematical Monthly* 112(1), 69–76

Williams, G. (1996). *Linear Algebra with Applications.* (3rd ed.) Dubuque, IA: Wm. C. Brown Publishers.

Yuster, T. (1984). "The Reduced Row Echelon Form of a Matrix is Unique: A Simple Proof," *Mathematics Magazine* 61(2), 93–94

Keyword Index

Index of Maple Commands and Packages

PURE AND APPLIED MATHEMATICS
A Wiley Series of Texts, Monographs, and Tracts

Founded by RICHARD COURANT
Editors Emeriti: MYRON B. ALLEN III, DAVID A. COX, PETER HILTON,
HARRY HOCHSTADT, PETER LAX, JOHN TOLAND

*Now available in a lower priced paperback edition in the Wiley Classics Library.
†Now available in paperback.